PROBABILITY
AND
STATISTICAL
INFERENCE

PROBABILITY
AND
STATISTICAL
INFERENCE

by

Robert V. Hogg

University of Iowa

and

Elliot A. Tanis

Hope College

MACMILLAN PUBLISHING CO., INC.
NEW YORK

COLLIER MACMILLAN PUBLISHERS
LONDON

Macmillan Publishing Co., Inc.
866 Third Avenue, New York, New York 10022

Collier Macmillan Canada, Ltd.

Library of Congress Cataloging in Publication Data

Hogg, Robert V
    Probability and statistical inference.

    Includes index.
    1. Probabilities.   2. Mathematical statistics.
I. Tanis, Elliot, joint author.   II. Title.
QA273.H694   519.2   75-35592
ISBN 0–02–355650–1

Printing: 1 2 3 4 5 6 7 8   Year: 7 8 9 0 1 2

preface

This book is designed for use in a course having from three to six semester hours of credit, such as a three-hour course for one semester, a two-quarter course, or a three-hour course for the full academic year. No previous study of statistics is assumed, and a standard two-semester course in calculus should provide an adequate mathematical background. For that matter, the material is organized in such a way that little multiple integration is needed until the last few chapters.

Our aim is to provide a book, at this mathematical level, that emphasizes fundamental concepts and presents them in a logical order. Probability and distributions of the discrete type are treated first. The usual descriptive statistics are found by computing the characteristics of a discrete-type empirical distribution. Histograms and ogives motivate the definitions of probability density and distribution functions of the continuous type. Certain basic sampling distribution theory is immediately used to make some elementary statistical inferences, the first of which is distribution-free and is based on the order statistics. The probabilities associated with these inferences are easy to determine using the binomial distribution and provide good applications of approximating distributions. After some of the standard parametric and nonparametric inferences involving one and two distributions, multivariate distributions are introduced. We feel that the student is better prepared to understand them at this stage, and they provide the necessary background for chi-square tests and the analysis of variance. The final chapter concerns certain interesting theoretical problems, most of which are treated more fully in an advanced course in mathematical statistics.

Although it is not necessary to have a computer available to study this text, we have included some computer output to make certain theories (like the central limit theorem) more plausible.

We are indebted to the *Biometrika* Trustees for permission to include Tables III and V, which are abridgments and adaptations of tables published in *Biometrika Tables for Statisticians*. We are also grateful to the Literary Executor of the late Sir Ronald A. Fisher, F.R.S., to Dr. Frank Yates, F.R.S., and to Longman Group Ltd, London, for permission to use Table III from their book *Statistical Tables for Biological, Agricultural, and Medical Research* (6th Edition, 1974), reproduced as our Table VI.

Finally, we wish to thank our colleagues and friends for many suggestions, Mrs. Mary DeYoung for her help with the typing, and our families for their patience and understanding during the preparation of this manuscript.

R. V. H.
E. A. T.

contents

10 CHI–SQUARE TESTS OF MODELS   329

11 ANALYSIS OF VARIANCE   359

12 A BRIEF THEORY OF STATISTICAL INFERENCE   381

PROBABILITY
AND
STATISTICAL
INFERENCE

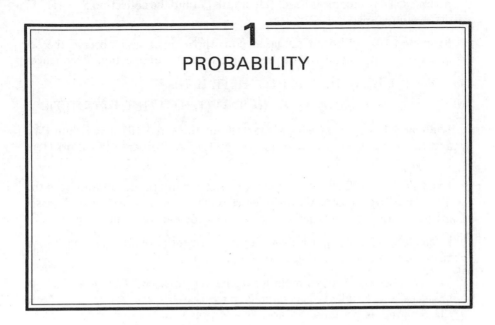

1
PROBABILITY

1.1 Random Experiments and Random Variables

Many decisions have to be made that involve uncertainties. In medical research, interest may center on the effectiveness of a new vaccine for mumps; an agronomist must decide if an increase in yield can be attributed to a new strain of wheat; a meteorologist is interested in predicting the probability of rain; the state legislature must decide whether decreasing speed limits will help prevent accidents; the admissions officer of a college must predict the college performance of an incoming freshman. Probability and statistics can provide the models that could help people make decisions such as these.

In the study of probability we shall consider *random experiments*. Each experiment ends in an *outcome* that cannot be determined with certainty before the performance of the experiment. However, the experiment is such that the collection of every possible outcome can be listed; and this collection of all outcomes is called the outcome space or, more frequently, the *sample space S*.

The following examples will help illustrate what we mean by random experiments, outcomes, and sample spaces:

Example 1.1–1. Consider the flip of an unbiased coin as a random experiment. The outcome is heads (H) or tails (T) and the collection $S = \{H, T\}$ is the sample space.

Example 1.1–2. Flip an unbiased coin three times and observe the sequence of heads and tails. Here the sample space is the collection of sequences

$$S = \{(H, H, H), (H, H, T), (H, T, H), (T, H, H),$$

$$(H, T, T), (T, H, T), (T, T, H), (T, T, T)\}.$$

Example 1.1–3. From a bowl containing three red (R), two white (W), and five blue (B) chips, draw one at random and observe its color. Here $S = \{R, W, B\}$.

Example 1.1–4. Each of six students selects an integer at random from the first 52 positive integers. We are interested in whether at least two of these six integers match (M) or whether they are all different (D). Thus, $S = \{M, D\}$.

Example 1.1–5. A light bulb is turned on continuously and we observe the time t until it burns out. Here $S = \{t : \ 0 \leq t\}$.

Example 1.1–6. A three-month-old chicken is selected from a flock of chickens and weighed. The sample space is $S = \{w : \ 0 < w \leq 9\}$. Note that 9 pounds is, perhaps, too large. In addition we are victims of the accuracy with which we can weigh chickens; thus, we might more realistically describe S as $S = \{w : \ w = 0.5, 0.6, \ldots, 5.0\}$. However, it is often easier to work with a mathematically "idealized" sample space rather than a more realistic one.

Let S denote a sample space, and let A be part of this collection S. Suppose the performance of the random experiment terminates so that the outcome is in A. Then we shall say that *event A* has occurred. Now consider the possibility of repeating the experiment a large number of times, say n. Then we can count the number of times that event A actually occurred throughout n performances; this number is called the frequency of event A and is denoted $N(A)$. The ratio $N(A)/n$ is called the *relative frequency* of event A in these n experiments. A relative frequency is usually very unstable for small values of n, but it tends to stabilize as n increases. Possibly you should check this by tossing a coin a large number of times, computing the relative frequency after each toss. This suggests that we associate with event A a number, say p, that is equal or approximately equal to the number about which the relative frequency seems to stabilize. This number p can then be taken as that number which the relative frequency of event A will be near in future performances of the experiment. Thus, although we cannot predict the outcome of a random experiment with certainty, we can, for a large value of n, predict fairly accurately the relative frequency associated with event A. The number p assigned to event A is called the *probability* of event A, and it is denoted by $P(A)$.

To illustrate some of these ideas, we performed the following two random experiments.

Example 1.1–7. Four unbiased coins are to be tossed and the number of heads observed. Here the sample space $S = \{0, 1, 2, 3, 4\}$. Let the event $A = \{0, 4\}$; that is, A occurs when the coins are all heads or all tails. This experiment was actually repeated a large number of times, and each time we recorded whether or not A had occurred. After 50 trials, the frequency of A was $N(A) = 4$; after 100 trials, $N(A) = 11$; after 500 trials, $N(A) = 49$; and after 1000 trials, $N(A) = 118$. These results provide the relative frequencies of 0.080, 0.110, 0.098, and 0.118, respectively. Accordingly, we believe that the probability $P(A)$ of the event A is close to 0.118. Later we will learn, if certain assumptions are fulfilled, that $P(A)$ equals 0.125.

Example 1.1–8. A pair of fair dice is to be cast and the sum of the dots on the top of the dice observed. Here

$$S = \{2, 3, 4, 5, 6, 7, 8, 9, 10, 11, 12\}.$$

Let $A = \{7\}$; that is, A occurs when the sum of the dots equals 7. After repeating the experiment a large number of times, we observed these combinations of the number n of trials and the frequency $N(A)$ of A:

$$n = 50, \qquad N(A) = 10;$$
$$n = 100, \qquad N(A) = 17;$$
$$n = 500, \qquad N(A) = 81;$$
$$n = 1000, \qquad N(A) = 175.$$

From these observations, the calculated respective relative frequencies $N(A)/n$ are 0.200, 0.170, 0.162, and 0.175. Other considerations later in the book allow us to assign the probability $P(A) = 1/6 = 0.167$ to the event A; this is near the relative frequency 0.175.

Note that a sample space S may be difficult to describe if the elements of S are not numbers. We shall now discuss how we can use a rule by which an element s of S may be associated with a number x. We begin the discussion with a simple example.

Example 1.1–9. In Example 1.1–1, we had the sample space $S = \{H, T\}$. Let X be a function defined on S such that $X(H) = 0$ and $X(T) = 1$. Thus, X is a real-valued function that has the sample space S as its domain and the space of real numbers $\{x: \ x = 0, 1\}$ as its range. We call X a random variable and, in this example, the space associated with X is $\{x: \ x = 0, 1\}$.

We now formulate the definition of a random variable.

DEFINITION 1.1–1. *Given a random experiment with a sample space S, a function X that assigns to each element s in S one and only one real number $X(s) = x$ is called a* random variable. *The space of X is the set of real numbers $\{x: \ x = X(s), s \in S\}$, where $s \in S$ means the element s belongs to the set S.*

It may be that the set S has elements that are themselves real numbers. In such an instance we could write $X(s) = s$ so that X is the identity function and the space of X is also S. This is illustrated in Example 1.1–10.

Example 1.1–10. Let the random experiment be the cast of a die. The sample space associated with this experiment is $S = \{1, 2, 3, 4, 5, 6\}$. For each $s \in S$, let $Y(s) = s$. The space of the random variable Y is then $\{1, 2, 3, 4, 5, 6\}$.

For notational purposes we shall denote the event $\{s: \ s \in S \text{ and } X(s) = a\}$ by $\{X = a\}$. That is, the event $\{X = a\}$ is the set of points in the sample space that are mapped onto the real number a by the function X. Similarly, $\{s: \ s \in S \text{ and } a < X(s) < b\}$ will be denoted by $\{a < X < b\}$. Now if we want to find the probabilities associated with events described in terms of X, such as $\{X = a\}$ and $\{a < X < b\}$, we use the probabilities of those events in the original space S, if they are known. That is, when we define the probability of these events, we shall let

$$P(X = a) = P(\{s: \ s \in S \text{ and } X(s) = a\})$$

and

$$P(a < X < b) = P(\{s: \ s \in S \text{ and } a < X(s) < b\}).$$

We say that probabilities are *induced* on the points of the space of X by the probabilities assigned to outcomes of the sample space S through the function X.

Example 1.1–11. If, in Example 1.1–10, we associate a probability of $1/6$ with each outcome, then, for example, $P(Y = 5) = 1/6$, $P(2 \leq Y \leq 5) = 4/6$, and $P(Y \leq 2) = 2/6$ seem to be reasonable assignments.

The student will no doubt recognize two major difficulties here:

(1) In many practical situations the probabilities assigned to the events A of the sample space S are unknown.
(2) Since there are many ways of defining a function X on S, which function do we want to use?

As a matter of fact, the solutions to these problems in particular cases are major concerns in applied statistics. In considering (2), statisticians try to determine what "measurement" (or measurements) should be taken on an outcome; that is, how best do we "mathematize" the outcome (which, for the anthropologist, might be a skull)? These measurement problems are most difficult and can only be answered by getting involved in a practical project. For (1), we need, through repeated observations (called sampling), to estimate these probabilities or "percentages." For example, what percentage of newborn girls in the University of Iowa Hospital weigh less than 7 pounds. Here a newborn baby girl is the outcome, and we have measured her one way (by weight); but obviously there are many other ways of measuring her. If we let X be the weight in pounds, we are interested in the probability $P(X < 7)$ and we can only estimate this by repeated observations. One obvious way of estimating this is by use of the relative frequency of $\{X < 7\}$ after a number of observations. If additional assumptions can be made, we will study, in this text, other ways of estimating this probability. It is this latter aspect with which mathematical statistics is concerned. That is, if we assume certain models, we find that the theory of statistics can explain how best to draw conclusions or make predictions. Now the construction of such a model does require some knowledge of probability, and most theories of probabilities are based on the concept of sets (or events). Accordingly, a basic review of the algebra of sets is given in Section 1.2.

One final remark should be made. In many instances, it is clear exactly what function X the experimenter wants to define on the sample space. For example, the caster in a dice game is concerned about the sum of the spots, say X, that are up on the pair of dice. Hence, we go directly to the space of X and sometimes even call this the sample space S, if there is no confusion. After all, in the dice game, the caster is directly concerned only with the probabilities associated with X. Hence, the reader can, in many instances, think of the space of X as being the sample space.

═══ *Exercises* ═══

1.1–1. In each of the following random experiments describe the sample space S. Use your intuition or any experience you may have had to assign a value to the probability p of each of the events A.

  (a)   The toss of an unbiased coin where the event A is heads.

  (b)   The cast of an honest die where the event A occurs if we observe a three, four, five, or six.

  (c)   The draw of a card from an ordinary deck of playing cards where the event A is a club.

  (d)   The choice of a point from a square with opposite vertices $(0, 0)$ and $(1, 1)$ where the event A occurs if the sum of the coordinates of the point is less than 3/4.

1.1–2. Describe the sample space for each of the following experiments.

  (a)   Toss a coin seven times and observe the number of heads.

  (b)   Toss a coin five times and observe the sequence of heads and tails.

(c)   Observe the number of tosses of a coin until the first head appears.

(d)   Draw five cards at random from a standard deck of cards and record each card in that five-card hand (order of drawing is not important).

1.1–3.   If a disk two inches in diameter is thrown at random on a tiled floor, where each tile is a square with sides four inches in length, assign a probability to the event that the disk will land entirely on one tile.

1.1–4.   Divide a line segment into two parts by selecting a point at random. Assign a probability to the event that the larger segment is at least two times longer than the shorter segment.

1.1–5.   Let the interval $[-r, r]$ be the base of a semicircle. If a point is selected at random from this interval, assign a probability to the event that the length of the perpendicular segment from this point to the semicircle is less than $r/2$.

1.1–6.   Consider the sequence of heads (H) and tails (T) if an unbiased coin is flipped four times.

(a)   List the 16 points in the sample space.

If X equals the number of observed heads, list which of these sample points correspond to

(b) $\{X = 3\}$,   (c) $\{0 \leq X \leq 1\}$.

1.1–7.   Let X equal the number of observed heads in two flips of an unbiased coin. If each point in the original sample space $S = \{HH, HT, TH, TT\}$ has probability 1/4, assign values to

(a) $P(X = 0)$,   (b) $P(X = 2)$,

(c) $P(X \leq 1)$.

1.1–8.   Let the random variable W equal a number selected at random from the closed interval from zero to one, that is $[0, 1]$. Describe the sample space S of W. Assign values to

(a) $P(0 \leq W \leq 1/3)$,   (b) $P(1/3 \leq W \leq 1)$,

(c) $P(W = 1/3)$,   (d) $P(1/2 < W < 5)$.

1.1–9.   Each of the numbers 1, 2, 3, 4, and 5 is written on a disk and placed in a hat. Two disks are drawn without replacement from the hat.

(a)   List the 10 possible outcomes for this experiment.

(b)   If the random variable Y is defined to be the sum of the two drawn numbers and each of the 10 outcomes has probability 1/10, assign values to $P(Y = 3)$, $P(Y = 5)$, and $P(6 \leq Y \leq 8)$.

1.2 Algebra of Sets

Before defining a probability set function in Section 1.3, we give some basic rules and definitions associated with set algebra. In addition, some terminology used in probability will be explained.

The totality of objects under consideration is called the *universal set* and is denoted S. Each object in S is called an *element* of S. If a set A is a collection of elements that are also in S, then A is said to be a *subset* of S. In applications in probability, S will usually denote the *sample space*. An *event A* will be a collection of possible outcomes of the experiment and will be a subset of S. We say that event A *has occurred* if the outcome of the experiment is an element of A. The set or event A may be described by listing all of its elements or by defining the properties that its elements must satisfy.

Example 1.2–1. Let $S = \{1, 2, 3, 4, 5, 6\}$. If A is the subset of S consisting of the even integers, we may write $A = \{2, 4, 6\}$ or $A = \{x: \ x \text{ is even}\}$. In order to emphasize that x is also in S we could write

$$A = \{x: \ x \text{ is in } S \text{ and } x \text{ is even}\}.$$

When a is an element in A, we write $a \in A$. When a is not an element in A, we write $a \notin A$. So, in Example 1.2–1, we have $2 \in A$ and $3 \notin A$. If every element of a set A is also an element in a set B, then A is a *subset* of B. We write $A \subset B$. In probability, if event B occurs whenever event A occurs, then $A \subset B$. The two sets A and B are equal, $A = B$, if $A \subset B$ and $B \subset A$. Note that it is always true that $A \subset A$ and $A \subset S$, where S is the universal set. We denote the subset that contains no elements by $\varnothing$. This set is called the *null* or *empty* set. For all sets A, $\varnothing \subset A$.

The set of elements in either A or B or possibly in both A and B is called the *union* of A and B and is denoted $A \cup B$. The set of elements in both A and B is called the *intersection* of A and B and is denoted $A \cap B$. The *complement* of a set A is the set of elements in the universal set S that are not in the set A and is denoted A'. In probability, if A and B are two events, the event that at least one of the two events has occurred is denoted by $A \cup B$, or the event that both events have occurred is denoted by $A \cap B$. The event that A has not occurred is denoted by A', and the event that A has not occurred but B has occurred is denoted by $A' \cap B$. If $A \cap B = \varnothing$, we say that A and B are *mutually exclusive*.

The operations of union and intersection may be extended to more than two sets. Let $A_1, A_2, \ldots, A_n$ be a finite collection of sets. Then the *union*

$$A_1 \cup A_2 \cup \cdots \cup A_n = \bigcup_{k=1}^{n} A_k$$

is the set of all elements that belong to at least one A_k, $k = 1, 2, \ldots, n$. The *intersection*

$$A_1 \cap A_2 \cap \cdots \cap A_n = \bigcap_{k=1}^{n} A_k$$

is the set of all elements that belong to every A_k, $k = 1, 2, \ldots, n$. Similarly, let $A_1, A_2, \ldots, A_n, \ldots$ be a denumerable collection of sets. Then x belongs to the *union*

$$A_1 \cup A_2 \cup A_3 \cup \cdots = \bigcup_{k=1}^{\infty} A_k$$

if x belongs to at least one A_k, $k = 1, 2, 3, \ldots$. Also x belongs to the *intersection*

$$A_1 \cap A_2 \cap A_3 \cap \cdots = \bigcap_{k=1}^{\infty} A_k$$

if x belongs to every A_k, $k = 1, 2, 3, \ldots$.

Example 1.2–2. Let S be the set of positive real numbers less than or equal to 6. Thus $S = \{x : 0 < x \le 6\}$. Let $A = \{x : 1 \le x \le 3\}$, $B = \{x : 2 \le x \le 6\}$, $C = \{x : 3 \le x < 5\}$, and $D = \{x : 0 < x < 2\}$. Then

$$A \cup B = \{x : 1 \le x \le 6\},$$
$$B \cup D = S,$$
$$B \cap D = \emptyset,$$
$$A \cap B = \{x : 2 \le x \le 3\},$$
$$B \cap C = C,$$
$$A' = \{x : 0 < x < 1 \text{ or } 3 < x \le 6\},$$
$$B' = \{x : 0 < x < 2\} = D.$$

Also

$$A \cup C \cup D = \{x : 0 < x < 5\}$$

and

$$A \cap B \cap C = \{x : x = 3\}.$$

Example 1.2–3. Let

$$A_k = \left\{ x : \frac{10}{k+1} \le x \le 10 \right\}, \qquad k = 1, 2, 3, \ldots.$$

Then

$$\bigcup_{k=1}^{\infty} A_k = \{x : 0 < x \le 10\};$$

note that the number zero is not in the union since it is not in one of the sets $A_1, A_2, A_3, \ldots$. Of course,

$$\bigcap_{k=1}^{\infty} A_k = \{x : 5 \le x \le 10\} = A_1$$

since $A_1 \subset A_k$, $k = 1, 2, 3, \ldots$.

A convenient way to illustrate operations on sets is with a *Venn* diagram. In Figure 1.2–1 the universal set S is represented by the rectangle and its interior and the subsets of S by the points enclosed by the circles. The sets under consideration are the shaded regions.

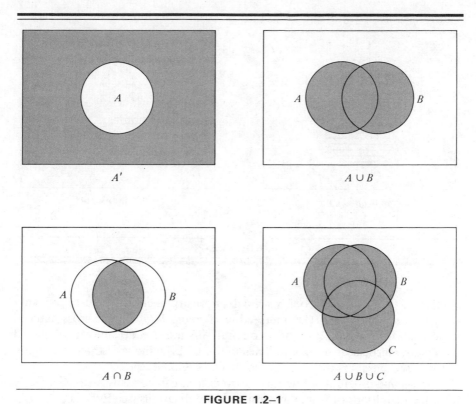

FIGURE 1.2–1

Set operations satisfy several properties. For example, if A, B, and C are subsets of S, we have the following:

Commutative laws:

$$A \cup B = B \cup A$$
$$A \cap B = B \cap A.$$

Associative laws:

$$(A \cup B) \cup C = A \cup (B \cup C)$$
$$(A \cap B) \cap C = A \cap (B \cap C).$$

Distributive laws:

$$A \cap (B \cup C) = (A \cap B) \cup (A \cap C)$$
$$A \cup (B \cap C) = (A \cup B) \cap (A \cup C).$$

De Morgan's laws:

$$(A \cup B)' = A' \cap B'$$
$$(A \cap B)' = A' \cup B'.$$

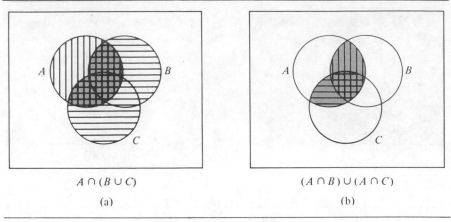

$$A \cap (B \cup C)$$

(a)

$$(A \cap B) \cup (A \cap C)$$

(b)

FIGURE 1.2–2

Rather than a formal proof, a Venn diagram argument will be used to justify the first distributive law. The intersection of A and $B \cup C$ is the cross-hatched region in (a) of Figure 1.2–2. The union of $A \cap B$ and $A \cap C$ is the region shaded with either vertical or horizontal lines in (b). Clearly, the cross-hatched region of the first diagram is the same as the shaded region of the second.

The second distributive law and the other laws can be easily justified directly from the definitions or from appropriate Venn diagrams (see Exercise 1.2–7).

Exercises

1.2–1. Let $S = \{1, 2, 3, 4, 5, 6\}$, $A = \{2, 4, 6\}$, $B = \{3, 4, 5\}$, and $C = \{1, 6\}$. Find the following sets.

(a) $A \cup C$,

(b) $A \cap B$,

(c) A',

(d) $A' \cap B$,

(e) $B \cap C'$,

(f) $A \cup B \cup C$,

(g) $(A \cup B)'$,

(h) $A' \cap B'$,

(i) $(C')'$,

(j) $[(A \cap B)' \cup C]'$.

1.2–2. Let $A_1 = (0, 1)$, $A_2 = (0, 1/2)$, $A_3 = (0, 1/3), \ldots, A_{10} = (0, 1/10)$, where (a, b) represents the open interval between a and b; that is, $(a, b) = \{x : \ a < x < b\}$. Find

$$\bigcup_{k=1}^{10} A_k \quad \text{and} \quad \bigcap_{k=1}^{10} A_k.$$

1.2–3. A box contains five radio tubes, two of which are defective. The tubes are tested, one at a time, until the second defective tube is found. Let X be the number tested.

(a) Describe the original sample space S and also that of X.
(b) What outcomes of S belong to the event $\{X = 4\} = A$?
(c) What outcomes of S belong to the event $\{X = 3\} = B$?
(d) What outcomes of S belong to (i) $A \cup B$? (ii) $A \cap B$? (iii) A'?

1.2–4. An urn contains three balls colored red, yellow, and green, respectively. An experiment consists of drawing the balls from the urn, one at a time without replacement and noting the order in which the colors appear. Consider the following events: $A = \{$red on first draw$\}$, $B = \{$green on second draw$\}$.
(a) Describe the sample space for this experiment.
(b) Find $A \cup B$ and $A \cap B$.
(c) Describe an event C such that A and C are mutually exclusive.

1.2–5. Let A, B, and C be three events for a given experiment. Use set notation to express the following statements.
(a) At least one of the three events occurs.
(b) All of the events occur.
(c) Exactly one of the three events occurs.
(d) None of the three events occurs.
(e) Events A and B occur but not C.

1.2–6. What numbers belong to the following unions and intersections of intervals of real numbers, where (a, b) and $[a, b]$ denote, respectively, the open and the closed intervals from a to b?

(a) $\displaystyle\bigcup_{n=2}^{\infty}\left(\frac{1}{n}, 1\right)$,

(b) $\displaystyle\bigcap_{n=1}^{\infty}\left(0, \frac{1}{n}\right)$,

(c) $\displaystyle\bigcup_{n=1}^{\infty}\left[\frac{1}{n}, 1\right]$,

(d) $\displaystyle\bigcap_{n=1}^{\infty}\left[0, 1 + \frac{1}{n}\right]$.

1.2–7. Using Venn diagrams, verify the second distributive law and De Morgan's laws.

1.2–8. Let S be the universal set and let $A \subset S$. Find each of the following.
(a) $A \cap S$,   (b) $A \cup S$,
(c) $\varnothing \cap A$,   (d) $\varnothing \cup A$,
(e) $A \cap A'$,   (f) $A \cup A'$.

1.2–9. If A and B are subsets of S, simplify

$$(A \cup B) \cap (A' \cup B) \cap (A \cup B').$$

1.3 Properties of Probability

Let the event A be a subset of the sample space S. Recall that we wish to associate with A a number $P(A)$ about which the relative frequency $N(A)/n$ of the event A tends to stabilize with large n. A function such as $P(A)$ or $N(A)$, which is evaluated for a set A and whose range is a collection of real numbers, is called a *set function*. Accordingly, in this section, we want to consider in general the probability set function $P(A)$ and discuss some of its properties. In succeeding sections we shall describe how the probability set function is defined for particular experiments.

REMARK. We stated earlier that an event A is a subset of the sample space S. However, it should be noted that it is not true in general that all subsets of a sample space S are events. Fortunately, in most applications, subsets that do not qualify as events do not arise very often. It is usually only in very theoretical discussions that these special subsets are considered. It is always true, however, when S is finite or denumerable, that all subsets of S can be treated as events. That is, if the number of elements of S can be put into a one-to-one correspondence with a subset of the positive integers, then every subset of S will be called an event. When S is nondenumerable, certain subsets exist that are not events. In this text, however, we consider only those subsets that can qualify as events. From the following definition and theorems, it can be observed that we want the union and the intersection of a number of events to be an event. In addition, it can be seen that the complement of an event should be an event.

DEFINITION 1.3–1. Probability *is a set function P that assigns to each event A in the sample space S a number $P(A)$, called the probability of the event A, such that the following properties are satisfied*:

(i) $P(A) \geq 0$,
(ii) $P(S) = 1$,
(iii) *If $A_1, A_2, A_3, \ldots$ are events and $A_i \cap A_j = \varnothing$, $i \neq j$, then*

$$P(A_1 \cup A_2 \cup A_3 \cup \cdots) = P(A_1) + P(A_2) + P(A_3) + \cdots.$$

Note that property (iii) implies that for any finite collection of events $A_1, A_2, A_3, \ldots, A_n$, we have

$$P(A_1 \cup A_2 \cup \cdots \cup A_n) = P(A_1) + P(A_2) + \cdots + P(A_n),$$

provided the events are mutually exclusive, that is, $A_i \cap A_j = \varnothing, i \neq j$.

The following theorems give some other important properties of the probability set function.

THEOREM 1.3–1. *For each event A,*

$$P(A) = 1 - P(A').$$

Proof: We have

$$S = A \cup A' \quad \text{and} \quad A \cap A' = \varnothing.$$

Thus, from properties (iii) and (ii), it follows that

$$1 = P(A) + P(A').$$

Hence,

$$P(A) = 1 - P(A').$$

THEOREM 1.3–2. $P(\varnothing) = 0.$

Proof: In Theorem 1.3–1, take $A = \varnothing$ so that $A' = S$. Thus,

$$P(\varnothing) = 1 - P(S) = 1 - 1 = 0.$$

THEOREM 1.3–3. *If events A and B are such that* $A \subset B$, *then* $P(A) \leq P(B)$.

Proof: Now

$$B = A \cup (B \cap A') \quad \text{and} \quad A \cap (B \cap A') = \varnothing.$$

Hence, from property (iii),

$$P(B) = P(A) + P(B \cap A') \geq P(A)$$

because from property (i),

$$P(B \cap A') \geq 0.$$

THEOREM 1.3–4. *For each event* $A, 0 \leq P(A) \leq 1$.

Proof: Since $\varnothing \subset A \subset S$, we have by Theorem 1.3–3 that

$$P(\varnothing) \leq P(A) \leq P(S).$$

But

$$P(\varnothing) = 0 \quad \text{and} \quad P(S) = 1,$$

which gives the desired result.

THEOREM 1.3–5. *If A and B are any two events, then*

$$P(A \cup B) = P(A) + P(B) - P(A \cap B).$$

Proof: The event $A \cup B$ can be represented as a union of disjoint sets, namely,

$$A \cup B = A \cup (A' \cap B).$$

Hence, by property (iii),

$$P(A \cup B) = P(A) + P(A' \cap B).$$

However,

$$B = (A \cap B) \cup (A' \cap B),$$

which is a union of disjoint sets. Thus,

$$P(B) = P(A \cap B) + P(A' \cap B)$$

so that

$$P(A' \cap B) = P(B) - P(A \cap B).$$

If this result is substituted in the equation involving $P(A \cup B)$, we obtain

$$P(A \cup B) = P(A) + P(B) - P(A \cap B),$$

which is the desired result.

THEOREM 1.3–6. *If A, B, and C are any three events, then*

$$P(A \cup B \cup C) = P(A) + P(B) + P(C) - P(A \cap B)$$
$$- P(A \cap C) - P(C \cap B) + P(A \cap B \cap C).$$

Proof: Write

$$A \cup B \cup C = A \cup (B \cup C)$$

and then apply Theorem 1.3–5. The details are left as an exercise.

Example 1.3–1.   Flip a coin twice and observe the sequence of heads and tails. Thus, the sample space may be represented as

$$S = \{(H, H), (H, T), (T, H), (T, T)\}.$$

Let the probability set function assign a probability of 1/4 to each point of S. Let

$$A_1 = \{(H, H), (H, T)\} \quad \text{and} \quad A_2 = \{(H, H), (T, H)\}.$$

Then

$$P(A_1) = P(A_2) = \frac{1}{2},$$

$$P(A_1 \cap A_2) = \frac{1}{4},$$

and in accordance with Theorem 1.3–5,

$$P(A_1 \cup A_2) = \frac{1}{2} + \frac{1}{2} - \frac{1}{4} = \frac{3}{4}.$$

Example 1.3–2. Let the sample space $S = A \cup B$, and let $P(A) = 0.6$ and $P(B) = 0.7$. Then

$$P(A \cup B) = P(A) + P(B) - P(A \cap B)$$

implies that

$$1 = 0.6 + 0.7 - P(A \cap B)$$

because $P(A \cup B) = 1$. Thus, $P(A \cap B) = 0.3$.

Let a probability set function $P(A)$ be defined on a sample space S. Let S be partitioned into k mutually disjoint subsets $A_1, A_2, \ldots, A_k$ in such a way that the union of these k mutually disjoint subsets is the sample space S. Thus, the events $A_1, A_2, \ldots, A_k$ are not only *mutually exclusive* but also *exhaustive* since their union equals S. Suppose that the random experiment is such that it may be *assumed* that each of the mutually exclusive and exhaustive events $A_1, A_2, \ldots, A_k$ has the same probability. Thus,

$$S = A_1 \cup A_2 \cup \cdots \cup A_k$$

and

$$P(S) = P(A_1) + P(A_2) + \cdots + P(A_k)$$

and, hence,

$$1 = kP(A_i)$$

since

$$P(A_1) = P(A_2) = \cdots = P(A_k).$$

That is,

$$P(A_i) = \frac{1}{k}, \qquad i = 1, 2, \ldots, k.$$

If the event A is the union of h of these mutually exclusive events, say

$$A = A_1 \cup A_2 \cup \cdots \cup A_h, \qquad h \le k$$

then

$$P(A) = P(A_1) + P(A_2) + \cdots + P(A_h) = \frac{h}{k}.$$

For this particular partition of S, the integer k is called the total number of ways in which the random experiment can terminate, and the integer h is called the number of ways that are favorable to the event A. Thus, $P(A)$ is equal to the number of ways favorable to the event A divided by the total number of ways in which the experiment can terminate. It should be emphasized that in order to assign the probability h/k to the event A, we must assume that each of the mutually exclusive and exhaustive events $A_1, A_2, \ldots, A_k$ has the same probability $1/k$. This assumption is then an important part of our probability model; if it is not realistic in an application, the probability of the event A cannot be computed this way.

Example 1.3–3. Let a card be drawn at random from an ordinary deck of 52 playing cards. The sample space S is the union of $k = 52$ outcomes, and it is reasonable to assume that each of these outcomes has the same probability, $1/52$. Accordingly, if A is the set of outcomes that are kings, $P(A) = 4/52 = 1/13$ because there are $h = 4$ kings in the deck. That is, $1/13$ is the probability of drawing a card that is a king provided the probability of each of the 52 outcomes has the same probability.

In Example 1.3–3, the computations are very easy because there is no difficulty in the determination of the appropriate values of h and k. However, instead of drawing only one card, suppose that 13 are taken at random and without replacement. We can think of each possible 13-card hand as being an outcome in a sample space, and it is reasonable to assume that each of these outcomes has the same probability. To use the above method to assign the probability of a hand, consisting of seven spades and six hearts, for illustration, we must be able to count the number h of all such hands as well as the number k of possible 13-card hands. In these more complicated situations, we need better methods of determining h and k. We discuss some of these counting techniques in Sections 1.4 and 1.5.

━━━ *Exercises* ━━━

1.3–1. Draw one card at random from a standard deck of cards. The sample space S is the collection of the 52 cards. Assume that the probability set function assigns $1/52$ to each of these 52 outcomes. Let

$$A = \{x: \quad x \text{ is a jack, queen, or king}\},$$
$$B = \{x: \quad x \text{ is a 9, 10, or jack and } x \text{ is red}\},$$
$$C = \{x: \quad x \text{ is a club}\},$$
$$D = \{x: \quad x \text{ is a diamond, a heart, or a spade}\}.$$

Find:

(a) $P(A)$, (b) $P(A \cap B)$,
(c) $P(A \cup B)$, (d) $P(C \cup D)$,
(e) $P(C \cap D)$.

1.3–2. A coin is tossed four times, and the sequence of heads and tails is observed.

(a) List each of the 16 sequences in the sample space S.
(b) Let events $A, B, C,$ and D be given by $A = \{$at least 3 heads$\}, B = \{$at most 2 heads$\}$,

C = {heads on the third toss}, and D = {1 head and 3 tails}. If the probability set function assigns 1/16 to each outcome in the sample space, find $P(A)$, $P(A \cap B)$, $P(B)$, $P(A \cap C)$, $P(D)$, $P(A \cup C)$, and $P(B \cap D)$.

1.3–3. If $P(A) = 0.4$, $P(B) = 0.5$, and $P(A \cap B) = 0.3$, find
   (a) $P(A \cup B)$,
   (b) $P(A \cap B')$,
   (c) $P(A' \cup B')$.

1.3–4. If $P(A) = 0.6$ and $P(A' \cap B) = 0.1$, find $P(A' \cap B')$.

1.3–5. If $P(A) = 0.4$, $P(B) = 0.5$, $P(C) = 0.7$, $P(A \cap B) = 0.2$, $P(A \cap C) = 0.2$, $P(B \cap C) = 0.4$, and $P(A \cap B \cap C) = 0.1$, find   (a) $P(A \cup B \cup C)$,   (b) $P(A \cup B \cup C')$.

1.3–6. If $P(A) = 0.2$, $P(B \cap A') = 0.4$, $P(C \cap A') = 0.2$, $P(A' \cap B \cap C) = 0.1$,   and $P(B \cap C) = 0.1$, find   (a) $P(A' \cap B' \cap C')$,   (b) $P(A \cup B \cup C)$.

1.3–7. If $S = A \cup B$, $P(A) = 0.7$, and $P(B) = 0.9$, find $P(A \cap B)$.

1.3–8. A poll of 100 persons determines that 82 like ice cream and 62 like cake. How many persons like both if each of the persons likes at least one of the two?

1.3–9. A report of a survey of 100 graduating seniors stated that the numbers who had studied the various languages were as follows: Spanish, 45; German, 36; French, 61; German and Spanish, 15; French and Spanish, 18; German and French, 12; all three languages, 7. The surveyor who turned in the report was fired. Why?

1.3–10. For each positive integer n, let $P(\{n\}) = (1/2)^n$. Let $A = \{n: \ 1 \le n \le 10\}$, $B = \{n: \ 1 \le n \le 20\}$ and $C = \{n: \ 11 \le n \le 20\}$. Find:
   (a) $P(A)$,   (b) $P(B)$,   (c) $P(A \cup B)$,
   (d) $P(A \cap B)$,   (e) $P(C)$,   (f) $P(B')$.

1.3–11. Prove Theorem 1.3–6.

1.4 Methods of Enumeration

In this section we develop counting techniques that are useful in determining the number of outcomes associated with the events of certain random experiments. We begin with a consideration of the multiplication principle.

   Multiplication Principle.   Suppose that an experiment E_1 has n_1 outcomes and for each of these possible outcomes an experiment E_2 has n_2 possible outcomes. The composite experiment $E_1 E_2$ that consists of performing first E_1 and then E_2 has $n_1 n_2$ possible outcomes.

   Example 1.4–1.   Let E_1 be the experiment of rolling a die and let E_2 be the experiment of tossing a coin. Then $n_1 = 6$ and $n_2 = 2$ and the composite experiment $E_1 E_2$ has $(6)(2) = 12$ possible outcomes. A *tree diagram* like that in Figure 1.4–1 is a useful device for illustrating the multiplication

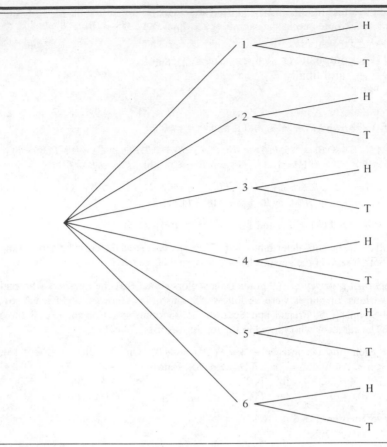

FIGURE 1.4–1

principle. This diagram shows that there are $n_1 = 6$ possibilities (branches) for the die and for each of these outcomes there are $n_2 = 2$ possibilities (branches) for the coin.

The outcome for the composite experiment can be denoted by an *ordered pair*, such as (3, H). In fact, the set of all possible outcomes can be denoted by the following rectangular array.

(1, H)	(2, H)	(3, H)	(4, H)	(5, H)	(6, H)
(1, T)	(2, T)	(3, T)	(4, T)	(5, T)	(6, T)

In the multiplication principle, the word "experiment" may be replaced by "procedure." Thus, if some procedure E_1 has n_1 possible outcomes, and for each of these a second procedure E_2 has n_2 possible outcomes, then the sequence of procedures $E_1 E_2$ has $n_1 n_2$ possible outcomes.

Clearly the multiplication principle can be extended to a sequence of more than two experiments or procedures. Suppose the experiment E_i has n_i, $i = 1, 2, \ldots, m$, possible outcomes after previous experiments have been performed. The composite experiment $E_1 E_2 \cdots E_m$, which consists of performing E_1, then $E_2, \ldots$, and finally E_m, has $n_1 n_2 \cdots n_m$ possible outcomes.

Example 1.4–2. A certain food service gives the following choices for dinner: E_1, soup or tomato juice; E_2, steak or shrimp; E_3, french fried potatoes or baked potatoes; E_4, corn or peas; E_5, jello, tossed salad, cottage cheese, or cole slaw; E_6, cake, cookies, pudding, brownie, vanilla ice cream, chocolate ice cream, or orange sherbet; E_7, coffee, tea, milk, or punch. How many different dinner selections are possible if one of the listed choices is made for each of $E_1, E_2, \ldots$, and E_7? By the multiplication principle there are

$$(2)(2)(2)(2)(4)(7)(4) = 1792 \text{ different combinations.}$$

Suppose that n positions are to be filled with n different objects. There are n choices for filling the first position, $n - 1$ for the second, $\ldots$, and 1 choice for the last position. So by the multiplication principle there are

$$n(n - 1) \cdots (2)(1) = n!$$

possible arrangements.

DEFINITION 1.4–1. *Each of the $n!$ arrangements (in a row) of n objects is called a* permutation *of the n objects. The symbol $n!$ is read n factorial. By definition, we take $0! = 1$; that is, we say that zero positions can be filled with zero objects in one way.*

Example 1.4–3. The number of permutations of the four letters a, b, c, and d is clearly $4! = 24$. However, the number of possible four-letter code words using the four letters a, b, c, and d, if letters may be repeated is $4^4 = 256$.

If only r positions are to be filled with objects selected from n different objects, $r \leq n$, the number of possible arrangements is

$$P(n, r) = n(n - 1)(n - 2) \cdots (n - r + 1).$$

That is, there are n ways to fill the first position, $(n - 1)$ ways to fill the second, and so on until there are $(n - r + 1)$ ways to fill the rth position.
Consequently,

$$P(n, r) = \frac{n(n - 1) \cdots (n - r + 1)(n - r) \cdots (3)(2)(1)}{(n - r) \cdots (3)(2)(1)} = \frac{n!}{(n - r)!}.$$

DEFINITION 1.4–2. *Each of the P(n, r) arrangements is called a* permutation *of n objects taken r at a time.*

Example 1.4–4. The number of possible four-letter code words in which all four letters are different is

$$P(26, 4) = (26)(25)(24)(23) = \frac{26!}{22!} = 358,800.$$

Students on a boat send signals back to shore by arranging five colored flags on a vertical flag pole. If the students have seven flags of different colors, they can send

$$P(7, 5) = \frac{7!}{2!} = 2520 \text{ different signals.}$$

Suppose now that the students have only five flags, three of which are red and two yellow. Although the flags can be placed on the pole in 5! different ways, these arrangements are not all distinguishable. To count the number of different signals, say C, that are now possible, consider the following. Suppose temporarily that we can tell the difference between the red and yellow flags. Then all 5! arrangements are possible. But we could get each of these 5! arrangements by first selecting one of the C distinguishable arrangements, then arranging the three red flags, and finally arranging the two yellow flags. This latter sequence could be achieved in

$$(C)(3!)(2!)$$

ways. Hence,

$$(C)(3!)(2!) = 5!$$

or, equivalently,

$$C = \frac{5!}{3!2!}.$$

Frequently, C is denoted by $C(5, 3)$ or $\binom{5}{3}$. That is,

$$C(5, 3) = \binom{5}{3} = \frac{5!}{3!2!}.$$

This example can obviously be generalized. If a set contains n objects of two types, r of one type and $n - r$ of the other type, the number of distinguishable arrangements of these n objects is

$$C(n, r) = \binom{n}{r} = \frac{n!}{r!(n - r)!}.$$

DEFINITION 1.4–3. *Each of the $C(n, r)$ arrangements is called a* distinguishable permutation *of these n objects, r of one type, $n - r$ of another type.*

Note that, in general,

$$C(n, r) = \binom{n}{r} = \frac{n!}{r!(n - r)!} = \frac{n!}{(n - r)!r!} = \binom{n}{n - r} = C(n, n - r).$$

Example 1.4–5. A coin is flipped ten times, and the sequence of heads and tails is observed. The number of possible ten-tuplets that result in four heads and six tails is

$$\binom{10}{4} = \frac{10!}{4!6!} = \frac{10!}{6!4!} = \binom{10}{6}.$$

Since two outcomes are possible on each flip, the total number of outcomes in the sample space S is 2^{10}. If A is the event that exactly four heads are observed, then there are $C(10, 4)$ arrangements of four heads and six tails so that

$$P(A) = \frac{N(A)}{N(S)} = \frac{C(10, 4)}{2^{10}} = \frac{10!}{4!6!} \left(\frac{1}{2}\right)^{10},$$

provided we can assume each of the 2^{10} outcomes has the same probability. Please note that the numerator $N(A) = C(10, 4)$ of $P(A)$ represents the number of elements in a certain subset of S. Other subsets could be defined by zero heads and ten tails, one head and nine tails; two heads and eight tails; and so on, which contain, respectively, $C(10, 0)$, $C(10, 1)$, $C(10, 2)$, ... elements. Of course, the sum of $C(10, r)$, $r = 0, 1, 2, \ldots, 10$, is equal to 2^{10} (see Exercise 1.5–16).

The above results can be extended. Suppose that in a set of n objects, n_1 are similar, n_2 are similar, ..., n_s are similar, where $n_1 + n_2 + \cdots + n_s = n$. The number of distinguishable permutations of the n objects is

$$\binom{n}{n_1, n_2, \ldots, n_s} = \frac{n!}{n_1! n_2! \cdots n_s!}.$$

Example 1.4–7. If the students on the boat have three red flags, four yellow flags, and two blue flags to arrange on a vertical flag pole, the number of possible signals is

$$\binom{9}{3, 4, 2} = \frac{9!}{3!4!2!} = 1260.$$

Exercises

1.4–1. How many different ordered outcomes are possible when rolling a red die, a blue die, and a white die?

1.4–2. How many different signals can be made using four flags of different colors on a vertical flag pole if exactly three flags are used for each signal?

1.4–3. In 1968 Michigan automobile license plates had two letters followed by a four-digit number. In 1974, there were three letters followed by a three-digit number. How many different plates were possible in   (a)   1968?   (b)   1974? Note which is larger.

1.4–4. In how many ways can six people be seated in a row of six chairs?

1.4–5. In how many ways can six students be seated in a row that has ten chairs?

1.4–6. A family of five is interested in a new seating arrangement for meals. How many different seating arrangements are possible?
   (a)   They sit on only three sides of a rectangular table because the fourth side is by a window, and the five chairs are fixed so that two, one, and two chairs are on the three sides, respectively.
   (b)   They have a round table and are only interested in relative positions of the family members.

1.4–7. How many four-letter code words are possible using the letters in HOPE if   (a)   the letters may not be repeated?   (b)   the letters may be repeated?

1.4–8. Show that the number of distinguishable permutations of n objects, if n_1 are of one kind, n_2 of a second kind, $\dots$, and n_s of the last kind (here $n = n_1 + n_2 + \cdots + n_s$) is $n!/(n_1!n_2!\cdots n_s!)$. HINT: First arrange the n_1 and the $n - n_1$ objects assuming the latter are all alike. Then, among the $n - n_1$, arrange n_2 and $n - n_1 - n_2$ objects assuming the latter are alike. Continue this and finally apply the multiplication principle.

1.4–9. Use the result of Exercise 1.4–8 to find the number of possible 11-letter code words (distinguishable permutations) using the letters in MISSISSIPPI.

1.4–10. A train of 100 cars has 40 box cars, 15 coal cars, 20 tank cars, and 25 flat cars. How many arrangements are distinguishable if only the type of car is taken into account?

1.4–11. A biology professor and his students are observing the breeding territory rights of the common gallinule, a bird that nests in a swampy area. They capture a bird and put three colored bands on each leg, choosing from red, white, yellow, and blue bands. How many gallinules can be banded differently?

1.4–12. One of the biology students from Exercise 1.4–11 is now studying ravens in northern Minnesota. After capturing a raven, the student puts two colored bands on each leg, choosing from red, white, blue, yellow, and green bands. How many ravens can be banded differently?

1.5 Sampling with and Without Replacement

Suppose a set contains n objects. Consider the problem of drawing r objects from this set. The order in which the objects are drawn may or may not be important. In addition, it is possible that a drawn object is replaced before the next object is drawn. Accordingly, we give some definitions and then some examples illustrating these possibilities.

DEFINITION 1.5–1. *If r objects are selected from a set of n objects and if the order of selection is considered, the selected set of r objects is called an* ordered sample *of size r.*

DEFINITION 1.5–2. Sampling with replacement *occurs when an object is selected and then replaced before the next object is selected.*

By the multiplication principle, the number of possible ordered samples of size r taken from a set of n objects is n^r when sampling with replacement.

DEFINITION 1.5–3. Sampling without replacement *occurs when an object is not replaced after it has been selected.*

By the multiplication principle, the number of possible ordered samples of size r taken from a set of n objects is

$$n(n-1)\cdots(n-r+1) = \frac{n!}{(n-r)!},$$

which is equivalent to $P(n, r)$, the number of permutations of n objects taken r at a time.

Example 1.5–1. A die is rolled five times. The number of possible ordered samples is 6^5. Note that rolling a die is equivalent to sampling with replacement.

Example 1.5–2. The number of ordered samples of five cards that can be drawn without replacement from a standard deck of 52 cards is

$$(52)(51)(50)(49)(48) = \frac{52!}{47!} = 311{,}875{,}200.$$

Often the order of selection is not important and interest centers only on the selected set of r objects. That is, we are interested in the number of subsets of size r that can be selected from a set of n different objects. In order to find the number of (unordered) subsets of size r, we count, in two different ways, the number of ordered subsets of size r that can be taken from the n distinguishable objects. By equating these two answers, we are able to count the number of (unordered) subsets of size r.

Let C denote the number of (unordered) subsets of size r selected from n different objects. We can obtain each of the $P(n, r)$ ordered subsets by first selecting one of the C unordered subsets of r objects and then ordering these r objects. Since the latter can be carried out in $r!$ ways, the multiplication principle yields $(C)(r!)$ ordered subsets, which must equal $P(n, r)$. Thus, we have

$$(C)(r!) = \frac{n!}{(n-r)!}$$

or

$$C = \frac{n!}{r!(n-r)!} = \binom{n}{r} = C(n, r).$$

Thus, a set of n different objects possesses

$$\binom{n}{r} = \frac{n!}{r!(n-r)!}$$

unordered subsets of size $r \leq n$. Note that this answer is the same as the number of distinguishable permutations of n objects, r of one kind and $n - r$ of another kind. We could also say that the number of ways in which r objects can be selected without replacement from n objects, when the order of selection is disregarded, is $\binom{n}{r}$. This motivates the following definition.

DEFINITION 1.5–4. *The number of* combinations *of n objects taken r at a time is*

$$C(n, r) = \binom{n}{r}.$$

Example 1.5–3. The number of possible five-card hands drawn from a deck of 52 playing cards is

$$\binom{52}{5} = \frac{52!}{5!47!} = 2,598,960.$$

The number of possible five-card hands that are all spades is

$$\binom{13}{5} = \frac{13!}{5!8!} = 1287.$$

Hence, if each five-card hand has the same probability, the probability of an all-spade hand is $1287/2,598,960$. Suppose now that the event A_1 is the set of outcomes in which exactly three cards are kings and exactly two cards are queens. We can select the three kings in any one of $\binom{4}{3}$ ways and the two queens in any one of $\binom{4}{2}$ ways. By the multiplication principle, the number of outcomes in A_1 is

$$h = \binom{4}{3}\binom{4}{2}.$$

Thus,

$$P(A_1) = \frac{\binom{4}{3}\binom{4}{2}}{\binom{52}{5}} = \frac{24}{2,598,960}.$$

Finally, let A_2 be the set of outcomes in which there are exactly two kings, two queens, and one jack. Then

$$P(A_2) = \frac{\binom{4}{2}\binom{4}{2}\binom{4}{1}}{\binom{52}{5}} = \frac{144}{2,598,960},$$

because the numerator of this fraction is the number of outcomes in A_2.

Example 1.5–4. Extending the last example, let B be the set of outcomes in which there are two pairs and one odd card. To determine $N(B)$, the number of outcomes in B, consider the following sequence: First select two of the 13 face values for the two pairs and take two of the four choices in each case. Now select one of the remaining 11 face values for the odd card and take one of the four choices. Thus, we have by the multiplication principle that

$$P(B) = \frac{N(B)}{N(S)} = \frac{\binom{13}{2}\binom{4}{2}\binom{4}{2}\binom{11}{1}\binom{4}{1}}{\binom{52}{5}} = \frac{123,552}{2,598,960} = 0.048.$$

We now present an example that will then be generalized to give what is commonly called the hypergeometric distribution.

Example 1.5–5. A lot, consisting of 100 fuses, is inspected by the following procedure. Five fuses are chosen at random and tested; if all five "blow" at the correct amperage, the lot is accepted. Suppose that the lot contains 20 defective fuses. If X is a random variable equal to the number of defective fuses in the sample of five, the probability of accepting the lot is

$$P(X = 0) = \frac{\binom{20}{0}\binom{80}{5}}{\binom{100}{5}} = 0.32,$$

approximately. More generally, if x is one of the numbers 0, 1, 2, 3, 4, or 5, then

$$P(X = x) = \frac{\binom{20}{x}\binom{80}{5-x}}{\binom{100}{5}}, \qquad x = 0, 1, 2, 3, 4, 5.$$

To generalize Example 1.5–5, consider a collection of $n = n_1 + n_2$ similar objects, n_1 of them belonging to one of two dichotomous classes (red chips, say) and n_2 of them belonging to the second class (blue chips, say). A collection of r objects is selected from these n objects at random and without replacement. Find the probability that exactly x (where the integer x satisfies $x \leq r$, $x \leq n_1$, and $r - x \leq n_2$) of these r objects are red (that is, x belong to the first class and $r - x$ belong to the second). Of course, we can select x red chips in any one of $\binom{n_1}{x}$ ways and $r - x$ blue chips in any one of $\binom{n_2}{r-x}$ ways. By the multiplication principle, the product $\binom{n_1}{x}\binom{n_2}{r-x}$ equals the number of ways the joint operation can be performed. If we assume that each of the $\binom{n}{r}$ ways of selecting r objects from $n = n_1 + n_2$ objects has the same probability, we have that the probability of selecting exactly x red chips is

$$P(X = x) = \frac{\binom{n_1}{x}\binom{n_2}{r-x}}{\binom{n}{r}},$$

where $x \leq r$, $x \leq n_1$, and $r - x \leq n_2$. We say that the random variable X has a *hypergeometric distribution*.

Example 1.5–6.   In a small pond there are 50 fish, ten of which have been tagged. If a fisherman's catch consists of seven fish, selected at random and without replacement, and X denotes the number of tagged fish, the probability that he has exactly two tagged fish is

$$P(X = 2) = \frac{\binom{10}{2}\binom{40}{5}}{\binom{50}{7}} = \frac{(45)(658,008)}{99,884,400} = 0.296,$$

approximately.

Let us now consider a generalization of the hypergeometric distribution. Instead of two classes, suppose that each of the n objects can be placed into one of s disjoint classes so that n_1 objects are in the first class, n_2 in the second, and so on until we find n_s in the sth class. Of course,

$$n = n_1 + n_2 + \cdots + n_s.$$

A collection of r objects is selected from these n at random and without replacement. Let the random variable X_i denote the number of observed objects in the sample belonging to the ith class. Find the probability that exactly x_i objects belong to the ith class, $i = 1, 2, \ldots, s$. Here

$$x_i \leq n_i \qquad \text{and} \qquad x_1 + x_2 + \cdots + x_s = r.$$

We can select x_i objects from the ith class in any one of $\binom{n_i}{x_i}$ ways, $i = 1, 2, \ldots, s$. By the multiplication principle, the product

$$\binom{n_1}{x_1}\binom{n_2}{x_2}\cdots\binom{n_s}{x_s}$$

equals the number of ways the joint operation can be performed. If we assume that each of the $\binom{n}{r}$ ways of selecting r objects from $n = n_1 + n_2 + \cdots + n_s$ objects has the same probability, we have that the probability of selecting exactly x_i objects from the ith class, $i = 1, 2, \ldots, s$, is

$$P(X_1 = x_1, X_2 = x_2, \ldots, X_s = x_s) = \frac{\binom{n_1}{x_1}\binom{n_2}{x_2}\cdots\binom{n_s}{x_s}}{\binom{n}{r}}.$$

Example 1.5–7. The probability that a 13-card bridge hand (selected at random and without replacement) contains two clubs, four diamonds, three hearts, and four spades is

$$\frac{\binom{13}{2}\binom{13}{4}\binom{13}{3}\binom{13}{4}}{\binom{52}{13}} = \frac{11,404,407,300}{635,013,559,600} = 0.018.$$

Example 1.5–8. The probability that a bridge hand contains two cards of one suit, three cards of a second suit, and four cards of the other two suits is

$$(4)(3)\frac{\binom{13}{2}\binom{13}{3}\binom{13}{4}\binom{13}{4}}{\binom{52}{13}} = 0.216.$$

The coefficient $(4)(3) = 12$ represents the number of ways that the suit with two cards can be selected, followed by the selection of the suit with three cards. Of course, each of the remaining two suits has four cards.

The numbers $\binom{n}{r}$ are frequently called *binomial coefficients* since they arise in the expansion of a binomial. We shall illustrate this by giving a justification of the binomial expansion

$$(a + b)^n = \sum_{r=0}^{n} \binom{n}{r} b^r a^{n-r}.$$

In the expansion of

$$(a + b)^n = (a + b)(a + b) \cdots (a + b),$$

either an a or a b is selected from each of the n factors. One possible product is then $b^r a^{n-r}$; this occurs when b is selected from each of r factors and a from each of the remaining $n - r$ factors. But this latter operation can be completed in $\binom{n}{r}$ ways, which then must be the coefficient of $b^r a^{n-r}$.

Perhaps you are familiar with Pascal's triangle, which gives a method for calculating the binomial coefficients.

$$
\begin{array}{ccccccccccccccc}
 & & & & & & & 1 \\
 & & & & & & 1 & & 1 \\
 & & & & & 1 & & 2 & & 1 \\
 & & & & 1 & & 3 & & 3 & & 1 \\
 & & & 1 & & 4 & & 6 & & 4 & & 1 \\
 & & 1 & & 5 & & 10 & & 10 & & 5 & & 1 \\
 & 1 & & 6 & & 15 & & 20 & & 15 & & 6 & & 1 \\
1 & & 7 & & 21 & & 35 & & 35 & & 21 & & 7 & & 1 \\
 & \vdots & & \vdots & & \vdots & & \vdots & & \vdots & & \vdots & & \vdots & & \vdots
\end{array}
$$

The nth row of this triangle gives the coefficients for $(a + b)^{n-1}$; for example, with $n = 5$,

$$(a + b)^4 = a^4 + 4a^3b + 6a^2b^2 + 4ab^3 + b^4.$$

To find an entry in the table other than a 1 on the boundary, add the two nearest numbers in the row directly above. The equation

$$\binom{n}{r} = \binom{n-1}{r} + \binom{n-1}{r-1}$$

explains why Pascal's triangle works.

This result can be proved easily by expressing the binomial coefficients in terms of factorials. This type of proof is left to the reader (see Exercise 1.5–5). However, consider the following alternate argument. Suppose a set of n objects has one distinguishable member, for example, 1 red ball among $n-1$ blue balls. The number of subsets of size r to which the red ball belongs is $\binom{n-1}{r-1}$ since the remaining $r-1$ balls are selected from the $n-1$ blue balls. The number of subsets of size r to which the red ball does not belong is $\binom{n-1}{r}$ since the r balls must be selected from the $n-1$ blue balls. Thus, $\binom{n-1}{r} + \binom{n-1}{r-1}$ must be the number of subsets of size r that can be selected out of n objects, namely, $\binom{n}{r}$. Hence

$$\binom{n}{r} = \binom{n-1}{r} + \binom{n-1}{r-1}.$$

This result supplies the reason why we can find the nth row of Pascal's triangle from the $(n-1)$st row. For example,

$$4 + 6 = \binom{4}{1} + \binom{4}{2} = \binom{5}{2} = 10.$$

The argument used in determining the binomial coefficients in the expansion of $(a + b)^n$ can be extended to find the expansion of $(a_1 + a_2 + \cdots + a_s)^n$. The coefficient of $a_1^{n_1} a_2^{n_2} \cdots a_s^{n_s}$ is

$$\binom{n}{n_1, n_2, \ldots, n_s} = \frac{n!}{n_1! n_2! \cdots n_s!};$$

This is sometimes called a multinomial coefficient.

Exercises

1.5–1. An urn contains 10 balls numbered from 1 to 10. How many ordered samples of size 3 can be drawn from the urn if the sampling is done (a) with replacement? (b) without replacement?

1.5–2. Ten people are to be grouped into two clubs, five in each club. Each club is then to elect a president and secretary. In how many ways can this be done?

1.5–3. Say that there are three defective items in a lot of 50 items. A sample of size 10 is taken at random and without replacement. Let X denote the number of defective items in the sample. Find the probability that the sample contains **(a)** exactly one defective item, **(b)** at most one defective item.

1.5–4. In a lot of 100 light bulbs, there are four bad bulbs. An inspector inspects ten bulbs selected at random. Find the probability of finding at least one defective bulb. HINT: First compute the probability of finding no defectives in the sample.

1.5–5. If r and n are integers such that $r \leq n - 1$, prove Pascal's rule

$$\binom{n}{r} = \binom{n-1}{r-1} + \binom{n-1}{r}.$$

HINT: Spell out the right-hand member by using factorials and show it is equal to the left-hand member.

1.5–6. Box 1 contains the letters B, C, D, E, F. Box 2 contains the letters W, X, Y, Z. How many five-letter code words are possible using three letters from box 1 and two letters from box 2?

1.5–7. Consider the birthdays of the students in a class of size r. Assume that the year consists of 365 days.
    (a) How many different ordered samples of birthdays are possible (r in sample) allowing repetitions (with replacements)?
    (b) The same as part (a) except requiring that all the students have different birthdays (without replacements)?
    (c) If we can assume that each ordered outcome in part (a) has the same probability, what is the probability that no two students have the same birthday?
    (d) For what value of r is the probability in part (c) about equal to 1/2? Is this number surprisingly small?

1.5–8. A bridge hand is defined as drawing 13 cards at random without replacement from a deck of 52 playing cards. How many different bridge hands are possible?

1.5–9. What is the probability of being dealt, at random, a bridge hand that does not contain a spade?

1.5–10. What is the probability that a bridge hand contains exactly five hearts?

1.5–11. Find the probability that a bridge hand contains three clubs, two diamonds, four hearts, and four spades.

1.5–12. Find the probability that a bridge hand consists of three cards of each of three suits and four cards of the fourth suit?

1.5–13. Find the probability that a bridge hand contains **(a)** exactly one ace, **(b)** at least one ace, **(c)** at most one ace, **(d)** all red cards.

1.5–14. A deal at bridge consists of dividing 52 cards among four players, 13 cards to each of them. In how many ways can this be done?

1.5–15. In a deal at bridge, what is the probability that each player will have exactly one ace?

1.5–16. Prove:

$$\sum_{r=0}^{n}(-1)^r\binom{n}{r}=0 \quad \text{and} \quad \sum_{r=0}^{n}\binom{n}{r}=2^n.$$

HINT: Consider $(1-1)^n$ and $(1+1)^n$.

1.5–17. Five cards are drawn at random and without replacement from an ordinary deck of cards. Compute the probability that this five-card hand contains exactly two aces.

1.5–18. A poker hand is defined as drawing five cards at random without replacement from a deck of 52 playing cards. Find the probability of each of the following poker hands.
 (a) Four of a kind (four cards of equal face values).
 (b) Full house (one pair and one triple of cards with equal face value).
 (c) Three of a kind (three equal face values plus two different cards).
 (d) Two pairs (two pairs of equal face values plus one other card).
 (e) One pair (one pair of equal face values plus three different cards).

1.5–19. An urn contains 10 red, 20 white, and 30 blue balls. A sample of size 8 is drawn from the urn, one at a time. Let X denote the number of red and Y the number of blue balls in the sample. Calculate $P(X=2, Y=3)$ if the sampling is done   (a)   with replacement, (b)   without replacement.

1.5–20. Suppose that there are 5 A, 10 B, and 10 C students in a class of 25 students. If seven students are selected at random and without replacement from this group, compute the probability of obtaining three A, two B, and two C students in this sample of seven.

1.5–21. Show that the multinomial expansion is

$$(a_1+a_2+\cdots+a_s)^n=\sum\frac{n!}{n_1!n_2!\cdots n_s!}a_1^{n_1}a_2^{n_2}\cdots a_s^{n_s},$$

where the summation is over all nonnegative n_i such that

$$n=n_1+n_2+\cdots+n_s.$$

HINT: Use an argument similar to that used to justify the binomial expansion.

1.5–22. Find the term involving $a^4b^3c^2d^3$ in $(a+b+c+d)^{12}$. HINT: Use the result of Exercise 1.5–21.

1.5–23. In a baseball team of size 30, 5 are freshmen, 10 are sophomores, 12 are juniors, and 3 are seniors. If eight players are selected at random, what is the probability that the sample includes exactly two students from each class?

1.6 Conditional Probability

Suppose that we are given 20 tulip bulbs that are very similar in appearance and told that 8 tulips will bloom early, 12 will bloom late, 13 will be red, and 7 will be yellow, in accordance with the various combinations of Table 1.6–1.

	EARLY (E)	LATE (L)	TOTALS
RED (R)	5	8	13
YELLOW (Y)	3	4	7
TOTALS	8	12	20

TABLE 1.6–1

If one bulb is selected at random, the probability that it will produce a red tulip (R) is given by $P(R) = 13/20$, under the assumption that each bulb is "equally likely." Suppose, however, that close examination of the bulb will reveal whether it will bloom early (E) or late (L). If we consider an outcome only if it results in a tulip bulb that will bloom early, only eight outcomes in the sample space are now of interest. Thus, it is natural to assign, under this limitation, the probability of 5/8 to R; that is, $P(R|E) = 5/8$, where $P(R|E)$ is read as the probability of R given that E has occurred. Note that

$$P(R|E) = \frac{5}{8} = \frac{N(R \cap E)}{N(E)} = \frac{N(R \cap E)/20}{N(E)/20} = \frac{P(R \cap E)}{P(E)}.$$

This example is illustrative of a number of common situations. That is, in some random experiments, we are interested only in those outcomes that are elements of a subset B of the sample space S. This means, for our purposes, that the sample space is effectively the subset B. We are now confronted with the problem of defining a probability set function with B as the "new" sample space. That is, for a given event A, we want to define $P(A|B)$, the probability of A considering only those outcomes of the random experiment that are elements of B. The previous example gives us the clue to that definition. That is, for experiments in which each outcome is equally likely, it makes sense to define $P(A|B)$ by

$$P(A|B) = \frac{N(A \cap B)}{N(B)}.$$

If we divide the numerator and the denominator of this fraction by $N(S)$, the number of outcomes in the sample space, we have

$$P(A|B) = \frac{N(A \cap B)/N(S)}{N(B)/N(S)} = \frac{P(A \cap B)}{P(B)}.$$

We are thus led to the following definition.

DEFINITION 1.6–1. *The* conditional probability *of an event A, given that event B has occurred, is defined by*

$$P(A|B) = \frac{P(A \cap B)}{P(B)}$$

provided that $P(B) > 0$.

A formal use of the definition is given in the following examples.

Example 1.6–1.   If $P(A) = 0.4$, $P(B) = 0.5$, and $P(A \cap B) = 0.3$, then $P(A|B) = 0.3/0.5 = 0.6$ and $P(B|A) = 0.3/0.4 = 0.75$.

Example 1.6–2.   Let five cards be dealt from an ordinary deck of playing cards at random and without replacement. Let A be the event that exactly two cards are spades and B be the event that exactly three cards are hearts. Then

$$P(A) = \frac{\binom{13}{2}\binom{39}{3}}{\binom{52}{5}}, \quad P(B) = \frac{\binom{13}{3}\binom{39}{2}}{\binom{52}{5}}, \quad P(A \cap B) = \frac{\binom{13}{2}\binom{13}{3}}{\binom{52}{5}}$$

so that

$$P(B|A) = \frac{\binom{13}{3}}{\binom{39}{3}} \quad \text{and} \quad P(A|B) = \frac{\binom{13}{2}}{\binom{39}{2}}.$$

It is interesting to note that conditional probability satisfies the axioms for a probability function, namely, with $P(B) > 0$,

(a) $P(A|B) \geq 0$.
(b) $P(B|B) = 1$.
(c) $P(A_1 \cup A_2 \cup \cdots | B) = P(A_1|B) + P(A_2|B) + \cdots$, provided the events $A_1, A_2, \ldots$ are mutually exclusive.

Properties (a) and (b) are evident because

$$P(A|B) = \frac{P(A \cap B)}{P(B)} \geq 0$$

since $P(A \cap B) \geq 0$ and $P(B) > 0$, and

$$P(B|B) = \frac{P(B \cap B)}{P(B)} = \frac{P(B)}{P(B)} = 1.$$

Property (c) holds because

$$P(A_1 \cup A_2 \cup \cdots | B) = \frac{P[(A_1 \cup A_2 \cup \cdots) \cap B]}{P(B)}$$

$$= \frac{P[(A_1 \cap B) \cup (A_2 \cap B) \cup \cdots]}{P(B)}.$$

But $(A_1 \cap B), (A_2 \cap B), \ldots$ are also mutually exclusive events; so

$$P(A_1 \cup A_2 \cup \cdots | B) = \frac{P(A_1 \cap B) + P(A_2 \cap B) + \cdots}{P(B)}$$

$$= \frac{P(A_1 \cap B)}{P(B)} + \frac{P(A_2 \cap B)}{P(B)} + \cdots$$

$$= P(A_1 | B) + P(A_2 | B) + \cdots.$$

From the definition of the conditional probability set function, we observe that

$$P(A \cap B) = P(B)P(A | B).$$

This relation is frequently called the *multiplication rule* for probabilities. Sometimes, after considering the nature of the random experiment, one can make reasonable assumptions so that it is easier to assign $P(B)$ and $P(A|B)$ rather than $P(A \cap B)$. Then $P(A \cap B)$ can be computed with these assignments. This will be illustrated in Examples 1.6–3 and 1.6–4.

Example 1.6–3. A bowl contains seven blue chips and three red chips. Two chips are to be drawn successively at random and without replacement. We want to compute the probability that the first draw results in a red chip (A) and the second draw results in a blue chip (B). It is reasonable to assign the following probabilities:

$$P(A) = \frac{3}{10} \quad \text{and} \quad P(B | A) = \frac{7}{9}.$$

Thus, the probability of red on the first draw and blue on the second draw is

$$P(A \cap B) = \left(\frac{3}{10}\right)\left(\frac{7}{9}\right) = \frac{7}{30}.$$

Example 1.6–4. From an ordinary deck of playing cards, cards are to be drawn successively at random and without replacement. The probability that the third spade appears on the sixth draw is computed as follows. Let A be the event of two spades in the first five cards drawn, and let B be the event of a spade on the sixth draw. Thus, the probability that we wish to compute is $P(A \cap B)$. It is reasonable to take

$$P(A) = \frac{\binom{13}{2}\binom{39}{3}}{\binom{52}{5}} \quad \text{and} \quad P(B | A) = \frac{11}{47}.$$

The desired probability $P(A \cap B)$ is then the product of these two numbers.

The multiplication rule can be extended to three or more events. In the case of three events, we have, by using the multiplication rule for two events,

$$P(A \cap B \cap C) = P[(A \cap B) \cap C]$$
$$= P(A \cap B)P(C \,|\, A \cap B).$$

But

$$P(A \cap B) = P(A)P(B \,|\, A).$$

Hence,

$$P(A \cap B \cap C) = P(A)P(B \,|\, A)P(C \,|\, A \cap B).$$

This type of argument can be used to extend the multiplication rule to more than three events, and the general formula for k events can be officially proved by mathematical induction.

Example 1.6–5. Four cards are to be dealt successively at random and without replacement from an ordinary deck of playing cards. The probability of receiving in order a spade, a heart, a diamond, and a club is

$$\left(\frac{13}{52}\right)\left(\frac{13}{51}\right)\left(\frac{13}{50}\right)\left(\frac{13}{49}\right),$$

a result that follows from the extension of the multiplication rule. In this computation, the assumptions involved seem clear.

We close this section with a different type of example.

Example 1.6–6. A grade school boy has five blue and four white marbles in his left pocket, four blue and five white marbles in his right pocket. If he transfers one marble at random from his left to his right pocket, what is the probability of then drawing a blue marble from his right pocket? For notation let BL, BR, and WL denote drawing blue from left pocket, blue from right pocket, and white from left pocket, respectively. Then

$$P(BR) = P(BL \cap BR) + P(WL \cap BR)$$

$$= P(BL)P(BR \,|\, BL) + P(WL)P(BR \,|\, WL)$$

$$= \left(\frac{5}{9}\right)\left(\frac{5}{10}\right) + \left(\frac{4}{9}\right)\left(\frac{4}{10}\right) = \frac{41}{90}.$$

====== *Exercises* ======

(In solving certain of these exercises, make the usual and rather natural assumptions.)

1.6–1. Two cards are drawn successively and without replacement from an ordinary deck of playing cards. Compute the probability of drawing **(a)** two hearts; **(b)** a heart on the first draw, a club on the second draw; **(c)** a heart on the first draw, an ace on the second draw. HINT: In part (c), note that a heart can be drawn by getting the ace of hearts or one of the other 12 hearts.

1.6–2. Compute the probability of drawing three hearts in succession from a deck of cards if the draws are made at random and without replacement.

1.6–3. A hand of 13 cards is to be dealt at random and without replacement from an ordinary deck of playing cards. Find the conditional probability that there are at least three kings in the hand given that the hand contains at least two kings.

1.6–4. Bowl A contains three red and two white chips and bowl B contains four red and three white chips. A chip is drawn at random from bowl A and transferred to bowl B. Compute the probability of then drawing a red chip from bowl B.

1.6–5. A drawer contains eight pairs of socks. If six socks are taken at random and without replacement, compute the probability that there is at least one matching pair among these six socks. HINT: Compute the probability that there is not a matching pair.

1.6–6. A bowl contains ten chips. Four of the chips are red, five are white, and one is blue. If three chips are taken at random and without replacement, compute the conditional probability that there is one chip of each color given that there is exactly one red chip among the three.

1.6–7. A bowl contains three red and nine blue chips. The chips are drawn at random, one at a time without replacement. Compute the probability that the second red chip is the fifth chip drawn.

1.6–8. A fish bowl contains 12 gray and 3 red fish, all the same size. If the fish are selected at random from the bowl, one at a time without replacement, what is the probability that the third red fish is the seventh fish selected?

1.6–9. In a string of eight Christmas tree light bulbs, three are defective. The bulbs are selected at random and tested, one at a time, until the third defective bulb is found. Compute the probability that the third defective bulb is the **(a)** third bulb tested, **(b)** fifth bulb tested, **(c)** eighth bulb tested.

1.6–10. An urn contains three chips, two marked Failure and one marked Success. Players X and Y take turns drawing a single chip from the urn, each chip being returned to the urn before the next player draws. The winner of the game is the first person to draw Success. The game continues until there is a winner. If X draws first, what is his or her probability of winning the game? Simplify your answer.

1.6–11. A small grocery store had ten cartons of milk, two of which were sour. If you are going to buy the sixth carton of milk sold that day at random, compute the probability of selecting a carton of sour milk.

1.6–12. You are a member of a class of 18 students. A bowl contains 18 chips, 1 blue and 17 red. Each student is to take one chip from the bowl without replacement. The student who draws the blue chip is guaranteed an A for the course.

    (a) If you have a choice of drawing first, fifth, or last, which position would you choose? Justify your choice using probability.

    (b) Suppose the bowl contains 2 blue and 16 red chips. What position would you now choose?

1.6–13. Draw two cards from the set of cards that consists of the four queens and four kings. Compute the probability that the cards are both queens, given that **(a)** at least one card is a queen, **(b)** at least one card is a black queen, **(c)** one card is the queen of spades. HINT: List the 28 points in the sample space.

1.6–14. A drawer contains four black, six brown, and eight olive socks. Two socks are selected at random from the drawer.

    (a) Compute the probability that both socks are the same color.

    (b) Compute the probability that both socks are olive if it is known that they are the same color.

1.7 Bayes' Formula

Let us begin this section with an example.

Example 1.7–1. Bowl B_1 contains two red and four white chips, bowl B_2 contains one red and two white chips, and bowl B_3 contains five red and four white chips. Say that the probabilities for selecting the bowls are not the same but are given by $P(B_1) = 1/3$, $P(B_2) = 1/6$, and $P(B_3) = 1/2$. The experiment consists of selecting a bowl with these probabilities and then drawing a chip at random from that bowl. Let us compute the probability of drawing a red chip, say $P(R)$. Note that $P(R)$ is dependent first of all which bowl is selected and then on the probability of drawing a red chip from the selected bowl. That is, the event R is the union of the mutually exclusive events $B_1 \cap R$, $B_2 \cap R$, and $B_3 \cap R$. Thus

$$P(R) = P(B_1 \cap R) + P(B_2 \cap R) + P(B_3 \cap R)$$

$$= P(B_1)P(R|B_1) + P(B_2)P(R|B_2) + P(B_3)P(R|B_3)$$

$$= \left(\frac{1}{3}\right)\left(\frac{2}{6}\right) + \left(\frac{1}{6}\right)\left(\frac{1}{3}\right) + \left(\frac{1}{2}\right)\left(\frac{5}{9}\right) = \frac{4}{9}.$$

Suppose now that the outcome of the experiment is a red chip, but we do not know from which bowl it was drawn. Accordingly, we compute the conditional probability that the chip was drawn from bowl B_1, namely, $P(B_1|R)$.

From the definition of conditional probability and the above result, we have that

$$P(B_1|R) = \frac{P(B_1 \cap R)}{P(R)}$$

$$= \frac{P(B_1)P(R|B_1)}{P(B_1)P(R|B_1) + P(B_2)P(R|B_2) + P(B_3)P(R|B_3)}$$

$$= \frac{(1/3)(2/6)}{(1/3)(2/6) + (1/6)(1/3) + (1/2)(5/9)} = \frac{2}{8}.$$

Similarly, we have that

$$P(B_2|R) = \frac{P(B_2 \cap R)}{P(R)} = \frac{(1/6)(1/3)}{4/9} = \frac{1}{8}$$

and

$$P(B_3|R) = \frac{P(B_3 \cap R)}{P(R)} = \frac{(1/2)(5/9)}{4/9} = \frac{5}{8}.$$

Note that the conditional probabilities $P(B_1|R)$, $P(B_2|R)$, and $P(B_3|R)$ have changed from the original probabilities $P(B_1)$, $P(B_2)$, and $P(B_3)$ in a way that agrees with your intuition. Namely, once the red chip was observed, the probability concerning B_3 seems more favorable than originally because B_3 has a larger percentage of red chips than do B_1 and B_2. The conditional probabilities of B_1 and B_2 decreased from their original ones once the red chip was observed. Frequently, the original probabilities are called *prior probabilities* and the conditional probabilities are the *posterior* probabilities.

We generalize the result of Example 1.7-1. Let $B_1, B_2, \ldots, B_m$ constitute a *partition* of the sample space S. That is,

$$S = B_1 \cup B_2 \cup \cdots \cup B_m \quad \text{and} \quad B_i \cap B_j = \varnothing, \quad i \neq j.$$

Of course, the events $B_1, B_2, \ldots, B_m$ are mutually exclusive and exhaustive (since the union of the disjoint sets equals the sample space S). Furthermore, suppose the (prior) probability of the event B_i is positive; that is $P(B_i) > 0$, $i = 1, \ldots, m$. If A is an event, then A is the union of m mutually exclusive events, namely,

$$A = (B_1 \cap A) \cup (B_2 \cap A) \cup \cdots \cup (B_m \cap A).$$

Thus

$$P(A) = \sum_{i=1}^{m} P(B_i \cap A)$$

$$= \sum_{i=1}^{m} P(B_i)P(A|B_i). \tag{1}$$

If $P(A) > 0$, we have that

$$P(B_k|A) = \frac{P(B_k \cap A)}{P(A)}, \qquad k = 1, 2, \ldots, m. \qquad (2)$$

Using equation (1) and replacing $P(A)$ in expression (2), we have *Bayes' formula*:

$$P(B_k|A) = \frac{P(B_k)P(A|B_k)}{\sum_{i=1}^{m} P(B_i)P(A|B_i)}, \qquad k = 1, 2, \ldots, m.$$

The conditional probability $P(B_k|A)$ is often called the posterior probability of B_k. The following illustrates one application of Bayes' formula.

Example 1.7–2. In a certain factory, machines I, II, and III are all producing springs of the same length. Of their production, machines I, II, and III produce 2, 1, and 3 % defective springs, respectively. Of the total production of springs in the factory, machine I produces 35%, machine II produces 25%, and machine III produces 40%. If one spring is selected at random from the total springs produced in a day, the probability that it is defective, in an obvious notation, equals

$$P(D) = P(I)P(D|I) + P(II)P(D|II) + P(III)P(D|III)$$

$$= \left(\frac{35}{100}\right)\left(\frac{2}{100}\right) + \left(\frac{25}{100}\right)\left(\frac{1}{100}\right) + \left(\frac{40}{100}\right)\left(\frac{3}{100}\right) = \frac{215}{10,000}.$$

If the selected spring is defective, the conditional probability that it was produced by machine III is, by Bayes' formula,

$$P(III|D) = \frac{P(III)P(D|III)}{P(D)} = \frac{(40/100)(3/100)}{215/10,000} = \frac{120}{215}.$$

Please note how the posterior probability of III increased from the prior probability of III after the defective spring was observed because III produces a larger percentage of defectives than do I and II.

══ *Exercises* ══

1.7–1. Bowl B_1 contains two red and five white chips and bowl B_2 contains four red and three white chips. A fair die is cast. If the outcome is a multiple of 3 (namely, 3 or 6), a chip is taken from bowl B_2, otherwise a chip is taken from bowl B_1.
  (a) Compute the probability that a red chip is taken.
  (b) Given that the selected chip is red, compute the conditional probability that it was taken from bowl B_1.

1.7–2. Using the information given in Exercise 1.7–1 and letting W represent the event associated with the draw of a white chip, compute (a) $P(W)$, (b) $P(B_2|W)$, (c) $P(B_1|W) + P(B_2|W)$.

1.7–3. Bowl B_1 contains three white and nine red chips, bowl B_2 contains eight white and four red chips, bowl B_3 contains ten white and two red chips. A card is drawn from an ordinary deck of playing cards. If a face card (king, queen, jack) is drawn, a chip is selected from bowl B_1; if an ace is drawn, a chip is selected from bowl B_2; if any other card is drawn, a chip is selected from bowl B_3.

    (a) Compute the probability of drawing a white chip.

    (b) Suppose a bowl is selected by another student in this manner and handed to you without telling you which bowl you hold. If you select a white chip from the bowl, evaluate the conditional probability that you have bowl B_2; that is, compute $P(B_2|W)$.

1.7–4. Bowl A contains two red chips, bowl B contains two white chips, and bowl C contains one red chip and one white chip. A bowl is selected at random (with equal probabilities), and one chip is taken at random from that bowl.

    (a) Compute the probability of selecting a white chip, say $P(W)$.

    (b) If the selected chip is white, compute the conditional probability that the other chip in the bowl is red. HINT: Compute $P(C|W)$.

1.7–5. Bowl C contains six red chips and four blue chips. Five of these ten chips are selected at random and without replacement and put in bowl D, which was originally empty. One chip is then drawn at random from bowl D. Given that this chip is blue, find the conditional probability that two red chips and three blue chips were transferred from bowl C to bowl D.

1.7–6. A package, say A, of 24 crocus bulbs contains 8 yellow, 8 white, and 8 purple crocus bulbs. A package, say B, of 24 crocus bulbs contains 6 yellow, 6 white, and 12 purple crocus bulbs. One of the two packages is selected at random.

    (a) If three bulbs from this package were planted and all three yielded purple flowers, compute the conditional probability that package B was selected.

    (b) If the three bulbs yielded one yellow flower, one white flower, and one purple flower, compute the conditional probability that package A was selected.

1.8 Independent Events

For certain pairs of events, the occurrence of one of them may or may not change the probability of the occurrence of the other. In the latter case they are said to be *independent events*. However, before giving the formal definition of independence, let us consider an example.

Example 1.8–1. Flip a coin twice and observe the sequence of heads and tails. The sample space is then

$$S = \{HH, HT, TH, TT\}.$$

It is reasonable to assign 1/4 to each of these four outcomes. Let

$$C = \{HH\},$$
$$B = \{\text{heads on first flip}\} = \{HH, HT\},$$
$$A = \{\text{tails on the second flip}\} = \{HT, TT\}.$$

Now $P(B) = 1/2$. However, if we are given that C has occurred, then $P(B|C) = 1$ because $C \subset B$. That is, the knowledge of the occurrence of C has changed the probability of B. On the other hand, if we are given that B has occurred,

$$P(A|B) = \frac{P(A \cap B)}{P(B)} = \frac{1/4}{1/2} = \frac{1}{2} = P(A).$$

So the occurrence of B has not changed the probability of A. That is, the probability of A does not depend upon knowledge about event B, so we say that A and B are independent events. That is, events A and B are independent if the occurrence of one of them does not affect the probability of the occurrence of the other. A more mathematical way of saying this is

$$P(A|B) = P(A) \qquad \text{or} \qquad P(B|A) = P(B),$$

provided that $P(B) > 0$ or, in the latter case, $P(A) > 0$. With the first of these equalities and the multiplication rule, we have

$$P(A \cap B) = P(B)P(A|B) = P(B)P(A).$$

The second of these equalities, namely $P(B|A) = P(B)$, gives us the same result:

$$P(A \cap B) = P(A)P(B|A) = P(A)P(B).$$

This example motivates the following definition of *independent* events.

DEFINITION 1.8–1. *Events A and B are* independent *if and only if*

$$P(A \cap B) = P(A)P(B).$$

Otherwise A and B are called dependent *events.*

Events that are independent are sometimes called *statistically independent, stochastically independent,* or *independent in a probability sense,* but in most instances we use independent without a modifier if there is no possibility of misunderstanding. It is interesting to note that the definition always holds if $P(A) = 0$ or $P(B) = 0$ because then $P(A \cap B) = 0$ since $(A \cap B) \subset A$ and $(A \cap B) \subset B$. Thus, the left-hand and right-hand members of $P(A \cap B) = P(A)P(B)$ both equal zero and thus are equal to each other.

Example 1.8–2.   Cast a die twice. Let

$$A = \{1 \text{ or } 2 \text{ on the first roll}\},$$
$$B = \{2, 3, \text{ or } 4 \text{ on the second roll}\}.$$

Of the 36 possible outcomes, 12 are favorable to A, 18 are favorable to B, and 6 are favorable to $A \cap B$. Thus, assigning equal probability to each of the 36 outcomes, we have

$$P(A)P(B) = \left(\frac{12}{36}\right)\left(\frac{18}{36}\right) = \frac{6}{36} = P(A \cap B).$$

Hence A and B are independent events by Definition 1.8–1.

THEOREM 1.8–1.   *If A and B are independent events, then the following pairs of events are also independent:*

(i)   *A and B';*
(ii)   *A' and B;*
(iii)   *A' and B'.*

Proof of (i):   With $P(A) > 0$, we have

$$P(B|A) + P(B'|A) = \frac{P(B \cap A)}{P(A)} + \frac{P(B' \cap A)}{P(A)} = \frac{P(A)}{P(A)} = 1.$$

$$= \frac{P[(B \cap A) \cup (B' \cap A)]}{P(A)}$$

That is, $P(B'|A) = 1 - P(B|A)$. Thus,

$$P(A \cap B') = P(A)P(B'|A) = P(A)[1 - P(B|A)]$$
$$= P(A)[1 - P(B)]$$
$$= P(A)P(B'),$$

since $P(B|A) = P(B)$ by hypothesis. Thus A and B' are independent events.

The proofs for parts (ii) and (iii) are left as exercises.

Before extending the definition of independent events to more than two events, we present the following example.

Example 1.8–3.   An urn contains four balls numbered 1, 2, 3, and 4. One ball is to be drawn at random from the urn. Let the events A, B, and C be defined by $A = \{1, 2\}$, $B = \{1, 3\}$, $C = \{1, 4\}$. Then $P(A) = P(B) = P(C) = 1/2$. Furthermore,

$$P(A \cap B) = \frac{1}{4} = P(A)P(B),$$

$$P(A \cap C) = \frac{1}{4} = P(A)P(C),$$

$$P(B \cap C) = \frac{1}{4} = P(B)P(C),$$

which implies that A, B, and C are independent in pairs (called *pairwise independence*). However,

$$P(A \cap B \cap C) = \frac{1}{4} \neq \frac{1}{8} = P(A)P(B)P(C).$$

That is, something seems to be lacking for the complete independence of A, B, and C.

This example illustrates the reason for the second condition in Definition 1.8–2.

DEFINITION 1.8–2. *Events A, B, and C are mutually independent if and only if the following two conditions hold:*

(i) *They are pairwise independent; that is, $P(A \cap B) = P(A)P(B)$, $P(A \cap C) = P(A)P(C)$, and $P(B \cap C) = P(B)P(C)$.*
(ii) $P(A \cap B \cap C) = P(A)P(B)P(C)$.

Definition 1.8–2 can be extended to mutual independence of four or more events. In this extension, each pair, triple, quartet, and so on must satisfy this type of multiplication rule.

Example 1.8–4. Toss a coin three times. Let the event A_i denote heads on the ith toss. The sample space is

$$S = \{HHH, HHT, HTH, THH, HTT, THT, TTH, TTT\}.$$

We assume that each element of this space has probability 1/8. Thus, we have

$$P(A_i) = \frac{4}{8}, \qquad i = 1, 2, 3;$$

$$P(A_i \cap A_j) = \frac{2}{8}, \qquad i \neq j;$$

$$P(A_1 \cap A_2 \cap A_3) = \frac{1}{8}.$$

Hence

$$P(A_i \cap A_j) = \frac{1}{4} = P(A_i)P(A_j), \qquad i \neq j,$$

and

$$P(A_1 \cap A_2 \cap A_3) = \frac{1}{8} = P(A_1)P(A_2)P(A_3).$$

Accordingly, A_1, A_2, and A_3 are mutually independent.

The proof and illustration of the following results are left as exercises. If A, B, and C are mutually independent events, then the following events are also independent:

(i) A and $(B \cap C)$
(ii) A and $(B \cup C)$
(iii) A' and $(B \cap C')$

In addition, A', B', and C' are mutually independent.

Many experiments consist of a sequence of trials that are independent in the following sense.

DEFINITION 1.8–3.   *Let an experiment E consist of a sequence of n trials, T_1, T_2, ..., T_n. The trials are independent if the outcome on any one trial does not affect the probabilities of outcomes on other trials. Moreover, if event A_i is associated with the ith trial, and $P(A_i)$ is its probability, $i = 1, 2, ..., n$, then*

$$P(A_1 \cap A_2 \cap \cdots \cap A_n) = P(A_1)P(A_2) \cdots P(A_n).$$

Trials that are not independent are said to be dependent.

The sample space for an experiment of n trials is a set of n-tuples, where the ith component denotes the outcome on the ith trial. For example, if a die is rolled five times,

$$S = \{(O_1, O_2, O_3, O_4, O_5): \quad O_i = 1, 2, 3, 4, 5, \text{ or } 6, \text{ for } i = 1, 2, 3, 4, 5\}.$$

That is, S is a set of five-tuples where each component is one of the first six positive integers.

If a coin is tossed three times,

$$S = \{(O_1, O_2, O_3): \quad O_i = \text{H or T}, i = 1, 2, 3\}.$$

We often drop the commas and parentheses and let, for illustration, (H, T, T) = HHT as in example 1.8–4.

Example 1.8–5.   Urn I contains two red balls and three white balls, urn II contains two red balls and one white ball, urn III contains one red ball and three white balls. At the ith trial a ball is drawn from urn i, $i = $ I, II, III. Assume that the trials are independent. Under reasonable assumptions, the probabilities of some of the possible outcomes are

$$P(\{(R, R, R)\}) = \left(\frac{2}{5}\right)\left(\frac{2}{3}\right)\left(\frac{1}{4}\right),$$

$$P(\{(W, R, W)\}) = \left(\frac{3}{5}\right)\left(\frac{2}{3}\right)\left(\frac{3}{4}\right),$$

and

$$P(\{(R, W, W)\}) = \left(\frac{2}{5}\right)\left(\frac{1}{3}\right)\left(\frac{3}{4}\right).$$

Example 1.8–6. An urn contains three red, two white, and four yellow balls. An ordered sample of size 3 is drawn from the urn. If the balls are drawn with replacement, so that one outcome does not affect the others, the trials are independent. Under reasonable assumptions, the probabilities of the two given outcomes (in an obvious notation) are

$$P(RWY) = \left(\frac{3}{9}\right)\left(\frac{2}{9}\right)\left(\frac{4}{9}\right) = \frac{8}{243}$$

and

$$P(\{YYR, RWW\}) = \left(\frac{4}{9}\right)\left(\frac{4}{9}\right)\left(\frac{3}{9}\right) + \left(\frac{3}{9}\right)\left(\frac{2}{9}\right)\left(\frac{2}{9}\right) = \frac{20}{243}.$$

If the balls are drawn without replacement, the trials are dependent. The probabilities, again under reasonable assumptions, of the two outcomes are

$$P(RWY) = \left(\frac{3}{9}\right)\left(\frac{2}{8}\right)\left(\frac{4}{7}\right) = \frac{1}{21}$$

and

$$P(\{YYR, RWW\}) = \left(\frac{4}{9}\right)\left(\frac{3}{8}\right)\left(\frac{3}{7}\right) + \left(\frac{3}{9}\right)\left(\frac{2}{8}\right)\left(\frac{1}{7}\right) = \frac{7}{84}.$$

===== *Exercises* =====

1.8–1. Roll a red die and a white die. Record the results so that there are 36 points in the sample space. Let $A = \{4$ on the red die$\}$ and $B = \{$sum of the numbers is odd$\}$. If we assume that each of the 36 points of the sample space has the same probability 1/36, are events A and B independent? Why?

1.8–2. Flip a coin and then independently cast a die. Compute the probability of observing heads on the coin and a 2 or 4 on the die.

1.8–3. Prove parts (ii) and (iii) of Theorem 1.8–1.

1.8–4. Let A and B be independent events with $P(A) = 1/4$ and $P(B) = 2/3$. Compute
(a) $P(A \cap B)$,　　　　　　(b) $P(A \cap B')$,　　　　　　(c) $P(A' \cap B')$,
(d) $P[(A \cup B)']$,　　　　(e) $P(A' \cap B)$.

1.8–5. If $P(A) = 0.8$, $P(B) = 0.5$, and $P(A \cup B) = 0.9$, are A and B independent events? Why?

1.8–6. If A, B, and C are mutually independent, show that the following pairs of events are independent: A and $(B \cap C)$, A and $(B \cup C)$, A' and $(B \cap C')$. Also show that A', B', and C' are mutually independent.

1.8–7. Each of three men fires one shot at a target. Let A_i denote the event that the target is hit by man i, $i = 1, 2, 3$. If we assume A_1, A_2, A_3 are mutually independent and if $P(A_1) = 0.7$, $P(A_2) = 0.9$, $P(A_3) = 0.8$, compute the probability that exactly two men hit the target (that is, one misses).

1.8–8. If the events A and B are mutually exclusive, are A and B always independent? If the answer is no, can they ever be independent? Explain.

1.8–9. If $A \subset B$, can A and B ever be independent events? Explain.

1.8–10. Flip an unbiased coin five independent times. Compute the probability of (a) HHTHT, (b) THHHT, (c) HTHTH, (d) three heads occurring in the five trials.

1.8–11. An urn contains two red balls and four white balls. Sample successively five times at random and with replacement so that the trials are independent. Compute the probability of (a) WWRWR, (b) RWWWR, (c) WRWRW, (d) three whites occurring in the five trials.

1.8–12. A rocket has a built-in redundant system. In this system if component A fails, it is by-passed and component B is used. If component B fails, it is by-passed and component C is used. The probability of failure of any one of these three components is 0.05, and the failures of these components are mutually independent events. What is the probability that this redundant system does *not* fail? HINT: Compute the probability of the complement, namely that A, B, and C all fail.

1.8–13. An urn contains seven white balls, three red balls, and five blue balls. Sample successively six times at random and with replacement so that the trials are independent. Compute (a) $P(\text{WRWBBW})$, (b) $P(\text{BWRBWW})$, (c) the probability of drawing three white balls, one red ball, and two blue balls.

2
DISTRIBUTIONS OF THE DISCRETE TYPE

2.1 Random Variables of the Discrete Type

Let X denote a random variable with space R. Suppose we know how the probability is distributed over the various subsets A of R; that is, we can compute $P(X \in A)$. In this sense, we speak of the distribution of the random variable X, meaning, of course, the distribution of probability associated with the space R of X.

Let X denote a random variable with one-dimensional space R, a subset of the real numbers. Suppose that the space R contains a countable number of points; that is, R contains only a finite number of points or the points of R can be put into a one-to-one correspondence with the positive integers. Such a set R is called a set of discrete points. Furthermore, the random variable X is called a random variable of the *discrete type*, and X is said to have a distribution of the discrete type.

For a random variable X of the discrete type, the induced probability $P(X = x)$ is frequently denoted by $f(x)$, and this function $f(x)$ is called the *probability density function*. Note that some authors refer to $f(x)$ as the probability function, the frequency function, or the probability mass function. We prefer probability density function, and it is hereafter abbreviated p.d.f.

Let $f(x)$ be the p.d.f. of the random variable X of the discrete type and let R be the space of X. The p.d.f. $f(x)$ is a real-valued function and satisfies the following properties since $f(x) = P(X = x), x \in R$:

(a) $f(x) > 0, \qquad x \in R;$

(b) $\sum_{x \in R} f(x) = 1;$

(c) $P(X \in A) = \sum_{x \in A} f(x), \qquad$ where $\quad A \subset R.$

We sometimes let $f(x) = 0$ when $x \notin R$ and thus the domain of $f(x)$ is the set of real numbers. When we define the p.d.f. $f(x)$ and do not say zero elsewhere, then we tacitly mean that $f(x)$ has been defined at all x's in the space R and it is assumed that $f(x) = 0$ elsewhere, namely, $f(x) = 0, x \notin R$. Since the probability $P(X = x) = f(x) > 0$ when $x \in R$ and since R contains all the probability associated with X, it is sometimes referred to as the *support* of X as well as the space of X.

Example 2.1–1. A young boy went fishing and caught $n = 10$ bluegills of which $n_1 = 4$ were under six inches in length. When he returned home he pulled two fish from the pail at random and without replacement so that each pair of fish has the same probability of being drawn. If X denotes the number of fish under six inches in this sample, the p.d.f. of X is defined by

$$f(x) = \frac{\binom{4}{x}\binom{6}{2-x}}{\binom{10}{2}}, \qquad x = 0, 1, 2.$$

Note that X has a hypergeometric distribution.

The graph of the p.d.f. of X would be a plot of the points $\{(x, f(x)): \ x \in R\}$, where R is the space of X. However, it is easier to visualize the corresponding probabilities if a vertical line segment is drawn from each $(x, f(x))$ to $(x, 0)$, $x \in R$, to form a *bar graph*. The p.d.f. defined in Example 2.1–1 is plotted as a bar graph in Figure 2.1–1.

Instead of considering events such as $(X = x)$ and the corresponding induced probabilities $f(x) = P(X = x)$, let us now consider events of the form $(X \le x)$ and the induced probabilities $P(X \le x)$. The following example will help the reader follow the more general discussion.

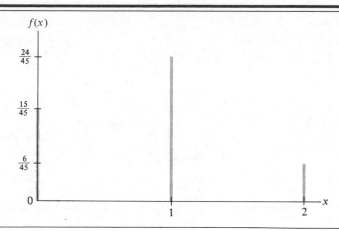

FIGURE 2.1–1

Example 2.1–2. Let the random variable X of the discrete type have the p.d.f. $f(x) = x/6$, $x = 1, 2, 3$. Then, for example,

$$P(X \leq 1) = f(1) = \frac{1}{6},$$

$$P(X \leq 3) = f(1) + f(2) + f(3) = \frac{1}{6} + \frac{2}{6} + \frac{3}{6} = 1,$$

$$P(X \leq 0) = 0,$$

$$P\left(X \leq \frac{3}{2}\right) = \frac{1}{6},$$

and

$$P\left(X \leq \frac{7}{3}\right) = \frac{1}{6} + \frac{2}{6} = \frac{1}{2}.$$

We let $F(x) = P(X \leq x)$. The function $F(x)$ is defined for each real number x, and its graph is depicted in Figure 2.1–2. Note that $F(x)$ is a nondecreasing and right-hand continuous function that has a minimum value of zero and a maximum value of one. The height of each jump at x corresponds to $P(X = x)$, $x = 1, 2, 3$.

Let X be a random variable of the discrete type with space R and p.d.f. $f(x) = P(X = x)$, $x \in R$. Now take x to be a real number and consider the set A of all points in R that are less than or equal to x. That is,

$$A = \{t: \quad t \leq x \quad \text{and} \quad t \in R\}.$$

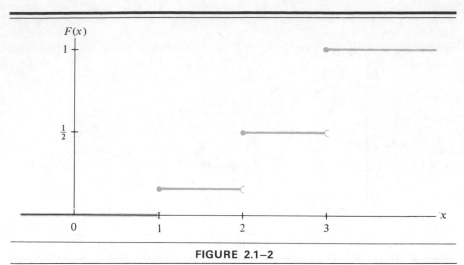

FIGURE 2.1–2

Define the function $F(x)$ by

$$F(x) = P(X \leq x) = \sum_{t \in A} f(t).$$

The function $F(x)$ is called the *distribution function* (sometimes, *cumulative distribution function*) of the discrete-type random variable X.

Several properties of a distribution function $F(x)$ can be listed as a consequence of the properties of the probability set function. (We use the symbols $F(\infty)$ and $F(-\infty)$ to mean

$$\lim_{x \to \infty} F(x) \quad \text{and} \quad \lim_{x \to -\infty} F(x),$$

respectively. In like manner, the symbols $\{x: \ x < \infty\}$ and $\{x: \ x < -\infty\}$ represent, respectively, the entire one-dimensional set and the null set.) The properties are

(a) $0 \leq F(x) \leq 1$ because $F(x)$ is a probability.
(b) $F(x)$ is a nondecreasing function of x. For if $x' < x''$, then

$$\{x: \ x \leq x'\} \subset \{x: \ x \leq x''\}.$$

Thus,

$$F(x) = P(X \leq x') \leq P(X \leq x'') = F(x''),$$

by Theorem 1.3–3, for the probability set function.
(c) $F(\infty) = 1$ and $F(-\infty) = 0$ because the set $\{x: \ x < \infty\}$ is the entire one-dimensional space and the set $\{x: \ x < -\infty\}$ is the null set.
(d) If X is a random variable of the discrete type, then $F(x)$ is a step function and the height of a step at x, $x \in R$, equals the probability $P(X = x)$.

It is clear that the probability distribution associated with the random variable X can be described by either the distribution function $F(x)$ or by the probability density function $f(x)$, and the function used is a matter of convenience.

══ *Exercises* ══

2.1–1. A bowl contains three red chips and five blue chips. Two chips are drawn at random and without replacement from the bowl. Let X denote the number of red chips in the sample.
   (a) Find the p.d.f. of X and draw the corresponding bar graph.
   (b) Draw the distribution function of X.

2.1–2. Let a chip be taken at random from a bowl that contains six white chips, three red chips, and one blue chip. Let the random variable $X = 1$ if the outcome is a white chip, let $X = 5$ if the outcome is a red chip, and let $X = 10$ if the outcome is a blue chip.
   (a) Find the p.d.f. of X.
   (b) Graph the p.d.f. as a bar graph.
   (c) Find the distribution function of X.
   (d) Graph the distribution function.

2.1–3. For each of the following, determine the constant c so that $f(x)$ satisfies the conditions of being a p.d.f. for a random variable X.
   (a) $f(x) = x/c$, $x = 1, 2, 3, 4$;
   (b) $f(x) = cx$, $x = 1, 2, 3, \ldots, 10$;
   (c) $f(x) = c(1/3)^x$, $x = 1, 2, 3, \ldots$;
   (d) $f(x) = c(x + 1)^2$, $x = 0, 1, 2, 3$.

2.1–4. For each of the probability density functions defined in Exercise 2.1–3, find $P(X = 1 \text{ or } 2)$, $P(X = 3.5)$, $P(X \le 2)$, $P(1/2 \le X \le 5/2)$.

2.1–5. Let a random experiment be the cast of a pair of unbiased dice, and let the random variable Y denote the sum of the dice.
   (a) Define the p.d.f. $g(y)$ of Y.
   (b) Find $P(Y \le 1)$, $P(Y \le 3)$, $P(Y \le 7/2)$.
   (c) Plot the distribution function $G(y)$ of Y.

2.1–6. Let $f(x)$ be the p.d.f. of a random variable X. Find the distribution function $F(x)$ of X and sketch its graph.
   (a) $f(x) = 1, x = 3$;   (b) $f(x) = 1/3, x = 1, 2, 3$;
   (c) $f(x) = x/15, x = 1, 2, 3, 4, 5$;   (d) $f(x) = (1/4)(3/4)^x, x = 0, 1, 2, \ldots$.

2.2 Mathematical Expectation

An extremely important concept in summarizing important characteristics of distributions of probability is that of mathematical expectation, which is introduced by an example.

Example 2.2–1. An enterprising young man who needs a little extra money devises a game of chance in which some of his friends might wish to participate. The game that he proposes is to let the participant cast an unbiased die and then receive a payment according to the following schedule: If the event $A = \{1, 2, 3\}$ occurs, he receives 1¢; if $B = \{4, 5\}$ occurs, he receives 5¢; and if $C = \{6\}$ occurs, he receives 35¢. The probabilities of the respective events are assumed to be 3/6, 2/6, and 1/6, since the die is unbiased. The problem that now faces the young man is the determination of the amount that should be charged for the opportunity of playing the game. He reasons, correctly, that if the game is played a large number of times, about 3/6 of the trials will require a payment of 1¢, about 2/6 of them will require one of 5¢, and about 1/6 of them will require one of 35¢. Thus, the approximate average payment is

$$(1)\left(\frac{3}{6}\right) + (5)\left(\frac{2}{6}\right) + (35)\left(\frac{1}{6}\right) = 8.$$

That is, he expects to pay 8¢ "on the average." Note that he never pays exactly 8¢; the payment is either 1¢, 5¢, or 35¢. However, the "weighted average" of 1, 5, and 35, in which the weights are the respective probabilities 3/6, 2/6, and 1/6, equals eight. Such a weighted average is called the *mathematical expectation* of payment. Thus, if the young man decides to charge 10¢ per play, he would make 2¢ per play "on the average." Since the most that a player would lose at the charge of 10¢ per play is 9¢, the young man might find that several players are attracted by the possible gain of 25¢.

A more mathematical way of formulating the preceding example would be to let X be the random variable defined by the outcome of the cast of the die. Thus, the p.d.f. of X is

$$f(x) = \frac{1}{6}, \qquad x = 1, 2, 3, 4, 5, 6.$$

In terms of the observed value x, the payment is given by the function

$$u(x) = \begin{cases} 1, & x = 1, 2, 3, \\ 5, & x = 4, 5, \\ 35, & x = 6. \end{cases}$$

The mathematical expectation of payment is then equal to

$$\sum_{x=1}^{6} u(x) f(x) = (1)\left(\frac{1}{6}\right) + (1)\left(\frac{1}{6}\right) + (1)\left(\frac{1}{6}\right)$$

$$+ (5)\left(\frac{1}{6}\right) + (5)\left(\frac{1}{6}\right) + (35)\left(\frac{1}{6}\right)$$

$$= (1)\left(\frac{3}{6}\right) + (5)\left(\frac{2}{6}\right) + (35)\left(\frac{1}{6}\right)$$

$$= 8.$$

This discussion suggests the more general definition of mathematical expectation of a function of X.

DEFINITION 2.2–1. *If $f(x)$ is the p.d.f. of the random variable X of the discrete type with space R and if the summation*

$$\sum_R u(x)f(x) = \sum_{x \in R} u(x)f(x)$$

exists, then the sum is called the mathematical expectation *or the* expected value *of the function $u(X)$, and it is denoted by $E[u(X)]$. That is,*

$$E[u(X)] = \sum_R u(x)f(x).$$

We can think of the expected value $E[u(X)]$ as a weighted mean of $u(x)$, $x \in R$, where the weights are the probabilities $f(x) = P(X = x)$, $x \in R$.

Before presenting additional examples, we list some useful facts about mathematical expectation in the following theorem.

THEOREM 2.2–1. *When it exists, mathematical expectation E satisfies the following properties:*

(i)   *If c is a constant, $E(c) = c$.*

(ii)   *If c is a constant and u is a function,*

$$E[cu(X)] = cE[u(X)].$$

(iii)   *If c_1 and c_2 are constants and u_1 and u_2 are functions, then*

$$E[c_1 u_1(X) + c_2 u_2(X)] = c_1 E[u_1(X)] + c_2 E[u_2(X)].$$

Proof: First, we have for the proof of (i) that

$$E(c) = \sum_R cf(x) = c \sum_R f(x) = c$$

because

$$\sum_R f(x) = 1.$$

Next, to prove (ii), we see that

$$E[cu(X)] = \sum_R cu(x)f(x)$$

$$= c \sum_R u(x)f(x)$$

$$= cE[u(X)].$$

Finally, the proof of (iii) is given by

$$E[c_1 u_1(X) + c_2 u_2(X)] = \sum_R [c_1 u_1(x) + c_2 u_2(x)] f(x)$$

$$= \sum_R c_1 u_1(x) f(x) + \sum_R c_2 u_2(x) f(x).$$

By applying (ii) we obtain

$$E[c_1 u_1(X) + c_2 u_2(X)] = c_1 E[u_1(X)] + c_2 E[u_2(X)].$$

Property (iii) can be extended to more than two terms by mathematical induction; that is, we have

(iii)' $\quad E\left[\sum_{i=1}^{k} c_i u_i(X)\right] = \sum_{i=1}^{k} c_i E[u_i(X)].$

Because of property (iii)', mathematical expectation E is called a *linear* or *distributive* operator.

Example 2.2–2.   Let X have the p.d.f.

$$f(x) = \frac{x}{10}, \qquad x = 1, 2, 3, 4.$$

Then

$$E(X) = \sum_{x=1}^{4} x\left(\frac{x}{10}\right) = (1)\left(\frac{1}{10}\right) + (2)\left(\frac{2}{10}\right) + (3)\left(\frac{3}{10}\right) + (4)\left(\frac{4}{10}\right) = 3,$$

$$E(X^2) = \sum_{x=1}^{4} x^2\left(\frac{x}{10}\right) = (1)^2\left(\frac{1}{10}\right) + (2)^2\left(\frac{2}{10}\right) + (3)^2\left(\frac{3}{10}\right) + (4)^2\left(\frac{4}{10}\right) = 10,$$

and

$$E[X(5 - X)] = 5E(X) - E(X^2) = (5)(3) - 10 = 5.$$

Example 2.2–3.   Let $u(x) = (x - b)^2$, where b is not a function of X, and suppose $E[(X - b)^2]$ exists. To find that value of b for which $E[(X - b)^2]$ is a minimum, we write

$$E[(X - b)^2] = E[X^2 - 2bX + b^2]$$
$$= E(X^2) - 2bE(X) + b^2$$

because $E(b^2) = b^2$. Thus, the derivative of $E[(X - b)^2]$ with respect to b is $-2E(X) + 2b$. This derivative is equal to zero when $b = E(X)$ and $E(X)$ is the value of b that minimizes $E[(X - b)^2]$. In the next section we learn that $E(X)$ and $E[(X - b)^2]$, when $b = E(X)$, have special names.

Sometimes we hear expressions such as the *odds in favor* of an event are 2 to 1. This simply means that the probability that the event will occur is 2/3, and the person betting on the event must put up $2 for each $1 against the event in order to create a fair bet. Likewise, if the *odds against* an event happening are 4 to 1, then the probability of the occurrence of that event is 1/5. Another not uncommon expression is that a bet on an event will earn *3 for 1*, meaning that a $1 bet will produce a return of $3 if the event occurs. Since the bettor has already given $1 of the $3, the result is equivalent to odds of 2 to 1 against the event or a probability of 1/3 that the event will occur. Although we will not use these expressions in this book, the student of probability should be aware of them.

═══ *Exercises* ═══

2.2–1. The number 3 appears on each of two chips in a bowl and the number 9 on a third chip in the bowl. Consider the following "game." A player draws one chip at random from the bowl and receives either $3 or $9, depending upon the number on that chip. If he or she plays this game a large number of times, how many dollars can the player expect to average per play.

2.2–2. Three chips are taken at random and without replacement from a bowl that has $n_1 = 5$ red chips and $n_2 = 4$ blue chips. If X equals the number of red chips among the three drawn, determine $E(X)$.

2.2–3. From a bowl containing five chips, four of which are marked $1 and one $24, a person selects two chips at random and without replacement. If the person is to receive the sum of the two respective amounts, compute the mathematical expectation of payment.

2.2–4. Let the random variable X have the p.d.f.

$$f(x) = \frac{(|x| + 1)^2}{9}, \qquad x = -1, 0, 1.$$

Compute $E(X)$, $E(X^2)$, and $E(3X^2 - 2X + 4)$.

2.2–5. Two unbiased dice are cast. A payment equal to the sum of the spots on the top sides is given the caster. Compute the expected value of the payment.

2.2–6. A bowl contains ten chips, eight of which are marked $1 and two $3. If three chips are selected at random and without replacement, compute the expected value of the payment that is equal to the sum of the values on the three chips in the sample.

2.2–7. In a particular lottery 3,000,000 tickets are sold each week for 50¢ apiece. Out of the 3,000,000 tickets, 12,006 are drawn and awarded prizes: 12,000 $25 prizes, four $10,000 prizes, one $50,000 prize, and one $200,000 prize. If you purchased a single ticket each week, what is the expected value of this game to you?

2.2–8. Each card in an ordinary deck is to have a value of zero except the jacks, queens, and kings, each of which has value one, and aces, each of which has value two. Let us select two cards at random and without replacement from this deck. Say the payment will equal the product of the values of those two cards. Compute the expected value of payment.

2.2–9. Let us select at random a number from the first n positive integers. If the payment is equal to the reciprocal of the number, find an expression for the expected payment. Evaluate this number when $n = 5$. Approximate this number when $n = 100$.

2.2–10. Let X be a random variable with support $\{1, 2, 3, 5, 15, 25, 50\}$, each point of which has the same probability $1/7$. Argue that $c = 5$ is the value that minimizes $E(|X - c|)$. Compare this to the value of b that minimizes $E[(X - b)^2]$.

2.2–11. Let X be the number selected at random from the first 10 positive integers; assuming equal probabilities on these 10 integers, compute $E[X(11 - X)]$.

2.2–12. A person bets \$1 to \$$z$ that he or she can draw, at random and without replacement, from an ordinary deck of cards two cards that will be of the same suit. Find z so that this is a fair bet.

2.3 The Mean and the Variance

Certain mathematical expectations are so important that they have special names. In this section we consider two of them: the mean and the variance.

If X is a random variable with p.d.f. $f(x)$ of the discrete type and space

$$R = \{b_1, b_2, b_3, \ldots\},$$

then, with $u(x) = x$,

$$E(X) = \sum_R xf(x)$$

$$= b_1 f(b_1) + b_2 f(b_2) + b_3 f(b_3) + \cdots$$

is the weighted mean of the numbers belonging to R, where the weights are given by the p.d.f. $f(x)$. We call $E(X)$ the *mean* of X (or the mean of the distribution) and denote it by μ. That is, $\mu = E(X)$.

REMARK. In mechanics, the weighted average of the points $b_1, b_2, b_3, \ldots$ in one-dimensional space is called the centroid of the system. Those readers without a mechanics background can think of the centroid as being the point of balance for the system in which the weights $f(b_1), f(b_2), f(b_3), \ldots$ are placed upon the points $b_1, b_2, b_3, \ldots$.

Example 2.3–1. Let X have the p.d.f.

$$f(x) = \begin{cases} \dfrac{1}{8}, & x = 0, 3, \\ \dfrac{3}{8}, & x = 1, 2. \end{cases}$$

The mean of X is

$$\mu = E(X) = 0\left(\frac{1}{8}\right) + 1\left(\frac{3}{8}\right) + 2\left(\frac{3}{8}\right) + 3\left(\frac{1}{8}\right) = \frac{3}{2}.$$

The next example shows that if the outcomes of X are equally likely (that is, have the same probability), the mean of X is the arithmetic average of these outcomes.

Example 2.3–2. Roll a fair die and let X denote the outcome. Thus, X has the p.d.f. $f(x) = 1/6$, $x = 1, 2, 3, 4, 5, 6$. Then

$$E(X) = \sum_{x=1}^{6} x\left(\frac{1}{6}\right) = \frac{1 + 2 + 3 + 4 + 5 + 6}{6} = \frac{7}{2},$$

which is the arithmetic average of the first six positive integers.

We have noted that the mean $\mu = E(X)$ is the centroid of a system of weights or a measure of the central location of the probability distribution of X. A measure of the dispersion or spread of a distribution is defined as follows. If $u(x) = (x - \mu)^2$ and $E[(X - \mu)^2]$ exists, the *variance*, frequently denoted by σ^2 or $\mathrm{Var}(X)$, of a random variable X of the discrete type (or variance of the distribution) is defined by

$$\sigma^2 = E[(X - \mu)^2] = \sum_{R} (x - \mu)^2 f(x).$$

The positive square root of the variance is called the *standard deviation* of X and is denoted by

$$\sigma = \sqrt{\mathrm{Var}(X)} = \sqrt{E[(X - \mu)^2]}.$$

Example 2.3–3. Let X denote the outcome when rolling a fair die. From Example 2.3–2 we know that $\mu = 7/2 = 3.5$. Thus,

$$\sigma^2 = E[(X - 3.5)^2] = \sum_{x=1}^{6} (x - 3.5)^2 \frac{1}{6}$$

$$= [(1 - 3.5)^2 + (2 - 3.5)^2 + \cdots + (6 - 3.5)^2]\left(\frac{1}{6}\right)$$

$$= \frac{35}{12}.$$

The standard deviation of X is $\sigma = \sqrt{35/12} = 1.708$, approximately.

It is worthwhile to note that the variance can be computed in another manner. We have

$$\sigma^2 = E[(X - \mu)^2] = E(X^2 - 2\mu X + \mu^2),$$

which, by the distributive property of E, is

$$\sigma^2 = E(X^2) - 2\mu E(X) + \mu^2 = E(X^2) - 2\mu^2 + \mu^2 = E(X^2) - \mu^2.$$

Sometimes $\sigma^2 = E(X^2) - \mu^2$ provides an easier way of computing $\text{Var}(X)$ than does $\sigma^2 = E[(X - \mu)^2]$. Thus, in Example 2.3–3 we could have first computed

$$E(X^2) = \sum_{x=1}^{6} x^2\left(\frac{1}{6}\right) = \frac{1^2 + 2^2 + \cdots + 6^2}{6} = \frac{91}{6}$$

and then

$$\sigma^2 = \text{Var}(X) = \frac{91}{6} - \left(\frac{7}{2}\right)^2 = \frac{35}{12}.$$

Although most students understand that $\mu = E(X)$ is, in some sense, a measure of the middle of the distribution of X, it is difficult to get much of a feeling for the variance and the standard deviation. Nevertheless, the following example illustrates that the standard deviation is a measure of dispersion or spread of the points belonging to the space R.

Example 2.3–4.   Let X have the p.d.f. $f(x) = 1/3$, $x = -1, 0, 1$. Here the mean is

$$\mu = \sum_{x=-1}^{1} xf(x) = (-1)\left(\frac{1}{3}\right) + (0)\left(\frac{1}{3}\right) + (1)\left(\frac{1}{3}\right) = 0.$$

Accordingly, the variance, here denoted by σ_X^2, is

$$\sigma_X^2 = E(X^2) = \sum_{x=-1}^{1} x^2 f(x) = (-1)^2\left(\frac{1}{3}\right) + (0)^2\left(\frac{1}{3}\right) + (1)^2\left(\frac{1}{3}\right)$$

$$= \frac{2}{3},$$

so the standard deviation is $\sigma_X = \sqrt{2/3}$. Next let the random variable Y have the p.d.f. $g(y) = 1/3$, $y = -2, 0, 2$. Its mean is also zero, and it is easy to show that $\text{Var}(Y) = 8/3$, so the standard deviation of Y is $\sigma_Y = 2\sqrt{2/3}$. Here the standard deviation of Y is twice that of X, reflecting the fact that the probability of Y is spread out twice as much as that of X.

Possibly Chebyshev's inequality will help give added significance to the standard deviation in terms of bounding certain probabilities. The inequality is valid for all distributions for which the standard deviation exists.

THEOREM 2.3–1. (Chebyshev's Inequality) *If the random variable X has a finite mean μ and finite variance σ^2, then for every $k \geq 1$,*

$$P(|X - \mu| \geq k\sigma) \leq \frac{1}{k^2}.$$

Proof: Let $f(x)$ denote the p.d.f. of X. Then

$$\sigma^2 = E[(X - \mu)^2] = \sum_{x \in R} (x - \mu)^2 f(x)$$

$$= \sum_{x \in A} (x - \mu)^2 f(x) + \sum_{x \in A'} (x - \mu)^2 f(x)$$

where

$$A = \{x: \quad |x - \mu| \geq k\sigma\}.$$

The second term in the right-hand member is the sum of nonnegative numbers and thus is greater than or equal to zero. Hence,

$$\sigma^2 \geq \sum_{x \in A} (x - \mu)^2 f(x).$$

However, in A, $|x - \mu| \geq k\sigma$; so

$$\sigma^2 \geq \sum_{x \in A} (k\sigma)^2 f(x) = k^2 \sigma^2 \sum_{x \in A} f(x).$$

But the latter summation equals $P(X \in A)$ and, thus,

$$\sigma^2 \geq k^2 \sigma^2 P(X \in A) = k^2 \sigma^2 P(|X - \mu| \geq k\sigma).$$

That is,

$$P(|X - \mu| \geq k\sigma) \leq \frac{1}{k^2}.$$

An obvious change in the theorem, namely, $\varepsilon = k\sigma$, gives

$$P(|X - \mu| \geq \varepsilon) \leq \frac{\sigma^2}{\varepsilon^2}.$$

In words, Chebyshev's inequality states that the probability that X differs from its mean by at least k standard deviations is less than or equal to $1/k^2$. It follows that the probability that X differs from its mean by less than k standard deviations is at least $1 - 1/k^2$. That is,

$$P(|X - \mu| < k\sigma) \geq 1 - \frac{1}{k^2}.$$

Thus, Chebyshev's inequality can be used as a bound for certain probabilities. However, in many instances, the bound $1/k^2$ is not very close to the true probability $P(|X - \mu| \geq k\sigma)$.

Example 2.3–5. If it is known that X has a mean of 25 and a variance of 16, then, since $\sigma = 4$, we have

$$P(17 < X < 33) = P(|X - 25| < 8)$$

$$= P(|X - \mu| < 2\sigma) \geq 1 - \frac{1}{4}$$

$$= 0.75$$

and

$$P(|X - 25| \geq 12) = P(|X - \mu| \geq 3\sigma) \leq \frac{1}{9}.$$

In a later chapter we shall use Chebyshev's inequality in certain theoretical discussions. The next example hints at such an application.

Example 2.3–6. Suppose that the variance of a random variable Y is d/n; that is, it is dependent on n in such a way that as n increases, the variance decreases. Now

$$P(|Y - \mu| \geq \varepsilon) \leq \frac{d}{n\varepsilon^2}.$$

Thus, for any fixed $\varepsilon > 0$, if n is sufficiently large so that $d/n\varepsilon^2$ is small, most of the probability of Y is concentrated in the interval $(\mu - \varepsilon, \mu + \varepsilon)$.

Let r be a positive integer. If

$$E(X^r) = \sum_R x^r f(x)$$

exists, it is called the rth *moment* of the distribution about the origin. The expression *moment* has its origin in the study of mechanics. In addition, the expectation

$$E[(X - b)^r] = \sum_R (x - b)^r f(x)$$

is called the rth moment of the distribution about b.

═══ *Exercises* ═══

2.3–1. Let the p.d.f. of X be given by $f(0) = 3/10$, $f(1) = 3/10$, $f(2) = 1/10$, and $f(3) = 3/10$. Compute the mean, variance, and standard deviation of X.

2.3–2. Find the mean and variance for the following discrete distributions.

(a) $f(x) = \dfrac{1}{5}$, $x = 5, 10, 15, 20, 25$;

(b) $f(x) = 1$, $x = 5$;

(c) $f(x) = \dfrac{3!}{x!(3-x)!}\left(\dfrac{1}{4}\right)^x\left(\dfrac{3}{4}\right)^{3-x}$, $x = 0, 1, 2, 3$.

2.3–3. Find $E(X)$ if the p.d.f. of X is defined by

$$f(x) = 2\left(\dfrac{1}{3}\right)^x, \qquad x = 1, 2, 3, \ldots.$$

HINT: Compare the series representations of $E(X)$ and $(1/3)E(X)$.

2.3–4. Select three chips at random and without replacement from a bowl that contains $n_1 = 6$ red chips and $n_2 = 4$ blue chips. Let X equal the number of red chips among the three. Compute the mean and the variance of X.

2.3–5. Find the mean and the variance of the distribution that has the distribution function

$$F(x) = \begin{cases} 0, & x < 10, \\[2mm] \dfrac{1}{4}, & 10 \le x < 15, \\[2mm] \dfrac{3}{4}, & 15 \le x < 20, \\[2mm] 1, & 20 \le x. \end{cases}$$

2.3–6. Given $E(X + 4) = 10$ and $E[(X + 4)^2] = 116$, determine μ and σ^2. HINT: First find $E(X)$ from the first equation and then $E(X^2)$ by considering

$$E[(X + 4)^2] = E(X^2 + 8X + 16).$$

2.3–7. Let μ and σ^2 denote the mean and variance of the random variable X. Determine $E[(X - \mu)/\sigma]$ and $E\{[(X - \mu)/\sigma]^2\}$.

2.3–8. If X is a random variable with mean 33 and variance 16, use Chebyshev's inequality to find a lower bound for $P(23 < X < 43)$.

2.3–9. If $E(X) = 17$ and $E(X^2) = 298$, use Chebyshev's inequality to determine a lower bound for $P(10 < X < 24)$.

2.3–10. Let X denote the outcome when rolling a fair die. Then $\mu = 7/2$ and $\sigma^2 = 35/12$. Note that the maximum deviation of X from μ equals 5/2. Express this deviation in terms of number of standard deviations; that is, find k where $k\sigma = 5/2$. Determine a lower bound for $P(|X - 3.5| < 2.5)$.

2.4 The Moment-Generating Function

In the last section, the moments of a random variable X were defined to be the expected values of powers of X. We now define a function of a real variable t that can be used to find the moments of a random variable and thus is called the moment-generating function.

DEFINITION 2.4–1.   *Let X be a random variable of the discrete type with p.d.f. $f(x)$ and space R. If there is a positive number h such that*

$$E(e^{tX}) = \sum_R e^{tx} f(x)$$

exists for $-h < t < h$, then the function of t defined by

$$M(t) = E(e^{tX})$$

is called the moment-generating function of X.

From the theory of mathematical analysis, it can be shown that the existence of $M(t)$, for $-h < t < h$, implies that derivatives of $M(t)$ of all orders exist at $t = 0$; moreover, it is permissible to interchange differentiation and summation. Thus,

$$M'(t) = \sum_R x e^{tx} f(x),$$

$$M''(t) = \sum_R x^2 e^{tx} f(x),$$

and, for each positive integer r,

$$M^{(r)}(t) = \sum_R x^r e^{tx} f(x).$$

Setting $t = 0$, we see that

$$M'(0) = \sum_R x f(x) = E(X),$$

$$M''(0) = \sum_R x^2 f(x) = E(X^2),$$

and, in general,

$$M^{(r)}(0) = \sum_R x^r f(x) = E(X^r).$$

In particular, if the moment-generating function exists,

$$\mu = M'(0) \quad \text{and} \quad \sigma^2 = M''(0) - [M'(0)]^2.$$

The above argument shows that we can find the moments of X by differentiating $M(t)$. It must be emphasized that in use we first evaluate the summation representing $M(t)$ and then differentiate that function of t. The following example should make this clear.

Example 2.4–1.   Let the p.d.f. of X be defined by

$$f(x) = 2\left(\frac{1}{3}\right)^x, \qquad x = 1, 2, 3, \dots .$$

The moment-generating function of X is

$$M(t) = \sum_{x=1}^{\infty} e^{tx} 2\left(\frac{1}{3}\right)^x = \sum_{x=1}^{\infty} 2\left(\frac{e^t}{3}\right)^x.$$

But the latter summation is that of an infinite geometric progression with first term $2(e^t/3)$ and common ratio $(e^t/3)$. Thus, we have

$$M(t) = \frac{2(e^t/3)}{1 - e^t/3}, \qquad \frac{e^t}{3} < 1.$$

That is,

$$M(t) = \frac{2e^t}{3 - e^t}, \qquad t < \ln 3 = h,$$

and thus we have found the moment-generating function of X. It is this function that we differentiate to get

$$M'(t) = \frac{(3 - e^t)2e^t - 2e^t(-e^t)}{(3 - e^t)^2}$$

and, thus,

$$\mu = M'(0) = \frac{3}{2}.$$

The next example provides a nice trick for finding μ and σ^2 in certain cases.

Example 2.4–2.   Let the moment-generating function $M(t)$ of X exist for $-h < t < h$. Consider now the function $R(t) = \ln M(t)$. The first two derivatives of $R(t)$ are, respectively,

$$R'(t) = \frac{M'(t)}{M(t)} \qquad \text{and} \qquad R''(t) = \frac{M(t)M''(t) - [M'(t)]^2}{[M(t)]^2}.$$

Setting $t = 0$, we have

$$R'(0) = \frac{M'(0)}{M(0)} \qquad \text{and} \qquad R''(0) = \frac{M(0)M''(0) - [M'(0)]^2}{[M(0)]^2}.$$

However, it is obvious from the definition of $M(t)$ that $M(0) = 1$. Thus,

$$R'(0) = M'(0) = \mu \quad \text{and} \quad R''(0) = M''(0) - [M'(0)]^2 = \sigma^2.$$

The results obtained in Example 2.4–2 frequently provide an easier way of computing μ and σ^2. For illustration, let us consider the moment-generating function of Example 2.4–1, namely, $M(t) = 2e^t/(3 - e^t)$, $t < \ln 3$. We have that

$$R(t) = \ln 2 + t - \ln(3 - e^t)$$

so that

$$R'(t) = 1 + \frac{e^t}{3 - e^t} \quad \text{and} \quad \mu = R'(0) = \frac{3}{2}.$$

In addition, it is easier to find $R''(t)$ than $M''(t)$. Thus,

$$R''(t) = \frac{(3 - e^t)e^t - e^t(-e^t)}{(3 - e^t)^2} \quad \text{and} \quad \sigma^2 = R''(0) = \frac{3}{4}.$$

Although the moment-generating function, when it exists, is a useful tool for determining moments, its major importance is the fact that it uniquely determines the distribution. Another way of stating this is that if two random variables X and Y have the same moment-generating function, then X and Y are identically distributed. Although the rigorous proof of this property is based on the theory of transforms in analysis, it seems fairly evident for distributions of the discrete type. First note clearly the form of the moment-generating function of a discrete random variable. If X has a p.d.f. $f(x)$ with support $\{b_1, b_2, \ldots\}$, then

$$M(t) = \sum_R e^{tx} f(x)$$

$$= f(b_1)e^{tb_1} + f(b_2)e^{tb_2} + \cdots.$$

Hence, the coefficient of e^{tb_i} is $f(b_i) = P(X = b_i)$. That is, if we write a moment-generating function of the discrete-type random variable in the above form, we can find the probabilities by considering the coefficients of e^{tb_i}.

Example 2.4–3. Let the moment-generating function of Y be defined by

$$M(t) = \frac{1}{15}e^t + \frac{2}{15}e^{2t} + \frac{3}{15}e^{3t} + \frac{4}{15}e^{4t} + \frac{5}{15}e^{5t}.$$

Then, for example, the coefficient of e^{2t} is $2/15 = f(2) = P(Y = 2)$. In general, we see that the p.d.f. of Y is $f(y) = y/15$, $y = 1, 2, 3, 4, 5$.

When the moment-generating function exists, derivatives of all orders exist at $t = 0$. Thus, it is possible to represent $M(t)$ as a Maclaurin's series, namely,

$$M(t) = M(0) + M'(0)\left(\frac{t}{1!}\right) + M''(0)\left(\frac{t^2}{2!}\right) + M'''(0)\left(\frac{t^3}{3!}\right) + \cdots.$$

That is, if the Maclaurin's series expansion of $M(t)$ can be found, the rth moment of X, $E(X^r)$, is the coefficient of $t^r/r!$. Or, if $M(t)$ exists and the moments are given, we can frequently sum the Maclaurin's series to obtain the closed form of $M(t)$. These points are illustrated in the next two examples.

Example 2.4–4. Suppose that the random variable Y has the moment-generating function $M(t) = (1 - t)^{-1}$, $t < 1$. The Maclaurin's series for this function is

$$M(t) = 1 + t + t^2 + t^3 + \cdots + t^r + \cdots$$

$$= 1 + (1!)\left(\frac{t}{1!}\right) + (2!)\left(\frac{t^2}{2!}\right) + (3!)\left(\frac{t^3}{3!}\right) + \cdots + (r!)\left(\frac{t^r}{r!}\right) + \cdots.$$

Hence,

$$E(X^r) = r!, \qquad r = 1, 2, 3, \ldots,$$

and, in particular,

$$\mu = 1! = 1 \qquad \text{and} \qquad \sigma^2 = 2! - 1^2 = 1.$$

Later we will discover that this $M(t)$ is associated with a random variable of the continuous type; however, this system for finding moments is quite valid in that case also.

Example 2.4–5. Let the moments of X be defined by

$$E(X^r) = 0.8, \qquad r = 1, 2, 3, \ldots.$$

The moment-generating function of X is then

$$M(t) = M(0) + \sum_{r=1}^{\infty} 0.8\left(\frac{t^r}{r!}\right) = 1 + 0.8 \sum_{r=1}^{\infty} \frac{t^r}{r!}$$

$$= 0.2 + 0.8 \sum_{r=0}^{\infty} \frac{t^r}{r!} = 0.2e^{0t} + 0.8e^{1t}.$$

Thus,

$$P(X = 0) = 0.2 \qquad \text{and} \qquad P(X = 1) = 0.8.$$

━━━━ *Exercises* ━━━━

2.4–1. Find the moment-generating function when the p.d.f. of X is defined by:

(a) $f(x) = \dfrac{1}{3}, \qquad x = 1, 2, 3$;

(b) $f(x) = 1, \qquad x = 5$;

(c) $f(x) = \dfrac{5!}{x!(5 - x)!}\left(\dfrac{1}{3}\right)^x\left(\dfrac{2}{3}\right)^{5-x}, \qquad x = 0, 1, \ldots, 5.$

2.4–2. Find the moment-generating function, mean, and variance of X if the p.d.f. of X is $f(x) = (1/2)(2/3)^x$, $x = 1, 2, 3, 4, \ldots$.

2.4–3. If the moment-generating function of X is

$$M(t) = \frac{2}{5}e^t + \frac{1}{5}e^{2t} + \frac{2}{5}e^{3t},$$

find the mean and variance of X, and define the p.d.f. of X.

2.4–4. If $E(X^r) = 5^r$, $r = 1, 2, 3, \ldots$, find $M(t)$, the moment-generating function of X, and define the p.d.f. of X.

2.4–5. If the moment-generating function of X is

$$M(t) = e^{t^2/2} = 1 + \frac{t^2/2}{1!} + \frac{(t^2/2)^2}{2!} + \frac{(t^2/2)^3}{3!} + \cdots, \qquad -\infty < t < \infty,$$

find $E(X^r)$. Note that we must consider two cases: r odd and r even. Again later we find that this $M(t)$ is that of a continuous-type random variable.

2.5 Bernoulli Trials and the Binomial Distribution

A *Bernoulli experiment* is a random experiment, the outcome of which can be classified in but one of two mutually exclusive and exhaustive ways, say, success or failure (for example, head or tail, life or death, nondefective or defective). A sequence of *Bernoulli trials* occurs when a Bernoulli experiment is performed several *independent* times so that the probability of success, say p, remains the *same* from trial to trial. That is, in such a sequence we let p denote the probability of success on each trial. In addition, we will frequently let $q = 1 - p$ denote the probability of failure; that is, we will use q and $1 - p$ interchangeably.

    Example 2.5–1. We flip a fair coin 10 independent times with heads being considered a success on each trial. This constitutes a sequence of 10 Bernoulli trials with $p = 1/2$.

    Example 2.5–2. An urn contains 10 red and 20 white balls. Draw five balls at random from the urn, one at a time and with replacement. Let the draw of a red ball be considered success. If the trials are independent, we have five Bernoulli trials with $p = 1/3$.

Let X be a random variable associated with a Bernoulli trial by defining it as follows:

$$X(\text{success}) = 1 \qquad \text{and} \qquad X(\text{failure}) = 0.$$

That is, the two outcomes, success and failure, are denoted by one and zero, respectively. The p.d.f. of X can be written as

$$f(x) = p^x(1 - p)^{1-x}, \qquad x = 0, 1.$$

We say that X has a *Bernoulli distribution*. The expected value of X is

$$\mu = E(X) = \sum_{x=0}^{1} xp^x(1 - p)^{1-x} = (0)(1 - p) + (1)(p) = p,$$

and the variance of X is

$$\sigma^2 = \text{Var}(X) = \sum_{x=0}^{1} (x - p)^2 p^x(1 - p)^{1-x}$$

$$= p^2(1 - p) + (1 - p)^2 p = p(1 - p).$$

Finally, the moment-generating function of X is

$$M(t) = E(e^{tX}) = \sum_{x=0}^{1} e^{tx} p^x(1 - p)^{1-x}$$

$$= (1 - p) + pe^t.$$

In a sequence of n Bernoulli trials, we shall let X_i denote the Bernoulli random variable associated with the ith trial. An observed sequence of n Bernoulli trials will then be an n-tuple of zeros and ones.

Example 2.5–3. A fair die is cast four independent times. Call the outcome a success if a six is rolled, all other outcomes being considered failures. A possible observed sequence would be $(0, 0, 1, 0)$, in which case a six would have been rolled on the third trial and a nonsix on each of the other three trials. Since the trials are independent, the probability of this particular outcome would be $(5/6)(5/6)(1/6)(5/6)$.

In a sequence of Bernoulli trials, we are often interested in the total number of successes and not in the order of their occurrence. If we let the random variable Y equal the number of observed successes in n Bernoulli trials, the possible values of Y are $0, 1, 2, \ldots, n$. If y successes occur, where $y = 0, 1, 2, 3, \ldots, n$, then $n - y$ failures occur. The number of ways of selecting y positions for the y successes in the n trials is

$$\binom{n}{y} = \frac{n!}{y!(n - y)!}.$$

Since the trials are independent and since the probabilities of success and failure on each trial are, respectively, p and $q = 1 - p$, the probability of each of these ways is $p^y(1 - p)^{n-y}$. Thus, the p.d.f. of Y, say $f(y)$, is the sum of the probabilities of these $\binom{n}{y}$ mutually exclusive events; that is,

$$f(y) = \binom{n}{y} p^y (1 - p)^{n-y}, \qquad y = 0, 1, 2, \ldots, n.$$

These probabilities are called *binomial probabilities* and the random variable Y is said to have a *binomial distribution*. A binomial distribution will be denoted by the symbol $b(n, p)$. The constants n and p are called the *parameters* of the binomial distribution; they correspond to the number n of trials and the probability p of success on each trial. Thus, if we say that Y is $b(12, 1/4)$, we mean that Y is the number of successes in $n = 12$ Bernoulli trials with probability $p = 1/4$ of success on each trial.

Example 2.5–4. The probability of rolling two sixes and three nonsixes in five independent casts of a fair die is

$$f(2) = \binom{5}{2}\left(\frac{1}{6}\right)^2\left(\frac{5}{6}\right)^3.$$

The random variable Y that denotes the number of sixes in the $n = 5$ casts is $b(5, 1/6)$.

Example 2.5–5. Cast a fair die 10 independent times. The probability of observing y sixes and $10 - y$ nonsixes is

$$f(y) = \binom{10}{y}\left(\frac{1}{6}\right)^y\left(\frac{5}{6}\right)^{10-y}, \qquad y = 0, 1, \ldots, 10,$$

which is the p.d.f. of the number Y of sixes in $n = 10$ independent casts of the die.

Values of the distribution function of a random variable X that is $b(n, p)$ are given in Appendix Table I for selected values of n and p. The use of this table is illustrated in the next example.

Example 2.5–6. The probability of observing exactly six heads when an unbiased coin is flipped ten independent times is

$$\binom{10}{6}\left(\frac{1}{2}\right)^6\left(\frac{1}{2}\right)^4 = P(X \le 6) - P(X \le 5)$$

$$= 0.8281 - 0.6230 = 0.2051,$$

since $P(X \le 6) = 0.8281$ and $P(X \le 5) = 0.6230$ from Appendix Table I.

The probability of observing at least six heads is

$$\sum_{y=6}^{10} \binom{10}{y}\left(\frac{1}{2}\right)^{y}\left(\frac{1}{2}\right)^{10-y} = 1 - P(X \le 5) = 1 - 0.6230 = 0.3770.$$

While probabilities for the binomial distribution $b(n, p)$ are given in Appendix Table I for selected p values less than or equal to 0.5, the next example demonstrates that this table can also be used for p values greater than 0.5.

Example 2.5–7. Let Y be $b(15, 0.7)$. Then the probability $P(Y = 11, 12, 13)$ is clearly equal to $P(Z = 2, 3, 4)$, where Z is equal to the number of failures in the $n = 15$ trials and thus is $b(15, 0.3)$. Hence

$$P(Y = 11, 12, 13) = P(Z = 2, 3, 4)$$
$$= P(Z \le 4) - P(Z \le 1)$$
$$= 0.5155 - 0.0353 = 0.4802.$$

Recall that if n is a positive integer, then

$$(a + b)^n = \sum_{y=0}^{n} \binom{n}{y} b^y a^{n-y}.$$

Thus, the sum of the binomial probabilities, if we use the above binomial expansion with $b = p$ and $a = 1 - p$, is

$$\sum_{y=0}^{n} \binom{n}{y} p^y (1 - p)^{n-y} = [(1 - p) + p]^n = 1,$$

a result that had to follow from the fact that $f(y)$ is a p.d.f.

To find the mean and variance of Y, we shall first find its moment-generating function. We have

$$M(t) = E[e^{tY}] = \sum_{y=0}^{n} e^{ty} \binom{n}{y} p^y (1 - p)^{n-y}$$

$$= \sum_{y=0}^{n} \binom{n}{y} (pe^t)^y (1 - p)^{n-y}.$$

Using the binomial expansion with $b = pe^t$ and $a = 1 - p$, we obtain

$$M(t) = [(1 - p) + pe^t]^n,$$

for all real t. Now

$$M'(t) = n[(1 - p) + pe^t]^{n-1}(pe^t)$$

and

$$M''(t) = n(n - 1)[(1 - p) + pe^t]^{n-2}(pe^t)^2$$
$$+ n[(1 - p) + pe^t]^{n-1}(pe^t).$$

Thus,

$$\mu = M'(0) = np$$

and

$$\sigma^2 = M''(0) - \mu^2 = n(n-1)p^2 + np - (np)^2 = np(1-p).$$

That is,

$$\mu = np \quad \text{and} \quad \sigma^2 = npq.$$

Note that when p is the probability of success on each trial, the expected number of successes in n trials is np, a result that agrees with most of our intuitions.

Example 2.5–8. Suppose that observation over a long period of time has disclosed that, on the average, one out of ten items produced by a process is defective. Select five items independently from the production line and test them. Let Y denote the number of defective items among the $n = 5$ items. Then Y is $b(5, 0.1)$. Furthermore,

$$E(Y) = 5(0.1) = 0.5, \quad \text{Var}(Y) = 5(0.1)(0.9) = 0.45.$$

For example, the probability of observing at most one defective item is

$$P(Y \le 1) = \binom{5}{0}(0.1)^0(0.9)^5 + \binom{5}{1}(0.1)^1(0.9)^4 = 0.9185.$$

If Y is the number of successes in n Bernoulli trials with probability p of success on each trial, then Y is $b(n, p)$. Furthermore Y/n gives the relative frequency of success, and, when p is unknown, Y/n can be used as an estimate of p. To gain some insight into the closeness of Y/n to p, we shall use Chebyshev's inequality. With $\varepsilon > 0$, we note that

$$P\left(\left|\frac{Y}{n} - p\right| \ge \varepsilon\right) = P(|Y - np| \ge n\varepsilon)$$

$$= P\left(|Y - np| \ge \frac{\sqrt{n\varepsilon}}{\sqrt{pq}}\sqrt{npq}\right).$$

However, $\mu = np$ and $\sigma = \sqrt{npq}$ are the mean and standard deviation of Y so that, with $k = \sqrt{n\varepsilon}/\sqrt{pq}$, we have

$$P\left(\left|\frac{Y}{n} - p\right| \ge \varepsilon\right) = P(|Y - \mu| \ge k\sigma) \le \frac{1}{k^2} = \frac{pq}{n\varepsilon^2}$$

or, equivalently,

$$P\left(\left|\frac{Y}{n} - p\right| < \varepsilon\right) \ge 1 - \frac{pq}{n\varepsilon^2}.$$

When p is completely unknown, we can use the fact that $pq = p(1 - p)$ is a maximum when $p = 1/2$ in order to bound these probabilities. That is,

$$1 - \frac{pq}{n\varepsilon^2} \geq 1 - \frac{(1/2)(1/2)}{n\varepsilon^2}.$$

For illustration, if $\varepsilon = 0.05$ and $n = 400$,

$$P\left(\left|\frac{Y}{400} - p\right| < 0.05\right) \geq 1 - \frac{(1/2)(1/2)}{400(0.0025)} = 0.75.$$

If, on the other hand, it is known that p is close to $1/10$, we would have

$$P\left(\left|\frac{Y}{400} - p\right| < 0.05\right) \geq 1 - \frac{(0.1)(0.9)}{400(0.0025)} = 0.91.$$

We shall see later that it is possible to give better bounds on these probabilities. Recall that Chebyshev's inequality is applicable to all distributions with a finite variance, and thus the bound is not always a tight one; that is, the bound is not necessarily close to the true probability.

In general, however, it should be noted that, with fixed $\varepsilon > 0$ and $0 < p < 1$, we have that

$$\lim_{n\to\infty} P\left(\left|\frac{Y}{n} - p\right| < \varepsilon\right) \geq \lim_{n\to\infty}\left(1 - \frac{pq}{n\varepsilon^2}\right) = 1.$$

But, since the probability of every event is less than or equal to one, it must be that

$$\lim_{n\to\infty} P\left(\left|\frac{Y}{n} - p\right| < \varepsilon\right) = 1.$$

That is, the probability that the relative frequency Y/n is within ε of p is close to one when n is large enough. This is one form of the *law of large numbers*, which we alluded to in our description of the relative frequency interpretation of probability in Section 1.1.

Example 2.5–9. In a close election, Smith is running against Jones. Supporters of Smith interview 500 voters selected at random and discover that 300 of them favor Smith. How confident can they be that Smith will win the election? If Y denotes the number of voters favoring Smith in the $n = 500$ interviews and if p is the proportion of total voters in the population favoring Smith, they know that

$$P\left(\left|\frac{Y}{500} - p\right| < 0.1\right) \geq 1 - \frac{(1/2)(1/2)}{500(0.01)} = 0.95.$$

Thus, they can be at least 95% confident that $0.5 < p < 0.7$, since the observed value of $Y/500$ equals $300/500 = 0.6$.

══ *Exercises* ══

2.5–1. Let X have a Bernoulli distribution. Find the mean and variance of X using the moment-generating function of X.

2.5–2. An urn contains 7 red and 11 white balls. Draw one ball at random from the urn. Let $X = 1$ if a red ball is drawn and let $X = 0$ if a white ball is drawn. Find the p.d.f., moment-generating function, mean, and variance of X.

2.5–3. Suppose that in Exercise 2.5–2, $X = 1$ if a red ball is drawn and $X = -1$ if a white ball is drawn. Find the p.d.f., moment-generating function, mean, and variance of X.

2.5–4. Suppose the moment-generating function of X is $M(t) = 1/3 + (2/3)e^t$. Determine the p.d.f., mean, and variance of X.

2.5–5. If Y is $b(5, 0.3)$, find the probability of   **(a)**   exactly two successes,   **(b)**   at least one success.   **(c)**   Find the mean, variance, and moment-generating function of Y.

2.5–6. Flip a pair of unbiased coins four independent times. Determine the probability of observing a pair of heads   **(a)**   exactly three times,   **(b)**   at least once.

2.5–7. Suppose that 2000 points are selected independently and at random from the unit square $S = \{(x, y): \ 0 \le x < 1, 0 \le y < 1\}$. Let W equal the number of points that fall in $A = \{(x, y): \ x^2 + y^2 \le 1\}$.
  (a)   How is W distributed?
  (b)   Give the mean, variance, and standard deviation of W.

2.5–8. An urn contains four failure balls and one success ball. A sample of size $n = 10$ is drawn from the urn, one at a time and with replacement. Let X denote the number of successes in the sample.
  (a)   How is X distributed?
  (b)   Give the probability of two successes, $P(X = 2)$.
  (c)   Give the probability of at most two successes.

2.5–9. If the moment-generating function of Y is $M(t) = (1/4 + 3e^t/4)^{12}$, find   **(a)** $E(Y)$,   **(b)**   Var(Y),   **(c)** $P(Y \ge 10)$.

2.5–10. An urn contains 10 red and 15 white balls. A sample of size 8 is drawn from the urn, one at a time. Let X denote the number of red balls in the sample. Calculate $P(X = 2)$ if the sampling is done   **(a)**   with replacement,   **(b)**   without replacement.

2.5–11. Let Y be $b(n, p)$.
  (a)   Show that $E(Y) = np$ by evaluating

$$E(Y) = \sum_{y=0}^{n} y \binom{n}{y} p^y (1 - p)^{n-y}.$$

HINT: Note that

$$E(Y) = \sum_{y=1}^{n} y \frac{n!}{y!(n - y)!} p^y (1 - p)^{n-y}$$

and

$$\frac{y}{y!} = \frac{1}{(y - 1)!}, \qquad y = 1, 2, 3, \ldots, n.$$

(b) Find $E[Y(Y - 1)]$ and then show that

$$\text{Var}(Y) = E[Y(Y - 1)] + E(Y) - [E(Y)]^2 = np(1 - p).$$

HINT: Note that

$$E[Y(Y - 1)] = \sum_{y=2}^{n} y(y - 1)\frac{n!}{y!(n - y)!}p^y(1 - p)^{n-y}$$

and

$$\frac{y(y - 1)}{y!} = \frac{1}{(y - 2)!}, \quad y = 2, 3, 4, \ldots, n.$$

2.5–12. It is claimed that for a particular lottery game, $1/10$ of the 50,000,000 tickets will win a prize. What is the probability of winning at least one prize if you purchase **(a)** 10 tickets? **(b)** 15 tickets?

2.5–13. For the lottery described in Exercise 2.5–12, find the smallest number of tickets that must be purchased so that the probability of winning at least one prize is greater than **(a)** 0.50, **(b)** 0.95.

2.5–14. The probability that a "10 ounce box" of corn flakes contains at least 10.1 ounces is 0.05. Compute the probability that a carton of 50 boxes of corn flakes, selected independently, has at most three boxes that contain at least 10.1 ounces of corn flakes.

2.5–15. If Y is $b(100, 1/2)$, give a lower bound for

$$P\left(\left|\frac{Y}{100} - \frac{1}{2}\right| < 0.10\right).$$

2.5–16. If Y is $b(500, 0.55)$, give a lower bound for

$$P\left(\left|\frac{Y}{500} - 0.55\right| < 0.05\right).$$

2.5–17. A random variable X has a binomial distribution with mean 6 and variance 3.6. Find $P(X = 4)$.

2.6 Geometric and Negative Binomial Distributions

Consider a sequence of Bernoulli trials with probability p of success. This sequence is observed until the first success occurs and we let Y denote the number of failures before this first success. For example, if the sequence starts, with F representing failure and S success, with $F, F, F, S, \ldots$, then $Y = 3$. Moreover, the probability of such a sequence is

$$P(Y = 3) = (q)(q)(q)(p) = q^3p = (1 - p)^3p.$$

In general, the p.d.f., $f(y) = P(Y = y)$, of Y is given by

$$f(y) = (1 - p)^y p, \qquad y = 0, 1, 2, \ldots.$$

We say that Y has a *geometric distribution*. Note that the first success occurs on the $(y + 1)$st trial.

Recall that the sum of the geometric series

$$\sum_{k=0}^{\infty} ar^k = \frac{a}{1 - r}$$

when $|r| < 1$. Thus,

$$\sum_{y=0}^{\infty} f(y) = \sum_{y=0}^{\infty} (1 - p)^y p = \frac{p}{1 - (1 - p)} = 1$$

so that $f(y)$ does satisfy the properties of a p.d.f.

From the sum of a geometric series we also note that

$$P(Y \geq k) = \sum_{y=k}^{\infty} (1 - p)^y p = \frac{(1 - p)^k p}{1 - (1 - p)} = (1 - p)^k,$$

and, thus,

$$P(Y < k) = \sum_{y=0}^{k-1} (1 - p)^y p = 1 - (1 - p)^k.$$

Example 2.6–1. In a sequence of independent rolls of a fair die, the probability that the first four is observed on the sixth trial is given by

$$P(Y = 5) = \left(\frac{5}{6}\right)^5 \left(\frac{1}{6}\right) = 0.067,$$

where Y denotes the number of nonfours before the occurrence of the first four. The probability that at least six trials are required to observe a four is given by

$$P(Y \geq 5) = \sum_{y=5}^{\infty} \left(\frac{5}{6}\right)^y \left(\frac{1}{6}\right) = \left(\frac{5}{6}\right)^5 = 0.402.$$

The probability that at most five trials are required to observe a four is given by

$$P(Y \leq 4) = \sum_{y=0}^{4} \left(\frac{5}{6}\right)^y \left(\frac{1}{6}\right) = 1 - \left(\frac{5}{6}\right)^5 = 0.598.$$

To find the mean and variance of the geometric distribution, we shall first find the moment-generating function. We have

$$M(t) = E[e^{tY}] = \sum_{y=0}^{\infty} e^{ty}(1 - p)^y p = \sum_{y=0}^{\infty} [(1 - p)e^t]^y p.$$

But this is the sum of a geometric series with $a = p$ and $r = (1 - p)e^t$. Thus,

$$M(t) = \frac{p}{1 - (1 - p)e^t}$$

when $(1 - p)e^t < 1$ or, equivalently, $t < -\ln(1 - p)$. It is left as an exercise (Exercise 2.6–3) to show that

$$\mu = M'(0) = \frac{1 - p}{p} = \frac{q}{p};$$

$$\sigma^2 = M''(0) - [M'(0)]^2 = \frac{1 - p}{p^2} = \frac{q}{p^2}.$$

Suppose now that the sequence of Bernoulli trials is observed until exactly r successes occur, where r is a fixed positive integer. Let the random variable Y denote the number of failures before the occurrence of that rth success. Then $y + r$ denotes the number of trials necessary to produce exactly r successes and y failures with the rth success occurring at the $(y + r)$th trial. By the multiplication rule of probabilities, the p.d.f. of Y, say $g(y)$, equals the product of the probability

$$\binom{y + r - 1}{r - 1} p^{r-1}(1 - p)^y$$

of obtaining exactly $r - 1$ successes in the first $y + r - 1$ trials and the probability p of a success on the $(y + r)$th trial. Thus,

$$g(y) = \binom{y + r - 1}{r - 1} p^r(1 - p)^y, \qquad y = 0, 1, 2, \ldots.$$

We say that Y has a *negative binomial distribution*.

Example 2.6–2. A fair die is cast on successive independent trials until the second six is observed. The probability of observing exactly ten nonsixes before the second six is cast is

$$\binom{10 + 2 - 1}{1} \left(\frac{1}{6}\right)^2 \left(\frac{5}{6}\right)^{10} = 0.049.$$

We shall now show that the sum of all the probabilities for the negative binomial distribution is equal to one, that is,

$$\sum_{y=0}^{\infty} g(y) = 1.$$

We shall use the Maclaurin's series (sometimes called the *negative binomial expansion*):

$$(1 - a)^{-n} = \sum_{k=0}^{\infty} (-1)^k \frac{(-n)(-n-1)\cdots(-n-k+1)}{k!} a^k \qquad (1)$$

$$= \sum_{k=0}^{\infty} \binom{n+k-1}{k} a^k, \qquad |a| < 1. \qquad (2)$$

The proof of these two expressions is left as an exercise (Exercise 2.6–13). Accordingly, using expression (2), we see that

$$\sum_{y=0}^{\infty} g(y) = \sum_{y=0}^{\infty} \binom{y+r-1}{r-1} p^r (1-p)^y$$

$$= p^r \sum_{y=0}^{\infty} \binom{r+y-1}{y} (1-p)^y$$

$$= p^r (1 - [1-p])^{-r} = 1.$$

The moment-generating function of Y is given by

$$M(t) = \sum_{y=0}^{\infty} e^{ty} \binom{y+r-1}{r-1} p^r (1-p)^y$$

$$= p^r \sum_{y=0}^{\infty} \binom{y+r-1}{y} [(1-p)e^t]^y.$$

But, from expression (2) with $a = (1-p)e^t$, we have

$$M(t) = p^r [1 - (1-p)e^t]^{-r}, \qquad t < -\ln(1-p).$$

It is left as an exercise (Exercise 2.6–14) to show that

$$\mu = \frac{r(1-p)}{p} = \frac{rq}{p};$$

$$\sigma^2 = \frac{r(1-p)}{p^2} = \frac{rq}{p^2}.$$

═══ *Exercises* ═══

2.6–1. When flipping an unbiased coin, determine the probability that **(a)** the first head occurs on the third trial, **(b)** at least three trials are necessary to observe a head, **(c)** at most three trials are necessary to observe a head.

2.6–2. When drawing balls at random and with replacement from an urn containing one red ball and three white balls, find the probability that **(a)** the first red ball is drawn on the fourth trial, **(b)** at most four trials are necessary to draw a red ball.

2.6–3. Verify that the mean and variance of the geometric distribution are $\mu = q/p$ and $\sigma^2 = q/p^2$.

2.6-4. Calculate the mean of the geometric distribution by evaluating the infinite series

$$E(Y) = \sum_{y=0}^{\infty} y(1 - p)^y p.$$

HINT: Write out the infinite series $E(Y)$ and then subtract $(1 - p)E(Y)$.

2.6-5. Flip an unbiased coin in a sequence of independent trials. Compute the probability that the first head is observed on the fifth trial, given that tails are observed on each of the first three trials.

2.6-6. Let Y have a geometric distribution. Show that

$$P(Y > k + j \mid Y \geq k) = P(Y > j),$$

where k and j are nonnegative integers.

2.6-7. The probability that a machine produces a defective item is 0.01. Each item is checked as it is produced. Assume that these are independent trials and compute the probability that at least 100 items must be checked to find one that is defective.

2.6-8. Compute the probability that a "300 hitter" in baseball (that is, a batter who gets a hit 3/10 of the time) requires at most four times at bat to get a hit. What assumptions did you make in working this problem? Are they realistic in practice?

2.6-9. If Y has a geometric distribution, then $Y + 1$ is the trial at which the first success occurs.

(a) Find $E(Y + 1)$.

(b) Let $Y + 1$ denote the number of independent trials needed to obtain the first three when casting an unbiased die. What does $E(Y + 1)$ equal? Does this answer make sense intuitively?

2.6-10. Five chips numbered 1, 2, 3, 4, and 5 are in a bowl. The chips are selected at random, one at a time and with replacement. The sampling is continued until each chip has been selected at least once. Given that $i - 1$ of the five chips have been selected, let X_i denote the number of additional draws to select a chip not drawn previously, $i = 1, 2, \ldots, 5$. Note that $X_1 = 1$.

(a) Show that $Y_i = X_i - 1, i = 1, 2, \ldots, 5$, has a geometric distribution.

(b) Find $E(X_i), i = 1, 2, \ldots, 5$.

(c) Let $W = X_1 + X_2 + \cdots + X_5$. Find $E(W)$, the expected number of draws to select each chip at least once.

2.6-11. Show that 63/512 is the probability that the fifth head is observed on the tenth independent flip of an unbiased coin.

2.6-12. An urn contains one red ball and three white balls. Sample successively and independently with replacement from this urn.

(a) Compute the probability that the second red ball occurs on the fourth trial.

(b) Compute the probability that at most four trials are necessary to observe two red balls.

(c) Compute the probability that at least four trials are necessary to observe two red balls.

2.6-13. Verify equations (1) and (2).

2.6-14. If Y has a negative binomial distribution, verify that $E(Y) = rq/p$ and $\mathrm{Var}(Y) = rq/p^2$.

2.7 The Poisson Distribution

Some experiments result in counting the numbers of times particular events occur in given times or on given physical objects. For example, we could count the number of phone calls arriving at a switchboard between 9 and 10 A.M., the number of flaws in 100 feet of wire, the number of customers that arrive at a ticket window between 12 noon and 2 P.M., or the number of defects in a 100 foot roll of aluminum screen that is 2 feet wide. Each count can be looked upon as a random variable associated with an approximate Poisson process provided the conditions in Definition 2.7–1 are satisfied. In this definition, the term *continuous interval* can represent a *given time* or *given object* and the term *change* can represent the *occurrence* of a certain event.

DEFINITION 2.7–1. *Let the number of changes that occur in a given continuous interval be counted. We have an* approximate Poisson process *with parameter* $\lambda > 0$ *if the following are satisfied:*

(i) *The numbers of changes occurring in nonoverlapping intervals are independent.*
(ii) *The probability of exactly one change in a sufficiently short interval of length h is approximately λh.*
(iii) *The probability of two or more changes in a sufficiently short interval is essentially zero.*

Suppose that an experiment satisfies the three points of an approximate Poisson process. Let X denote the number of changes in an interval of "length one" (where "length one" represents one unit of the quantity under consideration). We would like to find an approximation for $P(X = k)$, where k is a nonnegative integer. To achieve this, we partition the unit interval into n subintervals of equal length $1/n$. If n is sufficiently large (that is, much larger than k), we shall approximate the probability that k changes occur in this unit interval by finding the probability that one change occurs in each of exactly k of these n subintervals. The probability of one change occurring in any one subinterval of length $1/n$ is approximately $\lambda(1/n)$ by condition (ii). The probability of two or more changes in any one subinterval is essentially zero by condition (iii). So for each subinterval, exactly one change occurs with a probability of approximately $\lambda(1/n)$. Consider the occurrence of nonoccurrence of a change in each subinterval as a Bernoulli trial. By condition (i) we have a sequence of n Bernoulli trials with probability p approximately equal to $\lambda(1/n)$. Thus, an approximation for $P(X = k)$ is given by the binomial probability

$$\frac{n!}{k!(n-k)!}\left(\frac{\lambda}{n}\right)^k\left(1 - \frac{\lambda}{n}\right)^{n-k}.$$

In order to obtain a better approximation, choose a larger value for n. If n increases without bound we have that

$$\lim_{n \to \infty} \frac{n!}{k!(n-k)!} \left(\frac{\lambda}{n}\right)^k \left(1 - \frac{\lambda}{n}\right)^{n-k}$$

$$= \lim_{n \to \infty} \frac{n(n-1)\cdots(n-k+1)}{n^k} \frac{\lambda^k}{k!} \left(1 - \frac{\lambda}{n}\right)^n \left(1 - \frac{\lambda}{n}\right)^{-k}.$$

Now, for fixed k, we have

$$\lim_{n \to \infty} \frac{n(n-1)\cdots(n-k+1)}{n^k} = \lim_{n \to \infty} \left[1 \left(1 - \frac{1}{n}\right) \cdots \left(1 - \frac{k-1}{n}\right) \right] = 1,$$

$$\lim_{n \to \infty} \left(1 - \frac{\lambda}{n}\right)^n = e^{-\lambda},$$

and

$$\lim_{n \to \infty} \left(1 - \frac{\lambda}{n}\right)^{-k} = 1.$$

Thus,

$$\lim_{n \to \infty} \frac{n!}{k!(n-k)!} \left(\frac{\lambda}{n}\right)^k \left(1 - \frac{\lambda}{n}\right)^{n-k} = \frac{\lambda^k e^{-\lambda}}{k!} = P(X = k),$$

approximately. The distribution of probability associated with this process has a special name.

DEFINITION 2.7–2. *The random variable X has a* Poisson *distribution if its p.d.f. is of the form*

$$f(x) = \frac{\lambda^x e^{-\lambda}}{x!}, \qquad x = 0, 1, 2, \ldots,$$

where $\lambda > 0$.

It is easy to see that $f(x)$ enjoys the properties of a p.d.f. because clearly $f(x) \geq 0$ and

$$\sum_{x=0}^{\infty} \frac{\lambda^x e^{-\lambda}}{x!} = e^{-\lambda} \sum_{x=0}^{\infty} \frac{\lambda^x}{x!} = e^{-\lambda} e^{\lambda} = 1.$$

To discover the exact role of the parameter $\lambda > 0$, let us find some of the characteristics of the Poisson distribution. The moment-generating function of X is

$$M(t) = \sum_{x=0}^{\infty} e^{tx} \left(\frac{\lambda^x e^{-\lambda}}{x!}\right) = e^{-\lambda} \sum_{x=0}^{\infty} \frac{(\lambda e^t)^x}{x!}$$

$$= e^{-\lambda} e^{\lambda e^t} = e^{\lambda(e^t - 1)},$$

for all real values of t. To find the mean and variance of X, we have

$$R(t) = \ln M(t) = \lambda(e^t - 1),$$
$$R'(t) = \lambda e^t,$$

and

$$R''(t) = \lambda e^t.$$

Thus,

$$\mu = R'(0) = \lambda \quad \text{and} \quad \sigma^2 = R''(0) = \lambda.$$

That is, the parameter λ is equal to both the mean and the variance of the Poisson distribution.

Example 2.7–1. Let X have a Poisson distribution with a mean of $\lambda = 5$. Then

$$P(X = 3) = \frac{5^3 e^{-5}}{3!} = P(X \le 3) - P(X \le 2) = 0.140,$$

by Appendix Table II. The moment-generating function of X is

$$M(t) = e^{5(e^t - 1)}, \quad -\infty < t < \infty.$$

If events in a Poisson process occur at a mean rate of λ per unit, then the expected number of occurrences in an interval of length t is λt. For example, if phone calls arrive at a switchboard at a mean rate of three per minute, then the expected number of phone calls in a 5-minute period is $(3)(5) = 15$. Or if calls arrive at a mean rate of 22 in a 5-minute period, the expected number of calls per minute is $\lambda = 22(1/5) = 4.4$. Moreover, the number of occurrences, say X, in the interval of length t has the Poisson p.d.f.

$$f(x) = \frac{(\lambda t)^x e^{-\lambda t}}{x!}, \quad x = 0, 1, 2, \dots.$$

This is easily accepted by treating the interval of length t as if it were the "unit interval" with mean λt instead of λ.

Example 2.7–2. Flaws in a certain brand of tape occur on the average of one flaw per 1200 feet. If one assumes a Poisson distribution, what is the distribution of X, the number of flaws in a 4800 foot roll? The expected number of flaws in $4800 = 4(1200)$ feet is 4; that is, $\lambda = E(X) = 4$. Thus, the p.d.f. of X is

$$f(x) = \frac{4^x e^{-4}}{x!}, \quad x = 0, 1, 2, \dots$$

and, in particular,

$$P(X = 0) = \frac{4^0 e^{-4}}{0!} = e^{-4} = 0.018,$$

by Appendix Table II.

Example 2.7–3. Telephone calls enter a college switchboard on the average of two every 3 minutes. If one assumes an approximate Poisson process, what is the probability of five or more calls arriving in a 9-minute period? Let X denote the number of calls in a 9-minute period. We see that $\lambda = E(X) = 6$; that is, on the average, six calls will arrive during a 9-minute period. Thus,

$$P(X \geq 5) = 1 - P(X \leq 4) = 1 - \sum_{x=0}^{4} \frac{6^x e^{-6}}{x!}$$

$$= 1 - 0.285 = 0.715.$$

Not only is the Poisson distribution important in its own right, but it can also be used to approximate probabilities for a binomial distribution. If X has a Poisson distribution with parameter λ, we saw that

$$P(X = x) \approx \binom{n}{x}\left(\frac{\lambda}{n}\right)^x\left(1 - \frac{\lambda}{n}\right)^{n-x}.$$

Let $p = \lambda/n$ so that $\lambda = np$. That is, if X has the binomial distribution $b(n, p)$ with large n, then

$$\frac{(np)^x e^{-np}}{x!} \approx \binom{n}{x}p^x(1 - p)^{n-x}.$$

This approximation is reasonably good if n is "large." But since $\lambda = np$ was constant in the earlier argument, p should be "small." In particular, the approximation is quite accurate if $n \geq 20$ and $p \leq 0.05$ and it is very good if $n \geq 100$ and $np \leq 10$.

Example 2.7–4. A manufacturer of Christmas tree light bulbs knows that 2% of the bulbs are defective. Approximate the probability that a box of 100 of these bulbs contains at most three defective bulbs. Assuming independence, we have a binomial distribution with parameters $p = 0.02$ and $n = 100$. The Poisson distribution with $\lambda = 100\,(0.02) = 2$ gives

$$\sum_{x=0}^{3} \frac{2^x e^{-2}}{x!} = 0.857,$$

from Appendix Table II. Using the binomial distribution, we obtain, after some tedious calculations,

$$\sum_{x=0}^{3}\binom{100}{x}(0.2)^x(0.98)^{100-x} = 0.859.$$

Hence, in this case, the Poisson approximation is extremely close to the true value but much easier to find.

═══ *Exercises* ═══

2.7–1. Let X have a Poisson distribution with a mean of 4. Find (a) $P(2 \leq X \leq 5)$, (b) $P(X \geq 3)$, (c) $P(X \leq 3)$.

2.7–2. Let Y have a Poisson distribution with a variance of 3. Find $P(Y = 2)$.

2.7–3. If the moment-generating function of X is $e^{5(e^t-1)}$, find $P(3 \leq X < 5)$.

2.7–4. If X has a Poisson distribution so that $3P(X = 1) = P(X = 2)$, find $P(X = 4)$.

2.7–5. Flaws in a certain type of drapery material appear on the average of one in 150 square feet. If we assume the Poisson distribution, find the probability of at most one flaw in 225 square feet.

2.7–6. A certain type of aluminum screen that is 2 feet wide has on the average one flaw in a 100 foot roll. Find the probability that a 50 foot roll has no flaws.

2.7–7. Let X denote the number of alpha particles emitted by barium-133 during 1/10 of a second and counted by a Geiger counter in a fixed position. If $E(X) = 5.6$, calculate (a) $P(X \leq 2)$, (b) $P(2 \leq X < 6)$.

2.7–8. Suppose that the probability of suffering a side effect from a certain flu vaccine is 0.005. If 1000 persons are innoculated, find approximately the probability that (a) at most one person suffers; (b) four, five, or six people suffer.

2.7–9. A roll of a biased die results in a two only 1/10 of the time. Let X denote the number of two's in 100 rolls of this die. Approximate (a) $P(3 \leq X \leq 7)$, (b) $P(X \geq 5)$.

2.7–10. The moment-generating function of a random variable X is $e^{9(e^t-1)}$. Compute $P(\mu - 3\sigma < X < \mu + 3\sigma)$.

2.7–11. Let X have a Poisson distribution with mean λ.

(a) Show that $E(X) = \displaystyle\sum_{x=0}^{\infty} x\frac{\lambda^x e^{-\lambda}}{x!} = \lambda.$

(b) Show that $E[X(X - 1)] = \lambda^2$ and use this result to find $\mathrm{Var}(X)$.

3
EMPIRICAL DISTRIBUTIONS

3.1 Empirical Distribution Function

To elicit information about an unknown probability distribution the replication of the associated random experiment is useful in estimating the unknown probabilities and distribution characteristics or in checking the agreement between a physical random experiment and its mathematical model. Accordingly, we shall consider, in this section, repeating a random experiment a number, say n, of independent times in order to take the first step toward making these estimates and comparisons. Associated with these n independent random trials, suppose we have the respective random variables $X_1, X_2, \ldots, X_n$. We want these random variables to enjoy the properties of being mutually independent and identically distributed, properties that we now discuss.

Suppose, temporarily, that the random variables $X_1, X_2, \ldots, X_n$ are of the discrete type. Say x is a possible value in each of their spaces (or supports). If they are identically distributed according to a distribution with p.d.f. $f(x)$, then

$$P(X_1 = x) = P(X_2 = x) = \cdots = P(X_n = x) = f(x)$$

for all possible x values. Moreover, if $x_1, x_2, \ldots, x_n$ represent values in the spaces (supports) of $X_1, X_2, \ldots, X_n$, respectively, and if the n events $A_1 = (X_1 = x_1), A_2 = (X_2 = x_2), \ldots, A_n = (X_n = x_n)$ are mutually independent, then

$$P(A_1 \cap A_2 \cap \cdots \cap A_n) = P(A_1)P(A_2) \cdots P(A_n)$$

or, equivalently,

$$P(X_1 = x_1, X_2 = x_2, \ldots, X_n = x_n) = P(X_1 = x_1)P(X_2 = x_2) \cdots P(X_n = x_n).$$

If this latter statement is true for all possible $x_1, x_2, \ldots, x_n$ values, we say that $X_1, X_2, \ldots, X_n$ are *mutually independent random variables*. If the distributions of $X_1, X_2, \ldots, X_n$ are not the same and if $f_i(x_i)$ represents the p.d.f. of $X_i, i = 1, 2, \ldots, n$, it is interesting to note that this latter equality can be written as

$$P(X_1 = x_1, X_2 = x_2, \ldots, X_n = x_n) = f_1(x_1)f_2(x_2) \cdots f_n(x_n).$$

Frequently, in this case, the last product is called the *joint p.d.f.* of $X_1, X_2, \ldots, X_n$. Now if $X_1, X_2, \ldots, X_n$ actually are identically distributed with common p.d.f. $f(x)$, then

$$P(X_1 = x_1, X_2 = x_2, \ldots, X_n = x_n) = f(x_1)f(x_2) \cdots f(x_n).$$

In this case we say that the random variables $X_1, X_2, \ldots, X_n$ are mutually independent *and* identically distributed.

Example 3.1–1. Let the mutually independent random variables X_1, X_2, X_3 have the respective probability density functions

$$f_1(x_1) = \frac{x_1}{6}, \qquad x_1 = 1, 2, 3,$$

$$f_2(x_2) = \frac{1}{3}, \qquad x_2 = 1, 2, 3,$$

and

$$f_3(x_3) = \frac{4 - x_3}{6}, \qquad x_3 = 1, 2, 3.$$

Then, for illustration,

$$P(X_1 = 3, X_2 = 1, X_3 = 2) = f_1(3)f_2(1)f_3(2) = \left(\frac{3}{6}\right)\left(\frac{1}{3}\right)\left(\frac{2}{6}\right) = \frac{1}{18}$$

and

$$P(X_1 + X_2 + X_3 = 4) = P(X_1 = 2, X_2 = 1, X_3 = 1)$$
$$+ P(X_1 = 1, X_2 = 2, X_3 = 1)$$
$$+ P(X_1 = 1, X_2 = 1, X_3 = 2)$$

$$= \left(\frac{2}{6}\right)\left(\frac{1}{3}\right)\left(\frac{3}{6}\right) + \left(\frac{1}{6}\right)\left(\frac{1}{3}\right)\left(\frac{3}{6}\right) + \left(\frac{1}{6}\right)\left(\frac{1}{3}\right)\left(\frac{2}{6}\right)$$

$$= \frac{11}{108}.$$

Example 3.1–2. Let X_1, X_2, X_3, X_4 be four mutually independent and identically distributed random variables with the common Poisson p.d.f.

$$f(x) = \frac{2^x e^{-2}}{x!}, \qquad x = 0, 1, 2, \ldots.$$

Then, for illustration,

$$P(X_1 = 3, X_2 = 1, X_3 = 2, X_4 = 1) = f(3)f(1)f(2)f(1)$$
$$= \frac{2^{3+1+2+1}e^{-8}}{3!1!2!1!}$$
$$= \frac{2^7 e^{-8}}{12} = \frac{32}{3}e^{-8}.$$

Also, to compute the probability that exactly one of the X's equals zero, we treat zero as "success" and compute the probability of one success and three failures, namely,

$$\frac{4!}{1!3!}(e^{-2})(1 - e^{-2})^3,$$

because

$$P(X_i = 0) = e^{-2}, \qquad i = 1, 2, 3, 4.$$

We are now prepared to define a random sample.

DEFINITION 3.1–1. Say a random experiment that results in a random variable X having p.d.f. f(x) is repeated n independent times. Let $X_1, X_2, \ldots, X_n$ denote the n random variables associated with these outcomes. The collection of these random variables, which are mutually independent and identically distributed, is called a random sample from a distribution with p.d.f. f(x). The number n is called the sample size.

The common distribution of the random variables in the random sample is sometimes called the *population* from which the sample is taken. After the experiment has been performed these n times, we observe some definite numbers, say $X_1 = x_1, X_2 = x_2, \ldots, X_n = x_n$. These observed sample values, $x_1, x_2, \ldots, x_n$, will be used to elicit information about the unknown population (or common distribution).

DEFINITION 3.1–2. *Let* $x_1, x_2, \ldots, x_n$ *denote the observed values of the random sample* $X_1, X_2, \ldots, X_n$ *from a distribution. Let* $N(\{x_i : x_i \leq x\})$ *equal the number of these observed values that are less than or equal to x. Then the function*

$$F_n(x) = \frac{N(\{x_i : x_i \leq x\})}{n},$$

defined for each real number x, is called the empirical distribution function.

Please observe that $N(\{x_i : x_i \leq x\})$ is the frequency in n trials of the event that the X value is less than or equal to x. Hence, the empirical distribution function $F_n(x)$ is the relative frequency of that event, and, because of the law of large numbers, we expect $F_n(x)$ to be close to the corresponding true distribution function $F(x)$ when n is fairly large. Also, from Definition 3.1–2, we note that, in the empirical distribution, a weight (empirical probability) of $1/n$ is assigned to each observed x_i, $i = 1, 2, \ldots, n$. Hence, the empirical distribution is one of the discrete type. A simple example at this point will help us better understand this definition.

Example 3.1–3. Let the random experiment be the toss of ten coins. Repeat this experiment five independent times. Let X_i denote the number of heads on the ith toss. Suppose that the five observed values of the random sample $X_1, X_2, \ldots, X_5$ are $x_1 = 3, x_2 = 5, x_3 = 2, x_4 = 8, x_5 = 7$. Since there are two observed values less than or equal to 3, we have, for illustration, $F_5(3) = 2/5$. Similarly, $F_5(2) = 1/5$, $F_5(5.8) = 3/5$, $F_5(7.3) = 4/5$. It is easy to graph $F_5(x)$ after we observe that there is a jump of $1/5$ at each distinct observed value. Figure 3.1–1 depicts $F_5(x)$.

In general, to graph $F_n(x)$, note that there is a jump of $1/n$ at each distinct observed value. Moreover, if there are exactly k observed values equal to the same number, the jump equals k/n at that point.

The empirical distribution function $F_n(x)$ is a distribution function because

(a) $0 \leq F_n(x) \leq 1$ since $F_n(x) = N(\{x_i : x_i \leq x\})/n$ is a relative frequency.

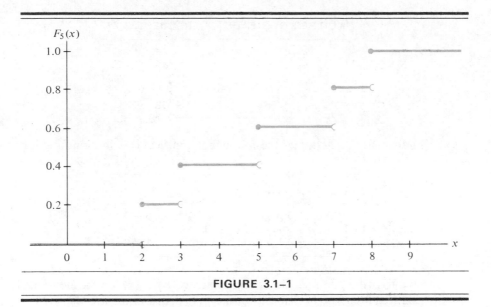

FIGURE 3.1–1

(b) $F_n(x)$ is a nondecreasing function since the number of observed values less than or equal to x does not decrease as x increases.

(c) For all values of x less than the smallest observed value of X, $F_n(x) = 0$, and for all values of x greater than or equal to the largest observed value of X, $F_n(x) = 1$.

(d) $F_n(x)$ is a step function, and the height of a step at each point x is equal to the relative frequency of the event $X = x$.

The function $F_n(x)$ is called an empirical distribution function because it is determined completely by the observed values of the random variable, and no assumptions about the underlying distribution of X are necessary. An empirical distribution function is one way of summarizing the information about the observed values of the random sample $X_1, X_2, \ldots, X_n$. Since $F_n(x)$ is the relative frequency of the event $X \leq x$, it is an approximation of the probability $P(X \leq x) = F(x)$, the distribution function of X evaluated at x. As a matter of fact, it can be proved in a more advanced probability course that $F_n(x)$ converges, in a certain probability sense that cannot be defined here, to $F(x)$ for all x, as n increases without bound.

Example 3.1–4. Let X denote the number of observed heads when four unbiased coins are tossed independently and at random. Recall that the distribution of X is $b(4, 1/2)$. One thousand repetitions of this experiment (actually simulated on the computer) yielded the following results.

NUMBER OF HEADS	FREQUENCY
0	65
1	246
2	358
3	272
4	59

This information determines the following empirical distribution function.

x	$F_{1000}(x)$	x	$F_{1000}(x)$
$(-\infty, 0)$	0.000	[2, 3)	0.669
[0, 1)	0.065	[3, 4)	0.941
[1, 2)	0.311	[4, ∞)	1.000

The graphs of the empirical distribution function $F_{1000}(x)$ and the theoretical distribution function $F(x)$ for the binomial distribution are depicted in Figure 3.1–2.

We will now introduce the expression *random number* by means of an example. Let a random variable have a distribution that assigns the probability of 1/10,000 to each of the numbers 0.0000, 0.0001, 0.0002, . . . , 0.9999. Let us determine a set, or a table, of four-digit numbers like these by repeating the random experiment associated with this random variable a large number of independent times. The resulting collection of numbers (or the observed values of the random sample) would constitute a table of four-digit random numbers. See Appendix Table VIII.

In general, consider a set of numbers that are observed values of a random variable for a large number of independent trials. Suppose the random variable has a distribution that assigns probability 10^{-k} to each of the numbers $0/10^k$, $1/10^k, 2/10^k, . . . , (10^k - 1)/10^k$. The resulting set of observations is then a table of k-digit random numbers. By choosing a suitably large value of k, the associated set of random numbers could be made arbitrarily close to a set of numbers selected at random from the interval [0, 1).

The production of a set of random numbers is not an easy task. Early in the history of statistics, some rather complicated and ingenious methods were employed to build tables of random numbers. Now, in the computer age, the standard software packages available with most digital computers contain programs that will generate numbers with the characteristics of random numbers. Such computer-produced random numbers are frequently called *pseudo-random numbers*. The prefix "pseudo" is used because the programs that produce them are such that if the starting number (the "seed" number) is known, all

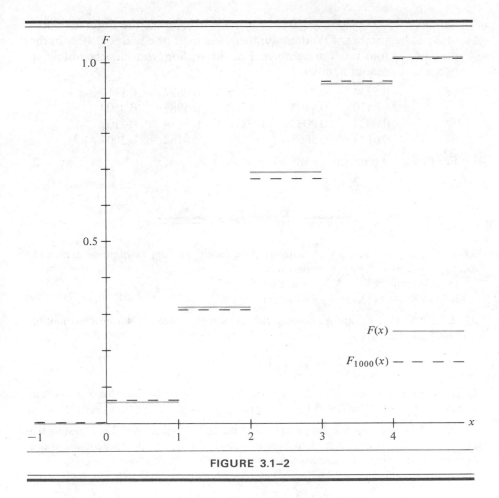

FIGURE 3.1–2

subsequent numbers in the sequence may be determined by simple arithmetical operations. Yet, despite their deterministic origin, these computer-produced numbers do behave as if they were truly randomly generated, and we shall not encumber our terminology by adding "pseudo."

Random numbers can be used to simulate many physical experiments. Although we will consider certain simulations throughout the text, the following is one simple example.

Example 3.1–5. In Example 3.1–4 we considered 1000 repetitions of the random experiment of tossing four coins. To simulate this experiment, consider five sets of four random numbers. Let X_i denote the number of heads on the ith toss, $i = 1, 2, 3, 4, 5$. If we associate heads with the random numbers $0.0000, 0.0001, \ldots, 0.4999$ and associate tails with the random numbers $0.5000, 0.5001, \ldots, 0.9999$, for $i = 1, 2, 3, 4, 5$, the observed value of X_i is the number of random numbers less than or equal to 0.4999 in the ith block of four random numbers. For illustration, consider the following five sets of random numbers:

0.5734	0.2395	0.7258	0.0832	0.1117
0.4420	0.8102	0.4562	0.5983	0.8509
0.0024	0.9033	0.3228	0.9644	0.7107
0.0127	0.5661	0.5261	0.8162	0.3004

For this set of numbers we would let $x_1 = 3, x_2 = 1, x_3 = 2, x_4 = 1, x_5 = 2$.

══ Exercises ══

3.1–1. Let X_1, X_2, and X_3 be mutually independent Poisson random variables with means $\mu_1 = 2, \mu_2 = 1, \mu_3 = 4$, respectively.
  (a) Compute $P(X_1 = 1, X_2 = 0, X_3 = 0)$.
  (b) Determine $P(X_1 + X_2 + X_3 = 1)$.

3.1–2. Let X_1, X_2, X_3, and X_4 denote a random sample of size $n = 4$ from a distribution with the geometric p.d.f.

$$f(x) = \left(\frac{2}{3}\right)\left(\frac{1}{3}\right)^x, \qquad x = 0, 1, 2, \ldots.$$

Let Y equal the maximum of X_1, X_2, X_3, and X_4. Find the probability that Y is less than or equal to 3. HINT: Note that $P(Y \le 3) = P(X_1 \le 3, X_2 \le 3, X_3 \le 3, X_4 \le 3)$.

3.1–3. A bowl contains five chips numbered 1 through 5, respectively. Let X denote the number on a chip drawn at random from this bowl. Ten independent repetitions of this experiment, sampling with replacement, yielded the following observations:

$x_1 = 3$	$x_5 = 3$	$x_8 = 3$
$x_2 = 3$	$x_6 = 5$	$x_9 = 4$
$x_3 = 1$	$x_7 = 5$	$x_{10} = 2$
$x_4 = 4$		

  (a) Define the empirical distribution function for this set of observations.
  (b) Graph the empirical distribution function along with the theoretical distribution function associated with this experiment.

3.1–4. Let X be the number of spots observed on a cast of a fair die. Cast this die 30 independent times to obtain the 30 observed values of X. Graph the empirical distribution function for your observations along with the graph of the distribution function of X.

3.1–5. The following numbers are a random sample of size 10 from some distribution.

$$-0.49, \quad 0.90, \quad 0.76, \quad -0.97, \quad -0.73, \quad 0.93, \quad -0.88, \quad -0.75, \quad 0.88, \quad 0.96.$$

  (a) Graph the empirical distribution function.
  (b) Use the empirical distribution function to estimate $P(X \le -0.50), P(X > 0)$, and $P(-0.50 \le X \le 0.50)$.

3.1–6. Toss three coins 50 times. Let X_i denote the number of heads observed on the ith toss, $i = 1, 2, \ldots, 50$. Sketch the graph of the empirical distribution function along with the theoretical distribution function. (These data will be used in Exercise 3.3–8.)

3.1–7. A salesman makes periodic visits to each of a number of customers who bottle soft drinks. The salesman sells them glass bottles by the case. He finds that the number of cases ordered by one of these regular customers is approximately uniformly distributed from 0 to 999 cases. That is, the probability is approximately 1/1000 that this customer orders k cases, $k = 0, 1, \ldots, 999$. The salesman makes a commission of 10¢ on each of the first 400 cases in an order and 12¢ on each case above the first 400.

    (a) How can we use a table of random numbers (or a random-number generator) to simulate the random variable that represents his commission in dollars on an order from this regular customer?

    (b) Simulate a random sample of 15 observations from this distribution and find the empirical distribution function.

3.1–8. The salesman in Exercise 3.1.–7 discovers that on any particular day he is able to visit one, two, or three regular customers with probabilities 0.3, 0.5, and 0.2, respectively. He also notes that the numbers of cases ordered by the customers are mutually independent and each has the given uniform distribution.

    (a) Show how to simulate the random variable that gives the salesman's commission *per day.*

    (b) Simulate a random sample of size 10 from this distribution and find the empirical distribution function.

3.1–9. Let X denote the number of baby gerbils in a litter. Four observations of X were $x_1 = 6$, $x_2 = 5$, $x_3 = 1$, and $x_4 = 2$.

    (a) Graph the empirical distribution function.

    (b) Give an estimate of $P(X > 4)$.

3.1–10. Let X denote the number of alpha particles emitted by barium-133 in 1/10 of a second and counted by a Geiger counter in a fixed position. Ten observations of X are 6, 3, 11, 8, 3, 7, 3, 6, 4, and 5.

    (a) Graph the empirical distribution function.

    (b) Give an estimate of $P(X \leq 5)$.

    (c) Calculate $P(X \leq 5)$ if X has a Poisson distribution with a mean equal to the average of these 10 observations.

3.2 Histograms and Ogives

In the preceding section we noted that the empirical distribution function $F_n(x)$ provides an estimate of the distribution function $F(x)$ of the underlying distribution. In this section we will discuss an estimate of the underlying p.d.f. We will also consider experiments for which there are many possible outcomes on each trial, in particular, cases in which a possible outcome might be any number in an interval. That is, we will observe for the first time random variables of the continuous type. For such random variables we shall describe, in addition to $F_n(x)$, another estimate of the distribution function.

For the moment, say that an experiment that has a finite number of possible outcomes and its corresponding random variable X is of the discrete type. One way to estimate $f(x) = P(X = x)$ is with the relative frequency of occurrences of x. This is illustrated in Example 3.2–1.

Example 3.2–1. In Example 3.1–4 we listed the frequency of observing x heads, $x = 0, 1, 2, 3, 4$, when four coins were tossed 1000 independent times. Using these data, we obtain the following estimate $f_{1000}(x)$ of the p.d.f. $f(x)$ of X. Recall, in this example, that we know the distribution of X is $b(4, 1/2)$ provided the coins are unbiased and are tossed independently. Thus, we are able to record $f(x)$.

x	$f_{1000}(x)$	$f(x)$
0	0.065	0.0625
1	0.246	0.2500
2	0.358	0.3750
3	0.272	0.2500
4	0.059	0.0625

We shall now consider experiments for which the theoretical set of possible outcomes forms an interval, as might occur in considerations of the heights of first grade children, the length of life of a light bulb, or the time required to run 100 yards. Note that such observations are often rounded off so that the set of observations may seem to come from a finite set. However, we shall consider for such experiments conceptual sample spaces that are intervals of finite or infinite length. A random variable whose space is an interval (or union of intervals) is called a *random variable of the continuous type*.

Suppose that a random experiment is such that the space of the random variable X associated with this experiment is $R = \{x : \quad a \leq x \leq b\}$. If this experiment is repeated n independent times with observations $x_1, x_2, \ldots, x_n$, an estimate of $P(c \leq X \leq d)$, where $a \leq c \leq d \leq b$, is the relative frequency $N(\{x_i : \quad c \leq x_i \leq d\})/n$. That is, an estimate is given by the proportion of outcomes that fall in the interval $[c, d]$.

We would like to extend this notion and obtain estimates of various probabilities associated with the distribution of the random variable X of the continuous type. In particular, we would like reasonably simple descriptive estimates of what we will define later as the p.d.f. $f(x)$ of the continuous type and the corresponding distribution function $F(x)$. One solution to this problem for distributions that have intervals of outcomes is now given.

Let us first concentrate on the spread of the sample by considering the smallest and the largest sample items. We will partition the interval determined by these two values into k intervals, where in practice k is often about 10 and the intervals are of equal length. Obviously, we do not always use $k = 10$; with a large sample size, the number k is frequently larger than 10 (and more than likely smaller with a small sample size). Practice with this type of procedure helps determine an appropriate value of k. As an example, consider a study of the weights of male college students. Suppose $n = 200$ and the sample values were recorded to the nearest pound so that 126 was the smallest weight and 242 the largest. Thus, the *range* of the sample is $242 - 126 = 116$. Accordingly, $k = 10$ intervals, each of length 12, would cover all the sample items. But so would $k = 12$ intervals, each of length 10. Another possibility would be $k = 9$ intervals, each of length 13. Obviously, we could list other possibilities; thus, we see that this problem has no unique solution, and experience will help determine an appropriate k value.

For illustration, suppose we take $k = 10$ intervals, each of length 12. Even that selection does not determine the 10 intervals exactly: the first one could begin at 126 and the last one end at 246, or the first begin at 124 and the last end at 244, and so on. What is done in practice? Usually, these intervals are selected as follows:

(1) Each interval begins and ends half way between two possible values of the measurements.
(2) The first interval begins about as much below the smallest item as the last interval ends above the largest.

Thus, a good selection for these 10 intervals would be given by the following boundaries.

(124.5, 136.5)	(184.5, 196.5)
(136.5, 148.5)	(196.5, 208.5)
(148.5, 160.5)	(208.5, 220.5)
(160.5, 172.5)	(220.5, 232.5)
(172.5, 184.5)	(232.5, 244.5)

These intervals are called *class intervals*, and the boundaries are *class boundaries*. In general, we will denote these k class intervals by

$$(c_0, c_1), \quad (c_1, c_2), \quad \ldots, \quad (c_{k-1}, c_k).$$

A frequency table is constructed that lists the class boundaries, a tabulation of the measurements in the various classes, the frequency f_i of each class, and the cumulative frequency. The probability estimates for the k classes are conveniently graphed in the form of a *relative frequency histogram*. Such a histogram is constructed by drawing a rectangle for each class having as its base the interval bounded by the class boundaries and an area equal to the relative frequency f_i/n of the observations for the class. That is, the function defined by

$$h(x) = \frac{f_i}{(n)(c_i - c_{i-1})} \qquad \text{for } c_{i-1} < x \le c_i, \quad i = 1, 2, \ldots, k,$$

is called a *relative frequency histogram*, where f_i is the frequency of the ith class and n is the total number of observations.

An example will help clarify some of these terms.

Example 3.2–2. Let us consider the study of the weights of male college students with $n = 200$. Say we record each weight in the appropriate class to obtain Table 3.2–1.

CLASS INTERVAL	TABULATION	FREQUENCY (f_i)	CUMULATIVE FREQUENCY
(124.5, 136.5)	ℳ //	7	7
(136.5, 148.5)	ℳ ℳ ///	13	20
(148.5, 160.5)	ℳ ℳ ℳ ℳ	20	40
(160.5, 172.5)	ℳ ℳ ℳ ℳ ℳ ℳ ℳ ////	39	79
(172.5, 184.5)	ℳ ℳ ℳ ℳ ℳ ℳ ℳ /	36	115
(184.5, 196.5)	ℳ ℳ ℳ ℳ ℳ ℳ ℳ //	37	152
(196.5, 208.5)	ℳ ℳ ℳ ///	18	170
(208.5, 220.5)	ℳ ℳ ℳ ////	19	189
(220.5, 232.5)	ℳ ///	8	197
(232.5, 244.5)	///	3	200

TABLE 3.2–1

For $124.5 < x \le 136.5$, the histogram is defined by

$$h(x) = \frac{7}{(200)(136.5 - 124.5)} = \frac{7}{(200)(12)} = 0.0029.$$

Continuing in this manner, we obtain the histogram given in Figure 3.2–1; the number within each rectangle is the area of that rectangle.

Note that the relative frequency histogram $h(x)$ satisfies the following properties.

(a) $h(x) \ge 0$ for all x.

(b) The total area bounded by the x axis and below $h(x)$ equals one; that is

$$\int_{c_0}^{c_k} h(x)\, dx = 1.$$

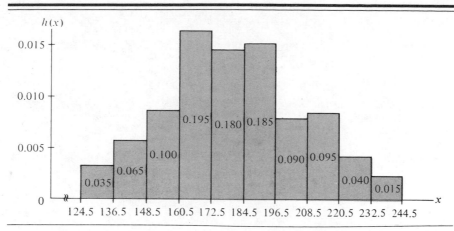

FIGURE 3.2–1

(c) The probability for an event A, which is composed of a union of class intervals, can be estimated by the area above A bounded by $h(x)$; that is,

$$P(A) \approx \int_A h(x)\, dx.$$

In the next chapter we shall define a random variable X of the continuous type. The properties that the p.d.f. $f(x)$ of X will satisfy are similar to those listed for $h(x)$.

We now return to an estimate of the distribution function $F(x)$ of X. With 200 observations, the empirical distribution function $F_{200}(x)$ is tedious to describe and graph. For all practical purposes, it amounts to listing 200 observations in order. So let us begin with the information given in Table 3.2–1, which certainly does not provide us with the individual observations. We then proceed as follows.

Let $F_n(x)$ denote the empirical distribution function. When a set of data has been classified in a frequency table like Table 3.2–1, we know the values of $F_n(x)$ at each of the class boundaries $c_0, c_1, \ldots, c_k$, namely,

$$F_n(c_i) = \frac{\text{cumulative frequency up to and including } c_i}{n}.$$

Graph the points

$$[c_0, F_n(c_0) = 0], \quad [c_1, F_n(c_1)], \quad \ldots, \quad [c_{k-1}, F_n(c_{k-1})], \quad [c_k, F_n(c_k) = 1].$$

Now draw a line segment between each pair of adjacent points; that is, a line segment is drawn between $[c_{i-1}, F_n(c_{i-1})]$ and $[c_i, F_n(c_i)]$, $i = 1, 2, \ldots, k$. The data given in Table 3.2–1 provide the plot shown in Figure 3.2-2, which is commonly called an *ogive*. We shall denote this function by $H(x)$.

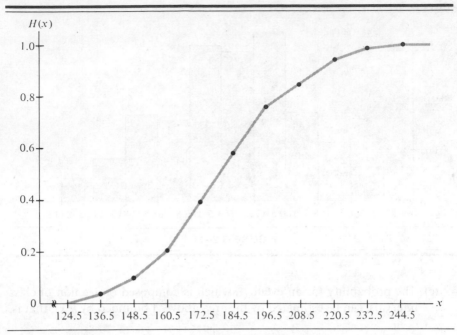

FIGURE 3.2–2

We make three observations about the ogive $H(x)$.

(a) Note that the ogive gives the cumulative area under the relative frequency histogram; that is,

$$H(x) = \int_{-\infty}^{x} h(t)\, dt = \int_{c_0}^{x} h(t)\, dt.$$

(b) $H(x)$ is a nondecreasing function with $0 \leq H(x) \leq 1$.

(c) The slope of the line segment that joins

$$[c_{i-1}, F_n(c_{i-1})] = [c_{i-1}, H(c_{i-1})] \qquad \text{and} \qquad [c_i, F_n(c_i)] = [c_i, H(c_i)]$$

is

$$\frac{H(c_i) - H(c_{i-1})}{c_i - c_{i-1}} = \frac{F_n(c_i) - F_n(c_{i-1})}{c_i - c_{i-1}} = \frac{f_i/n}{c_i - c_{i-1}}$$

$$= h(x), \qquad c_{i-1} < x \leq c_i.$$

That is, for $c_{i-1} < x < c_i$, $h(x) = H'(x)$, the derivative of the ogive function. We shall see in the next chapter that a similar relationship holds between the p.d.f. $f(x)$ and the distribution function $F(x)$ of a random variable of the continuous type.

══ *Exercises* ══

3.2–1. A random sample of 50 college-bound high school seniors yielded the following high school cumulative grade point averages (GPA's).

3.77	2.78	3.40	2.20	3.26
3.00	2.85	2.65	3.08	2.92
3.69	2.83	2.75	3.97	2.74
2.90	3.38	2.38	2.71	3.31
3.92	3.29	4.00	3.50	2.80
3.57	2.84	3.18	3.66	2.86
2.81	3.10	2.84	2.89	2.59
2.95	2.77	3.90	2.82	3.89
2.83	2.28	3.20	2.47	3.00
3.78	3.48	3.52	3.20	3.30

(a) Construct a frequency table for these 50 GPA's using 10 intervals of equal length. Use $c_0 = 2.005$ and $c_{10} = 4.005$.

(b) Construct a relative frequency histogram and an ogive for the grouped data.

3.2–2. Construct a relative frequency histogram and an ogive for the following set of data, which is a random sample of 75 Scholastic Aptitude Test scores in mathematics for college-bound high school students.

545	558	564	649	381	525	525
650	518	542	654	406	663	520
598	391	446	683	578	499	466
683	551	584	512	492	535	446
569	486	521	571	683	532	528
659	689	698	584	379	445	462
621	486	492	466	659	551	460
499	597	512	353	499	436	424
478	535	354	507	597	512	758
610	499	400	615	597	525	564
479	571	650	446	709		

3.2–3. A random sample of 48 production line workers under age 40 in a large food-processing firm showed the following weekly wages.

134.49	124.97	134.50	129.11	127.34
111.86	125.14	138.20	139.71	105.41
127.71	104.71	153.87	106.71	103.61
140.08	134.27	107.52	91.50	181.34
110.61	128.68	137.58	137.33	128.41
167.69	112.85	117.34	109.76	126.73
181.82	176.19	128.36	117.14	145.62
135.37	98.64	117.39	127.84	109.63
153.49	125.66	139.75	106.40	88.42
119.70	131.55	104.75		

(a)   Construct a relative frequency histogram and an ogive for these wages using 10 intervals of equal length. Use $c_0 = 85.005$ and $c_{10} = 185.005$.

(b)   Construct another histogram and ogive for the same set of numbers using seven intervals of equal length with $c_0 = 86.005$ and $c_7 = 184.005$.

3.2–4.   A sample of 100 numbers selected at random from the interval $(0, 1)$ yielded the following:

0.441	0.886	0.587	0.316	0.438
0.893	0.642	0.143	0.500	0.287
0.670	0.588	0.704	0.754	0.291
0.876	0.149	0.184	0.522	0.813
0.327	0.216	0.898	0.186	0.590
0.412	0.128	0.896	0.427	0.517
0.606	0.531	0.733	0.750	0.835
0.906	0.060	0.152	0.731	0.181
0.649	0.302	0.760	0.217	0.530
0.147	0.196	0.593	0.373	0.962
0.297	0.735	0.475	0.973	0.567
0.010	0.159	0.293	0.975	0.210
0.875	0.967	0.347	0.652	0.921
0.556	0.215	0.375	0.322	0.205
0.943	0.137	0.242	0.921	0.099
0.496	0.003	0.855	0.464	0.510
0.381	0.015	0.195	0.315	0.647
0.564	0.855	0.067	0.482	0.051
0.394	0.510	0.041	0.320	0.444
0.729	0.916	0.673	0.557	0.812

(a)   Construct a relative frequency histogram $h(x)$ and an ogive $H(x)$ for these numbers. Use 10 classes of equal length with $c_0 = 0.0005$ and $c_{10} = 1.0005$.

(b)   Can you guess at the functions $f(x)$ and $F(x)$ associated with this distribution of the continuous type that $h(x)$ and $H(x)$ are estimating, respectively? Consider both the experiment and the observations in defining $f(x)$ and $F(x)$.

3.2–5.   A random sample of size 100 yielded the following:

1.43	1.42	0.82	1.38	1.89	1.15	1.01	1.49
1.60	0.05	1.63	1.15	1.77	1.64	1.85	1.04
0.87	1.08	1.53	0.47	0.65	1.70	1.89	1.54
1.80	1.62	0.44	0.76	1.25	1.43	1.29	1.60
0.66	0.38	1.20	0.61	1.92	1.84	0.86	1.25
1.21	1.49	0.98	1.43	0.95	0.82	0.86	1.75
1.90	1.88	1.21	1.72	1.46	1.81	1.31	0.60
1.91	0.97	1.76	1.69	1.95	1.99	1.89	0.32
1.30	0.76	1.55	1.25	0.99	1.53	1.87	1.01
0.79	1.51	1.66	1.66	1.41	1.37	1.06	1.26
1.88	1.73	1.92	1.68	0.87	0.76	1.75	1.07
1.11	1.65	1.77	1.33	0.76	1.12	0.28	1.74
1.79	1.13	1.79	1.77				

(a)   Group these data into 10 classes with $c_0 = 0.005$ and $c_{10} = 2.005$.
(b)   Construct a relative frequency histogram $h(x)$ and ogive $H(x)$.
(c)   If it is known that the underlying p.d.f. $f(x)$ is a linear function for $0 \le x \le 2$, can you guess the continuous-type $f(x)$, based on these data?
(d)   What is the corresponding distribution function $F(x)$?

3.2–6. Let X denote the number of alpha particles emitted by barium-133 in 1/10 of a second and counted by a Geiger counter in a fixed position. One hundred observations of X are

7	5	7	7	4	6	5	6	6	2
4	5	4	3	8	5	6	3	4	2
3	3	4	2	9	12	6	8	5	8
6	3	11	8	3	6	3	6	4	5
4	4	9	6	10	9	5	9	5	4
4	4	6	7	7	8	6	9	4	6
5	3	8	4	7	4	7	8	5	8
3	3	4	2	9	1	6	8	5	8
5	7	5	9	3	4	8	5	3	7
3	6	4	8	10	5	6	5	4	4

The average of these numbers is 5.59.
(a)   Find $f_{100}(x)$, the estimate of the unknown p.d.f. of the discrete type.
(b)   Do these data seem to represent a random sample from a Poisson distribution with mean the same as the average of these 100 observations, namely 5.6? (At this stage, we do not have any statistical tests available. You should just compare $f_{100}(x)$ with the Poisson p.d.f. with $\mu = 5.6$. Finding an appropriate p.d.f. $f(x)$ that seemingly fits the data comes under the general heading of model building.)

3.3 The Mean and the Variance of a Sample

In the first two sections of this chapter we considered the estimation of the distribution of probability. That is, we showed how a set of data can be used to construct an empirical distribution function, an ogive, or a relative frequency histogram. These are used to estimate various probabilities and, in particular, the distribution function and p.d.f. We now consider the estimation of certain characteristics or parameters of the distribution from the data.

Since the empirical distribution assigns probability $1/n$ to each X_i, the mean of this distribution is given by

$$\sum_{i=1}^{n} X_i \left(\frac{1}{n}\right)$$

and is denoted by the symbol $\overline{X}$. That is,

$$\overline{X} = \frac{1}{n} \sum_{i=1}^{n} X_i$$

is the mean of the empirical distribution. But since this empirical distribution is determind by the sample, $\overline{X}$ is more frequently called simply the *mean of the sample*.

The variance of the empirical distribution is given by

$$\sum_{i=1}^{n} (X_i - \overline{X})^2 \left(\frac{1}{n} \right);$$

this quantity is denoted by S^2 and is called the *variance of the sample*. That is,

$$S^2 = \frac{1}{n} \sum_{i=1}^{n} (X_i - \overline{X})^2$$

but, from an earlier result about the computation of the variance of a distribution, namely $\sigma^2 = E(X^2) - \mu^2$, we know that the sample variance can be computed by

$$S^2 = \frac{1}{n} \sum_{i=1}^{n} X_i^2 - \overline{X}^2.$$

That is, $\overline{X}$ and S^2 are the mean and the variance of a distribution (the empirical distribution), and all the rules concerning means and variances would apply to them also. The positive square root S of the variance S^2 is called the *standard deviation of the sample*. The lowercase symbols $\bar{x}$ and s^2 will denote observed values of $\overline{X}$ and S^2. Since $F_n(x)$ is close to $F(x)$ for large n, the observed mean $\bar{x}$ and variance s^2 of the sample can be used as estimates of the corresponding distributional characteristics μ and σ^2.

REMARK.   Many writers define the variance of the sample as

$$\sum_{i=1}^{n} \frac{(X_i - \overline{X})^2}{n - 1}.$$

As we will see later, there is a good reason for doing this. However, we prefer the definition with $1/n$ rather than $1/(n-1)$ because, with $1/n$, S^2 is a variance of a distribution, namely, the empirical distribution. This is not the case with $1/(n-1)$; thus, our definition seems more consistent.

Example 3.3–1.   An unbiased die is case five independent times. Say the five random variables associated with the outcomes of this experiment take on the values $x_1 = 6, x_2 = 2, x_3 = 4, x_4 = 1, x_5 = 2$. The observed sample mean is

$$\bar{x} = \frac{6 + 2 + 4 + 1 + 2}{5} = 3$$

and the observed sample variance is

$$s^2 = \frac{(6 - 3)^2 + (2 - 3)^2 + (4 - 3)^2 + (1 - 3)^2 + (2 - 3)^2}{5} = 3.2.$$

The mean and the variance of the underlying distribution, for which a probability of 1/6 is assigned to each of the numbers 1, 2, 3, 4, 5, 6, are, respectively,

$$\mu = \frac{1 + 2 + 3 + 4 + 5 + 6}{6} = 3.5$$

and

$$\sigma^2 = \sum_{x=1}^{6} \frac{(x - 3.5)^2}{6} = \frac{35}{12} = 2.92.$$

Suppose that a random experiment has only k possible outcomes and that several of these outcomes are repeated in n independent trials of the experiment. For illustration, consider the cast of a die $n = 100$ times, so that we must have some repetitions. Or suppose that data for an experiment have been grouped in a frequency table with k classes. In such a case we could assume that the possible outcomes are the midpoints of the classes, denoted by u_i and called the *class marks*. Say that these outcomes occur with respective frequencies f_i, $i = 1, 2, \ldots, k$. In these cases we have k points $u_1, u_2, \ldots, u_k$ to which we assign the respective relative frequencies $f_1/n, f_2/n, \ldots, f_k/n$ as empirical probabilities. The mean and the variance of this distribution are, respectively,

$$\bar{u} = \sum_{i=1}^{k} u_i \left(\frac{f_i}{n}\right) = \frac{1}{n} \sum_{i=1}^{k} f_i u_i$$

and

$$s_u^2 = \sum_{i=1}^{k} (u_i - \bar{u})^2 \left(\frac{f_i}{n}\right) = \frac{1}{n} \sum_{i=1}^{k} f_i (u_i - \bar{u})^2 = \frac{1}{n} \sum_{i=1}^{k} f_i u_i^2 - \bar{u}^2.$$

For grouped "continuous data," these statistics can be used as approximations of $\bar{x}$ and s_x^2, which are computed before grouping. But in the discrete cases, we have that $\bar{x} = \bar{u}$ and $s_x^2 = s_u^2$.

Example 3.3–2. An urn contains 15 balls, one numbered 1, two numbered 2, three numbered 3, four numbered 4, and five numbered 5. Let X denote the outcome when a ball is drawn at random from this urn with replacement. One hundred repetitions of this experiment yielded the outcomes in Table 3.3–1. The sample mean of these data is

$$\bar{u} = \bar{x} = \frac{371}{100} = 3.71.$$

u_i	f_i	$f_i u_i$	$f_i u_i^2$
1	8	8	8
2	9	18	36
3	20	60	180
4	30	120	480
5	33	165	825
Totals	100	371	1529

TABLE 3.3–1

The sample variance of these data is

$$s_u^2 = s_x^2 = \frac{1529}{100} - (3.71)^2 = 1.53.$$

The sample standard deviation is

$$s = \sqrt{s^2} = 1.24.$$

Since X has the p.d.f. $f(x) = x/15$, $x = 1, 2, 3, 4, 5$, it is easy to show that $\mu = 3.67$, $\sigma^2 = 1.56$, and $\sigma = 1.25$ are the mean, the variance, and the standard deviation of X, respectively. Thus, the sample characteristics are reasonably close to those of the underlying distribution in this case.

Just as μ and σ^2 were of value in locating the center and measuring the degree of dispersion of a distribution, $\bar{X}$ and S^2 are of value in locating the center and measuring the dispersion of the sample values or, equivalently, of the empirical distribution.

To emphasize the fact that the sample standard deviation s does measure spread, let us apply Chebyshev's inequality to the distribution of numbers $x_1, x_2, \ldots, x_n$. Here probability will be empirical probability (or relative frequency), and the mean $\bar{x}$ and the variance s^2 will be those of the sample (or of the empirical distribution). Referring to Chebyshev's inequality (Theorem 2.3–1), we have, for every $k \geq 1$,

$$\frac{N(\{x_i: \ |x_i - \bar{x}| \geq ks\})}{n} \leq \frac{1}{k^2}$$

or, equivalently,

$$\frac{N(\{x_i: \ |x_i - \bar{x}| < ks\})}{n} \geq 1 - \frac{1}{k^2}.$$

Note that this is simply a restatement of Chebyshev's inequality as applied to an empirical distribution so no additional proof is necessary.

In words, the second inequality says: "The proportion of the sample values that lie within k standard deviations of the sample mean $\bar{x}$ is at least $1 - 1/k^2$."

Thus, we see that Chebyshev's inequality can be thought of as a statement about a collection of numbers.

A function of the items of a random sample that does not depend on any unknown parameters of the distribution is called a *statistic*. Thus, the mean $\overline{X}$ and the variance S^2 of the sample are statistics. A good statistic is one that helps summarize the information contained in a sample. Much of mathematical statistics is devoted to finding standards by which the quality of a statistic is measured and then constructing statistics that are optimal by these standards.

Of course, $\overline{X}$ and S are only two of a great many statistics that are useful functions of the sample values. Three others, which are easy to compute and, in a sense, are competitors of S and $\overline{X}$, are

(a) The range of the sample. The *sample range* indicates dispersion and is equal to the difference of the largest and smallest items of the sample. That is,

$$\text{Range} = \text{Max}(X_1, X_2, \ldots, X_n) - \text{Min}(X_1, X_2, \ldots, X_n).$$

This statistic, like the sample standard deviation, can be used to estimate the degree of dispersion of the underlying distribution and to measure the dispersion within the sample values. Recall that it was used in the determination of the length of the class intervals in Section 3.2.

(b) The midrange and the median of the sample. The *midrange* is defined as the average of the smallest and the largest items of the sample, that is,

$$\frac{\text{Max}(X_1, X_2, \ldots, X_n) + \text{Min}(X_1, X_2, \ldots, X_n)}{2}.$$

The *median* is the sample item that is middle in magnitude; or, in the case of an even number of sample items, it is the average of the two middle items. These statistics, like the sample mean, are sometimes useful in estimating the location of the "center" of a distribution of probability.

Example 3.3–3. The random variables X_1, X_2, X_3, X_4, X_5 are a random sample from a uniform distribution on the 10 digits $0, 1, 2, \ldots, 9$. The sampling yields the values $x_1 = 4$, $x_2 = 9$, $x_3 = 3$, $x_4 = 1$, $x_5 = 6$. Five statistics that may be computed from this sample are

$$\bar{x} = \frac{4 + 9 + 3 + 1 + 6}{5} = \frac{23}{5} = 4.6,$$

$$s^2 = \frac{4^2 + 9^2 + 3^2 + 1^2 + 6^2}{5} - (4.6)^2 = 7.44,$$

$$\text{Range} = 9 - 1 = 8,$$

$$\text{Median} = 4,$$

$$\text{Midrange} = \frac{9 + 1}{2} = 5.$$

It is interesting to note the corresponding characteristics of the underlying distribution, which has p.d.f. $f(x) = 1/10, \quad x = 0, 1, 2, \ldots, 9$, namely,

$$\mu = \sum_{x=0}^{9} x\left(\frac{1}{10}\right) = \frac{9(10)}{2}\left(\frac{1}{10}\right) = 4.5,$$

$$\sigma^2 = \sum_{x=0}^{9} x^2\left(\frac{1}{10}\right) - (4.5)^2 = \frac{(9)(10)(19)}{6}\left(\frac{1}{10}\right) - (4.5)^2 = 8.25,$$

$$\text{Range} = 9 - 0 = 9,$$

$$\text{Median} = \frac{4+5}{2} = 4.5,$$

$$\text{Midrange} = \frac{9+0}{2} = 4.5.$$

Exercises

3.3–1. Find the mean, variance, standard deviation, range, median, and midrange for the following test scores: 78, 93, 81, 95, and 83.

3.3–2. A spot check on traffic conducted during the month of September revealed that 3, 5, 2, 2, 7, and 3 cars were waiting at a particular stop sign at six randomly chosen times during the noon hours. Find the median, midrange, variance, and range of these sample values.

3.3–3. Denote a sample of n numbers by $x_1, x_2, \ldots, x_n$. Suppose that we transform each observation x_i into a new observation z_i using the transformation $z_i = ax_i + b$, $i = 1, 2, \ldots, n$. Show that the relationships between the means and the variances of $x_1, x_2, \ldots, x_n$ and $z_1, z_2, \ldots, z_n$ are

$$\bar{z} = a\bar{x} + b \qquad \text{and} \qquad s_z^2 = a^2 s_x^2.$$

3.3–4. Consider the following sample of five numbers: 1.0, 2.3, 3.0, 4.2, and 1.5. Use the results of Exercise 3.3–3 to find five observations:
  (a) With the same variance as the above observations but with a mean three units larger.
  (b) With the same mean but a variance four times as large as the variance of the given sample.
  (c) With a mean three units larger and a variance four times as large.

3.3–5. Hot billets of steel are rolled into long round bars for shipment to customers. Fissures (hairline cracks) in the bars can affect the quality of a customer's product so the bars are inspected before shipping. A random sample of 12 bars were inspected, yielding the following data.

NUMBER OF FISSURES (u_i)	NUMBER OF BARS (f_i)
0	0
1	4
2	5
3	2
4	1

Find the mean and variance of the number of fissures per bar in this sample of size $n = 12$.

3.3–6. Find the mean $\bar{u}$ and variance s_u^2 for the 50 high school grade point averages given in Exercise 3.2–1, after the data have been grouped in a frequency table. Compare $\bar{u}$ and s_u^2 with $\bar{x} = 3.116$ and $s^2 = 0.208$, the mean and variance for the ungrouped data.

3.3–7. Find the mean $\bar{u}$ and variance s_u^2 for the 100 random numbers listed in Exercise 3.2–4, after the numbers have been grouped in a frequency table. Compare $\bar{u}$ and s_u^2 with the theoretical values of the underlying uniform distribution, namely, $\mu = 1/2$ and $\sigma^2 = 1/12$.

3.3–8. Find the mean and variance for the 50 outcomes when you tossed three coins 50 times for Exercise 3.1–6. Compare $\bar{x}$ and s^2 with the mean and variance of the underlying distribution $b(3, 1/2)$.

3.3–9. Find the sample mean and sample variance of the data given in Exercise 3.2–6. Are they approximately equal? (Recall that the mean and the variance of a Poisson distribution are equal; and, if these are sample observations of a Poisson random variable, these two sample characteristics should be about equal.)

3.3–10. The mean and variance of the test scores for 900 students were $\bar{x} = 83$ and $s^2 = 36$, respectively. At least how many students received test scores between 71 and 95?

4
DISTRIBUTIONS OF THE
CONTINUOUS TYPE

4.1 Random Variables of the Continuous Type

In Section 3.2 we considered random variables whose spaces were not composed of a countable number of points but were intervals or a union of intervals. Such random variables are said to be of the *continuous type*. Also recall that the relative frequency histogram $h(x)$ associated with n observations of a random variable of that type is a nonnegative function defined so that the total area between its graph and the x axis equals one. In addition, $h(x)$ is constructed so that the integral

$$\int_a^b h(x)\, dx$$

is an estimate of the probability $P(a < X < b)$, where the interval (a, b) is a subset of the space R of the random variable X.

Let us now consider what happens to the function $h(x)$ in the limit, as n increases without bound and as the lengths of the class intervals decrease to zero. It is to be hoped that $h(x)$ will become closer and closer to some function, say $f(x)$, that gives the true probabilities, such as $P(a < X < b)$, through the integral

$$P(a < X < b) = \int_a^b f(x)\, dx.$$

That is, $f(x)$ should be a nonnegative function such that the total area between its graph and the x axis equals one. Moreover, the probability $P(a < X < b)$ is the area bounded by the graph of $f(x)$, the x axis, and the lines $x = a$ and $x = b$. This suggests the following formal definition.

DEFINITION 4.1–1. *The probability density function (p.d.f.) of a random variable X of the continuous type, with space R that is an interval or union of intervals, is an integrable function $f(x)$ satisfying the following conditions:*

(i) $f(x) > 0$, $x \in R$.

(ii) $\displaystyle\int_R f(x)\, dx = 1.$

(iii) *The probability of the event $X \in A$ is*

$$P(X \in A) = \int_A f(x)\, dx.$$

The corresponding distribution of probability is said to be one of the continuous type.

Example 4.1–1. Let the random variable X be the length of life of an electron tube, with space $R = \{x: \ 0 \le x < \infty\}$. Suppose that a reasonable probability model for X is given by the p.d.f.

$$f(x) = \frac{1}{100}\, e^{-x/100}, \qquad x \in R.$$

Note that $f(x) > 0$, $x \in R$, and

$$\int_R f(x)\, dx = 1.$$

The probability that this electron tube lasts more than 100 hours is given by

$$P(X > 100) = \int_{100}^\infty \frac{1}{100}\, e^{-x/100}\, dx = e^{-1} = 0.368.$$

The p.d.f. and the area of interest are depicted in Figure 4.1–1.

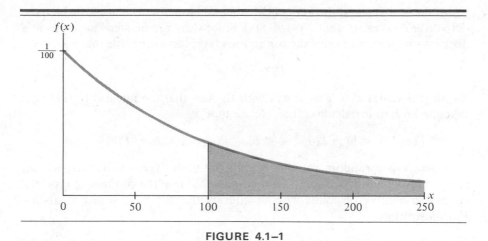

FIGURE 4.1–1

So that we can avoid repeated references to the space (or support) R of the random variable X, we shall adopt the following convention when describing probability density functions of both the discrete and continuous types. We extend the definition of the p.d.f. $f(x)$ to the entire set of real numbers by letting it equal zero when $x \notin R$. For example, it seems clear that

$$f(x) = \begin{cases} \dfrac{x}{10}, & x = 1, 2, 3, 4, \\ 0 \text{ elsewhere,} \end{cases}$$

is a p.d.f. of a discrete-type random variable X having support

$$\{x : \quad x = 1, 2, 3, 4\}.$$

On the other hand,

$$f(x) = \begin{cases} \dfrac{1}{100} e^{-x/100}, & 0 \le x < \infty, \\ 0 \text{ elsewhere,} \end{cases}$$

has the properties of a p.d.f. of a continuous-type random variable X having support $\{x : \quad 0 \le x < \infty\}$. It will always be understood that $f(x) = 0$ when $x \notin R$, even when this is not explicitly written out.

The *distribution function* of a random variable X of the continuous type, defined in terms of the p.d.f. of X, is given by

$$F(x) = P(X \le x) = \int_{-\infty}^{x} f(t)\, dt.$$

From the fundamental theorem of calculus we have, for x values for which the derivative $F'(x)$ exists, that $F'(x) = f(x)$. Since there are no steps or jumps in a distribution function $F(x)$ of the continuous type, it must be true that

$$P(X = b) = 0$$

for all real values of b. This agrees with the fact that the integral $\int_b^b f(x)\,dx$ is taken to be zero in calculus. Thus, we see that

$$P(a < X < b) = P(a < X \le b) = P(a \le X < b) = F(b) - F(a),$$

provided X is a random variable of the continuous type. Moreover, we can change the definition of a p.d.f. of a random variable of the continuous type at a countable number of points without altering the distribution of probability. For illustration,

$$f(x) = \begin{cases} 1, & 0 < x < 1, \\ 0 \text{ elsewhere,} \end{cases}$$

and

$$f(x) = \begin{cases} 1, & 0 \le x \le 1, \\ 0 \text{ elsewhere,} \end{cases}$$

are equivalent in the computation of probabilities involving a random variable X with the uniform distribution on the unit interval. This distribution is called uniform because the p.d.f. is constant on the support.

Example 4.1–2. Let Y be a continuous random variable with p.d.f. $g(y) = 2y$, $0 < y < 1$. The distribution function of Y is defined by

$$G(y) = \begin{cases} 0, & y < 0, \\ \int_0^y 2t\,dt = y^2, & 0 \le y < 1, \\ 1, & 1 \le y. \end{cases}$$

Figure 4.1–2 gives the graph of the p.d.f. $g(y)$ and the graph of the distribution function $G(y)$. For examples of computations of probabilities, consider

$$P\left(\frac{1}{2} < Y \le \frac{3}{4}\right) = G\left(\frac{3}{4}\right) - G\left(\frac{1}{2}\right) = \left(\frac{3}{4}\right)^2 - \left(\frac{1}{2}\right)^2 = \frac{5}{16}$$

and

$$P\left(\frac{1}{4} \le Y < 2\right) = G(2) - G\left(\frac{1}{4}\right) = 1 - \left(\frac{1}{4}\right)^2 = \frac{15}{16}.$$

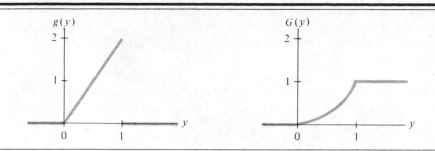

FIGURE 4.1–2

REMARK. It should be noted that the p.d.f. of a random variable X of the continuous type does not need to be a continuous function. For example,

$$f(x) = \begin{cases} \dfrac{1}{2}, & 0 < x < 1 \quad \text{or} \quad 2 < x < 3, \\ 0 \text{ elsewhere,} \end{cases}$$

enjoys the properties of a p.d.f. of a distribution of the continuous type, and yet $f(x)$ has discontinuities at $x = 0, 1, 2,$ and 3. However, the distribution function associated with a distribution of the continuous type is always a continuous function.

For continuous-type random variables, the definitions associated with mathematical expectation are the same as those in the discrete case except that integrals replace summations. For illustration, let X be a continuous random variable with a p.d.f. $f(x)$.

The *expected value of X* or *mean of X* is

$$\mu = E(X) = \int_{-\infty}^{\infty} x f(x) \, dx.$$

The *variance of X* is

$$\sigma^2 = \text{Var}(X) = \int_{-\infty}^{\infty} (x - \mu)^2 f(x) \, dx.$$

The *standard deviation of X* is

$$\sigma = \sqrt{\text{Var}(X)}.$$

The *moment-generating function*, if it exists, is

$$M(t) = \int_{-\infty}^{\infty} e^{tx} f(x) \, dx, \qquad -h < t < h.$$

Moreover, results such as $\sigma^2 = E(X^2) - \mu^2$ and $\mu = R'(0)$, $\sigma^2 = R''(0)$, where $R(t) = \ln M(t)$, are still valid. And again it is important to note that the moment-generating function completely determines the distribution.

Example 4.1–3. Let X have a uniform distribution on the interval $[0, 1]$; that is, the p.d.f. of X is $f(x) = 1, 0 \leq x \leq 1$. Then

$$\mu = E(X) = \int_0^1 (x)(1)\, dx = \frac{1}{2} = 0.5$$

and

$$\sigma^2 = \text{Var}(X) = \int_0^1 (x - 0.5)^2(1)\, dx = \left[\frac{(x - 0.5)^3}{3}\right]_0^1 = \frac{1}{12}.$$

The moment-generating function is

$$M(t) = \int_0^1 e^{tx}(1)\, dx = \left[\frac{e^{tx}}{t}\right]_0^1 = \frac{e^t - 1}{t}, \qquad t \neq 0.$$

In general, $M(t)$ equals one when $t = 0$; that is, $M(0) = 1$.

Example 4.1–4. Let X have the p.d.f.

$$f(x) = \begin{cases} xe^{-x}, & 0 \leq x < \infty, \\ 0 & \text{elsewhere.} \end{cases}$$

Then

$$M(t) = \int_0^\infty e^{tx}xe^{-x}\, dx = \int_0^\infty xe^{-(1-t)x}\, dx$$

$$= \left[-\frac{xe^{-(1-t)x}}{1-t} - \frac{e^{-(1-t)x}}{(1-t)^2}\right]_0^\infty$$

$$= \frac{1}{(1-t)^2},$$

provided $t < 1$. Hence,

$$R(t) = -2[\ln(1 - t)], \qquad R'(t) = \frac{2}{1-t}, \qquad R''(t) = \frac{2}{(1-t)^2}$$

and

$$\mu = R'(0) = 2, \qquad \sigma^2 = R''(0) = 2.$$

═══ *Exercises* ═══

4.1-1. Let the random variable Y have the p.d.f. $g(y) = cy^2$, $-1 < y < 1$, zero elsewhere.
(a) Determine the value of the constant c so that $g(y)$ satisfies the properties of a p.d.f.
(b) Define the distribution function of Y.
(c) Find $P(0 < Y < 1)$, $P(0 < Y \le 3)$, $P(Y = 1/2)$, and $P(-1 < Y \le 1/2)$.

4.1-2. The p.d.f. of X is $f(x) = c/x^2$, $1 < x < \infty$, zero elsewhere.
(a) Calculate the value of c so that $f(x)$ is a p.d.f.
(b) Show that $E(X)$ does not exist.

4.1-3. The p.d.f. of Y is $g(y) = d/y^3$, $1 < y < \infty$, zero elsewhere.
(a) Calculate the value of d so that $g(y)$ is a p.d.f.
(b) Find $E(Y)$.
(c) Show that $\text{Var}(Y)$ does not exist.

4.1-4. Let $f(x)$ be the p.d.f. of a random variable X. Define the distribution function $F(x)$ of X and sketch its graph if
(a) $f(x) = 4x^3$, $\quad 0 < x < 1$;
(b) $f(x) = e^{-x}$, $\quad 0 < x < \infty$.

4.1-5. Sketch the graphs of the following probability density functions. Also find and sketch the graphs of the distribution functions associated with these distributions.

(a) $f(x) = \left(\dfrac{3}{2}\right)x^2$, $\quad -1 < x < 1$, $\quad$ zero elsewhere;

(b) $f(x) = \dfrac{1}{2}$, $\quad -1 < x < 1$, $\quad$ zero elsewhere:

(c) $f(x) = \begin{cases} x + 1, & -1 < x < 0, \\ 1 - x, & 0 < x < 1. \end{cases}$

4.1-6. Find the mean and variance for each of the distributions in Exercise 4.1-5.

4.1-7. Find the moment-generating function $M(t)$ of the distribution with p.d.f. $f(x) = (1/10)e^{-x/10}$, $0 < x < \infty$. Use $M(t)$ or $R(t) = \ln M(t)$ to determine the mean μ and the variance σ^2.

4.1-8. The logistic distribution is associated with the distribution function $F(x) = (1 + e^{-x})^{-1}$, $-\infty < x < \infty$. Find the p.d.f. of the logistic distribution and show that its graph is symmetric about $x = 0$.

4.1-9. The *(100p)th percentile* is a number π_p such that the area under $f(x)$ to the left of π_p is p. That is,

$$p = \int_{-\infty}^{\pi_p} f(x)\, dx = F(\pi_p).$$

Let $f(x) = 1/2$, $-1 < x < 1$. Find (a) $\pi_{0.5}$, (b) $\pi_{0.25}$, (c) $\pi_{0.90}$.

4.1-10. Let $f(x) = (x + 1)/2$, $-1 < x < 1$. Find (a) $\pi_{0.64}$, (b) $\pi_{0.25}$, (c) $\pi_{0.81}$.

4.2 The Uniform Distribution

Let the random variable X denote the outcome when a point is selected at random from an interval $[a, b]$, $-\infty < a < b < \infty$. If the experiment is performed in a "fair" manner, it is reasonable to assume that the probability that the point is selected from the interval $[a, x]$, $a \leq x < b$ is $(x - a)/(b - a)$. That is, the probability is proportional to the length of the interval so that the distribution function of X is

$$F(x) = \begin{cases} 0, & x < a, \\ \dfrac{x - a}{b - a}, & a \leq x < b, \\ 1, & b \leq x. \end{cases}$$

Because X is a continuous-type random variable, $F'(x)$ is equal to the p.d.f. of X whenever $F'(x)$ exists; thus, when $a < x < b$, we have $F'(x) = 1/(b - a)$.

The random variable X has a *uniform distribution* if its p.d.f. is equal to a constant on its support. In particular, if the support is the interval $[a, b]$, then

$$f(x) = \frac{1}{b - a}, \qquad a \leq x \leq b.$$

Moreover, we shall say that X is $U(a, b)$. This distribution is also referred to as *rectangular* because the graph of $f(x)$ suggests that name. Note that we could have taken $f(a) = 0$ or $f(b) = 0$ without altering the probabilities since this is a continuous-type distribution, and we will do this in some cases.

The mean, variance, and moment-generating function of X are easy to calculate (see Exercise 4.2–1); they are

$$\mu = \frac{a + b}{2}, \qquad \sigma^2 = \frac{(b - a)^2}{12},$$

$$M(t) = \begin{cases} \dfrac{e^{tb} - e^{ta}}{t(b - a)}, & t \neq 0, \\ 1, & t = 0. \end{cases}$$

In Section 3.1 we discussed the concept of "random numbers." Conceptually, a random number from the interval $(0, 1)$ is the observed value of a random variable X with a distribution $U(0, 1)$. The values of a random sample from a distribution $U(0, 1)$ can be simulated by (a) spinning a balanced spinner with a scale from 0 to 1, (b) using a table of random numbers (see Appendix Table VIII), or (c) using a random number generator on a computer. In turn, values of a random sample from a uniform distribution can be used to simulate the values of a random sample from both discrete and continuous distributions. The following example illustrates the use of random numbers to simulate a sample from the geometric distribution.

Example 4.2–1. Toss a pair of unbiased coins until a pair of heads is observed. The event of a pair of heads is called "success"; so the number of failures, say Y, before the first success occurs has a geometric distribution. Here, $p = 1/4$ is the probability of observing a pair of heads on a single trial. To simulate four observations of Y, we shall use the random numbers in Appendix Table VIII.

In that table, we let the numbers 0.0000 through 0.2499 correspond to success. Beginning in the upper left-hand corner, we read 0.3407, 0.5044, 0.0045. Since the first random number that corresponds to success occurs on the third trial, $y_1 = 2$ failures were observed before the first success. Continuing down the column of random numbers, we read 0.7536, 0.7653, 0.6157, 0.6593, 0.3187, 0.4780, 0.9414, 0.1948; thus, $y_2 = 7$ failures were observed before the next success. Continuing, we find $y_3 = 0$ and $y_4 = 4$. Note that the sample mean

$$\bar{y} = \frac{2 + 7 + 0 + 4}{4} = 3.25$$

is close to

$$\mu = \frac{3/4}{1/4} = 3,$$

the mean of the geometric distribution.

The following theorem enables us to simulate a random sample from a continuous-type distribution.

THEOREM 4.2–1. *Let Y have a distribution $U(0, 1)$. Let $F(x)$ have the properties of a distribution function of the continuous type. Let X be a function of Y defined by $Y = F(X)$. Then X is a continuous-type random variable with distribution function $F(x)$.*

Proof: To simplify the proof, assume that $y = F(x)$ is a strictly increasing function when $0 < y < 1$. Then, because of this property, the events $X \leq x$ and $F(X) \leq F(x)$ are equivalent, and the distribution function of X is

$$P(X \leq x) = P[F(X) \leq F(x)] = P[Y \leq F(x)].$$

However, $Y = F(X)$ has the uniform distribution with distribution function $P(Y \leq y) = y, 0 \leq y \leq 1$. Thus, we have that

$$P(X \leq x) = P[Y \leq F(x)] = F(x), \qquad 0 \leq F(x) \leq 1.$$

That is, the distribution function $P(X \leq x)$ is $F(x)$, which completes the proof.

Theorem 4.2–1 holds, however, even though $F(x)$ is not strictly increasing. In those cases, while the events $X \leq x$ and $F(X) \leq F(x)$ are not necessarily identical, they do have the same probability. Hence, the proof is actually satisfactory in this more general case. However, when $F(x)$ is strictly increasing, we can frequently solve $Y = F(X)$ for $X = F^{-1}(Y)$, say, and this makes the simulation of observations of the random variable X from observations of the uniform random variable Y rather simple. This is illustrated in the next example.

Example 4.2–2. Say we want to simulate three observations of the random variable X that has the p.d.f.

$$f(x) = \frac{1}{2\sqrt{x}}, \quad 0 < x < 1,$$

and distribution function

$$F(x) = \sqrt{x}, \qquad 0 < x < 1.$$

Let Y be $U(0, 1)$ and define X by $Y = F(X) = \sqrt{X}$. Then $X = Y^2$ has the given p.d.f. and distribution function. If, now, y is a uniform random number on $[0, 1)$, then

$$x = F^{-1}(y) = y^2$$

is an observed value of X. For example, given the three random numbers $y_1 = 0.1440$, $y_2 = 0.9859$, and $y_3 = 0.4999$, the corresponding observations of X are

$$x_1 = (0.1440)^2 = 0.0207,$$
$$x_2 = (0.9859)^2 = 0.9720,$$

and

$$x_3 = (0.4999)^2 = 0.2499.$$

====== *Exercises* ======

(Use Appendix Table VIII for the simulations.)

4.2–1. Show that the mean, variance, and moment-generating function of the uniform distribution are as given in this section.

4.2–2. Let $f(x) = 1/2$, $-1 \leq x \leq 1$, be the p.d.f. of X. Graph the p.d.f. and distribution function, and record the mean and variance, of X.

4.2–3. An urn contains one red ball, one white ball, one green ball, and one yellow ball. A random sample of size 10 is drawn from the urn, one at a time and with replacement. Describe how this experiment can be simulated. Then simulate this experiment.

4.2–4. A coin is tossed two independent times. One point is won if both tosses are heads; otherwise zero is won.
  (a) If X denotes the outcome of this game, what is the distribution of X?
  (b) Simulate four independent repetitions of this game.
  (c) How is the sum of these four independent repetitions distributed?

4.2–5. Simulate a random sample of size 6 from a distribution $b(4, 1/4)$.

4.2–6. Simulate a random sample of size 5 from a geometric distribution for which $p = 1/3$.

4.2–7. Simulate a random sample of size 4 from a negative binomial distribution with parameters $p = 2/3$ and $r = 3$.

4.2–8. The probabilities that a manufactured item is good, fair, or defective are $6/10$, $3/10$, and $1/10$, respectively. Simulate the production of seven independent items.

4.2–9. The p.d.f. of X is $f(x) = 2x, 0 < x < 1$.
  (a) Find the distribution function of X.
  (b) Describe how an observation of X can be simulated.
  (c) Simulate 10 observations of X.
  (d) Graph the empirical and theoretical distribution functions associated with X.

4.2–10. Repeat Exercise 4.2–9 when $f(x) = 1/(1 - x)^2, 0 < x < 1/2$.

4.2–11. Repeat Exercise 4.2–9 when $f(x)$ is that of the logistic distribution (Exercise 4.1–8).

4.2–12. Let Y have a distribution $U(0, 1)$, and let $W = a + (b - a)Y$.
  (a) Find the distribution function of W. HINT: Find $P[a + (b - a)Y \le w]$.
  (b) How is W distributed?

4.2–13. If the moment-generating function of X is

$$M(t) = \frac{e^{5t} - e^{4t}}{t}, \qquad t \ne 0, \qquad \text{and} \qquad M(0) = 1,$$

find   (a) $E(X)$,   (b) $\text{Var}(X)$,   (c) $P(4.2 < X \le 4.7)$.

4.2–14. Suppose X is a continuous random variable with distribution function $F(x)$. Assume that $F(a) = 0$, $F(b) = 1$, and $F(x)$ is strictly increasing for $a < x < b$. Show that $Y = F(X)$ is $U(0, 1)$. HINT: Consider

$$P(Y \le y) = P[F(X) \le y] = P[X \le F^{-1}(y)].$$

4.2–15. The p.d.f. of X is defined by $f(x) = (3/2)x^2, -1 < x < 1$. Show that $Y = (X^3 + 1)/2$ has a distribution $U(0, 1)$. HINT: See Exercise 4.2–14.

4.3 The Exponential Distribution

When previously observing a process of the (approximate) Poisson type, we counted the number of changes occurring in a given interval. This number was a discrete-type random variable with a Poisson distribution. But not only is the number of changes a random variable; the waiting times between successive changes are also random variables. However, the latter are of the continuous type, since each of them can assume any positive value. In particular, let W denote the waiting time until the first change occurs when observing a Poisson process in which the mean number of changes in the unit interval is λ. Then W is a continuous-type random variable, and we proceed to find its distribution function.

Clearly, this waiting time is nonnegative. Thus, the distribution function $F(w) = 0$, $w < 0$. For $w \geq 0$,

$$
\begin{aligned}
F(w) = P(W \leq w) &= 1 - P(W > w) \\
&= 1 - P \text{ (no changes in } [0, w]) \\
&= 1 - e^{-\lambda w},
\end{aligned}
$$

since we previously discovered that $e^{-\lambda w}$ equals the probability of no changes in an interval of length w. Thus, when $w > 0$, the p.d.f. of W is given by

$$
F'(w) = \lambda e^{-\lambda w}.
$$

We usually say that the random variable X has an *exponential distribution* if its p.d.f. is defined by

$$
f(x) = \begin{cases} \dfrac{1}{\theta} e^{-x/\theta}, & 0 \leq x < \infty, \\[2mm] 0, & x < 0, \end{cases}
$$

where the parameter $\theta > 0$. Accordingly, the waiting time W has an exponential distribution with $\theta = 1/\lambda$. To determine the exact meaning of the parameter θ, we first find the moment-generating function. It is

$$
M(t) = \int_0^\infty e^{tx} \left(\frac{1}{\theta}\right) e^{-x/\theta} \, dx = \int_0^\infty \left(\frac{1}{\theta}\right) e^{-(1-\theta t)x/\theta} \, dx
$$

$$
= \left[\frac{e^{-(1-\theta t)x/\theta}}{1-\theta t} \right]_0^\infty = \frac{1}{1-\theta t}, \qquad t < \frac{1}{\theta}.
$$

Thus,

$$
R(t) = -\ln(1 - \theta t),
$$

$$
R'(t) = \frac{\theta}{1 - \theta t},
$$

and

$$
R''(t) = \frac{\theta^2}{(1 - \theta t)^2}.
$$

Hence, for an exponential distribution, we have

$$\mu = R'(0) = \theta \qquad \text{and} \qquad \sigma^2 = R''(0) = \theta^2.$$

So if λ is the mean number of changes in the unit interval, then $\theta = 1/\lambda$ is the mean waiting time for the first change. In particular, suppose $\lambda = 7$ is the mean number of changes per minute, then the mean waiting time for the first change is $1/7$ of a minute.

Example 4.3–1. Let X have an exponential distribution with a mean of 100. The p.d.f. of X is

$$f(x) = \frac{1}{100} e^{-x/100}, \qquad 0 \le x < \infty.$$

The probability that X is less than 90 is

$$P(X < 90) = \int_0^{90} \frac{1}{100} e^{-x/100} \, dx = 1 - e^{-90/100} = 0.593.$$

Example 4.3–2. Customers arrive in a certain shop according to an approximate Poisson process at a mean rate of 20 per hour. What is the probability that the shopkeeper will have to wait more than 5 minutes for his first customer to arrive? Let X denote the waiting time *in minutes* until the first customer arrives and note that $\lambda = 1/3$ is the expected number of arrivals per minute. Thus,

$$\theta = \frac{1}{\lambda} = 3$$

and

$$f(x) = \frac{1}{3} e^{-(1/3)x}, \qquad 0 \le x < \infty.$$

Hence,

$$P(X > 5) = \int_5^\infty \frac{1}{3} e^{-(1/3)x} \, dx = e^{-5/3} = 0.189.$$

Example 4.3–3. Suppose that the life of a certain type of electron tube has an exponential distribution with a mean life of 500 hours. If X denotes the life of a tube (or the time to failure of a tube), then

$$P(X > x) = \int_x^\infty \frac{1}{500} e^{-t/500} \, dt = e^{-x/500}.$$

Suppose the tube has been in operation for 300 hours. The conditional probability that it will last for another 600 hours is

$$P(X > 900 | X > 300) = \frac{P(X > 900)}{P(X > 300)} = \frac{e^{-900/500}}{e^{-300/500}} = e^{-6/5}.$$

It is extremely important to note that this conditional probability is exactly equal to $P(X > 600) = e^{-6/5}$. That is, the probability that it will last an additional 600 hours, given that it has operated 300 hours, is the same as the probability that it would last 600 hours when first put into operation. Thus, for such tubes, an old tube is as good as a new one, and we say that the failure rate is constant. Certainly, with constant failure rate, there is no advantage in replacing tubes that are operating satisfactorily. Obviously, this is not true in practice because most would have an increasing failure rate; hence, the exponential distribution is probably not the best model for the probability distribution of such a life.

REMARK. In Exercise 4.3–5, the result of Example 4.3–3 is generalized; namely, if the life has an exponential distribution, the probability that it will last a time of at least $x + y$ units, given that it has lasted at least x units, is exactly the same as the probability that it will last at least y units when first put into operation. In effect, this states that the exponential distribution has a forgetfulness (or no memory) property. It is also interesting to observe that, for continuous random variables whose support is $(0, \infty)$, the exponential distribution is the only distribution with this forgetfulness property. Recall, however, that when we considered distributions of the discrete type, we noted that the geometric distribution has this property.

Observations of an exponential random variable can be simulated using observations of a random variable Y, which is $U(0, 1)$. This is illustrated in the next example.

Example 4.3–4. Let X have an exponential distribution with mean $\theta = 1$. That is, the p.d.f. of X is

$$f(x) = e^{-x}, \qquad 0 < x < \infty;$$

thus, the distribution function is

$$F(x) = \int_0^x e^{-w} \, dw = 1 - e^{-x}, \qquad 0 < x < \infty.$$

Setting $y = F(x) = 1 - e^{-x}$, we have that $x = -\ln(1 - y)$. Thus, if Y has a distribution $U(0, 1)$, then $X = -\ln(1 - Y)$ has the given exponential distribution. Hence, if we want to generate a random sample X_1, X_2, X_3, X_4 of size 4 from this distribution, we observe four random numbers, say y_1, y_2, y_3, y_4, and then compute

$$x_i = -\ln(1 - y_i), \qquad i = 1, 2, 3, 4.$$

We do this by using the first four numbers in the last column of Appendix Table VIII, which are

$$y_1 = 0.1514,$$
$$y_2 = 0.6697,$$
$$y_3 = 0.0527,$$
$$y_4 = 0.4749.$$

From these we compute

$$x_1 = -\ln(1 - 0.1514) = 0.1642,$$
$$x_2 = 1.1078,$$
$$x_3 = 0.0541,$$
$$x_4 = 0.6442.$$

In this sample of size 4 from this exponential distribution, we observe that one of the values is greater than the mean $\theta = 1$ and the other three are less than the mean. We might ask ourselves what is the probability of this occurring. If we call an observation greater than one a "success," then we want the probability of one success and three failures in four independent trials. Since the items of the random sample have the same exponential distribution, the probability of success on each trial is

$$p = P(X > 1) = \int_1^\infty e^{-x}\, dx = e^{-1}.$$

The desired probability is then

$$\frac{4!}{1!3!} p(1 - p)^3 = 4e^{-1}(1 - e^{-1})^3 = 0.3717.$$

═══ *Exercises* ═══

4.3–1. Let X have an exponential distribution with a mean of $\theta = 20$. Compute
  (a) $P(10 < X < 30)$, (b) $P(X > 30)$,
  (c) $P(X > 40 | X > 10)$.

4.3–2. Let the p.d.f. of X be $f(x) = (1/2)e^{-x/2}, 0 \le x < \infty$.
  (a) What are the mean, variance, and moment-generating function of X?
  (b) Calculate $P(X > 3)$.
  (c) Calculate $P(X > 5 | X > 2)$.

4.3–3. What is the p.d.f. of X if the moment-generating function of X is given by the following?

  (a) $M(t) = \dfrac{1}{1 - 3t}$, $t < \dfrac{1}{3}$;   (b) $M(t) = \dfrac{3}{3 - t}$, $t < 3$.

4.3–4. Telephone calls enter a college switchboard according to a Poisson process on the average of two every 3 minutes. Let X denote the waiting time until the first call that arrives after 10 A.M.
   (a)   What is the p.d.f. of X?
   (b)   Find $P(X > 2)$.

4.3–5. Let X have an exponential distribution with mean $\theta > 0$. Show that

$$P(X > x + y \mid X > x) = P(X > y).$$

4.3–6. (a)   Simulate a random sample of size $n = 10$ from an exponential distribution with mean $\theta = 3$.
   (b)   Sketch the empirical and theoretical distribution functions on the same set of axes.

4.3–7. Let $F(x)$ be the distribution function of the continuous-type random variable X and assume that $F(x) = 0$ for $x \leq 0$ and $0 < F(x) < 1$ for $0 < x$. Prove that if

$$P(X > x + y \mid X > x) = P(X > y),$$

then

$$F(x) = 1 - e^{-\lambda x}, \qquad 0 < x.$$

HINT: Show that $g(x) = 1 - F(x)$ satisfies the functional equation $g(x + y) = g(x)g(y)$, which implies that $g(x) = a^{cx}$.

4.3–8. A certain type of aluminum screen two feet in width has on the average three flaws in a 100-foot roll.

   (a)   What is the probability that the first 40 feet in a roll contain no flaws?
   (b)   What assumption did you make to solve part (a)?

4.3–9. Find the median of the exponential distribution with mean θ. If a random sample of size $n = 5$ is taken from this distribution, determine the probability that at least two of the sample items are less than the median of the distribution. Does the latter probability depend upon the fact that the distribution is exponential? Or, would it hold for every distribution of the continuous type?

4.4 The Gamma and Chi-Square Distributions

In the (approximate) Poisson process with mean λ, we have seen that the waiting time until the first change has an exponential distribution. In this section we let W denote the waiting time until the hth change occurs and find the distribution of W.

The distribution function of W, when $w > 0$, is given by

$$F(w) = P(W \le w) = 1 - P(W > w)$$

$$= 1 - P \text{ (less than } h \text{ changes occur in } [0, w])$$

$$= 1 - \sum_{k=0}^{h-1} \frac{(\lambda w)^k e^{-\lambda w}}{k!},$$

since the number of changes in the interval $[0, w]$ has a Poisson distribution with mean λw. Because W is a continuous-type random variable, $F'(w)$ is equal to the p.d.f. of W whenever this derivative exists. We have, provided $w > 0$, that

$$F'(w) = \lambda e^{-\lambda w} - e^{-\lambda w} \sum_{k=1}^{h-1} \left[\frac{k(\lambda w)^{k-1} \lambda}{k!} - \frac{(\lambda w)^k \lambda}{k!} \right]$$

$$= \lambda e^{-\lambda w} - e^{-\lambda w} \left[\lambda - \frac{\lambda(\lambda w)^{h-1}}{(h-1)!} \right]$$

$$= \frac{\lambda(\lambda w)^{h-1}}{(h-1)!} e^{-\lambda w}.$$

Of course, if $w < 0$, then $F(w) = 0$ and $F'(w) = 0$. A p.d.f. of this form is said to be one of the gamma type and the random variable W is said to have a *gamma distribution*.

Before determining the characteristics of the gamma distribution, let us consider the gamma function for which the distribution was named. The *gamma function* is defined by

$$\Gamma(t) = \int_0^\infty y^{t-1} e^{-y} \, dy, \qquad 0 < t.$$

This integral is obviously positive for $0 < t$, and values of it are often given in a table of integrals. If $t > 1$, integration of the gamma function of t by parts yields

$$\Gamma(t) = [-y^{t-1} e^{-y}]_0^\infty + \int_0^\infty (t-1) y^{t-2} e^{-y} \, dy$$

$$= (t-1) \int_0^\infty y^{t-2} e^{-y} \, dy = (t-1)\Gamma(t-1).$$

For example, $\Gamma(6) = 5\Gamma(5)$ and $\Gamma(3) = 2\Gamma(2) = (2)(1)\Gamma(1)$. Whenever $t = n$, a positive integer, we have, by repeated application of $\Gamma(t) = (t-1)\Gamma(t-1)$, that

$$\Gamma(n) = (n-1)\Gamma(n-1) = (n-1)(n-2)\cdots(2)(1)\Gamma(1).$$

However,

$$\Gamma(1) = \int_0^\infty e^{-y} \, dy = 1.$$

Thus, when n is a positive integer, we have that

$$\Gamma(n) = (n - 1)!;$$

and, for this reason, the gamma function is called the generalized factorial. Incidentally, $\Gamma(1)$ corresponds to $0!$, and we have noted that $\Gamma(1) = 1$, which is consistent with earlier definitions.

Let us now formally define the p.d.f. of the gamma distribution and find its characteristics. The random variable X has a *gamma distribution* if its p.d.f. is defined by

$$f(x) = \begin{cases} \dfrac{1}{\Gamma(\alpha)\theta^{\alpha}} x^{\alpha - 1} e^{-x/\theta}, & 0 \leq x < \infty, \\ 0, & x < 0. \end{cases}$$

Hence, W, the waiting time until the hth change in a Poisson process, has a gamma distribution with parameters $\alpha = h$ and $\theta = 1/\lambda$. To see that $f(x)$ actually has the properties of a p.d.f., note that $f(x) \geq 0$ and

$$\int_{-\infty}^{\infty} f(x) \, dx = \int_{0}^{\infty} \frac{x^{\alpha - 1} e^{-x/\theta}}{\Gamma(\alpha)\theta^{\alpha}} \, dx,$$

which, by the change of variables $y = x/\theta$, equals

$$\int_{0}^{\infty} \frac{(\theta y)^{\alpha - 1} e^{-y}}{\Gamma(\alpha)\theta^{\alpha}} \theta \, dy = \frac{1}{\Gamma(\alpha)} \int_{0}^{\infty} y^{\alpha - 1} e^{-y} \, dy = \frac{\Gamma(\alpha)}{\Gamma(\alpha)} = 1.$$

The moment-generating function of X is

$$M(t) = \int_{0}^{\infty} \frac{e^{tx} x^{\alpha - 1} e^{-x/\theta}}{\Gamma(\alpha)\theta^{\alpha}} \, dx$$

$$= \int_{0}^{\infty} \frac{x^{\alpha - 1} e^{-(1 - \theta t)x/\theta}}{\Gamma(\alpha)\theta^{\alpha}} \, dx.$$

Let $y = (1 - \theta t)x/\theta$, when $1 - \theta t > 0$ or, equivalently, $t < 1/\theta$, to obtain

$$M(t) = \int_{0}^{\infty} \frac{[\theta y/(1 - \theta t)]^{\alpha - 1} e^{-y}}{\Gamma(\alpha)\theta^{\alpha}} \left(\frac{\theta}{1 - \theta t} \right) dy$$

$$= \frac{1}{(1 - \theta t)^{\alpha}\Gamma(\alpha)} \int_{0}^{\infty} y^{\alpha - 1} e^{-y} \, dy = \frac{1}{(1 - \theta t)^{\alpha}}.$$

Accordingly, with $R(t) = \ln M(t) = -\alpha \ln(1 - \theta t)$, we have

$$R'(t) = \frac{\alpha\theta}{1 - \theta t} \quad \text{and} \quad R''(t) = \frac{\alpha\theta^{2}}{(1 - \theta t)^{2}}.$$

Thus, the mean and the variance are, respectively,

$$\mu = R'(0) = \alpha\theta \quad \text{and} \quad \sigma^{2} = R''(0) = \alpha\theta^{2}.$$

Example 4.4–1. Suppose that an average of 30 customers per hour arrive at a shop in accordance with a Poisson process. That is, if a minute is our unit, then $\lambda = 1/2$. What is the probability that the shopkeeper will wait more than 5 minutes before both of the first two customers arrive? If X denotes the waiting time in minutes until the second customer arrives, then X has a gamma distribution with $\alpha = h = 2$, $\theta = 1/\lambda = 2$. Hence,

$$P(X > 5) = \int_5^\infty \frac{x^{2-1}e^{-x/2}}{\Gamma(2)2^2}\,dx = \int_5^\infty \frac{xe^{-x/2}}{4}\,dx$$

$$= \frac{1}{4}[(-2)xe^{-x/2} - 4e^{-x/2}]_5^\infty$$

$$= \frac{7}{2}e^{-5/2} = 0.287.$$

Example 4.4–2. Telephone calls arrive at a switchboard at a mean rate of $\lambda = 2$ per minute according to a Poisson process. Let X denote the waiting time in minutes until the fifth call arrives. The p.d.f. of X, with $\alpha = h = 5$ and $\theta = 1/\lambda = 1/2$, is

$$f(x) = \frac{2^5 x^4}{4!}e^{-2x}, \qquad 0 \le x < \infty.$$

The mean and the variance of X are, respectively, $\mu = 5/2$ and $\sigma^2 = 5/4$.

We now consider a special case of the gamma distribution that plays an important role in statistics. Let X have a gamma distribution with $\theta = 2$ and $\alpha = r/2$, where r is a positive integer. The p.d.f. of X is

$$f(x) = \begin{cases} \dfrac{1}{\Gamma(r/2)2^{r/2}}x^{r/2-1}e^{-x/2}, & 0 \le x < \infty, \\[2mm] 0, & x < 0. \end{cases}$$

We say that X has a *chi-square distribution with r degrees of freedom*, which we abbreviate by saying X is $\chi^2(r)$. The mean and the variance of this chi-square distribution are

$$\mu = \alpha\theta = r \qquad \text{and} \qquad \sigma^2 = \alpha\theta^2 = 2r.$$

That is, the mean equals the number of degrees of freedom and the variance equals twice the number of degrees of freedom. Incidentally, at this time, it is impossible to explain why this parameter r is called the number of degrees of freedom. From the results concerning the more general gamma distribution, we see that its moment-generating function is

$$M(t) = (1 - 2t)^{-r/2}, \qquad t < 1/2.$$

Because the chi-square distribution is so important in applications, tables have been prepared giving the values of the distribution function

$$F(x) = \int_0^x \frac{1}{\Gamma(r/2)2^{r/2}} \, w^{r/2-1} e^{-w/2} \, dw$$

for selected values of r and x. For an example, see Appendix Table III.

Example 4.4–3.   Let X have a chi-square distribution with $r = 5$ degrees of freedom. Then

$$P(1.145 \leq X \leq 12.83) = F(12.83) - F(1.145) = 0.975 - 0.050 = 0.925$$

and

$$P(X > 15.09) = 1 - F(15.09) = 1 - 0.99 = 0.01.$$

Example 4.4–4.   If X is $\chi^2(7)$, two constants, a and b, such that

$$P(a < X < b) = 0.95$$

are $a = 1.690$ and $b = 16.01$. Other constants a and b can be found, and we are only restricted in our choices by the limited table.

Example 4.4–5.   If customers arrive at a shop on the average of 30 per hour in accordance with a Poisson process, what is the probability that the shopkeeper will have to wait longer than 9.390 minutes for the first nine customers to arrive? Note that the mean rate of arrivals per minute is $\lambda = 1/2$. Thus, $\theta = 2$ and $h = \alpha = r/2 = 9$. Thus, if X denotes the waiting time until the ninth arrival, X is $\chi^2(18)$. Hence,

$$P(X > 9.390) = 1 - 0.05 = 0.95.$$

Example 4.4–6.   If X has an exponential distribution with a mean of 2, the p.d.f. of X is

$$f(x) = \frac{1}{2} e^{-x/2} = \frac{x^{2/2-1} e^{-x/2}}{\Gamma(2/2)2^{2/2}}, \qquad 0 \leq x < \infty.$$

That is, X is $\chi^2(2)$. Thus, for illustration,

$$P(0.051 < X < 7.378) = 0.975 - 0.025 = 0.95.$$

In the next section we shall give a relationship between the chi-square and normal distributions. Then in later sections we shall illustrate the importance of the chi-square distribution in applications.

Exercises

4.4–1. Telephone calls enter a college switchboard at a mean rate of 2/3 call per minute according to a Poisson process. Let X denote the waiting time until the tenth call arrives.
    (a) What is the p.d.f. of X?
    (b) What are the moment-generating function, mean, and variance of X?

4.4–2. If X has a gamma distribution with $\theta = 4$ and $\alpha = 2$, find $P\{X < 5\}$.

4.4–3. Cars arrive at a toll booth at a mean rate of five cars every 10 minutes according to a Poisson process. Find the probability that the toll collector will have to wait longer than 26.30 minutes before collecting the eighth toll.

4.4–4. If the moment-generating function of a random variable W is $M(t) = (1 - 7t)^{-20}$, find the p.d.f., mean, and variance of W.

4.4–5. Let X denote the number of alpha particles emitted by barium-133 and observed by a Geiger counter in a fixed position. Assume that $\lambda = 14.7$ is the mean number of counts per second. Let W denote the waiting time to observe 100 counts. Twenty-five independent observations of W were

6.9	7.3	6.7	6.4	6.3
5.9	7.0	7.1	6.5	7.6
7.2	7.1	6.1	7.3	7.6
7.6	6.7	6.3	5.7	6.7
7.5	5.3	5.4	7.4	6.9

    (a) Give the p.d.f., mean, and variance of W.
    (b) Calculate the sample mean and sample variance of the 25 observations of W.
    (c) Use the relative frequency of the event $\{W \le 6.6\}$ to approximate $P(W \le 6.6)$.

4.4–6. If the moment-generating function of W is $M(t) = (1 - 5t)^{-1}$, $t < 1/5$, find **(a)** $E(W)$, **(b)** $\text{Var}(W)$, **(c)** $P(0 < W < 5)$.

4.4–7. If X is $\chi^2(17)$, find **(a)** $P(X < 7.564)$, **(b)** $P(X > 27.59)$, **(c)** $P(6.408 < X \le 27.59)$.

4.4–8. If X is $\chi^2(12)$, find constants a and b such that

$$P(a < X < b) = 0.90 \quad \text{and} \quad P(X < a) = 0.05.$$

4.4–9. If X is $\chi^2(23)$, find the following.
    (a) $P(10.20 < X < 35.17)$.
    (b) Constants a and b such that $P(a < X < b) = 0.95$ and $P(X < a) = 0.025$.
    (c) The mean and variance of X.

4.4–10. If the moment-generating function of X is $M(t) = (1 - 2t)^{-12}$, $t < 1/2$, find **(a)** $E(X)$, **(b)** $\text{Var}(X)$, **(c)** $P(15.66 < X < 42.98)$.

4.4–11. Let $X_1, X_2, \ldots, X_{10}$ be a random sample of size $n = 10$ from a distribution $\chi^2(19)$. Find the probability that exactly two of the 10 sample items exceed 30.14.

4.5 The Normal Distribution

The normal distribution is perhaps the most important distribution in statistical applications since many measurements have (approximate) normal distributions. One explanation of this fact is the role of the normal distribution in the Central Limit Theorem. One form of this theorem will be considered in Section 5.3.

We give the definition of the p.d.f. for the normal distribution, verify that it is a p.d.f., and then justify the use of μ and σ^2 in this formula. That is, we will show that μ and σ^2 are actually the mean and the variance of this distribution. The random variable X has a *normal distribution* if its p.d.f. is defined by

$$f(x) = \frac{1}{\sigma\sqrt{2\pi}} \exp\left[-\frac{(x-\mu)^2}{2\sigma^2}\right], \quad -\infty < x < \infty,$$

where μ and σ are parameters satisfying $-\infty < \mu < \infty, 0 < \sigma < \infty$, and where $\exp[v]$ means e^v. Briefly, we say that X is $N(\mu, \sigma^2)$.

Clearly $f(x) > 0$. We now evaluate the integral

$$I = \int_{-\infty}^{\infty} \frac{1}{\sigma\sqrt{2\pi}} \exp\left[-\frac{(x-\mu)^2}{2\sigma^2}\right] dx.$$

In I, change variables of integration by letting $z = (x - \mu)/\sigma$. Thus,

$$I = \int_{-\infty}^{\infty} \frac{1}{\sqrt{2\pi}} e^{-z^2/2} \, dz.$$

Since $I > 0$, if $I^2 = 1$, then $I = 1$. Now

$$I^2 = \frac{1}{2\pi} \left[\int_{-\infty}^{\infty} e^{-x^2/2} \, dx\right]\left[\int_{-\infty}^{\infty} e^{-y^2/2} \, dy\right],$$

or, equivalently,

$$I^2 = \frac{1}{2\pi} \int_{-\infty}^{\infty} \int_{-\infty}^{\infty} \exp\left(-\frac{x^2+y^2}{2}\right) dx \, dy.$$

Letting $x = r \cos \theta$, $y = r \sin \theta$ (that is, using polar coordinates), we have

$$I^2 = \frac{1}{2\pi} \int_0^{2\pi} \int_0^{\infty} e^{-r^2/2} r \, dr \, d\theta$$

$$= \frac{1}{2\pi} \int_0^{2\pi} d\theta = \frac{1}{2\pi} 2\pi = 1.$$

Thus, $I = 1$, and we have shown that $f(x)$ has the properties of a p.d.f.

The moment-generating function of X is

$$M(t) = \int_{-\infty}^{\infty} \frac{e^{tx}}{\sigma\sqrt{2\pi}} \exp\left[-\frac{(x-\mu)^2}{2\sigma^2}\right] dx$$

$$= \int_{-\infty}^{\infty} \frac{1}{\sigma\sqrt{2\pi}} \exp\left\{-\frac{1}{2\sigma^2}[x^2 - 2(\mu + \sigma^2 t)x + \mu^2]\right\} dx.$$

To evaluate this integral, we complete the square in the exponent

$$x^2 - 2(\mu + \sigma^2 t)x + \mu^2 = [x - (\mu + \sigma^2 t)]^2 - 2\mu\sigma^2 t - \sigma^4 t^2.$$

Thus,

$$M(t) = \exp\left(\frac{2\mu\sigma^2 t + \sigma^4 t^2}{2\sigma^2}\right)\int_{-\infty}^{\infty} \frac{1}{\sigma\sqrt{2\pi}} \exp\left\{-\frac{1}{2\sigma^2}[x - (\mu + \sigma^2 t)]^2\right\} dx.$$

Note that the integrand in the last integral is like the p.d.f. of a normal distribution with μ replaced by $\mu + \sigma^2 t$. However, the normal p.d.f. integrates to one for all real μ, in particular when it equals $\mu + \sigma^2 t$. Thus,

$$M(t) = \exp\left(\frac{2\mu\sigma^2 t + \sigma^4 t^2}{2\sigma^2}\right) = \exp\left(\mu t + \frac{\sigma^2 t^2}{2}\right).$$

With $R(t) = \ln M(t) = \mu t + \sigma^2 t^2/2$, we have

$$R'(t) = \mu + \sigma^2 t \quad \text{and} \quad R''(t) = \sigma^2.$$

Thus,

$$R'(0) = \mu \quad \text{and} \quad R''(0) = \sigma^2.$$

That is, the parameters μ and σ^2 in the p.d.f. of X are the mean and the variance of X.

Example 4.5–1. If the p.d.f. of X is

$$f(x) = \frac{1}{\sqrt{32\pi}} \exp\left[-\frac{(x+7)^2}{32}\right], \quad -\infty < x < \infty,$$

then X is $N(-7, 16)$. That is, X has a normal distribution with a mean $\mu = -7$, variance $\sigma^2 = 16$, and moment-generating function $M(t) = \exp(-7t + 8t^2)$.

Example 4.5–2. If the moment-generating function of X is

$$M(t) = \exp(5t + 12t^2),$$

then X is $N(5, 24)$ and its p.d.f. is

$$f(x) = \frac{1}{\sqrt{48\pi}} \exp\left[-\frac{(x-5)^2}{48}\right], \quad -\infty < x < \infty.$$

If Z is $N(0, 1)$, we shall say that Z has a standard normal distribution. Moreover, the distribution function of Z is

$$\Phi(z) = P(-\infty < Z \le z) = \int_{-\infty}^{z} \frac{1}{\sqrt{2\pi}} e^{-w^2/2} \, dw.$$

It is not possible to evaluate this integral by finding an antiderivative that can be expressed as an elementary function. However, numerical approximations for integrals of this type have been tabulated and are given in Appendix Table IV. The bell-shaped curve in Figure 4.5–1 represents the graph of the p.d.f. of Z and the shaded area equals $\Phi(z)$. Because of the symmetry of the standard p.d.f., we have that $\Phi(-z) = 1 - \Phi(z)$ for all real z. Thus, we list in the table only those areas $\Phi(z)$ associated with positive values of z.

Example 4.5–3.   If Z is $N(0, 1)$, then

$$P(0 \le Z \le 2) = \Phi(2) - \Phi(0) = 0.9772 - 0.5000 = 0.4772,$$
$$P(1.25 \le Z \le 2.75) = \Phi(2.75) - \Phi(1.25) = 0.9970 - 0.8944 = 0.1026,$$

and

$$P(-1.65 \le Z \le 0.70) = \Phi(0.70) - \Phi(-1.65)$$
$$= \Phi(0.70) - [1 - \Phi(1.65)] = 0.7580 - 1 + 0.9505$$
$$= 0.7085.$$

Example 4.5–4.   If Z is $N(0, 1)$, find constants $a, b,$ and c such that

$$P(0 \le Z \le a) = 0.4147, \quad P(Z > b) = 0.05, \quad \text{and} \quad P(|Z| \le c) = 0.95.$$

These three equations are equivalent to

$$P(Z \le a) = 0.9147, \qquad P(Z \le b) = 0.95, \qquad \text{and} \qquad P(Z \le c) = 0.975,$$

respectively. Thus, using Appendix Table IV we see that $a = 1.37, b = 1.645,$ and $c = 1.96$.

If X is $N(\mu, \sigma^2)$, the next theorem shows that the random variable $(X - \mu)/\sigma$ is $N(0, 1)$. Thus, Appendix Table IV can be used to find probabilities concerning X.

THEOREM 4.5–1.   *If X is $N(\mu, \sigma^2)$, then $Z = (X - \mu)/\sigma$ is $N(0, 1)$.*

Proof: The distribution function of Z is

$$P(Z \le z) = P\left(\frac{X - \mu}{\sigma} \le z\right) = P(X \le z\sigma + \mu)$$

$$= \int_{-\infty}^{z\sigma + \mu} \frac{1}{\sigma\sqrt{2\pi}} \exp\left[-\frac{(x - \mu)^2}{2\sigma^2}\right] dx.$$

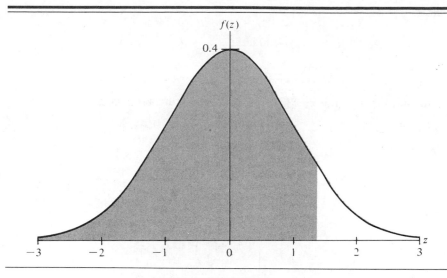

FIGURE 4.5–1

In the integral representating $P(Z \leq z)$, use the change of variable of integration given by $w = (x - \mu)/\sigma$ (that is, $x = w\sigma + \mu$) to obtain

$$P(Z \leq z) = \int_{-\infty}^{z} \frac{1}{\sqrt{2\pi}} e^{-w^2/2} \, dw.$$

But this is the expression for $\Phi(z)$, the distribution function of a standardized normal random variable. Hence, Z is $N(0, 1)$.

This theorem can be used to find probabilities about X, which is $N(\mu, \sigma^2)$, as follows:

$$P(a \leq X \leq b) = P\left(\frac{a - \mu}{\sigma} \leq \frac{X - \mu}{\sigma} \leq \frac{b - \mu}{\sigma}\right) = \Phi\left(\frac{b - \mu}{\sigma}\right) - \Phi\left(\frac{a - \mu}{\sigma}\right)$$

since $(X - \mu)/\sigma$ is $N(0, 1)$.

Example 4.5–5.   If X is $N(3, 16)$, then

$$P(4 \leq X \leq 8) = P\left(\frac{4 - 3}{4} \leq \frac{X - 3}{4} \leq \frac{8 - 3}{4}\right)$$

$$= \Phi(1.25) - \Phi(0.25) = 0.8944 - 0.5987 = 0.2957,$$

$$P(0 \leq X \leq 5) = P\left(\frac{0 - 3}{4} \leq Z \leq \frac{5 - 3}{4}\right)$$

$$= \Phi(0.5) - \Phi(-0.75) = 0.4649,$$

and

$$P(-2 \le X \le 1) = P\left(\frac{-2-3}{4} \le Z \le \frac{1-3}{4}\right)$$

$$= \Phi(-0.5) - \Phi(-1.25) = 0.2029.$$

Example 4.5–6. If X is $N(25, 36)$, we find a constant c such that

$$P(|X - 25| \le c) = 0.9544.$$

We want

$$P\left(\frac{-c}{6} \le \frac{X - 25}{6} \le \frac{c}{6}\right) = 0.9544.$$

Thus,

$$\Phi\left(\frac{c}{6}\right) - \left[1 - \Phi\left(\frac{c}{6}\right)\right] = 0.9544$$

and

$$\Phi\left(\frac{c}{6}\right) = 0.9772.$$

Hence, $c/6 = 2$ and $c = 12$. That is, the probability that X falls within two standard deviations of its mean is the same as the probability that the standard normal variable Z falls within two units of zero.

In the next theorem we give a relationship between the chi-square and normal distributions.

THEOREM 4.5–2. *If the random variable X is $N(\mu, \sigma^2)$, $\sigma^2 > 0$, then the random variable $V = (X - \mu)^2/\sigma^2$ is $\chi^2(1)$.*

Proof: Because $V = Z^2$, where $Z = (X - \mu)/\sigma$ is $N(0, 1)$, the distribution function $G(v)$ of V is, for $v \ge 0$,

$$G(v) = P(Z^2 \le v) = P(-\sqrt{v} \le Z \le \sqrt{v}).$$

That is, with $v \ge 0$,

$$G(v) = \int_{-\sqrt{v}}^{\sqrt{v}} \frac{1}{\sqrt{2\pi}} e^{-z^2/2} \, dz = 2 \int_{0}^{\sqrt{v}} \frac{1}{\sqrt{2\pi}} e^{-z^2/2} \, dz.$$

If we change the variable of integration by writing $z = \sqrt{y}$, then, since $z' = 1/(2\sqrt{y})$, we have

$$G(v) = \int_{0}^{v} \frac{1}{\sqrt{2\pi y}} e^{-y/2} \, dy, \qquad 0 \le v.$$

Of course, $G(v) = 0$, when $v < 0$. Hence, the p.d.f. $g(v) = G'(v)$ of the continuous-type random variable V is, by one form of the fundamental theorem of calculus,

$$g(v) = \begin{cases} \dfrac{1}{\sqrt{\pi}\sqrt{2}}\, v^{1/2-1}e^{-v/2}, & 0 < v < \infty, \\ 0 \text{ elsewhere.} \end{cases}$$

Since $g(v)$ is a p.d.f., it must be that

$$\int_0^\infty \frac{1}{\sqrt{\pi}\sqrt{2}}\, v^{1/2-1}e^{-v/2}\, dv = 1.$$

The change of variables $x = v/2$ yields

$$1 = \frac{1}{\sqrt{\pi}} \int_0^\infty x^{1/2-1}e^{-x}\, dx = \frac{1}{\sqrt{\pi}}\, \Gamma\!\left(\frac{1}{2}\right).$$

Hence, $\Gamma(1/2) = \sqrt{\pi}$, and, thus, V is $\chi^2(1)$.

Example 4.5–7. If Z is $N(0, 1)$, then

$$P(|Z| < 1.96 = \sqrt{3.842}) = 0.95$$

and, of course,

$$P(Z^2 < 3.842) = 0.95$$

from the chi-square table with $r = 1$.

===== *Exercises* =====

4.5–1. If Z is $N(0, 1)$, find
   (a) $P(0.53 < Z \le 2.06)$,   (b) $P(-0.79 \le Z < 1.52)$,
   (c) $P(-2.63 < Z \le -0.51)$,   (d) $P(Z > -1.77)$,
   (e) $P(Z > 2.89)$,   (f) $P(|Z| < 1.96)$,
   (g) $P(|Z| < 1)$,   (h) $P(|Z| < 2)$,
   (i) $P(|Z| < 3)$.

4.5–2. If X is normally distributed with a mean of 6 and a variance of 25, find
   (a) $P(6 \le X \le 12)$,   (b) $P(0 \le X \le 8)$,
   (c) $P(-2 < X \le 0)$,   (d) $P(X > 21)$,
   (e) $P(|X - 6| < 5)$,   (f) $P(|X - 6| < 10)$,
   (g) $P(|X - 6| < 15)$.

4.5–3. If Z is $N(0, 1)$, find values of c such that
   (a) $P(|Z| \ge c) = 0.05$,   (b) $P(|Z| \le c) = 0.90$,
   (c) $P(Z > c) = 0.05$.

4.5–4. If the moment-generating function of X is $M(t) = \exp(-6t + 32t^2)$, find **(a)** $P(-4 \le X < 16)$, **(b)** $P(-10 < X \le 0)$.

4.5–5. If X is $N(650, 625)$, find **(a)** $P(600 \le X < 660)$, **(b)** a constant $c > 0$ such that $P(|X - 650| \le c) = 0.9544$.

4.5–6. If the moment-generating function of X is $M(t) = \exp(166t + 200t^2)$, find
  (a) the mean of X,   **(b)** the variance of X,
  (c) $P(170 < X < 200)$,   **(d)** $P(148 \le X \le 172)$.

4.5–7. If X is $N(7, 4)$, find $P(15.36 \le (X - 7)^2 \le 20.08)$.

4.5–8. If X is $N(\mu, \sigma^2)$, show that $Y = aX + b$ is $N(a\mu + b, a^2\sigma^2)$, $a \ne 0$. HINT: Find the distribution function $P(Y \le y)$ of Y and, in the resulting integral, let $w = ax + b$ or, equivalently, $x = (w - b)/a$.

4.5–9. Scores are often assumed to be normally distributed. Let X denote the score on an SAT mathematics examination. The following data are 100 observations of X selected at random from a college freshman class.

500	561	383	683	481	433	453	499	449	545
499	598	519	610	521	577	493	479	446	473
512	571	367	551	591	431	552	407	663	545
400	492	528	597	611	466	525	556	532	584
445	538	407	590	440	448	663	517	543	659
492	545	475	407	532	394	650	571	558	518
579	608	535	584	543	431	625	492	637	374
545	507	353	597	420	677	569	497	577	466
486	427	549	335	321	683	597	432	565	551
506	709	499	413	466	558	453	542	493	683

(a) Group these data into 10 classes and draw a relative frequency histogram.
(b) Do these data seem to come from a normal distribution with mean μ about equal to $\bar{x} = 519.59$ and variance σ^2 about equal to $s^2 = 6978.44$?
(c) Approximate $P(X \ge 560)$ using the relative frequency of the event $\{X \ge 560\}$ and compare this approximation with the probability calculated assuming that X is $N(519.59, 6978.44)$.

4.6 Mixed Distributions

Thus far we have considered random variables that are either discrete or continuous. In most applications these are the types that are encountered. However, on some occasions combinations of the two types of random variables are found. That is, in some experiments, positive probability is assigned to each of certain points and also is spread over an interval of outcomes, each point of which has zero probability. An illustration will help clarify these remarks.

Example 4.6–1. A bulb for a slide projector is tested by turning it on, letting it burn for 1 hour, and then turning it off. Let X equal the length of time that the bulb performs satisfactorily during this test. There is positive probability that the bulb will burn out when it is turned on; hence,

$$P(X = 0) > 0.$$

It could also burn out during the 1 hour time period; thus,

$$P(0 < X < 1) > 0$$

with $P(X = x) = 0$ when $x \in (0, 1)$. In addition, $P(X = 1) > 0$.

The distribution function for a distribution of the mixed type will be a combination of those for the discrete and continuous types. That is, at each point of positive probability the distribution function will be discontinuous so that the height of the step there equals the corresponding probability; at all other points the distribution function will be continuous.

Example 4.6–2. Let X have a distribution function $F(x)$ defined by

$$F(x) = \begin{cases} 0, & x < 0, \\ \dfrac{x^2}{4}, & 0 \le x < 1, \\ \dfrac{1}{2}, & 1 \le x < 2, \\ \dfrac{x}{3}, & 2 \le x < 3, \\ 1, & 3 \le x. \end{cases}$$

This distribution function is depicted in Figure 4.6–1. Probabilities can be computed using $F(x)$; for illustration consider:

$$P(0 < X < 1) = \frac{1}{4},$$

$$P(0 < X \le 1) = \frac{1}{2},$$

$$P(X = 1) = \frac{1}{4},$$

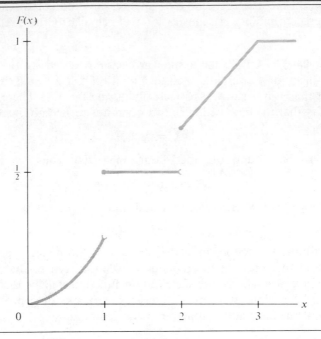

FIGURE 4.6–1

and

$$P(1 \le X \le 2) = \frac{2}{3} - \frac{1}{4} = \frac{5}{12}.$$

Example 4.6–3. Consider the following "game." An unbiased coin is tossed. If the outcome is heads, the player receives $2. If the outcome is tails, the player spins a balanced spinner that has a scale from 0 to 1 and receives that fraction of a dollar associated with the point selected by the spinner. If X denotes the amount received, the space of X is $R = [0, 1) \cup \{2\}$. The distribution function of X is defined by

$$F(x) = \begin{cases} 0, & x < 0, \\ \dfrac{x}{2}, & 0 \le x < 1, \\ \dfrac{1}{2}, & 1 \le x < 2, \\ 1, & 2 \le x. \end{cases}$$

The graph of the distribution function $F(x)$ is given in Figure 4.6–2.

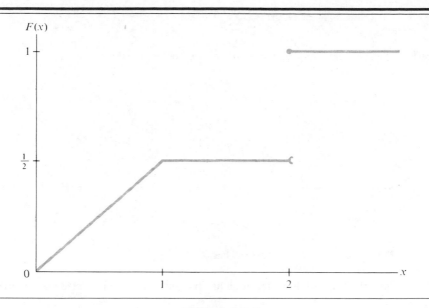

FIGURE 4.6–2

Suppose the random variable X has a distribution of the mixed type. To find the expectation of the function $u(X)$ of X, a combination of a sum and a Riemann integral is used, as shown in Example 4.6–4.

Example 4.6–4. We shall find the mean and variance for the random variable given in Example 4.6–2. Note that there $F'(x) = x/2$ when $0 < x < 1$, and $F'(x) = 1/3$ when $2 < x < 3$; also $P(X = 1) = 1/4$ and $P(X = 2) = 1/6$. Accordingly, we have

$$\mu = E(x) = \int_0^1 x\left(\frac{x}{2}\right) dx + 1\left(\frac{1}{4}\right) + 2\left(\frac{1}{6}\right) + \int_2^3 x\left(\frac{1}{3}\right) dx$$

$$= \left[\frac{x^3}{6}\right]_0^1 + \frac{1}{4} + \frac{1}{3} + \left[\frac{x^2}{6}\right]_2^3 = \frac{19}{12}$$

and

$$\sigma^2 = E(X^2) - [E(X)]^2$$

$$= \int_0^1 x^2\left(\frac{x}{2}\right) dx + 1^2\left(\frac{1}{4}\right) + 2^2\left(\frac{1}{6}\right) + \int_2^3 x^2\left(\frac{1}{3}\right) dx - \left(\frac{19}{12}\right)^2 = \frac{1}{48}.$$

═══ *Exercises* ═══

4.6–1. From the graph of the distribution function determine the indicated probabilities.

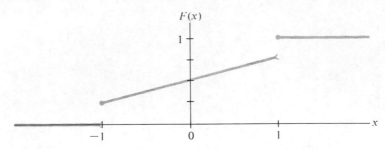

(a) $P(X < 0)$,   **(b)** $P(X < -1)$,   **(c)** $P(X \le -1)$,

(d) $P(X < 1)$,   **(e)** $P\left(-1 \le X < \dfrac{1}{2}\right)$.

4.6–2. From the graph of the distribution function determine the indicated probabilities.

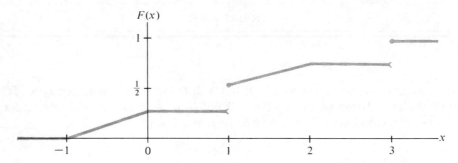

(a) $P\left(-\dfrac{1}{2} \le X \le \dfrac{1}{2}\right)$,   **(b)** $P\left(\dfrac{1}{2} < X < 1\right)$,   **(c)** $P\left(\dfrac{3}{4} < X < 2\right)$,

(d) $P(X > 1)$,   **(e)** $P(2 < X < 3)$.

4.6–3. Let X be a random variable of the mixed type having the distribution function

$$F(x) = \begin{cases} 0, & x < 0, \\[2mm] \dfrac{x^2}{4}, & 0 \le x < 1, \\[2mm] \dfrac{x+1}{4}, & 1 \le x < 2, \\[2mm] 1, & 2 \le x. \end{cases}$$

(a) Carefully sketch the graph of $F(x)$.
(b) Find the mean and the variance of X.
(c) Find (i) $P(1/4 < X < 1)$, (ii) $P(X = 1)$, (iii) $P(X = 1/2)$, (iv) $P(1/2 \le X < 2)$.

4.6–4. Find the mean and variance of X if the distribution function of X is

$$F(x) = \begin{cases} 0, & x < 0, \\ 1 - \left(\dfrac{2}{3}\right)e^{-x}, & 0 \leq x. \end{cases}$$

4.6–5. (a) Find the expected value of X in the game described in Example 4.6–3.
(b) Simulate 10 plays of this game. Use the random numbers in Appendix Table VIII for the outcome on the spinner.
(c) Is the average of your 10 plays close to the expected value calculated in part (a)?

4.6–6. Consider the following "game." A fair die is rolled. If the outcome is even, the player receives a number of dollars equal to the outcome on the die. If the outcome is odd, a number is selected at random from the interval $[0, 1)$ with a balanced spinner and the player receives that fraction of a dollar associated with the point selected.
(a) Define and sketch the distribution function of X, the amount received.
(b) Find the expected value of X.
(c) Simulate 10 plays of this game.
(d) Is the average of your 10 plays close to the expected value calculated in part (b)?

5

BASIC SAMPLING DISTRIBUTION THEORY

5.1 Distributions of Functions of Random Variables

Functions of random variables are usually of interest in statistical applications. In Chapter 3, for example, we considered functions, called statistics, of the items of a random sample; two important statistics are the mean $\bar{X}$ and the variance S^2 of the sample $X_1, X_2, \ldots, X_n$. Although in a particular sample, say $x_1, x_2, \ldots, x_n$, we observed definite values of these statistics, $\bar{x}$ and s^2, we should recognize that each value is only one observation of the respective random variables $\bar{X}$ and S^2. That is, each $\bar{X}$ or S^2 (or, more generally, any statistic) is also a random variable with its own distribution. In this chapter, we will determine the distributions of some of the important statistics (and, more generally, distributions of functions of random variables). These determinations fall in that area of statistics usually referred to as *sampling distribution theory.*

REMARK.   Throughout this text we will assume that the functions of random variables under consideration are such that they are also random variables. That is, although it is possible to construct *unusual* functions that do not induce probabilities from the original spaces to the spaces of the functions, we will not consider any of them in this course.

We begin with a simple illustration.

Example 5.1–1. Let us cast an unbiased die two independent times and observe the number of spots that turn up on these trials, say X_1 and X_2. Note that this is equivalent to saying that X_1 and X_2 are items of a random sample of size $n = 2$ from a distribution with p.d.f.

$$f(x) = \frac{1}{6}, \qquad x = 1, 2, 3, 4, 5, 6.$$

As in many games, suppose that we are interested in the sum $Y = X_1 + X_2$. We immediately recognize that the space of Y is $R = \{2, 3, 4, \ldots, 11, 12\}$. To determine $g(y) = P(Y = y)$, $y \in R$, let us look at some special cases. Say $y = 3$. We note that the event $\{Y = 3\}$ can occur in two mutually exclusive ways: $\{X_1 = 1, X_2 = 2\}$ and $\{X_1 = 2, X_2 = 1\}$. Thus, recalling that X_1 and X_2 are independent, we obtain

$$P(Y = 3) = P(X_1 = 1, X_2 = 2) + P(X_1 = 2, X_2 = 1)$$

$$= \left(\frac{1}{6}\right)\left(\frac{1}{6}\right) + \left(\frac{1}{6}\right)\left(\frac{1}{6}\right) = \frac{2}{36}.$$

Similarly, we can find

$$P(Y = 10) = P(X_1 = 4, X_2 = 6) + P(X_1 = 5, X_2 = 5)$$

$$+ P(X_1 = 6, X_2 = 4)$$

$$= \left(\frac{1}{6}\right)\left(\frac{1}{6}\right) + \left(\frac{1}{6}\right)\left(\frac{1}{6}\right) + \left(\frac{1}{6}\right)\left(\frac{1}{6}\right) = \frac{3}{36}.$$

If this process is continued, we finally arrive at the p.d.f. $g(y)$ of Y, which can be given by Table 5.1–1 or the formula

$$g(y) = \frac{6 - |y - 7|}{36}, \qquad y = 2, 3, \ldots, 12.$$

We note that we can compute $E(Y)$ from the p.d.f. $g(y)$ of Y to obtain

$$E(Y) = \sum_{y=2}^{12} y g(y)$$

$$= 2\left(\frac{1}{36}\right) + 3\left(\frac{2}{36}\right) + \cdots + 10\left(\frac{3}{36}\right) + 11\left(\frac{2}{36}\right) + 12\left(\frac{1}{36}\right) = 7.$$

Let X_1 and X_2 be the items of a random sample of size $n = 2$ from a distribution with p.d.f. $f(x)$. We recall that the joint p.d.f. of X_1 and X_2 is $f(x_1)f(x_2)$ and define the *expected value* of a function $u(X_1, X_2)$ of the two random variables by

$$E[u(X_1, X_2)] = \sum\sum_{(x_1, x_2)} u(x_1, x_2) f(x_1) f(x_2),$$

where the summation is taken over all possible pairs (x_1, x_2).

y	g(y)	y	g(y)
2	1/36	8	5/36
3	2/36	9	4/36
4	3/36	10	3/36
5	4/36	11	2/36
6	5/36	12	1/36
7	6/36		

TABLE 5.1–1

REMARK. An extension of this definition to a function of more than two random variables is made in an obvious way, namely,

$$E[u(X_1, X_2, \ldots, X_n)] = \sum \sum \cdots \sum_{(x_1, x_2, \ldots, x_n)} u(x_1, x_2, \ldots, x_n) f_1(x_1) f_2(x_2) \cdots f_n(x_n)$$

where the summation is over all possible n-tuples $(x_1, x_2, \ldots, x_n)$ and $f_1(x_1) f_2(x_2) \cdots f_n(x_n)$ is the joint p.d.f. of the n mutually independent random variables. In the continuous case, integrals replace the summations. Later, when we consider dependent random variables, the definition remains the same with the joint p.d.f. in that case replacing $f_1(x_1) f_2(x_2) \cdots f_n(x_n)$.

Since, in Example 5.1–1, $x_1 = 1, 2, \ldots, 6$ and $x_2 = 1, 2, \ldots, 6$, we have 36 pairs (x_1, x_2) over which to sum to compute $E[u(X_1, X_2)]$. Moreover, if

$$u(X_1, X_2) = X_1 + X_2,$$

we have

$$E(X_1 + X_2) = \sum \sum_{(x_1, x_2)} (x_1 + x_2) f(x_1) f(x_2)$$

$$= \sum \sum_{(x_1, x_2)} x_1 f(x_1) f(x_2) + \sum \sum_{(x_1, x_2)} x_2 f(x_1) f(x_2)$$

$$= E(X_1) + E(X_2).$$

That is, again the expected value of the sum is equal to the sum of the expected values. We have seen that this is true for sums involving only one random variable. It is also true for more general situations involving sums of functions of several random variables and is proved by exhibiting a display similar to the preceding one. This means that again the expectation E is a linear operator.

For Example 5.1–1, we have that

$$E(X_1) = E(X_2) = \frac{7}{2},$$

because both X_1 and X_2 have the uniform distribution over the numbers 1, 2, 3, 4, 5, 6. Hence,

$$E(X_1 + X_2) = E(X_1) + E(X_2) = \frac{7}{2} + \frac{7}{2} = 7,$$

which is the same as $E(Y)$ computed using the p.d.f. of Y. The equality

$$E(Y) = E(X_1 + X_2)$$

illustrates a more general theorem that we will use but will *not* prove. Its proof will be left to a more advanced course in mathematical statistics.

THEOREM 5.1–1. *Let* $Y = u(X_1, X_2, \ldots, X_n)$ *be a function of the random variables* $X_1, X_2, \ldots, X_n$. *If* $E(Y)$ *exists, it is equal to* $E[u(X_1, X_2, \ldots, X_n)]$. *That is*

$$E(Y) = E[u(X_1, X_2, \ldots, X_n)].$$

REMARK. In this theorem, it is understood that $E(Y)$ is computed from the distribution of Y (if it can be found) and that $E[u(X_1, X_2, \ldots, X_n)]$ is computed using the joint distribution of $X_1, X_2, \ldots, X_n$. Certainly, it would be hoped that the two would be equal since $Y = u(X_1, X_2, \ldots, X_n)$.

To illustrate the theorem further and also observe an example of another important theorem, consider the product $Z = X_1 X_2$, where X_1 and X_2 have been defined in Example 5.1–1. Then

$$E(Z) = E(X_1 X_2) = \sum_{(x_1, x_2)} \sum x_1 x_2 f(x_1) f(x_2)$$

$$= \sum_{x_1=1}^{6} \sum_{x_2=1}^{6} x_1 x_2 f(x_1) f(x_2)$$

$$= \left[\sum_{x_1=1}^{6} x_1 f(x_1) \right] \left[\sum_{x_2=1}^{6} x_2 f(x_2) \right]$$

$$= E(X_1) E(X_2) = \left(\frac{7}{2} \right) \left(\frac{7}{2} \right) = \frac{49}{4}.$$

That is, if X_1 and X_2 are independent random variables, then

$$E(X_1 X_2) = E(X_1) E(X_2),$$

which is a special case of Theorem 5.1–2, the proof of which will be left until later in the course.

THEOREM 5.1–2. *Let* $X_1, X_2, \ldots, X_n$ *be mutually independent random variables such that* $E[u_i(X_i)]$, $i = 1, 2, \ldots, n$, *exist. Then*

$$E[u_1(X_1) u_2(X_2) \cdots u_n(X_n)] = E[u_1(X_1)] E[u_2(X_2)] \cdots E[u_n(X_n)].$$

REMARK. Sometimes students recognize that $X^2 = X \cdot X$ and thus believe that $E(X^2)$ is equal to $[E(X)][E(X)] = [E(X)]^2$ because the above theorem states that the expected value of the product is the product of the expected values. However, please note the hypothesis of independence in the theorem, and certainly X is not independent of itself. Incidentally, if $E(X^2)$ did equal $[E(X)]^2$, then the variance of X

$$\sigma^2 = E(X^2) - [E(X)]^2$$

would always equal zero. This really happens only in the case of degenerate (one point) distributions.

Example 5.1–2. It is interesting to note that these two theorems allow us to determine the mean, the variance, and the moment-generating function of a function such as $Y = X_1 + X_2$, where X_1 and X_2 have been defined in Example 5.1–1. We have seen that

$$\mu_Y = E(Y) = E(X_1 + X_2) = E(X_1) + E(X_2) = \frac{7}{2} + \frac{7}{2} = 7.$$

The variance of Y is

$$\sigma_Y^2 = E[(Y - \mu_Y)^2] = E[(X_1 + X_2 - \mu_1 - \mu_2)^2],$$

where $\mu_i = E(X_i) = 7/2$, $i = 1, 2$. Thus,

$$\sigma_Y^2 = E[(X_1 - \mu_1) + (X_2 - \mu_2)]^2$$
$$= E[(X_1 - \mu_1)^2 + 2(X_1 - \mu_1)(X_2 - \mu_2) + (X_2 - \mu_2)^2].$$

But, from an earlier comment about the expected value being a linear operator, we have

$$\sigma_Y^2 = E[(X_1 - \mu_1)^2] + 2E[(X_1 - \mu_1)(X_2 - \mu_2)] + E[(X_2 - \mu_2)^2].$$

However, since X_1 and X_2 are independent, then

$$E[(X_1 - \mu_1)(X_2 - \mu_2)] = [E(X_1 - \mu_1)][E(X_2 - \mu_2)]$$
$$= (\mu_1 - \mu_1)(\mu_2 - \mu_2) = 0.$$

Thus,

$$\sigma_Y^2 = \sigma_1^2 + \sigma_2^2,$$

where

$$\sigma_i^2 = E[(X_i - \mu_i)^2], \qquad i = 1, 2.$$

In the case in which X_1 and X_2 are the spots on two independent casts of a die, we have $\sigma_i^2 = 35/12$, $i = 1, 2$, and, hence,

$$\sigma_Y^2 = \frac{35}{12} + \frac{35}{12} = \frac{35}{6}.$$

Finally, the moment-generating function is

$$M_Y(t) = E(e^{tY}) = E[e^{t(X_1 + X_2)}] = E(e^{tX_1}e^{tX_2}).$$

The independence of X_1 and X_2 implies that

$$M_Y(t) = E(e^{tX_1})E(e^{tX_2}).$$

For our example in which X_1 and X_2 have the same p.d.f.

$$f(x) = 1/6, \qquad x = 1, 2, \ldots, 6,$$

and thus the same moment-generating function

$$M_X(t) = \frac{1}{6}e^t + \frac{1}{6}e^{2t} + \frac{1}{6}e^{3t} + \frac{1}{6}e^{4t} + \frac{1}{6}e^{5t} + \frac{1}{6}e^{6t},$$

we have that

$$M_Y(t) = [M_X(t)]^2 = \frac{1}{36}e^{2t} + \frac{2}{36}e^{3t} + \frac{3}{36}e^{4t} + \frac{4}{36}e^{5t} + \frac{5}{36}e^{6t} + \frac{6}{36}e^{7t}$$

$$+ \frac{5}{36}e^{8t} + \frac{4}{36}e^{9t} + \frac{3}{36}e^{10t} + \frac{2}{36}e^{11t} + \frac{1}{36}e^{12t}.$$

Note that the coefficient of e^{bt} is equal to the probability $P(Y = b)$; for illustration, $3/36 = P(Y = 10)$. This agrees with the result found in Example 5.1–1, and thus we see that we could find the distribution of Y by determining its moment-generating function.

A careful consideration of Example 5.1–2 will make the next section about sums of random variables easier to read. We now close this section with two examples.

Example 5.1–3. Let X_1 and X_2 be two independent random variables with respective means μ_1 and μ_2 and variances σ_1^2 and σ_2^2. Let us determine the mean and the variance of $Y = X_1 X_2$. We have

$$\mu_Y = E(X_1 X_2) = E(X_1)E(X_2) = \mu_1 \mu_2$$

and

$$\sigma_Y^2 = E(X_1^2 X_2^2) - \mu_1^2 \mu_2^2 = E(X_1^2)E(X_2^2) - \mu_1^2 \mu_2^2.$$

However, $E(X_i^2) - \mu_i^2 = \sigma_i^2$, $i = 1, 2$; so

$$\sigma_Y^2 = (\sigma_1^2 + \mu_1^2)(\sigma_2^2 + \mu_2^2) - \mu_1^2 \mu_2^2 = \sigma_1^2 \sigma_2^2 + \mu_1^2 \sigma_2^2 + \mu_2^2 \sigma_1^2.$$

Example 5.1–4. Let X_1 and X_2 be a random sample from a distribution $N(\mu, \sigma^2)$. If $Y = X_1 + X_2$, then

$$E(e^{tY}) = E[e^{t(X_1 + X_2)}] = E(e^{tX_1})E(e^{tX_2}).$$

But since X_1 and X_2 have the same normal distribution, we have

$$E(e^{tX_i}) = e^{\mu t + \sigma^2 t^2/2}, \qquad i = 1, 2.$$

Accordingly,

$$E(e^{tY}) = e^{\mu t + \sigma^2 t^2/2} e^{\mu t + \sigma^2 t^2/2}$$
$$= e^{(2\mu)t + (2\sigma^2)t^2/2}.$$

However, this is the moment-generating function of a random variable that is $N(2\mu, 2\sigma^2)$. Thus, from the uniqueness of the moment-generating function, Y has that distribution.

═══ *Exercises* ═══

5.1–1. Let X_1 and X_2 be items of a random sample of size $n = 2$ from a distribution with p.d.f. $f(x) = x/6$, $x = 1, 2, 3$. Find the p.d.f. of $Y = X_1 + X_2$. Determine the mean and the variance of the sum in two ways.

5.1–2. Let X_1 and X_2 be a random sample of size $n = 2$ from a distribution with p.d.f. $f(x) = 6x(1 - x)$, $0 < x < 1$. Find the mean and the variance of $Y = X_1 + X_2$.

5.1–3. Let X_1 and X_2 be two independent random variables with normal distributions $N(1, 9)$ and $N(-3, 16)$, respectively. Find the moment-generating function of $Y = X_1 + X_2$. What distribution is associated with this moment-generating function?

5.1–4. Let X_1, X_2, and X_3 be three mutually independent random variables, which are $\chi^2(r_1)$, $\chi^2(r_2)$, and $\chi^2(r_3)$, respectively. Find the moment-generating function of

$$Y = X_1 + X_2 + X_3.$$

What distribution has this moment-generating function?

5.1–5. Let X_1 and X_2 be independent with normal distributions $N(5, 50)$ and $N(1, 50)$, respectively. Compute $P(X_2 < X_1)$. HINT: Find the moment-generating function of $Y = X_1 - X_2$ and then compute $P(Y > 0) = P(X_2 < X_1)$.

5.1–6. Let X_1 and X_2 have distributions $b(n_1, p)$ and $b(n_2, p)$, respectively. Find the moment-generating function of $Y = X_1 + X_2$. How is Y distributed?

5.2 Sums of Independent Random Variables

In this section we shall restrict attention to those functions that are linear combinations of random variables. We shall first prove an important theorem about the mean and the variance of such a linear combination.

THEOREM 5.2–1. *Let $X_1, X_2, \ldots, X_n$ be n independent random variables with respective means $\mu_1, \mu_2, \ldots, \mu_n$ and variances $\sigma_1^2, \sigma_2^2, \ldots, \sigma_n^2$. Then the mean and the variance of $Y = \sum_{i=1}^{n} a_i X_i$, where $a_1, a_2, \ldots, a_n$ are real constants, are*

$$\mu_Y = \sum_{i=1}^{n} a_i \mu_i \qquad \text{and} \qquad \sigma_Y^2 = \sum_{i=1}^{n} a_i^2 \sigma_i^2,$$

respectively.

Proof: We have that

$$\mu_Y = E(Y) = E\left(\sum_{i=1}^{n} a_i X_i\right) = \sum_{i=1}^{n} a_i E(X_i) = \sum_{i=1}^{n} a_i \mu_i$$

because the expected value of the sum is the sum of the expected values (that is, E is a linear operator). Also,

$$\sigma_Y^2 = E[(Y - \mu_Y)^2] = E\left[\left(\sum_{i=1}^{n} a_i X_i - \sum_{i=1}^{n} a_i \mu_i\right)^2\right]$$

$$= E\left\{\left[\sum_{i=1}^{n} a_i (X_i - \mu_i)\right]^2\right\} = E\left[\sum_{i=1}^{n} \sum_{j=1}^{n} a_i a_j (X_i - \mu_i)(X_j - \mu_j)\right].$$

Again using the fact that E is a linear operator, we obtain

$$\sigma_Y^2 = \sum_{i=1}^{n} \sum_{j=1}^{n} a_i a_j E[(X_i - \mu_i)(X_j - \mu_j)].$$

However, if $i \neq j$, then from the independence of X_i and X_j we have

$$E[(X_i - \mu_i)(X_j - \mu_j)] = E(X_i - \mu_i)E(X_j - \mu_j) = (\mu_i - \mu_i)(\mu_j - \mu_j) = 0.$$

Thus, the variance can be written as

$$\sigma_Y^2 = \sum_{i=1}^{n} a_i^2 E[(X_i - \mu_i)^2] = \sum_{i=1}^{n} a_i^2 \sigma_i^2.$$

We give two illustrations of the theorem.

Example 5.2–1. Let the independent random variables X_1 and X_2 have respectively means $\mu_1 = -4$ and $\mu_2 = 3$ and variances $\sigma_1^2 = 4$ and $\sigma_2^2 = 9$. The mean and the variance of $Y = 3X_1 - 2X_2$ are, respectively,

$$\mu_Y = (3)(-4) + (-2)(3) = -18$$

and

$$\sigma_Y^2 = (3)^2(4) + (-2)^2(9) = 72.$$

Example 5.2–2. Let $X_1, X_2, \ldots, X_n$ be a random sample of size n from a distribution with mean μ and variance σ^2. First let $Y = X_1 - X_2$; then

$$\mu_Y = \mu - \mu = 0 \quad \text{and} \quad \sigma_Y^2 = (1)^2\sigma^2 + (-1)^2\sigma^2 = 2\sigma^2.$$

Now consider the sample mean

$$\bar{X} = \frac{X_1 + X_2 + \cdots + X_n}{n},$$

which is a linear function with each $a_i = 1/n$. Then

$$\mu_{\bar{X}} = \sum_{i=1}^{n} \left(\frac{1}{n}\right)\mu = \mu \quad \text{and} \quad \sigma_{\bar{X}}^2 = \sum_{i=1}^{n} \left(\frac{1}{n}\right)^2 \sigma^2 = \frac{\sigma^2}{n}.$$

That is, the mean of $\bar{X}$ is that of the distribution from which the sample arose, but the variance of $\bar{X}$ is that of the underlying distribution divided by n.

In some applications it is sufficient to know the mean and variance of a linear combination of random variables, say Y. However, it is often helpful to know exactly how Y is distributed. The next theorem can frequently be used to find the distribution of a linear combination of independent random variables.

THEOREM 5.2–2. *Let $X_1, X_2, \ldots, X_n$ be independent random variables with respective moment-generating functions $M_{X_i}(t)$, $i = 1, 2, 3, \ldots, n$. Then the moment-generating function of $Y = \sum_{i=1}^{n} a_i X_i$ is*

$$M_Y(t) = \prod_{i=1}^{n} M_{X_i}(a_i t).$$

Proof: The moment-generating function of Y is given by

$$M_Y(t) = E[e^{tY}] = E[e^{t(a_1 X_1 + a_2 X_2 + \cdots + a_n X_n)}]$$
$$= E[e^{a_1 t X_1} e^{a_2 t X_2} \cdots e^{a_n t X_n}]$$
$$= E[e^{a_1 t X_1}] E[e^{a_2 t X_2}] \cdots E[e^{a_n t X_n}]$$

using Theorem 5.1–2. However, since

$$E(e^{tX_i}) = M_{X_i}(t),$$

then

$$E(e^{a_i t X_i}) = M_{X_i}(a_i t).$$

Thus, we have that

$$M_Y(t) = M_{X_1}(a_1 t) M_{X_2}(a_2 t) \cdots M_{X_n}(a_n t)$$

$$= \prod_{i=1}^{n} M_{X_i}(a_i t).$$

A corollary follows immediately, and it will be used in some important examples.

COROLLARY. Let $X_1, X_2, \ldots, X_n$ be a random sample from a distribution with moment-generating function $M(t)$.

(i) The moment-generating function of $Y = \sum_{i=1}^{n} X_i$ is

$$M_Y(t) = \prod_{i=1}^{n} M(t) = [M(t)]^n.$$

(ii) The moment-generating function of $\overline{X} = \sum_{i=1}^{n} (1/n)X_i$ is

$$M_{\overline{X}}(t) = \prod_{i=1}^{n} M\left(\frac{t}{n}\right) = \left[M\left(\frac{t}{n}\right)\right]^n.$$

Proof: For (i), let $a_i = 1$, $i = 1, 2, \ldots, n$, in Theorem 5.2–2. For (ii), take $a_i = 1/n$, $i = 1, 2, \ldots, n$.

The following examples and the exercises give some important applications of Theorem 5.2–2 and its Corollary.

Example 5.2–3. Let $X_1, X_2, \ldots, X_n$ denote the outcomes on n Bernoulli trials. The moment-generating function of X_i, $i = 1, 2, \ldots, n$, is

$$M(t) = q + pe^t.$$

If

$$Y = \sum_{i=1}^{n} X_i,$$

then

$$M_Y(t) = \prod_{i=1}^{n} (q + pe^t) = (q + pe^t)^n.$$

Thus, we again see that Y is $b(n, p)$.

Example 5.2–4. Let $X_1, X_2, \ldots, X_n$ be a random sample from a distribution $N(\mu, \sigma^2)$. Since the moment-generating function of each X is

$$M_X(t) = \exp\left(\mu t + \frac{\sigma^2 t^2}{2}\right),$$

the moment-generating function of

$$\overline{X} = \frac{1}{n} \sum_{i=1}^{n} X_i$$

is, from the Corollary of Theorem 5.2–2, equal to

$$M_{\bar{X}}(t) = \left\{ \exp\left[\mu\left(\frac{t}{n}\right) + \frac{\sigma^2(t/n)^2}{2} \right] \right\}^n$$

$$= \exp\left[\mu t + \frac{(\sigma^2/n)t^2}{2} \right].$$

However, the moment-generating function uniquely determines the distribution of the random variable. Since this one is that associated with the normal distribution $N(\mu, \sigma^2/n)$, the sample mean $\bar{X}$ is $N(\mu, \sigma^2/n)$.

The previous example shows that if $X_1, X_2, \ldots, X_n$ is a random sample from a distribution $N(\mu, \sigma^2)$, then the probability distribution of $\bar{X}$ is also normal with the same mean μ but a variance σ^2/n. This means that $\bar{X}$ has a greater probability of falling in an interval containing μ than does a single observation, say X_1. For example, if $\mu = 50, \sigma^2 = 16, n = 64$, then $P(49 < \bar{X} < 51) = 0.9544$, whereas $P(49 < X_1 < 51) = 0.1974$.

We close this section with an important example about the sum of mutually independent chi-square random variables, a result that is used often in statistics.

Example 5.2–5. Let $X_1, X_2, \ldots, X_n$ be n mutually independent chi-square random variables with $r_1, r_2, \ldots, r_n$ degrees of freedom, respectively. The moment-generating function of

$$Y = X_1 + X_2 + \cdots + X_n$$

is, by Theorem 5.2–2, with $a_i = 1, i = 1, 2, \ldots, n,$

$$M_Y(t) = M_{X_1}(t)M_{X_2}(t)\cdots M_{X_n}(t).$$

But,

$$M_{X_i}(t) = (1 - 2t)^{-r_i/2}, \qquad t < \frac{1}{2}, \quad i = 1, 2, \ldots, n;$$

so

$$M_Y(t) = (1 - 2t)^{-r_1/2}(1 - 2t)^{-r_2/2}\cdots(1 - 2t)^{-r_n/2}$$

$$= (1 - 2t)^{-(r_1 + r_2 + \cdots + r_n)/2}, \qquad t < \frac{1}{2}.$$

That is, Y has the moment-generating function of a chi-square distribution with $r_1 + r_2 + \cdots + r_n$ degrees of freedom. Thus, from the uniqueness of the moment-generating function, Y is $\chi^2(r_1 + r_2 + \cdots + r_n)$.

══ *Exercises* ══

5.2–1. Let X_1 and X_2 be two independent random variables with respective means 3 and 7 and variances 9 and 25. Compute the mean and the variance of $Y = -2X_1 + X_2$.

5.2–2. Let $X_1, X_2, \ldots, X_{16}$ be a random sample from a distribution $N(77, 25)$. Compute (a) $P(77 < \bar{X} < 79.5)$, (b) $P(74.2 < \bar{X} < 78.4)$.

5.2–3. Let X be $N(50, 36)$. Using the same set of axes, sketch the graphs of the probability density functions of (a) X; (b) $\bar{X}$, the mean of a random sample of size 9 from this distribution; (c) $\bar{X}$, the mean of a random sample of size 36 from this distribution.

5.2–4. Let X_1, X_2, X_3 be mutually independent random variables with Poisson distributions having means $2, 1, 4$, respectively. Find the moment-generating function and then the distribution of the sum $Y = X_1 + X_2 + X_3$ and compute $P(3 \le Y \le 9)$.

5.2–5. Let $X_1, X_2, \ldots, X_n$ be n mutually independent normal variables with means $\mu_1, \mu_2, \ldots, \mu_n$ and variances $\sigma_1^2, \sigma_2^2, \ldots, \sigma_n^2$, respectively. Show that the linear function $Y = \sum c_i X_i$ is $N(\sum c_i \mu_i, \sum c_i^2 \sigma_i^2)$, where $c_1, c_2, \ldots, c_n$ are real constants.

5.2–6. Use the result of Exercise 5.2–5 to compute

$$P(X < Y) = P(0 < Y - X),$$

where X and Y are independent random variables with normal distributions $N(4, 9)$ and $N(7, 16)$, respectively.

5.2–7, Let $\bar{X}$ and $\bar{Y}$ represent the means of random samples, each of size 4, from the respective normal distributions of Exercise 5.2–6. Compute $P(\bar{X} < \bar{Y})$. Compare this to the answer for $P(X < Y)$ found in Exercise 5.2–6.

5.2–8. Generalize Exercise 5.2–4 by showing that the sum of n independent Poisson random variables with respective means $\mu_1, \mu_2, \ldots, \mu_n$ is Poisson with mean

$$\mu_1 + \mu_2 + \cdots + \mu_n.$$

5.2–9. Let $W = X_1 + X_2 + \cdots + X_h$, a sum of h mutually independent and identically distributed exponential random variables with mean θ. Show that W has a gamma distribution with mean $h\theta$.

5.2–10. Let X and Y, with respective p.d.f.'s $f(x)$ and $g(y)$, be independent discrete random variables, each of whose support is a subset of the nonnegative integers $0, 1, 2, \ldots$. Show that the p.d.f. of $W = X + Y$ is given by the *convolution formula*

$$h(w) = \sum_{x=0}^{w} f(x)g(w - x).$$

HINT: Argue that $h(w) = P(W = w)$ is the probability of the $w + 1$ mutually exclusive events $(x, w - x)$, $x = 0, 1, \ldots, w$ or find the moment-generating function of W.

5.2–11. Let X_1, X_2, X_3, X_4 be a random sample from a distribution having p.d.f. $f(x) = (x + 1)/6$, $x = 0, 1, 2$.
    (a) Use Exercise 5.2–10 to find the p.d.f. of $W_1 = X_1 + X_2$.
    (b) What is the p.d.f. of $W_2 = X_3 + X_4$?
    (c) Now find the p.d.f. of $W = W_1 + W_2 = X_1 + X_2 + X_3 + X_4$.

5.2–12. Roll a die eight times and denote the outcomes by $X_1, X_2, \ldots, X_8$. Use Exercise 5.2–10 to find the p.d.f.'s of

(a) $X_1 + X_2$,　　　　　　(b) $\displaystyle\sum_{i=1}^{4} X_i$,　　　　　　(c) $\displaystyle\sum_{i=1}^{8} X_i$.

5.3 The Central Limit Theorem

In Section 5.2, we found that the mean $\overline{X}$ of a random sample of size n from a distribution with mean μ and variance $\sigma^2 > 0$ is a random variable with the properties that

$$E(\overline{X}) = \mu \quad\text{and}\quad \text{Var}(\overline{X}) = \frac{\sigma^2}{n}.$$

Thus, as n increases, the variance of $\overline{X}$ decreases. Consequently, the distribution of $\overline{X}$ clearly depends on n, and we see that we are dealing with sequences of distributions. To denote this we will place the subscript n on the random variables in this section. For example, in the case of the mean of a random sample of size n, we write $\overline{X}_n$.

We can easily obtain our first result about $\overline{X}_n$ by using Chebyshev's inequality (Theorem 2.3–1). Consider, for every $\varepsilon > 0$,

$$P[|\overline{X}_n - \mu| \geq \varepsilon] = P\left[|\overline{X}_n - \mu| \geq \left(\frac{\varepsilon\sqrt{n}}{\sigma}\right)\left(\frac{\sigma}{\sqrt{n}}\right)\right].$$

But the standard deviation of $\overline{X}$ is $\sigma/\sqrt{n}$. Hence, using Chebyshev's inequality with $k = \varepsilon\sqrt{n}/\sigma$, we have

$$P[|\overline{X}_n - \mu| \geq \varepsilon] \leq \frac{\sigma^2}{\varepsilon^2 n}.$$

Since probability is nonnegative, it follows that

$$0 \leq \lim_{n\to\infty} P(|\overline{X}_n - \mu| \geq \varepsilon) \leq \lim_{n\to\infty} \frac{\sigma^2}{\varepsilon^2 n} = 0.$$

This implies that

$$\lim_{n\to\infty} P(|\overline{X}_n - \mu| \geq \varepsilon) = 0,$$

or, equivalently,

$$\lim_{n\to\infty} P(|\overline{X}_n - \mu| < \varepsilon) = 1.$$

The preceding discussion shows that the probability associated with the distribution of $\overline{X}_n$ becomes concentrated in an arbitrary small interval about μ as n increases. This is one form of the *law of large numbers*.

Example 5.3–1. Let $X_1, X_2, \ldots, X_n$ denote a random sample from a distribution $b(1, p)$, $0 < p < 1$; this distribution has mean p and variance $p(1 - p) > 0$. We know that

$$Y_n = X_1 + X_2 + \cdots + X_n$$

is $b(n, p)$. From the preceding result, with $\overline{X}_n = Y_n/n$, we see that

$$\lim_{n \to \infty} P\left[\left| \frac{Y_n}{n} - p \right| < \varepsilon \right] = 1.$$

That is, *in this sense*, the relative frequency Y_n/n gets close to p, which is another form of the law of large numbers.

We have seen that, in some probability sense, $\overline{X}_n$ converges to μ in the limit, or, equivalently, $\overline{X}_n - \mu$ converges to zero. Let us multiply the difference $\overline{X}_n - \mu$ by some function of n so that the result will not converge to zero. In our search for such a function, it is natural to consider

$$W_n = \frac{\overline{X}_n - \mu}{\sigma/\sqrt{n}} = \frac{\sqrt{n}(\overline{X}_n - \mu)}{\sigma} = \frac{Y_n - n\mu}{\sqrt{n}\,\sigma},$$

where Y_n is the sum of the items of the random sample. The reason for this is that W_n is a standardized random variable and has mean 0 and variance 1 for each positive integer n. That is,

$$E(W_n) = E\left[\frac{\overline{X}_n - \mu}{\sigma/\sqrt{n}} \right] = \frac{E(\overline{X}_n) - \mu}{\sigma/\sqrt{n}} = \frac{\mu - \mu}{\sigma/\sqrt{n}} = 0$$

and

$$\mathrm{Var}(W_n) = E(W_n^2) = E\left[\frac{(\overline{X}_n - \mu)^2}{\sigma^2/n} \right] = \frac{E[(\overline{X}_n - \mu)^2]}{\sigma^2/n} = \frac{\sigma^2/n}{\sigma^2/n} = 1.$$

Thus, while $\overline{X}_n - \mu$ "degenerates" to zero, the factor $\sqrt{n}/\sigma$ in $\sqrt{n}(\overline{X}_n - \mu)/\sigma$ "spreads out" the probability enough to prevent this degeneration. What then is the distribution of W_n as n increases? One observation that might shed some light on the answer to this question can be made immediately. If the sample arises from a normal distribution then, from Example 5.2–4, we know that $\overline{X}_n$ is $N(\mu, \sigma^2/n)$ and, hence, W_n is $N(0, 1)$ for each positive n. Thus, in the limit, the distribution of W_n must be $N(0, 1)$. So if the solution of the question does not depend on the underlying distribution (that is, it is unique), the answer must be $N(0, 1)$. As we will see, this is exactly the case, and this result is so important it is called the *Central Limit Theorem*.

THEOREM 5.3–1.   (Central Limit Theorem) *Let* $\bar{X}_n$ *be the mean of a random sample* $X_1, X_2, \ldots, X_n$ *of size n from a distribution with a finite mean* μ *and a finite positive variance* σ^2. *Then the distribution of*

$$W_n = \frac{\bar{X}_n - \mu}{\sigma/\sqrt{n}} = \frac{\sum_{i=1}^{n} X_i - n\mu}{\sqrt{n}\,\sigma}$$

is $N(0, 1)$ *in the limit as* $n \to \infty$.

REMARK.   The student must recognize that a *limiting distribution* or *convergence in distribution* has not been officially defined. At this level, we believe an intuitive approach is better. Moreover, in the proof, we even ask the student to accept the result that if a sequence of moment-generating functions approaches a given one, then the limit of the distributions associated with this sequence must be getting close to the distribution corresponding to that given moment-generating function. This result does appeal to one's intuition. As a matter of fact, in an advanced course, we do not need the existence of the moment-generating function but would use the characteristic function instead. However, assuming that such does exist, we will show that the limit of the sequence of moment-generating functions of W_n is that of the $N(0, 1)$ distribution, namely, $\exp(t^2/2)$.

Proof: We first consider

$$E[\exp(tW_n)] = E\left\{\exp\left[\left(\frac{t}{\sqrt{n}\,\sigma}\right)\left(\sum_{i=1}^{n} X_i - n\mu\right)\right]\right\}$$

$$= E\left\{\exp\left[\left(\frac{t}{\sqrt{n}}\right)\left(\frac{X_1 - \mu}{\sigma}\right)\right]\cdots\exp\left[\left(\frac{t}{\sqrt{n}}\right)\left(\frac{X_n - \mu}{\sigma}\right)\right]\right\}$$

$$= E\left\{\exp\left[\left(\frac{t}{\sqrt{n}}\right)\left(\frac{X_1 - \mu}{\sigma}\right)\right]\right\}\cdots E\left\{\exp\left(\frac{t}{\sqrt{n}}\right)\left(\frac{X_n - \mu}{\sigma}\right)\right]\right\},$$

which follows from the mutual independence of $X_1, X_2, \ldots, X_n$. Then

$$E[\exp(tW_n)] = \left[m\left(\frac{t}{\sqrt{n}}\right)\right]^n, \qquad -h < \frac{t}{\sqrt{n}} < h,$$

where

$$m(t) = E\left\{\exp\left[t\left(\frac{X_i - \mu}{\sigma}\right)\right]\right\}, \qquad -h < t < h,$$

is the common moment-generating function of each

$$Y_i = \frac{X_i - \mu}{\sigma}, \qquad i = 1, 2, \ldots, n.$$

Since $E(Y_i) = 0$ and $E(Y_i^2) = 1$, it must be that

$$m(0) = 1, \qquad m'(0) = E\left(\frac{X_i - \mu}{\sigma}\right) = 0, \qquad m''(0) = E\left[\left(\frac{X_i - \mu}{\sigma}\right)^2\right] = 1.$$

Hence, using Taylor's formula with a remainder, we can find a number t_1 between 0 and t such that

$$m(t) = m(0) + m'(0)t + \frac{m''(t_1)t^2}{2} = 1 + \frac{m''(t_1)t^2}{2}.$$

By adding and subtracting $t^2/2$, we have that

$$m(t) = 1 + \frac{t^2}{2} + \frac{[m''(t_1) - 1]t^2}{2}.$$

Using this expression of $m(t)$ in $E[\exp(tW_n)]$, we can represent the moment-generating function of W_n by

$$E[\exp(tW_n)] = \left\{1 + \frac{1}{2}\left(\frac{t}{\sqrt{n}}\right)^2 + \frac{1}{2}[m''(t_1) - 1]\left(\frac{t}{\sqrt{n}}\right)^2\right\}^n$$

$$= \left\{1 + \frac{t^2}{2n} + \frac{[m''(t_1) - 1]t^2}{2n}\right\}^n, \qquad -\sqrt{n}h < t < \sqrt{n}h,$$

where now t_1 is between 0 and $t/\sqrt{n}$. Since $m''(t)$ is continuous at $t = 0$ and $t_1 \to 0$ as $n \to \infty$, we have that

$$\lim_{n \to \infty} [m''(t_1) - 1] = 1 - 1 = 0.$$

Thus, using a result from advanced calculus, we have that

$$\lim_{n \to \infty} E[\exp(tW_n)] = \lim_{n \to \infty}\left\{1 + \frac{t^2}{2n} + \frac{[m''(t_1) - 1]t^2}{2n}\right\}^n$$

$$= \lim_{n \to \infty}\left\{1 + \frac{t^2/2}{n}\right\}^n = e^{t^2/2},$$

which completes the proof.

    A practical use of the Central Limit Theorem is approximating, when n is "sufficiently large," the distribution function of W_n, namely,

$$P(W_n \le w) \approx \int_{-\infty}^{w} \frac{1}{\sqrt{2\pi}} e^{-z^2/2} \, dz.$$

We present some illustrations of this application, discuss "sufficiently large," and try to give an intuitive feeling for the Central Limit Theorem.

Example 5.3–2. Let $X_1, X_2, \ldots, X_{20}$ denote a random sample of size 20 from the uniform distribution $U(0, 1)$. Here $E(X_i) = 1/2$ and $\text{Var}(X_i) = 1/12$, $i = 1, 2, \ldots, 20$. If $Y_{20} = X_1 + X_2 + \cdots + X_{20}$, then

$$P(Y_{20} \leq 9.1) = P\left(\frac{Y_{20} - 20(1/2)}{\sqrt{20/12}} \leq \frac{9.1 - 10}{\sqrt{20/12}}\right)$$

$$\approx \Phi(-0.70)$$

$$= 0.2420.$$

Also,

$$P(8.5 \leq Y_{20} \leq 11.7) = P\left(\frac{8.5 - 10}{\sqrt{5/3}} \leq \frac{Y_{20} - 10}{\sqrt{5/3}} \leq \frac{11.7 - 10}{\sqrt{5/3}}\right)$$

$$\approx \Phi(1.32) - \Phi(-1.16)$$

$$= 0.7836.$$

Example 5.3–3. Let $\bar{X} = \bar{X}_{25}$ denote the mean of a random sample of size 25 from the distribution whose p.d.f. is $f(x) = x^3/4, 0 < x < 2$. It is easy to show that $\mu = 8/5 = 1.6$ and $\sigma^2 = 8/75$. Thus,

$$P(1.5 \leq \bar{X} \leq 1.65) = P\left(\frac{1.5 - 1.6}{\sqrt{8/75}/\sqrt{25}} \leq \frac{\bar{X} - 1.6}{\sqrt{8/75}/\sqrt{25}} \leq \frac{1.65 - 1.6}{\sqrt{8/75}/\sqrt{25}}\right)$$

$$\approx \Phi(0.77) - \Phi(-1.53)$$

$$= 0.7164.$$

Example 5.3–4. Let $\bar{X} = \bar{X}_{15}$ denote the mean of a random sample of size $n = 15$ from the distribution whose p.d.f. is $f(x) = (3/2)x^2, -1 < x < 1$. Here $\mu = 0$ and $\sigma^2 = 3/5$. Thus,

$$P(0.03 \leq \bar{X} \leq 0.15) = P\left(\frac{0.03 - 0}{\sqrt{3/5}/\sqrt{15}} \leq \frac{\bar{X} - 0}{\sqrt{3/5}/\sqrt{15}} \leq \frac{0.15 - 0}{\sqrt{3/5}/\sqrt{15}}\right)$$

$$\approx \Phi(0.75) - \Phi(0.15)$$

$$= 0.2138.$$

These examples have shown how the Central Limit Theorem can be used for approximating certain probabilities concerning the mean $\bar{X}_n$ of a random sample. That is, $\bar{X}_n$ is approximately $N(\mu, \sigma^2/n)$ when n is "sufficiently large," where μ and σ^2 are the mean and the variance of the underlying distribution from which the sample arose. Generally if n is greater than 25 or 30, these approximations will be good. However, if the underlying distribution is symmetric, unimodal, and of the continuous type, a value of n as small as 4 or 5 can yield a very adequate approximation. Moreover, if the original distribution

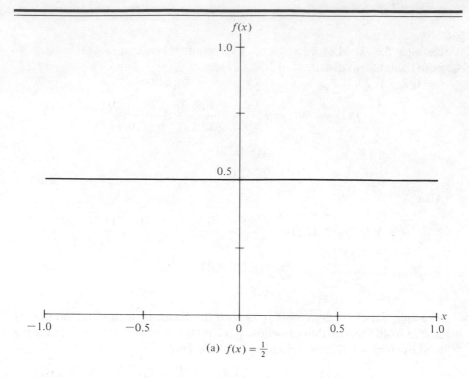

(a) $f(x) = \frac{1}{2}$

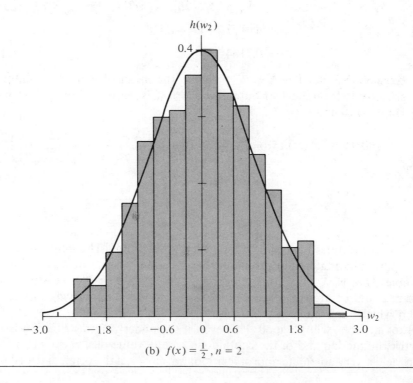

(b) $f(x) = \frac{1}{2}$, $n = 2$

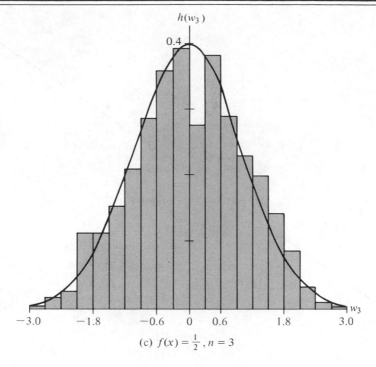

(c) $f(x) = \frac{1}{2}$, $n = 3$

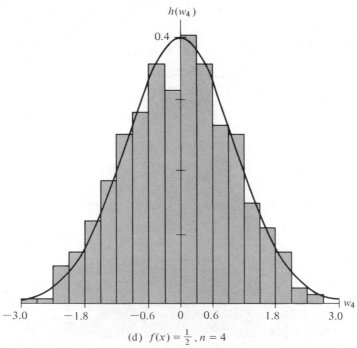

(d) $f(x) = \frac{1}{2}$, $n = 4$

FIGURE 5.3–1

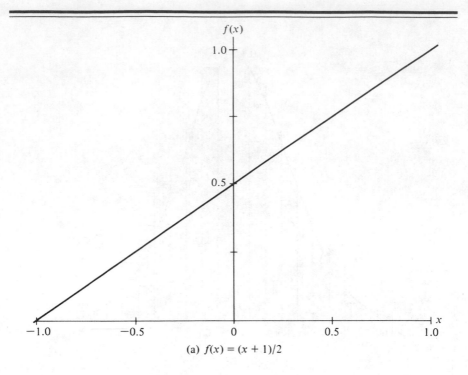

(a) $f(x) = (x + 1)/2$

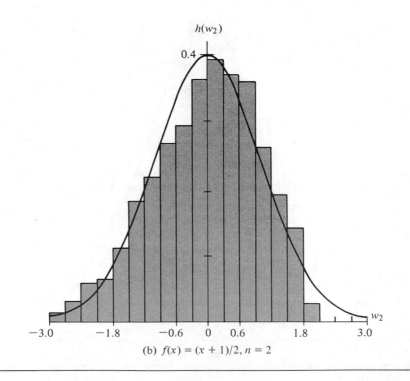

(b) $f(x) = (x + 1)/2, n = 2$

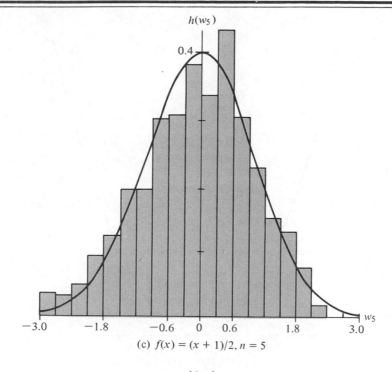

(c) $f(x) = (x + 1)/2, n = 5$

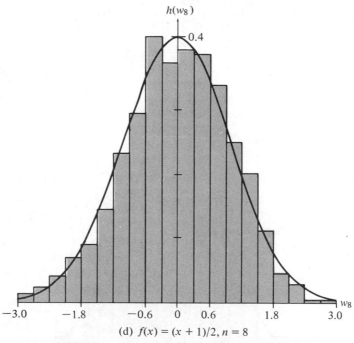

(d) $f(x) = (x + 1)/2, n = 8$

FIGURE 5.3–2

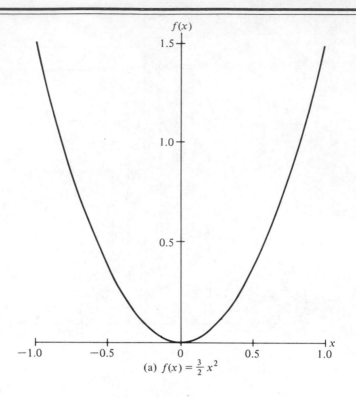

(a) $f(x) = \frac{3}{2}x^2$

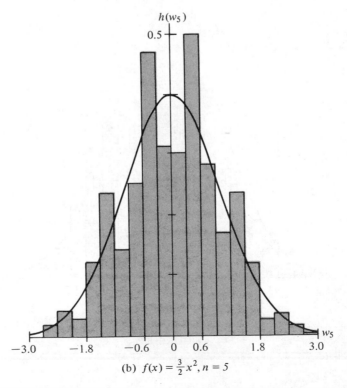

(b) $f(x) = \frac{3}{2}x^2, n = 5$

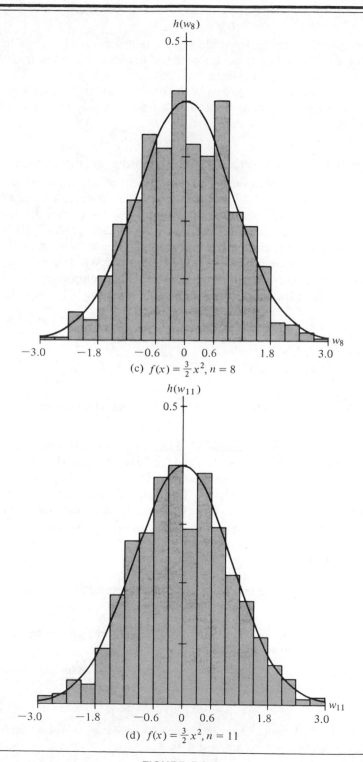

(c) $f(x) = \frac{3}{2}x^2, n = 8$

(d) $f(x) = \frac{3}{2}x^2, n = 11$

FIGURE 5.3–3

is approximately normal, $\overline{X}_n$ would have a distribution very close to normal when n equals 2 or 3. In fact, we know that if the sample is taken from $N(\mu, \sigma^2)$, $\overline{X}_n$ is exactly $N(\mu, \sigma^2/n)$ for every $n = 1, 2, 3, \ldots$.

The next example will help to give the reader a better intuitive feeling about the Central Limit Theorem. We shall also see how the size of n affects the distribution of $W_n = (\overline{X}_n - \mu)/(\sigma/\sqrt{n})$.

Example 5.3–5. It is often difficult to find the exact distribution of $W_n = (\overline{X}_n - \mu)/(\sigma/\sqrt{n})$. However, let us consider some empirical evidence about the distribution of W_n by simulating random samples on the computer. Let $X_1, X_2, \ldots, X_n$ denote a random sample of size n from the distribution with p.d.f. $f(x)$, mean μ, and variance σ^2. We shall generate 1000 random samples of size n from this distribution and compute a value of W_n for each sample, thus obtaining 1000 observed values of W_n. A histogram of these 1000 values is constructed using 20 intervals of equal length. We depict the results of this experiment, along with the underlying p.d.f., as follows:

(1) In Figure 5.3–1, $f(x) = 1/2, -1 < x < 1, \mu = 0, \sigma^2 = 1/3$, for $n = 2, 3, 4$.
(2) In Figure 5.3–2, $f(x) = (x + 1)/2, -1 < x < 1, \mu = 1/3, \sigma^2 = 2/9$, for $n = 2, 5, 8$.
(3) In Figure 5.3–3, $f(x) = (3/2)x^2, -1 < x < 1, \mu = 0, \sigma^2 = 3/5$, for $n = 5, 8, 11$.

The $N(0, 1)$ p.d.f. has been superimposed on each histogram.

Note very clearly that Example 5.3–5 has not proved anything. It is presented to give empirical evidence of the truth of the Central Limit Theorem. And so far all the illustrations have concerned distributions of the continuous type. However, the hypothesis for the Central Limit Theorem does not require the distribution to be continuous. We shall consider applications of the Central Limit Theorem for discrete-type distributions in the next section.

══ *Exercises* ══

5.3–1. Let $\overline{X}_{12}$ be the mean of a random sample of size 12 from the uniform distribution on the interval $(0, 1)$. Approximate $P(1/2 \le \overline{X}_{12} \le 2/3)$.

5.3–2. Let $Y_{15} = X_1 + X_2 + \cdots + X_{15}$ be the sum of a random sample of size 15 from the distribution whose p.d.f. is $f(x) = (3/2)x^2, -1 < x < 1$. Approximate

$$P(-0.3 \le Y_{15} \le 1.5).$$

5.3–3. Let $\overline{X}_{36}$ be the mean of a random sample of size 36 from an exponential distribution with mean 3. Approximate $P(2.5 \le \overline{X}_{36} \le 4)$.

5.3–4. Approximate $P(39.75 \leq \bar{X}_{32} \leq 41.25)$, where $\bar{X}_{32}$ is the mean of a random sample of size 32 from a distribution with mean $\mu = 40$ and variance $\sigma^2 = 8$.

5.3–5. Let $X_1, X_2, \ldots, X_{18}$ be a random sample of size 18 from a chi-square distribution with $r = 1$. Recall that $\mu = 1$, $\sigma^2 = 2$.

   (a) How is $Y_{18} = \sum_{i=1}^{18} X_i$ distributed?

   (b) Using the result of part (a), we see from Appendix Table III that

$$P(Y_{18} \leq 9.390) = 0.05 \quad \text{and} \quad P(Y_{18} \leq 34.80) = 0.99.$$

Compare these two probabilities with the approximations found using the Central Limit Theorem.

5.4 Approximations for Discrete Distributions

In this section we illustrate how the normal distribution can be used to approximate probabilities for certain discrete-type distributions. One of the most important discrete distributions is the binomial distribution. To see how the Central Limit Theorem can be applied, recall that a binomial random variable can be described as the sum of Bernoulli random variables. That is, let $X_1, X_2, \ldots, X_n$ be a random sample from a Bernoulli distribution with a mean $\mu = p$ and a variance $\sigma^2 = p(1 - p)$, where $0 < p < 1$. Then $Y_n = \sum_{i=1}^{n} X_i$ is $b(n, p)$. The Central Limit Theorem states that the distribution of

$$W_n = \frac{Y_n - np}{\sqrt{np(1 - p)}} = \frac{\bar{X}_n - p}{\sqrt{p(1 - p)/n}}$$

is $N(0, 1)$ in the limit as $n \to \infty$. Thus, if n is "sufficiently large," probabilities for $b(n, p)$ can be approximated using $N[np, np(1 - p)]$. A rule often stated is that n is "sufficiently large" if $np \geq 5$ and $n(1 - p) \geq 5$.

Note that we will be approximating probabilities for a discrete distribution with probabilities for a continuous distribution. Let us discuss a reasonable procedure in this situation. For example, if V is $N(\mu, \sigma^2)$, $P(a < V < b)$ is equivalent to the area bounded by the p.d.f. of V, the v axis, $v = a$, and $v = b$. If Y is $b(n, p)$, we shall define a "probability histogram" for Y as follows. For $k - 1/2 < y < k + 1/2$, let

$$f(y) = \frac{n!}{k!(n - k)!} p^k(1 - p)^{n-k}, \quad k = 0, 1, 2, \ldots, n.$$

Then $P(Y = k)$ can be represented by the area of the rectangle with a height of $P(Y = k)$ and a base of length 1 centered at k. Figure 5.4–1 shows the graph of the probability histogram for the binomial distribution $b(3, 1/3)$. When using

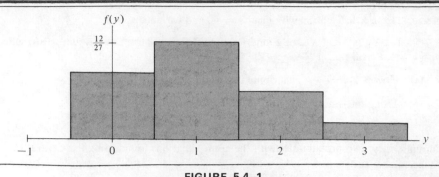

FIGURE 5.4–1

the normal distribution to approximate probabilities for the binomial distribution, areas under the p.d.f. for the normal distribution will be used to approximate areas of rectangles in the probability histogram for the binomial distribution.

Example 5.4–1. Let $Y = Y_{10}$ be $b(10, 1/2)$. Then, by the Central Limit Theorem, $P(a < Y < b)$ can be approximated using the normal distribution with mean $10(1/2) = 5$ and variance $10(1/2)(1/2) = 5/2$. Figure 5.4–2 shows the graph of the probability histogram for $b(10, 1/2)$ and the graph of the p.d.f. of $N(5, 5/2)$. Note that the area of the rectangle whose base is

$$\left(k - \frac{1}{2}, \quad k + \frac{1}{2} \right)$$

and the area under the normal curve between $k - 1/2$ and $k + 1/2$ are approximately equal for each integer k.

Example 5.4–2. Let Y_{18} be $b(18, 1/6)$. Because $np = 18(1/6) = 3 < 5$, the normal approximation is not too good here. Figure 5.4–3 illustrates this by depicting the probability histogram for $b(18, 1/6)$ and the p.d.f. of $N(3, 5/2)$.

Example 5.4–3. Let Y_{10} have the binomial distribution of Example 5.4–1 and Figure 5.4–2, namely, $b(10, 1/2)$. Then

$$P(3 \leq Y_{10} < 6) = P(2.5 \leq Y_{10} \leq 5.5)$$

because $P(Y_{10} = 6)$ is not in the desired answer. But the latter equals

$$P\left(\frac{2.5 - 5}{\sqrt{10/4}} \leq \frac{Y_{10} - 5}{\sqrt{10/4}} \leq \frac{5.5 - 5}{\sqrt{10/4}} \right) \approx \Phi(0.32) - \Phi(-1.58)$$

$$= 0.5685.$$

Using the binomial formula, we find that $P(3 \leq Y_{10} < 6) = 0.5683$.

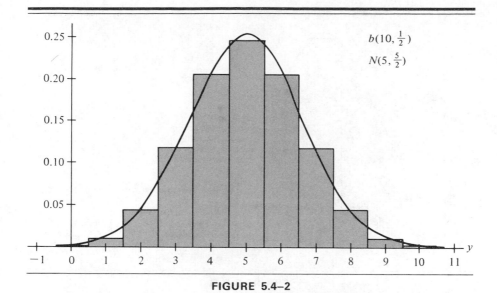

FIGURE 5.4–2

Example 5.4–4. Let Y_{36} be $b(36, 1/2)$. Then

$$P(12 < Y_{36} \leq 18) = P(12.5 \leq Y_{36} \leq 18.5)$$

$$= P\left(\frac{12.5 - 18}{\sqrt{9}} \leq \frac{Y_{36} - 18}{\sqrt{9}} \leq \frac{18.5 - 18}{\sqrt{9}}\right)$$

$$\approx \Phi(0.167) - \Phi(-1.833)$$

$$= 0.5329.$$

Note that 12 was increased to 12.5 because $P(Y_{36} = 12)$ is not included in the desired probability. Using the binomial formula, we find that

$$P(12 < Y_{36} \leq 18) = P(13 \leq Y_{36} \leq 18) = 0.5334.$$

Also,

$$P(Y_{36} = 20) = P(19.5 \leq Y_{36} \leq 20.5)$$

$$= P\left(\frac{19.5 - 18}{\sqrt{9}} \leq \frac{Y_{36} - 18}{\sqrt{9}} \leq \frac{20.5 - 18}{\sqrt{9}}\right)$$

$$\approx \Phi(0.833) - \Phi(0.5)$$

$$= 0.1060.$$

Using the binomial formula, we have $P(Y_{36} = 20) = 0.1063$.

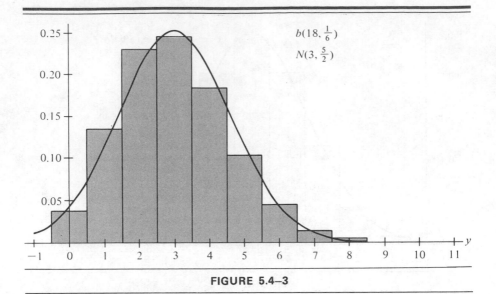

$$b(18, \tfrac{1}{6})$$
$$N(3, \tfrac{5}{2})$$

FIGURE 5.4–3

Note that in general, if Y is $b(n, p)$,

$$P(Y \leq k) \approx \Phi\left(\frac{k + 1/2 - np}{\sqrt{npq}}\right)$$

and

$$P(Y < k) \approx \Phi\left(\frac{k - 1/2 - np}{\sqrt{npq}}\right).$$

We now show how the Poisson distribution with large enough mean can be approximated using a normal distribution.

Example 5.4–5. A random variable having a Poisson distribution with mean 20 can be thought of as the sum Y_{20} of the items of a random sample of size 20 from a Poisson distribution with mean 1. Thus,

$$W_{20} = \frac{Y_{20} - 20}{\sqrt{20}}$$

has a distribution that is approximately $N(0, 1)$. So, for illustration,

$$P(16 < Y_{20} \leq 21) = P(16.5 \leq Y_{20} \leq 21.5)$$

$$= P\left(\frac{16.5 - 20}{\sqrt{20}} \leq \frac{Y_{20} - 20}{\sqrt{20}} \leq \frac{21.5 - 20}{\sqrt{20}}\right)$$

$$\approx \Phi(0.335) - \Phi(-0.783)$$

$$= 0.4142.$$

Note that 16 is increased to 16.5 because $Y_{20} = 16$ is not included in the event $16 < Y_{20} \le 21$. The answer using the Poisson formula is 0.4226.

In general, if Y has a Poisson distribution with mean λ, $W = (Y - \lambda)/\sqrt{\lambda}$ is approximately $N(0, 1)$ when λ is sufficiently large. The exercises illustrate applications of the Central Limit Theorem to other discrete distributions. However, we would like to close this section by showing that, in some cases, the binomial distribution can be approximated by the Poisson distribution.

To see that the binomial probabilities can be approximated by those of the Poisson when n is sufficiently large and p fairly small, let us consider the moment-generating function of Y_n, which is $b(n, p)$. We will take the limit of this as $n \to \infty$ such that $np = \lambda$ is a constant; thus, $p \to 0$. The moment-generating function of Y_n, written as $M(t; n)$ to emphasize that it depends on n, is

$$M(t; n) = (1 - p + pe^t)^n.$$

Because $p = \lambda/n$, we have that

$$M(t; n) = \left[1 - \frac{\lambda}{n} + \frac{\lambda}{n} e^t\right]^n$$

$$= \left[1 + \frac{\lambda(e^t - 1)}{n}\right]^n.$$

Thus, since, in calculus,

$$\lim_{n \to \infty} \left(1 + \frac{b}{n}\right)^n = e^b,$$

we have

$$\lim_{n \to \infty} M(t; n) = e^{\lambda(e^t - 1)},$$

which exists for all real t. But this is the moment-generating function of a Poisson random variable with mean λ. Hence this Poisson distribution seems like a reasonable approximation to the binomial one when n is large and p is small. This approximation is usually found to be fairly successful if $n \ge 20$ and $p \le 0.05$ and very successful if $n \ge 100$ and $np \le 10$. Obviously, it could be used in other situations too; we only want to stress that the approximation becomes better with larger n and smaller p.

Example 5.4–6. Let Y_{50} be $b(50, 1/25)$. Then

$$P(Y_{50} \le 1) = \left(\frac{24}{25}\right)^{50} + 50\left(\frac{1}{25}\right)\left(\frac{24}{25}\right)^{49} = 0.400.$$

Since $\lambda = np = 2$, the Poisson approximation is

$$P(Y_{50} \le 1) \approx 0.406$$

from Appendix Table II.

═══ *Exercises* ═══

5.4–1. Let Y_{18} be $b(18, 1/3)$. Use the normal distribution to approximate
(a) $P(4 < Y_{18} \leq 6)$,
(b) $P(5 \leq Y_{18} < 8)$,
(c) $P(Y_{18} = 7)$.

5.4–2. A die is rolled independently 240 times. Approximate the probability that (a) more than 40 rolls are fives, (b) the number of two's and three's is from 75 to 83, inclusive.

5.4–3. Let $X_1, X_2, \ldots, X_{30}$ be a random sample of size 30 from a Poisson distribution with a mean of 2/3. Approximate

(a) $P\left(15 < \sum_1^{30} X_i \leq 22 \right)$,   (b) $P\left(21 \leq \sum_1^{30} X_i < 27 \right)$,

(c) $P\left(\dfrac{1}{2} \leq \bar{X}_{30} \leq \dfrac{3}{4} \right)$.

5.4–4. Let $X_1, X_2, \ldots, X_{48}$ be a random sample of size 48 from the distribution with p.d.f. $f(x) = 1/x^2, 1 < x < \infty$. Approximate the probability that at most 10 of these random variables have values greater than 4. HINT: Let the ith trial be a success if $X_i > 4, i = 1, 2, \ldots, 48$.

5.4–5. Let $X_1, X_2, \ldots, X_{36}$ be a random sample of size 36 from the geometric distribution with p.d.f. $f(x) = (1/4)^x(3/4)$, $x = 0, 1, 2, \ldots$. Approximate

(a) $P\left(10 \leq \sum_1^{36} X_i \leq 13 \right)$,   (b) $P\left(\dfrac{1}{4} \leq \bar{X}_{36} \leq \dfrac{1}{2} \right)$.

HINT: Observe that the distribution of the sum is of the discrete type.

5.4–6. A die is rolled 24 independent times. Let Y_{24} be the sum of the 24 resulting values. Approximate (a) $P(Y_{24} \geq 86)$, (b) $P(Y_{24} < 86)$, (c) $P(70 < Y_{24} \leq 86)$, recalling that Y_{24} is a random variable of the discrete type.

5.4–7. Let Y_{50} be the number of defectives in a box of 50 articles taken from the output of a machine. Each article is defective with probability 0.01. What is the probability that $Y_{50} = 0, 1, 2$, or 3 (a) using the binomial distribution? (b) using the Poisson approximation?

5.4–8. The probability that a certain type of inoculation takes effect is 0.995. Use the Poisson distribution to approximate the probability that at most two out of 400 people given the inoculation find that it has not taken effect? HINT: Let $p = 1 - 0.995 = 0.005$.

5.5 Order Statistics

In recent years, the importance of order statistics has increased owing to the more frequent use of nonparametric inferences and robust procedures. However, order statistics have always been prominent because, among other things, they are needed to determine rather simple statistics such as the sample median, the sample range, and the empirical distribution function. The *order statistics* are the items of the random sample arranged, or ordered, in magnitude from the smallest to the largest. We will consider certain interesting aspects about their distributions in this section; and in Chapter 6 we will demonstrate how they can be used to make certain statistical inferences.

In most of our discussions about order statistics, we will assume that the random sample arises from a continuous-type distribution. This means, among other things, that the probability of any two sample items being equal is zero. That is, the probability is *one* that the items can be ordered from smallest to largest without having two equal values. Of course, in practice, we do frequently observe *ties*; but if the probability of this is small, the following distribution theory will hold approximately. Thus, in the discussion here, we are assuming that the probability of ties is zero.

Example 5.5–1. The values $x_1 = 0.62$, $x_2 = 0.98$, $x_3 = 0.31$, $x_4 = 0.81$, and $x_5 = 0.53$ are the $n = 5$ observed values of five independent trials of an experiment with p.d.f. $f(x) = 2x$, $0 < x < 1$. The observed order statistics are

$$y_1 = 0.31 < y_2 = 0.53 < y_3 = 0.62 < y_4 = 0.81 < y_5 = 0.98.$$

Frequently, the middle item, here $y_3 = 0.62$, is called the sample median and the difference of the largest and the smallest, here

$$y_5 - y_1 = 0.98 - 0.31 = 0.67,$$

is called the sample range.

If $X_1, X_2, \ldots, X_n$ are items of a random sample of size n from a continuous-type distribution, we let the random variables

$$Y_1 < Y_2 < \cdots < Y_n$$

denote the order statistics of that sample. That is,

$$Y_1 = \text{smallest of } X_1, X_2, \ldots, X_n,$$
$$Y_2 = \text{second smallest of } X_1, X_2, \ldots, X_n,$$
$$\vdots$$
$$Y_n = \text{largest of } X_1, X_2, \ldots, X_n.$$

There is a very simple method for determining the distribution function of the rth order statistic, Y_r. This procedure depends on the binomial distribution and is illustrated in Example 5.5–2.

Example 5.5–2. Let $Y_1 < Y_2 < Y_3 < Y_4 < Y_5$ be the order statistics of a random sample X_1, X_2, X_3, X_4, X_5 of size $n = 5$ from the distribution with p.d.f. $f(x) = 2x$, $0 < x < 1$. Consider $P(Y_4 \leq 1/2)$. For the event $Y_4 \leq 1/2$ to occur, at least four of the items X_1, X_2, X_3, X_4, X_5 must be less than $1/2$ because Y_4 is the fourth smallest among the five items. Thus if the event $X_i \leq 1/2$, $i = 1, 2, \ldots, 5$, is called "success," we must have at least four successes in the five mutually independent trials each of which has probability of success

$$P\left(X_i \leq \frac{1}{2}\right) = \int_0^{1/2} 2x \, dx = \left(\frac{1}{2}\right)^2 = \frac{1}{4}.$$

Thus,

$$P\left(Y_4 \leq \frac{1}{2}\right) = \binom{5}{4}\left(\frac{1}{4}\right)^4\left(\frac{3}{4}\right) + \left(\frac{1}{4}\right)^5 = 0.0156.$$

In general, if $0 < y < 1$, then the distribution function of Y_4 is

$$G(y) = P(Y_4 \leq y) = \binom{5}{4}(y^2)^4(1 - y^2) + (y^2)^5$$

since this represents the probability of at least four "successes" in five independent trials, each of which has probability of success

$$P(X_i \leq y) = \int_0^y 2x \, dx = y^2.$$

The p.d.f. of Y_4 is therefore, for $0 < y < 1$,

$$g(y) = G'(y) = \binom{5}{4}4(y^2)^3(2y)(1 - y^2) + \binom{5}{4}(y^2)^4(-2y) + 5(y^2)^4(2y)$$

$$= \frac{5!}{3!1!}(y^2)^3(1 - y^2)(2y), \qquad 0 < y < 1.$$

Please note that in this example the distribution function of each X is $F(x) = x^2$ when $0 < x < 1$. Thus,

$$g(y) = \frac{5!}{3!1!}[F(y)]^3[1 - F(y)]f(y), \qquad 0 < y < 1.$$

The preceding example should make the following generalization easier to read. Let $Y_1 < Y_2 < \cdots < Y_n$ be the order statistics of a random sample of size n from a distribution of the continuous type with distribution function $F(x)$ and p.d.f. $F'(x) = f(x)$. Note that the event that the rth order statistic $Y_r \le y$ can occur if and only if at least r of the n items are less than or equal to y. That is, here the probability of "success" on each trial is $F(y)$ and we must have at least r successes. Thus,

$$G(y) = P(Y_r \le y) = \sum_{k=r}^{n} \binom{n}{k} [F(y)]^k [1 - F(y)]^{n-k}.$$

That is, rewriting this slightly, we have

$$G(y) = \sum_{k=r}^{n-1} \binom{n}{k} [F(y)]^k [1 - F(y)]^{n-k} + [F(y)]^n.$$

Thus, the p.d.f. of Y_r is

$$g(y) = G'(y) = \sum_{k=r}^{n-1} \binom{n}{k} (k) [F(y)]^{k-1} f(y) [1 - F(y)]^{n-k}$$

$$+ \sum_{k=r}^{n-1} \binom{n}{k} [F(y)]^k (n - k) [1 - F(y)]^{n-k-1} [-f(y)]$$

$$+ n[F(y)]^{n-1} f(y).$$

But since

$$\binom{n}{k} (k) = \frac{n!}{(k-1)!(n-k)!} \quad \text{and} \quad \binom{n}{k} (n-k) = \frac{n!}{k!(n-k-1)!},$$

we have that the p.d.f. of Y_r is

$$g(y) = \frac{n!}{(r-1)!(n-r)!} [F(y)]^{r-1} [1 - F(y)]^{n-r} f(y),$$

which is the first term of the first summation in $g(y) = G'(y)$. The remaining terms in $g(y) = G'(y)$ sum to zero because the second term of the first summation (when $k = r + 1$) equals the negative of the first term in the second summation (when $k = r$), and so on. Finally, the last term of the second summation equals the negative of $n[F(y)]^{n-1} f(y)$. To see this clearly, the student is urged to write out a number of terms in these summations (see Exercise 5.5-4).

REMARK. There is one very satisfactory way to construct heuristically the expression for the p.d.f. of Y_r. To do this, we must recall the multinomial probability and then consider the probability element $g(y)(\Delta y)$ of Y_r. If the length Δy is *very* small, $g(y)(\Delta y)$ represents approximately the probability

$$P(y < Y_r \le y + \Delta y).$$

Thus, we want the probability $g(y) \Delta y$ that $(r - 1)$ items fall less than y, $(n - r)$ items are greater than $y + \Delta y$, and one item falls between y and $y + \Delta y$. Recall that the probabilities on a single trial are

$$P(X \le y) = F(y)$$
$$P(X > y + \Delta y) = 1 - F(y + \Delta y) \approx 1 - F(y)$$
$$P(y < X \le y + \Delta y) \approx f(y) \Delta y.$$

Thus, the multinomial probability is approximately

$$g(y) \Delta y = \frac{n!}{(r - 1)!\,1!\,(n - r)!} [F(y)]^{r-1} [1 - F(y)]^{n-r} [f(y) \Delta y].$$

If we divide each member by the length Δy, the formula for $g(y)$ results.

Example 5.5–3. Let $Y_1 < Y_2 < Y_3 < Y_4$ be the order statistics of a random sample of size $n = 4$ from a distribution with the uniform p.d.f. $f(x) = 1$, $0 < x < 1$. Since $F(x) = x$, when $0 \le x < 1$, the p.d.f. of Y_3 is

$$g(y) = \frac{4!}{2!\,1!} y^2(1 - y), \qquad 0 < y < 1.$$

Thus, for illustration,

$$P\left(\frac{1}{3} < Y_3 < \frac{2}{3}\right) = \int_{1/3}^{2/3} 12y^2(1 - y) \, dy$$

$$= [4y^3 - 3y^4]_{1/3}^{2/3} = \frac{13}{27}.$$

Example 5.5–4. Let $Y_1 < Y_2 < \cdots < Y_7$ be the order statistics of a random sample of size $n = 7$ from a distribution with p.d.f. $f(x) = 3(1 - x)^2$, $0 < x < 1$. Compute the probability that the sample median is less than $1 - \sqrt[3]{0.6}$; that is, find $P(Y_4 < 1 - \sqrt[3]{0.6})$. Note that the probability of a single item being less than $1 - \sqrt[3]{0.6}$ is

$$\int_0^{1 - \sqrt[3]{0.6}} 3(1 - x)^2 \, dx = [-(1 - x)^3]_0^{1 - \sqrt[3]{0.6}}$$

$$= 1 - (\sqrt[3]{0.6})^3 = 0.4.$$

Thus,

$$P(Y_4 < 1 - \sqrt[3]{0.6}) = \sum_{k=4}^{7} \binom{7}{k}(0.4)^k(0.6)^{7-k} = 0.2898.$$

Now, in Example 5.5–4, it is easy enough to look up the resulting probability in Appendix Table I. Suppose, however, the sample size is much larger and we wish to know certain probabilities associated with a given order statistic, such as $P(Y_r < y)$. Of course, these probabilities can be expressed in terms of binomial probabilities, which are difficult to evaluate but in turn can be approximated by normal probabilities, provided n is large. This is illustrated in Example 5.5–5.

Example 5.5–5. Let $Y_1 < Y_2 < \cdots < Y_{100}$ be the order statistics of a random sample of size $n = 100$ from a continuous distribution having median $\pi_{0.5} = 68.1$ (that is, the probability that a single item X is less than 68.1 is 1/2). What then is the probability that the 55th order statistic, Y_{55}, is less than $\pi_{0.5} = 68.1$? Since a binomial distribution with $n = 100$ and $p = 1/2$ has mean 50 and standard deviation 5, we have that

$$P(Y_{55} < 68.1) = \sum_{k=55}^{100} \binom{100}{k} \left(\frac{1}{2}\right)^k \left(\frac{1}{2}\right)^{100-k}$$

$$= 1 - \sum_{k=0}^{54} \binom{100}{k} \left(\frac{1}{2}\right)^k \left(\frac{1}{2}\right)^{100-k}$$

$$= 1 - \Phi\left(\frac{54.5 - 50}{5}\right) = 1 - \Phi(0.9)$$

$$= 1 - 0.8159 = 0.1841.$$

Recall that, in Section 4.2, we noted that if X had a distribution function $F(x)$ of the continuous type, then $F(X)$ has a uniform distribution on the interval zero to one. If $Y_1 < Y_2 < \cdots < Y_n$ are the order statistics of a random sample $X_1, X_2, \ldots, X_n$ of size n, then

$$F(Y_1) < F(Y_2) < \cdots < F(Y_n)$$

because F is a nondecreasing function and the probability of an equality is again zero. Note that this last display could be looked upon as an ordering of the mutually independent random variables $F(X_1), F(X_2), \ldots, F(X_n)$, each of which is $U(0, 1)$. That is,

$$Z_1 = F(Y_1) < Z_2 = F(Y_2) < \cdots < Z_n = F(Y_n)$$

can be thought of as the order statistics of a random sample of size n from that uniform distribution. Since the distribution function of $U(0, 1)$ is $G(w) = w$, $0 < w < 1$, the p.d.f. of the rth order statistics $Z_r = F(Y_r)$ is

$$h(z) = \frac{n!}{(r-1)!(n-r)!} z^{r-1}(1-z)^{n-r}, \qquad 0 < z < 1.$$

Of course the mean, $E(Z_r) = E[F(Y_r)]$ of $Z_r = F(Y_r)$, is given by the integral

$$E(Z_r) = \int_0^1 z \frac{n!}{(r-1)!(n-r)!} z^{r-1}(1-z)^{n-r} \, dz.$$

This can be evaluated by integrating by parts several times, but it is easier to obtain the answer if we rewrite it as follows:

$$E(Z_r) = \left(\frac{r}{n+1}\right) \int_0^1 \frac{(n+1)!}{r!(n-r)!} z^r (1-z)^{n-r} \, dz.$$

The integrand in this last expression can be thought of as the p.d.f. of the $(r + 1)$st order statistic of a random sample of size $n + 1$ from a distribution $U(0, 1)$. Hence, the integral in the latter form of $E(Z_r)$ must equal one; thus,

$$E(Z_r) = \frac{r}{n+1}, \qquad r = 1, 2, \ldots, n.$$

There is an extremely interesting interpretation of $Z_r = F(Y_r)$. Note that $F(Y_r)$ is the cumulated probability up to and including Y_r—or, equivalently, the area under $f(x) = F'(x)$ but less than Y_r. Hence, $F(Y_r)$ can be treated as a random area. Since $F(Y_{r-1})$ is also a random area, $F(Y_r) - F(Y_{r-1})$ is the random area under $f(x)$ between Y_{r-1} and Y_r. The expected value of the random area between any two adjacent order statistics is then

$$E[F(Y_r) - F(Y_{r-1})] = E[F(Y_r)] - E[F(Y_{r-1})]$$

$$= \frac{r}{n+1} - \frac{r-1}{n+1} = \frac{1}{n+1}.$$

Also, it is easy to show (see Exercise 5.5–6)

$$E[F(Y_1)] = \frac{1}{n+1} \quad \text{and} \quad E[1 - F(Y_n)] = \frac{1}{n+1}.$$

That is, the order statistics $Y_1 < Y_2 < \cdots < Y_n$ partition the support of X into $n + 1$ parts and thus create $n + 1$ areas under $f(x)$ and above the x axis as depicted in Figure 5.5–1 with $n = 5$. "On the average," the $n + 1$ areas equal $1/(n + 1)$.

If we recall that the $(100p)$th percentile π_p is such that the area under $f(x)$ to the left of π_p is p, the preceding discussion suggests that we let Y_r be an estimator of π_p, where $p = r/(n + 1)$. For this reason, we define the $(100p)th$ *percentile of the sample* as Y_r, where $r = (n + 1)p$. In case $(n + 1)p$ is not an integer, we use a weighted average of the two adjacent order statistics Y_r and Y_{r+1}, where r is the greatest integer $[(n + 1)p]$ in $(n + 1)p$. In particular, the sample median is

$$\begin{cases} Y_{(n+1)/2} & \text{when } n \text{ is odd,} \\[2mm] \dfrac{Y_{n/2} + Y_{(n/2)+1}}{2} & \text{when } n \text{ is even.} \end{cases}$$

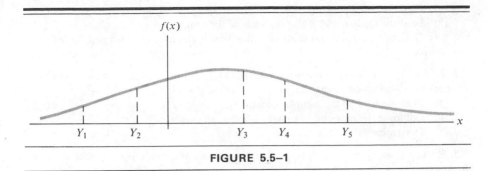

FIGURE 5.5–1

Example 5.5–6. Let the following be the observed order statistics of a random sample of size $n = 7$ from a continuous-type distribution:

$$35.2, \quad 46.1, \quad 59.7, \quad 62.4, \quad 66.7, \quad 71.2, \quad 72.3.$$

Since $(n + 1)/2 = 4$, the sample median is

$$y_4 = 62.4.$$

The 25th and 75th percentiles of the sample (sometimes called the first and third *quartiles*) are, respectively,

$$y_2 = 46.1 \quad \text{and} \quad y_6 = 71.2$$

because $(n + 1)(1/4) = 2$ and $(n + 1)(3/4) = 6$. Since $(n + 1)(0.6) = 4.8$, the 60th percentile of the sample would be taken to be the weighted average

$$(0.2)y_4 + (0.8)y_5 = (0.2)(62.4) + (0.8)(66.7) = 65.84.$$

Note that this is just a linear interpolation between y_4 and y_5.

REMARK. There is not universal agreement upon the definition of the percentiles of the sample. The one given here is satisfactory and will essentially agree with any other definition, particularly when the sample size is large. Certainly, for a large sample size, we want, for illustration, about 60% of the sample items below the 60th percentile (and about 40% above). If you use our definition when n is large, you will see that this is essentially the situation.

===== *Exercises* =====

5.5–1. Let the following $n = 9$ numbers represent an observed random sample from a distribution of the continuous type:

$$13.8, \quad 26.1, \quad 19.6, \quad 17.8, \quad 28.1, \quad 21.7, \quad 14.5, \quad 12.4, \quad 18.3$$

(a) Find the order statistics.

(b) Find the median and 80th percentile of the sample.

(c) Determine the first and third quartiles (that is, the 25th and 75th percentiles) of the sample.

5.5-2. Let $Y_1 < Y_2 < Y_3$ be the order statistics of a random sample of size $n = 3$ from the exponential distribution with p.d.f. $f(x) = e^{-x}$, $0 < x < \infty$.

(a) Find the p.d.f. of the sample median Y_2.

(b) Compute the probability that Y_3 is less than two.

(c) Determine $P(1 < Y_1)$.

5.5-3. Let $Y_1 < Y_2 < \cdots < Y_8$ be the order statistics of a random sample of size $n = 8$ from a continuous-type distribution with 70th percentile $\pi_{0.7} = 27.3$.

(a) Determine $P(Y_7 < 27.3)$.

(b) Find $P(Y_5 < 27.3 < Y_8)$ by recognizing that the event $Y_5 < 27.3 < Y_8$ happens if and only if there are at least five but less than eight "successes." That is, the number of successes must equal 5, 6, or 7.

5.5-4. In the expression for $g(y) = G'(y)$ on page 173, let $n = 6, r = 3$, and write out the summations, showing that the "telescoping" suggested in the text is achieved.

5.5-5. Let $Y_1 < Y_2 < \cdots < Y_{72}$ be the order statistics of a random sample of size $n = 72$ from a distribution of the continuous type having the $100(1/3)$th percentile $\pi_{1/3} = 7.2$.

(a) Approximate the probability that $Y_{20} < \pi_{1/3} = 7.2$.

(b) Find approximately the probability $P(Y_{18} < \pi_{1/3} < Y_{30})$ by first noting that the event $Y_{18} < \pi_{1/3} < Y_{30}$ means that we must have at least 18 but less that 30 "successes."

5.5-6. Let $Z_1 < Z_2 < \cdots < Z_n$ be the order statistics of a random sample of size n from a distribution $U(0, 1)$.

(a) Find the p.d.f. of Z_1 and that of Z_n.

(b) Use the results of (a) to verify that $E(Z_1) = 1/(n + 1)$ and $E(Z_n) = n/(n + 1)$.

5.5-7. Let $Z_1 < Z_2 < \cdots < Z_n$ be the order statistics of a random sample of size n from a distribution $U(0, 1)$.

(a) Show that $E(Z_r^2) = r(r + 1)/(n + 1)(n + 2)$ using a technique similar to that used in determining that $E(Z_r) = r/(n + 1)$.

(b) Find the variance of Z_r.

6

DISTRIBUTION–FREE
CONFIDENCE INTERVALS

6.1 Confidence Intervals for Percentiles

In the earlier chapters, we have alluded to estimating distribution characteristics from certain characteristics of the sample, hoping that the latter would be reasonably close to the former. For example, the sample percentiles can be thought of as estimates of the corresponding distribution percentiles and the sample mean $\bar{x}$ as an estimate of μ, the mean of the distribution. Even the histogram associated with a sample can be taken as an estimate of the p.d.f. of the underlying distribution. Naturally, the question of the accuracy of these estimates arises. In this chapter we will begin to answer that question by finding *confidence intervals* for the unknown characteristics. That is, we will be able to say something about the *error structure* of our estimates. The process of arguing from a part to the whole (in this case, from the sample to the distribution or population) is often called an inference, in particular, a *statistical inference* since it is based on statistics. Moreover, since little will be assumed about the underlying distribution in the construction of these confidence intervals, they are often called *distribution-free* confidence intervals. That is, in this chapter, we will assume little about our models. However, as we will see later, if more is known about the underlying distribution, other techniques can be used to make statistical inferences.

In Section 5.5 we considered order statistics and their distributions. Moreover, it was suggested there that order statistics could be used to estimate the percentiles of an unknown distribution. For example, if $Y_1 < Y_2 < Y_3 < Y_4 < Y_5$ are the order statistics of a random sample of size $n = 5$ from a continuous-type distribution, then the sample median Y_3 could be thought of as an estimator of the distribution median $\pi_{0.5}$. However, we are certain that all of us recognize that, with only a sample of size 5, we would be quite lucky if the observed $Y_3 = y_3$ were very close to $\pi_{0.5}$. Maybe a more realistic position is described by the following.

Instead of simply using Y_3 as an estimator of $\pi_{0.5}$, let us also compute the probability that the random interval (Y_1, Y_5) includes $\pi_{0.5}$. That is, let us determine $P(Y_1 < \pi_{0.5} < Y_5)$. This is extremely easy if we follow the procedure given in Section 5.5. Again say that we have success if an individual item, say X, is less than $\pi_{0.5}$; thus, the probability of success on one of the independent trials is $P(X < \pi_{0.5}) = 0.5$. In order for the first order statistic Y_1 to be less than $\pi_{0.5}$ and the last order statistic Y_5 to be greater than $\pi_{0.5}$, we must have at least one success but not five successes. That is,

$$P(Y_1 < \pi_{0.5} < Y_5) = \sum_{k=1}^{4} \binom{5}{k}\left(\frac{1}{2}\right)^k\left(\frac{1}{2}\right)^{5-k}$$

$$= 1 - \left(\frac{1}{2}\right)^5 - \left(\frac{1}{2}\right)^5 = \frac{15}{16}.$$

So the probability that the random interval (Y_1, Y_5) includes $\pi_{0.5}$ is $15/16 \approx 0.94$. Suppose this random sample is actually taken and the order statistics are observed to equal $y_1 < y_2 < y_3 < y_4 < y_5$, respectively. Then, of course, either the known interval (y_1, y_5) includes $\pi_{0.5}$ or it does not. If $\pi_{0.5}$ is unknown, however, we do not know whether $\pi_{0.5}$ is in the interval (y_1, y_5) or not. Thus, we continue to refer to the probability 0.94 by saying that the known interval (y_1, y_5) is a 94% *confidence interval* for the unknown distribution median $\pi_{0.5}$. In this context, the probability $0.94 = 94\%$ is often called a *confidence coefficient*.

It is interesting to note what happens as the sample size increases. Let $Y_1 < Y_2 < \cdots < Y_n$ be the order statistics of a random sample of size n from a distribution of the continuous type. Thus, $P(Y_1 < \pi_{0.5} < Y_n)$ is the probability that there is at least one "success" but not n successes, where the probability of success on each trial is $P(X < \pi_{0.5}) = 0.5$. Consequently,

$$P(Y_1 < \pi_{0.5} < Y_n) = \sum_{k=1}^{n-1} \binom{n}{k}\left(\frac{1}{2}\right)^k\left(\frac{1}{2}\right)^{n-k}$$

$$= 1 - \left(\frac{1}{2}\right)^n - \left(\frac{1}{2}\right)^n = 1 - \left(\frac{1}{2}\right)^{n-1}.$$

This probability increases as n increases so that the corresponding confidence interval (y_1, y_n) would have a very large confidence coefficient, $1 - (1/2)^{n-1}$. Unfortunately, the interval (y_1, y_n) tends to get wider as n increases and thus we are not "pinning down" $\pi_{0.5}$ very well. However, if we used the interval (y_2, y_{n-1}) or (y_3, y_{n-2}), we would obtain shorter intervals but also smaller confidence coefficients. Let us investigate this possibility further.

With the order statistics $Y_1 < Y_2 < \cdots < Y_n$ associated with a random sample of size n from a continuous-type distribution, consider $P(Y_i < \pi_{0.5} < Y_j)$, where $i < j$. For example, we might want

$$P(Y_2 < \pi_{0.5} < Y_{n-1}) \quad \text{or} \quad P(Y_3 < \pi_{0.5} < Y_{n-2}).$$

On each of the n independent trials we say that we have success if that X is less than $\pi_{0.5}$; thus, the probability of success on each trial is $P(X < \pi_{0.5}) = 0.5$. Consequently, to have the ith order statistic Y_i less than $\pi_{0.5}$ and the jth order statistic greater than $\pi_{0.5}$, we must have at least i successes but fewer than j successes (or else $Y_j < \pi_{0.5}$). That is,

$$P(Y_i < \pi_{0.5} < Y_j) = \sum_{k=i}^{j-1} \binom{n}{k}\left(\frac{1}{2}\right)^k\left(\frac{1}{2}\right)^{n-k} = \gamma.$$

For particular values of n, i, and j, this probability, say γ, which is the sum of probabilities from a binomial distribution, can be calculated directly or approximated by an area under the normal p.d.f. provided n is large enough. The observed interval (y_i, y_j) could then serve as a $(100\gamma)\%$ confidence interval for the unknown distribution median $\pi_{0.5}$.

Example 6.1–1. Let us take $n = 9$ values at random from a large collection of test scores (suppose these were scored so that any number between 0 and 100 could result). Say the observed order statistics are

$$43.8, \quad 55.2, \quad 61.3, \quad 66.7, \quad 72.1, \quad 74.0, \quad 79.5, \quad 83.1, \quad 90.6.$$

Before the sample is drawn we know that

$$P(Y_2 < \pi_{0.5} < Y_8) = \sum_{k=2}^{7} \binom{9}{k}\left(\frac{1}{2}\right)^k\left(\frac{1}{2}\right)^{9-k} = 0.9610,$$

from Appendix Table I. Thus, the confidence interval $(y_2 = 55.2, y_8 = 83.1)$ for $\pi_{0.5}$, the median of all the scores, has a 96.1% confidence coefficient.

So that the student need not compute many of these probabilities, we give in Table 6.1–1 the necessary information for constructing confidence intervals of the form (y_i, y_{n+1-i}) for the unknown $\pi_{0.5}$ for sample sizes $n = 5, 6, \ldots, 20$.

n	$(i, n + 1 - i)$	$P(Y_i < \pi_{0.5} < Y_{n+1-i})$	n	$(i, n + 1 - i)$	$P(Y_i < \pi_{0.5} < Y_{n+1-i})$
5	(1, 5)	0.9376	13	(3, 11)	0.9776
6	(1, 6)	0.9688	14	(4, 11)	0.9426
7	(1, 7)	0.9844	15	(4, 12)	0.9648
8	(2, 7)	0.9296	16	(5, 12)	0.9232
9	(2, 8)	0.9610	17	(5, 13)	0.9510
10	(2, 9)	0.9786	18	(5, 14)	0.9692
11	(3, 9)	0.9346	19	(6, 14)	0.9364
12	(3, 10)	0.9614	20	(6, 15)	0.9586

TABLE 6.1–1

The subscript i is selected so that the confidence coefficient $P(Y_i < \pi_{0.5} < Y_{n+1-i})$ is greater than 90% and as close to 95% as possible.

For sample sizes larger than 20, we approximate those binomial probabilities with areas under the normal curve. To illustrate how good these approximations are, we compute the probability corresponding to $n = 16$ in Table 6.1–1. Here,

$$\gamma = P(Y_5 < \pi_{0.5} < Y_{12}) = \sum_{k=5}^{11} \binom{16}{k}\left(\frac{1}{2}\right)^k\left(\frac{1}{2}\right)^{16-k}$$

$$= P(W = 5, 6, \ldots, 11) = 0.9232,$$

where W is $b(16, 1/2)$. Thus,

$$\gamma = P(4.5 < W < 11.5) = P\left(\frac{4.5 - 8}{2} < \frac{W - 8}{2} < \frac{11.5 - 8}{2}\right)$$

because W has mean $np = 8$ and variance $np(1 - p) = 4$. The standardized variable $Z = (W - 8)/2$ has an approximate normal distribution. Thus,

$$\gamma \approx \Phi\left(\frac{3.5}{2}\right) - \Phi\left(\frac{-3.5}{2}\right) = 2\Phi(1.75) - 1 = 0.9198.$$

This compares very favorably with the probability 0.9232 recorded in Table 6.1–1.

It is quite clear that the argument used to find a confidence interval for the median $\pi_{0.5}$ of a distribution of the continuous type can be applied to any percentile π_p. In this case we say that we have success on a single trial if that X is less than π_p. Thus, the probability of success on each of the independent trials is $P(X < \pi_p) = p$. Accordingly, with $i < j$, $\gamma = P(Y_i < \pi_p < Y_j)$ is the probability that we have at least i successes but no more than j successes. Thus,

$$\gamma = P(Y_i < \pi_p < Y_j) = \sum_{k=i}^{j-1} \binom{n}{k} p^k (1 - p)^{n-k}.$$

Once the sample is observed and the order statistics determined, the known interval (y_i, y_j) could serve as a $(100\gamma)\%$ confidence interval for the unknown distribution percentile π_p.

Example 6.1–2. Let the following numbers represent the order statistics of the $n = 27$ observations obtained in a random sample from a certain population of incomes (measured in hundreds of dollars).

61	80	92	105	129	164
69	83	93	113	141	191
71	84	96	121	143	217
74	86	100	122	156	276
79	87	104			

Say we are interested in estimating the 25th percentile $\pi_{0.25}$ of the population. Of course, since $(n + 1)p = 28(1/4) = 7$, the 7th order statistic, namely, $y_7 = 83$, would be a point estimate of $\pi_{0.25}$. To get a confidence interval for $\pi_{0.25}$ let us move up and down a few order statistics from y_7, say to y_4 and y_{10}. What is the confidence coefficient associated with the interval (y_4, y_{10})? Of course, before the sample is drawn, we had

$$\gamma = P(Y_4 < \pi_{0.25} < Y_{10}) = \sum_{k=4}^{9} \binom{27}{k}(0.25)^k(0.75)^{27-k}$$

$$= P(3.5 < W < 9.5),$$

where W is $b(27, 1/4)$ with mean $27/4 = 6.75$ and variance $81/16$. Hence,

$$\gamma \approx \Phi\left(\frac{9.5 - 6.75}{9/4}\right) - \Phi\left(\frac{3.5 - 6.75}{9/4}\right) = \Phi\left(\frac{11}{9}\right) - \Phi\left(-\frac{13}{9}\right) = 0.8149.$$

Thus, $(y_4 = 74, y_{10} = 87)$ serves as an 81.49% confidence interval for $\pi_{0.25}$. It should be noted that we could choose other intervals, such as $y_3 = 71$ and $y_{11} = 92$, and these would have different confidence coefficients. The persons involved in the study must select the desired confidence coefficient, and then the appropriate order statistics are taken, usually fairly symmetrically about the $(n + 1)p$th order statistic.

Obviously, in these procedures, it is fairly important to be able to determine rather easily the order statistics. One scheme that has some appeal is John Tukey's "stem and leaf" system. It will be illustrated using the 50 GPA's found in Exercise 3.2–1 as summarized in Table 6.1–2. The first number, 3.77, has been recorded as the "leaf" 7 after the "stem" 3.7; the second number, 3.00, as the "leaf" 0 after the "stem" 3.0; . . . ; the last number, 3.30, as the last "leaf" 0 after the "stem" 3.3. Obviously, this does give us a "histogram-type" configuration and thus provides some picture of how those 50 GPA's are distributed. Moreover, with such a display it is easy to find the order statistics; the reader should verify

STEM	LEAVES	STEM	LEAVES
		3.1	08
2.2	80	3.2	9006
2.3	8	3.3	810
2.4	7	3.4	80
2.5	9	3.5	720
2.6	5	3.6	96
2.7	87514	3.7	78
2.8	1353449206	3.8	9
2.9	052	3.9	207
3.0	080	4.0	0

TABLE 6.1–2

that $y_{13} = 2.81$ and $y_{42} = 3.66$. Also, the interval $(y_{22} = 2.90, y_{29} = 3.18)$ serves as a confidence interval for the unknown $\pi_{0.5}$ of the population from which the sample arose. Its confidence coefficient is

$$P(Y_{22} < \pi_{0.5} < Y_{29}) = \sum_{k=22}^{28} \binom{50}{k}\left(\frac{1}{2}\right)^k\left(\frac{1}{2}\right)^{50-k}$$

$$\approx \Phi\left(\frac{28.5 - 25}{5/\sqrt{2}}\right) - \Phi\left(\frac{21.5 - 25}{5/\sqrt{2}}\right)$$

$$= \Phi(0.99) - \Phi(-0.99) = 0.6778.$$

Exercises

6.1–1. Let $Y_1 < Y_2 < Y_3 < Y_4 < Y_5 < Y_6$ be the order statistics of a random sample of size $n = 6$ from a distribution of the continuous type having $(100p)$th percentile π_p. Compute
(a) $P(Y_2 < \pi_{0.5} < Y_5)$, (b) $P(Y_1 < \pi_{0.25} < Y_4)$, (c) $P(Y_4 < \pi_{0.9} < Y_6)$.

6.1–2. Let the following be $n = 12$ test scores, which were selected at random from a large collection of such scores and then ordered:

$$41, \quad 52, \quad 58, \quad 64, \quad 66, \quad 69, \quad 74, \quad 75, \quad 80, \quad 83, \quad 83, \quad 97.$$

(a) Find a 96.14% confidence interval for $\pi_{0.5}$.
(b) The interval $(y_1 = 41, y_7 = 74)$ could serve as a confidence interval for $\pi_{0.3}$. What is its confidence coefficient?

6.1–3. Using the $n = 100$ numbers given in Exercise 3.2–4, use "stem and leaf" to determine y_{11}, y_{33}, and y_{71}. Here let the stems be $0.0, 0.1, 0.2, \ldots, 0.9$. Thus, the first number 0.441 has the "leaf" 41 and the "stem" 0.4. Find (y_i, y_{n+1-i}) so that this confidence interval for $\pi_{0.5}$ has a confidence coefficient about equal to 0.95.

6.1–4. Using the $n = 100$ values given in Exercise 3.2–5, use stems $0.2, 0.3, \ldots, 1.9$ and construct a "stem and leaf" display. Determine an approximate 90% confidence interval for each of $\pi_{0.2}, \pi_{0.5}, \pi_{0.8}$.

6.1–5. A biologist who studies spiders selected a random sample of 20 male green lynx spiders (a spider that does not weave a web but chases and leaps on its prey) and measured in millimeters the lengths of one of the front legs of the 20 spiders. Use the following measurements to construct a confidence interval for $\pi_{0.5}$ that has a confidence coefficient about equal to 0.95.

15.10	13.55	15.75	20.00	15.45
13.60	16.45	14.05	16.95	19.05
16.40	17.05	15.25	16.65	16.25
17.75	15.40	16.80	17.55	19.05

6.1–6. The biologist (Exercise 6.1–5) also selected a random sample of 20 female green lynx spiders and measured in millimeters the lengths of one of their front legs. Use the following data to construct a confidence interval for $\pi_{0.5}$ that has a confidence coefficient about equal to 0.95.

15.85	18.00	11.45	15.60	16.10
18.80	12.85	15.15	13.30	16.65
16.25	16.15	15.25	12.10	16.20
14.80	14.60	17.05	14.15	15.85

6.2 Confidence Intervals for Means

In Section 5.2, we saw that the mean $\overline{X}$ of a random sample of size n from the normal distribution $N(\mu, \sigma^2)$ has the normal distribution $N(\mu, \sigma^2/n)$. This means that the standardized variable $(\overline{X} - \mu)/(\sigma/\sqrt{n})$ is $N(0, 1)$. Moreover, in Section 5.3, we discovered that this same ratio $(\overline{X} - \mu)/(\sigma/\sqrt{n})$ has, provided n is large enough, the approximate normal distribution $N(0, 1)$ even though the underlying distribution is *not* normal. We will use this latter result to construct confidence intervals for unknown means of distributions.

Let us first consider the situation in which the random sample arose from a distribution with unknown mean μ but known variance σ^2. That is, we assume that we know σ^2, the variance of the underlying distribution. Then, for the probability γ and for n sufficiently large, we can find a number z_0 from Appendix Table IV such that

$$P\left(-z_0 \leq \frac{\overline{X} - \mu}{\sigma/\sqrt{n}} \leq z_0\right) \approx \gamma.$$

For example, if $\gamma = 0.95$, then $z_0 = 1.96$, and if $\gamma = 0.90$, then $z_0 = 1.645$. Now recalling that $\sigma > 0$, we see that the following inequalities are equivalent:

$$-z_0 \leq \frac{\overline{X} - \mu}{\sigma/\sqrt{n}} \leq z_0,$$

$$-z_0\left(\frac{\sigma}{\sqrt{n}}\right) \leq \overline{X} - \mu \leq z_0\left(\frac{\sigma}{\sqrt{n}}\right),$$

$$-\overline{X} - z_0\left(\frac{\sigma}{\sqrt{n}}\right) \leq -\mu \leq -\overline{X} + z_0\left(\frac{\sigma}{\sqrt{n}}\right),$$

and

$$\overline{X} + z_0\left(\frac{\sigma}{\sqrt{n}}\right) \geq \mu \geq \overline{X} - z_0\left(\frac{\sigma}{\sqrt{n}}\right).$$

Thus, since the probability of the first of these is approximately γ, the probability of the last must also be approximately γ because the latter is true if and only if the former is true. That is, we have

$$P\left[\overline{X} - z_0\left(\frac{\sigma}{\sqrt{n}}\right) \leq \mu \leq \overline{X} + z_0\left(\frac{\sigma}{\sqrt{n}}\right)\right] \approx \gamma.$$

So the probability that the random interval

$$\left[\overline{X} - z_0\left(\frac{\sigma}{\sqrt{n}}\right), \overline{X} + z_0\left(\frac{\sigma}{\sqrt{n}}\right)\right]$$

includes the unknown mean μ is approximately γ.

Once the sample is observed and the sample mean computed to equal $\bar{x}$, the interval $[\bar{x} - z_0(\sigma/\sqrt{n}), \bar{x} + z_0(\sigma/\sqrt{n})]$ is a known interval. Since the probability that the *random* interval covers μ before the sample was drawn is approximately equal to γ, we now call the computed interval, $\bar{x} \pm z_0(\sigma/\sqrt{n})$ (for brevity), an approximate $(100\gamma)\%$ confidence interval for the unknown mean μ. For illustration, $\bar{x} \pm 1.96(\sigma/\sqrt{n})$ is a 95% confidence interval for μ.

The closeness of the approximate probability γ to the exact probability γ depends on both the underlying distribution and the sample size. When the underlying distribution is unimodal (has only one mode) and continuous, the approximation is usually quite good for even small n, such as $n = 5$. As the underlying distribution becomes "less normal" (that is, badly skewed or discrete), a larger sample size might be required to keep a reasonably accurate approximation. But, in all cases, usually an n of 25 or 30 is quite adequate.

Example 6.2–1. Say it is known that the distribution of salaries of a certain group of persons has a standard deviation about equal to $2000. To make a statistical inference about the unknown mean μ, a random sample of $n = 16$ is taken and the resulting mean is found to be $\bar{x} = \$23,412$. Then

$$23,412 \pm 1.96\left(\frac{2000}{\sqrt{16}}\right) \quad \text{or} \quad [22,432; 24,392]$$

is an approximate 95% confidence interval for μ.

If the random sample arose from a normal distribution, the argument is exactly the same as that above except that now the probability γ is exact for any sample size.

Example 6.2–2. Let us assume that some examination scores have a distribution with mean μ and standard deviation $\sigma = 15$. A random sample of size 25 is taken and it is found that $\bar{x} = 69.2$. Then

$$\bar{x} \pm 1.654\left(\frac{\sigma}{\sqrt{n}}\right) \quad \text{or} \quad 69.2 \pm 1.645\left(\frac{15}{\sqrt{25}}\right) \quad \text{or} \quad [64.265, 74.135]$$

is an approximate 90% confidence interval for μ.

Students frequently raise the question "If you do not know the mean μ, why do you believe that you know the standard deviation σ as in Examples 6.2–1 and 6.2–2?" Obviously, this is an excellent question because in most instances we do *not* know the standard deviation. In some cases, however, we might have an excellent idea about the value of the standard deviation. For illustration, in Example 6.2–2, an experienced teacher might very well have a good notion of what the standard deviation will be on a certain type of examination. But certainly, most of the time, the investigator will not have any more idea about the standard deviation than about the mean—and frequently less. Let us consider how we should proceed under those circumstances.

Assuming that σ^2 is now unknown, let us reconsider the standardized variable

$$\frac{\bar{X} - \mu}{\sigma/\sqrt{n}} = \frac{\bar{X} - \mu}{\sqrt{\sigma^2/n}}.$$

We would like to replace the unknown σ^2 by some estimator that can be calculated from the sample. A natural choice is the sample variance

$$S^2 = \frac{1}{n}\sum_{i=1}^{n}(X_i - \bar{X})^2 = \frac{1}{n}\left[\sum_{i=1}^{n}X_i^2 - n\bar{X}^2\right].$$

Many statisticians, however, prefer to use a slight modification of this statistic, namely one that is an *unbiased* estimator of σ^2. That is, they use one with an expected value σ^2. To see how we must modify S^2 to do this, let us consider

$$E(S^2) = \frac{1}{n}\left[\sum_{i=1}^{n} E(X_i^2) - nE(\overline{X}^2) \right].$$

Recalling that if W is a random variable, then

$$\sigma_W^2 = E(W^2) - \mu_W^2 \qquad \text{or} \qquad E(W^2) = \mu_W^2 + \sigma_W^2,$$

we have

$$E(S^2) = \frac{1}{n}\left[\sum_{i=1}^{n} (\mu^2 + \sigma^2) - n\left(\mu^2 + \frac{\sigma^2}{n} \right) \right]$$

$$= \frac{1}{n} [n\mu^2 + n\sigma^2 - n\mu^2 - \sigma^2]$$

$$= \frac{1}{n}(n - 1)\sigma^2.$$

Hence,

$$E\left(\frac{nS^2}{n - 1} \right) = \sigma^2;$$

that is,

$$\frac{nS^2}{n - 1} = \frac{1}{n - 1} \sum_{i=1}^{n} (X_i - \overline{X})^2$$

is an unbiased estimator of σ^2. Incidentally, this is the reason that so many statisticians define the sample variance with $n - 1$ replacing n. We prefer using n in the definition of S^2 and calling $nS^2/(n - 1)$ an unbiased estimator of σ^2.

If this unbiased estimator is substituted for σ^2 in $(\overline{X} - \mu)/\sqrt{\sigma^2/n}$, we obtain

$$\frac{\overline{X} - \mu}{\sqrt{[nS^2/(n - 1)]/n}} = \frac{\overline{X} - \mu}{S/\sqrt{n - 1}}.$$

This ratio plays an extremely important role in statistical inference as will be seen in Section 7.4. At the moment, we accept the fact that if n is large, say 30 or greater, this ratio still has the approximate normal distribution $N(0, 1)$. This statement is true whether or not the underlying distribution is normal. However, if the underlying distribution is badly skewed or contaminated with occasional outliers, most statisticians would prefer to have a sample size somewhat larger than 30 (or even find confidence intervals for the distribution median $\pi_{0.5}$ as described in Section 6.1).

Example 6.2–3. Lake Macatawa, an inlet lake on the east side of Lake Michigan, is divided into an east basin and a west basin. To measure the effect on the lake of salting city streets in the winter, students took 32 samples of water from the west basin and measured the amount of sodium in parts per million in order to make a statistical inference about the unknown mean μ. They obtained the following data:

13.0	18.5	16.4	14.8	19.4
17.3	23.2	24.9	20.8	19.3
18.8	23.1	15.2	19.9	19.1
18.1	25.1	16.8	20.4	17.4
25.2	23.1	15.3	19.4	16.0
21.7	15.2	21.3	21.5	16.8
15.6	17.6			

For these data $\bar{x} = 19.07$ and $s^2 = 10.27$. Thus, an approximate 95% confidence interval for μ is

$$\bar{x} \pm 1.96\left(\frac{s}{\sqrt{n-1}}\right) \quad \text{or} \quad 19.07 \pm 1.96\sqrt{\frac{10.27}{31}} \quad \text{or} \quad [17.94, 20.20].$$

In Sections 7.2–7.4 some of the concepts discussed in this section are reviewed. In addition, you will find there discussions of confidence intervals for variances, ratio of two variances, and differences of two means. In particular, the importance of the ratio $(\bar{X} - \mu)/(S/\sqrt{n-1})$ is highlighted in the discussion of confidence intervals for μ with small values of the sample size n.

===== *Exercises* =====

6.2–1. A pet store sells gerbil food in "2-pound" bags that are weighed on the platform of an old 25-pound scale. Suppose it is known that the standard deviation of weights is $\sigma = 0.12$ pound. If a sample of $n = 16$ bags of gerbil food were weighed carefully in a laboratory and the average weight was $\bar{x} = 2.09$ pounds, find a 95% confidence interval for μ, the mean weight of gerbil food in the "2-pound" bags sold by the pet store.

6.2–2. Students took $n = 35$ samples of water from the east basin of Lake Macatawa (see Example 6.2–3) and measured the amount of sodium in parts per million. For their data they calculated $\bar{x} = 24.11$ and $s^2 = 24.44$. Find an approximate 90% confidence interval for μ, the mean of the amount of sodium in parts per million.

6.2–3.   As a clue to the amount of organic waste in Lake Macatawa (see Example 6.2–3), a count was made of the number of bacteria colonies in 100 milliliters of water. The number of colonies, in hundreds, for $n = 30$ samples of water from the east basin yielded

93	140	8	120	3	120
33	70	91	61	7	100
19	98	110	23	14	94
57	9	66	53	28	76
58	9	73	49	37	92

Find a 90% confidence interval for μ, the mean number of colonies in 100 milliliters of water.

6.2–4.   To determine whether bacteria count was lower in the west basin of Lake Macatawa (Exercise 6.2–3) than in the east basin, $n = 37$ samples of water were taken from the west basin and the number of bacteria colonies in 100 milliliters of water was counted. The sample mean and sample standard deviation were $\bar{x} = 11.95$ and $s = 11.80$, measured in hundreds of colonies. Find a 95% confidence interval for the mean number of colonies in 100 milliliters of water.

6.2–5.   A sample of $n = 35$ welfare clients was taken in order to make a statistical inference about medical expenses for all welfare clients during a given six-month period. Use the following data to find a 95% confidence interval for the mean μ of medical expenses.

24.74	67.46	12.80	20.16	0.00
55.96	0.00	2.40	29.04	193.24
19.12	0.00	67.25	65.68	108.44
107.26	324.03	0.00	0.00	11.57
28.81	0.00	4.60	22.95	0.00
359.61	53.18	675.29	105.47	70.64
0.00	0.00	13.38	132.60	11.99

Compare this interval to an approximate 95% confidence interval for $\pi_{0.5}$ using the method given in Section 6.1.

6.2–6.   Find a 98% confidence interval for the mean μ of the Scholastic Aptitude Test scores in mathematics for a large freshman class if a sample of $n = 37$ students had a sample mean of $\bar{x} = 536.86$ and a sample standard deviation of $s = 93.73$.

6.3 Confidence Intervals for Percentages

In Sections 6.1 and 6.2 we found confidence intervals for the percentiles and the mean of a distribution using the order statistics, the mean, and the variance of the sample. Since little was assumed about the underlying distribution; we call this distribution-free statistical inference. Another important distribution-free inference can be made by recalling, from Chapter 3, that the histogram is also an excellent description of how the items of a sample are distributed. We might naturally inquire about the accuracy of those relative frequencies (or percentages) associated with the various classes. For illustration, in Example 3.2–2 concerning $n = 200$ weights, we found that the relative frequency of the class interval (172.5, 184.5) was 0.18, that is, 18 %. If we think of this collection of 200 weights as a random sample observed from a larger population of weights, how close is 18 % to the true percentage of weights in that class interval throughout the entire population?

In considering this problem, we generalize it somewhat by treating the class interval (172.5, 184.5) as "success." That is, there is some true probability of success p, namely the fraction of the population in that interval. Let Y equal the frequency of measurements in the interval out of the n observations so that, under the assumptions of independence and constant probability p, Y has the binomial distribution $b(n, p)$. Thus, the problem is to determine the accuracy of the relative frequency Y/n as an estimator of p. We solve this by finding for the unknown p a confidence interval based on Y/n.

In Section 5.4, we noted that

$$\frac{Y - np}{\sqrt{np(1 - p)}} = \frac{(Y/n) - p}{\sqrt{p(1 - p)/n}}$$

has an approximate normal distribution $N(0, 1)$, provided n is large enough. This means that, for a given probability γ, we can find a z_0 from Appendix Table IV such that

$$P\left[-z_0 \leq \frac{(Y/n) - p}{\sqrt{p(1 - p)/n}} \leq z_0\right] \approx \gamma.$$

The inequality

$$-z_0 \leq \frac{(Y/n) - p}{\sqrt{p(1 - p)/n}} \leq z_0$$

is equivalent to

$$K(p) = \left(\frac{Y}{n} - p\right)^2 - z_0^2 p\left(\frac{1 - p}{n}\right) \leq 0.$$

But $K(p)$ is a quadratic expression in p, and it is easy to find those p values for which $K(p) \leq 0$ by determining the two zeros of $K(p)$, say $p_1(Y/n)$ and $p_2(Y/n)$. These zeros have been denoted as functions of Y/n, which they are (as well as of z_0), and for $\gamma = 0.9544$ and $z_0 = 2$ the reader is asked to find $p_1(Y/n)$ and $p_2(Y/n)$ in Exercise 6.3–1. There it is clear that the preceding inequalities are equivalent to

$$p_1\left(\frac{Y}{n}\right) \leq p \leq p_2\left(\frac{Y}{n}\right)$$

and thus

$$P\left[p_1\left(\frac{Y}{n}\right) \leq p \leq p_2\left(\frac{Y}{n}\right)\right] \approx \gamma.$$

The observed value of Y/n, say y/n, then provides an approximate $(100\gamma)\%$ confidence interval $[p_1(y/n), p_2(y/n)]$ for p.

Rather than proceed by solving the inequality $K(p) \leq 0$, most statisticians understand better the situation by making yet another additional approximation, namely, replacing p by Y/n in $p(1-p)/n$. That is, it is still approximately true, if n is large enough, that

$$P\left[-z_0 \leq \frac{(Y/n) - p}{\sqrt{(Y/n)(1 - Y/n)/n}} \leq z_0\right] \approx \gamma.$$

This probability statement can easily be written as

$$P\left[\frac{Y}{n} - z_0\sqrt{\frac{(Y/n)(1 - Y/n)}{n}} \leq p \leq \frac{Y}{n} + z_0\sqrt{\frac{(Y/n)(1 - Y/n)}{n}}\right] \approx \gamma.$$

Thus, for large n, if the observed Y equals y, the interval

$$\left[\frac{y}{n} - z_0\sqrt{\frac{(y/n)(1 - y/n)}{n}}, \frac{y}{n} + z_0\sqrt{\frac{(y/n)(1 - y/n)}{n}}\right]$$

serves as an approximate $(100\gamma)\%$ confidence interval for p. Frequently, this is written as

$$\frac{y}{n} \pm z_0\sqrt{\frac{(y/n)(1 - y/n)}{n}},$$

for brevity. This clearly notes, as does $\bar{x} \pm z_0(s/\sqrt{n-1})$ in Section 6.2, the reliability of the estimate y/n, namely, we are $(100\gamma)\%$ confident that p is within $z_0\sqrt{(y/n)(1 - y/n)/n}$ of y/n.

Example 6.3–1. Let us return to the example of the histogram with $n = 200$ and $y/n = 0.18$. If $\gamma = 0.95$ so that $z_0 = 1.96$, then

$$0.18 \pm 1.96\sqrt{\frac{(0.18)(0.82)}{200}}$$

serves as an approximate 95% confidence interval for the true fraction p. That is, $[0.127, 0.233]$, which is the same as $[12.7\%, 23.3\%]$, is an approximate 95% confidence interval for the percentage of weights of the entire population in the interval $(172.5, 184.5)$.

Example 6.3–2. In a certain political campaign, one candidate has a poll taken at random among the voting population. The results are $n = 112$ and $y = 59$. Even though $y/n = 59/112 = 0.527$, should the candidate feel very confident of winning? An approximate 95% confidence interval for the fraction p of the voting population for this candidate is

$$0.527 \pm 1.96\sqrt{\frac{(0.527)(0.473)}{112}}$$

or, equivalently, $[0.435, 0.619]$. Thus, there is a good possibility that p is less than 50%, and the candidate should certainly take this into account in campaigning.

Frequently, there are two (or more) possible independent ways of performing an experiment; suppose these have probabilities of success p_1 and p_2, respectively. Let n_1 and n_2 be the number of independent trials associated with these two methods, and let us say they result in Y_1 and Y_2 successes, respectively. In order to make a statistical inference about the difference $p_1 - p_2$, we proceed as follows.

Since the independent random variables Y_1/n_1 and Y_2/n_2 have respective means p_1 and p_2 and variances $p_1(1 - p_1)/n_1$ and $p_2(1 - p_2)/n_2$, we know from Section 5.2 that the difference $Y_1/n_1 - Y_2/n_2$ must have mean $p_1 - p_2$ and variance

$$\frac{p_1(1 - p_1)}{n_1} + \frac{p_2(1 - p_2)}{n_2}.$$

(Recall that the variances are added to get the variance of a difference of two independent random variables.) Moreover, the fact that Y_1/n_1 and Y_2/n_2 have approximate normal distributions would suggest that the difference

$$\frac{Y_1}{n_1} - \frac{Y_2}{n_2}$$

would have an approximate normal distribution with the above mean and variance (see Exercise 5.2–5). That is,

$$\frac{(Y_1/n_1) - (Y_2/n_2) - (p_1 - p_2)}{\sqrt{p_1(1 - p_1)/n_1 + p_2(1 - p_2)/n_2}}$$

has an approximate normal distribution $N(0, 1)$. If we now replace p_1 and p_2 in the denominator of this ratio by Y_1/n_1 and Y_2/n_2, respectively, it is still true, for large enough n_1 and n_2, that the new ratio will be approximately $N(0, 1)$. Thus, for a given γ, we can find z_0 from Appendix Table IV such that

$$P\left[-z_0 \leq \frac{(Y_1/n_1) - (Y_2/n_2) - (p_1 - p_2)}{\sqrt{(Y_1/n_1)(1 - Y_1/n_1)/n_1 + (Y_2/n_2)(1 - Y_2/n_2)/n_2}} \leq z_0\right] \approx \gamma.$$

It is easy to solve this to obtain, once Y_1 and Y_2 are observed to be y_1 and y_2, an approximate $(100\gamma)\%$ confidence interval

$$\frac{y_1}{n_1} - \frac{y_2}{n_2} \pm z_0 \sqrt{\frac{(y_1/n_1)(1 - y_1/n_1)}{n_1} + \frac{(y_2/n_2)(1 - y_2/n_2)}{n_2}}$$

for the unknown difference $p_1 - p_2$. Note again how this form indicates the reliability of the estimate $y_1/n_1 - y_2/n_2$ of the difference $p_1 - p_2$.

Example 6.3–3. Two detergents were tested for their ability to remove stains of a certain type. An inspector judged the first one to be successful on 63 out of 91 independent trials and the second one to be successful on 42 out of 79 independent trials. The respective relative frequencies of success are 0.692 and 0.532. An approximate 90% confidence interval for the difference $p_1 - p_2$ of the two detergents is

$$0.692 - 0.532 \pm 1.645 \sqrt{\frac{(0.692)(0.308)}{91} + \frac{(0.532)(0.468)}{79}},$$

or, equivalently, [0.038, 0.282]. Accordingly, it seems that the first detergent is definitely better than the second one for removing these stains.

═══ *Exercises* ═══

6.3–1. In the notation of this section solve $K(p) \leq 0$ when $\gamma = 0.9544$ and $z_0 = 2$ by showing that the zeros of $K(p)$ are $[Y + 2 \pm 2\sqrt{Y(n - Y)/n + 1}]/(n + 4)$.

6.3–2. In the past, 10% of all automobile rear view mirrors manufactured by a particular company were defective. After a new profit-sharing plan was developed, a sample of $n = 100$ mirrors was selected at random and $y = 5$ of these mirrors were defective. Find an approximate 95% confidence interval for p, the new proportion of defective mirrors.

6.3–3. In order to estimate the proportion p of beet seeds in a large bin that will germinate, $n = 200$ seeds were selected at random and planted. Of these 200 seeds, $y = 168$ germinated.
(a) Find a 95% confidence interval for p.
(b) Would the seed company be justified in claiming that at least 80% of their seeds will germinate?

6.3–4. A student loaded a die by drilling holes in two spots and filling each hole with a steel rod. To estimate the probability p of rolling a four with this loaded die, the student rolled the die $n = 600$ times and observed a four $y = 87$ times.
(a) Construct an approximate 90% confidence interval for p.
(b) Was the student successful in decreasing the probability of rolling a four?

6.3–5. A vendor was interested in determining whether a new, more expensive brand of coffee was preferred over the present brand. If $y = 53$ out of a random sample of $n = 90$ customers prefer the new brand, give a 95% confidence interval for p, the percentage of customers that prefer the new brand.

6.3–6. In order to estimate the percentage of a large class of college freshmen that had high school GPA's from 3.2 to 3.6 inclusive, a sample of $n = 50$ students was taken and $y = 9$ students fell in this class. Give a 95% confidence interval for the percentage of this freshman class having a high school GPA of 3.2 to 3.6.

6.3–7. A tire dealer estimates that 9 out of 100 belted tires of Brand A are defective and 5 out of 100 tires of Brand B are defective before the tread is worn. Use 9/100 and 5/100 as estimates of p_A and p_B, respectively, to find an approximate 90% confidence interval for the difference of p_A and p_B, the probabilities of defective tires for each of the two brands.

6.3–8. We shall compare the percentages, say p_1 and p_2, of students in the classes of 1973 and 1979 at a particular college who received SAT verbal scores greater than 640. Such a score was received by 16 out of a random sample of 100 students and by 6 out of a random sample of 50 students from the classes of 1973 and 1979, respectively. Use these data to construct a 90% confidence interval for $p_1 - p_2$.

6.4 Sample Size

In statistical consulting, the first question frequently asked is "How large should the sample size be?" In order to convince the inquirer that the answer will depend upon the variation associated with the random variable under observation, the statistician could correctly respond, "Only one item is needed, provided the standard deviation of the distribution is zero." That is, if σ equals zero, then the value of that one item would necessarily equal the unknown mean of the distribution. This, of course, is an extreme case and one that is not met in practice; however, it should help convince persons that the smaller the variance, the smaller the sample size needed to achieve a given degree of accuracy. This will become clearer as we consider several examples. Let us begin with a problem that involves a statistical inference about the unknown mean of a distribution.

Example 6.4–1. Suppose we wish to evaluate a new method of teaching an introductory mathematics course, say by programmed materials. At the end of the course, the evaluation will be made on the basis of scores of the participating students on some standard test. We desire to determine the number n of students who are to be selected at random from a larger group of students to take the course taught by this new method. Since the programmed materials are relatively expensive, assume that we cannot afford to let all the students take the course this new way. In addition, we might have reason to question the value of this approach and hence not want to expose every student to this procedure. So let us find the sample size n such that we are fairly confident that $\bar{x} \pm 1$ contains the unknown mean μ. From past experience assume that we know that the standard deviation associated with this type of test is about 15. Accordingly, using the fact that $\bar{X}$ is approximately $N(\mu, \sigma^2/n)$, we see that the interval given by $\bar{x} \pm 2(15/\sqrt{n})$ will serve as an approximate 95.44% confidence interval for μ. That is, we want

$$2\left(\frac{15}{\sqrt{n}}\right) = 1$$

or, equivalently,

$$\sqrt{n} = 30 \qquad \text{and thus} \qquad n = 900.$$

It is quite likely that it had not been anticipated that as many as 900 students would be needed in this study. If that is the case, the statistician must discuss, with those involved in the experiment, whether or not the accuracy and the confidence level could be relaxed some. For illustration, rather than requiring $\bar{x} \pm 1$ to be a 95.44% confidence interval for μ, possibly $\bar{x} \pm 2$ would be a satisfactory 80% one. If this modification is acceptable, we now have

$$1.282\left(\frac{15}{\sqrt{n}}\right) = 2$$

or, equivalently,

$$\sqrt{n} = 9.615 \quad \text{and} \quad n \approx 92.4.$$

Since n must be an integer, we would use 92 or 93 in practice. Most likely, the persons involved in this project would find this a more reasonable sample size. Of course, any sample size greater than 92 could be used. Then either the length of the confidence interval could be decreased from that of $\bar{x} \pm 2$ or the confidence coefficient could be increased from 80% or a combination of both. Also, since there might be some question of whether the standard deviation σ actually equals 15, the sample standard deviation s would no doubt be used in the construction of the interval. For example, suppose that the sample characteristics observed are

$$n = 145, \quad \bar{x} = 77.2, \quad s = 13.2,$$

then

$$\bar{x} \pm \frac{1.282s}{\sqrt{n-1}} \quad \text{or} \quad 77.2 \pm 1.41$$

provides an 80% confidence interval for μ.

In general, if we want the $(100\gamma)\% = 100(1-\alpha)\%$ confidence interval for μ to be no longer than that given by $\bar{x} \pm \varepsilon$, the sample size n is the solution of

$$\varepsilon = \frac{z_0\sigma}{\sqrt{n}}, \quad \text{where} \quad \Phi(z_0) = 1 - \frac{\alpha}{2}.$$

That is,

$$n = \frac{z_0^2\sigma^2}{\varepsilon^2},$$

where it is assumed that σ^2 is known.

The type of statistic we see most often in newspapers and magazines is an estimate of a proportion p. We might, for example, want to know the percentage of the labor force that is unemployed or the percentage of voters favoring a certain candidate. Sometimes extremely important decisions are made on the basis of these estimates. If this is the case, we would most certainly desire short confidence intervals for p with large confidence coefficients. We recognize that these conditions will require a large sample size. On the other hand, if the fraction p being estimated is not too important, an estimate associated with a longer confidence interval with a smaller confidence coefficient is satisfactory; and thus a smaller sample size can be used.

Example 6.4-2. Suppose we know that the unemployment rate has been about 5 % (0.05). However, we wish to update our estimate in order to make an important decision about the national economic policy. Accordingly, let us say we wish to be 99 % confident that the new estimate of p is within 0.001 of the true p. If we assume Bernoulli trials (an assumption that might be questioned), the relative frequency y/n, based upon a large sample size n, provides the 99 % confidence interval

$$\frac{y}{n} \pm 2.576 \sqrt{\frac{(y/n)(1 - y/n)}{n}}.$$

Although we do not know y/n exactly before sampling, we do know, since y/n will be near 0.05, that

$$2.576 \sqrt{\frac{(y/n)(1 - y/n)}{n}} \approx 2.576 \sqrt{\frac{(0.05)(0.95)}{n}}$$

and we want this number to equal 0.001. That is,

$$2.576 \sqrt{\frac{(0.05)(0.95)}{n}} = 0.001$$

or, equivalently,

$$\sqrt{n} = 2576 \sqrt{0.0475} \quad \text{and} \quad n \approx 315{,}199.$$

That is, under our assumptions, such a sample size is needed in order to achieve the reliability and the accuracy desired.

We hope from the preceding examples that the student will recognize how important it is to know the sample size (or length of the confidence interval and confidence coefficient) before he or she can place much weight on a statement such as 51 % of the voters seem to favor candidate A, 46 % favor candidate B, and 3 % are undecided. Is this statement based upon a sample of 100 or 2000 or 10,000 voters? If we assume Bernoulli trials, the approximate 95.44 % confidence intervals for the fraction of voters favoring candidate A in these cases are, respectively, [0.41, 0.61], [0.49, 0.53], and [0.50, 0.52]. Quite obviously, the first interval, with $n = 100$, does not assure candidate A that at least half the voters support him or her, while the interval with $n = 10{,}000$ is more convincing.

Thus we see that there must be some advanced planning if the resulting estimates are to achieve the anticipated accuracy and reliability. Suppose we want an estimate of p that is within ε of the unknown p with

$$(100\gamma)\% = 100(1 - \alpha)\%$$

confidence. If it is known that p is about equal to p^*, the necessary sample size n is the solution of

$$\varepsilon = \frac{z_0\sqrt{p^*(1 - p^*)}}{\sqrt{n}}, \qquad \text{where} \quad \Phi(z_0) = 1 - \frac{\alpha}{2}.$$

That is,

$$n = \frac{z_0^2 p^*(1 - p^*)}{\varepsilon^2}.$$

It is often true, however, that we do not have a strong prior idea about p, as we did in Example 6.4–2 about the rate of unemployment. It is interesting to observe that no matter what value p takes between zero and one, it is always true that $p(1 - p) \leq 1/4$. Hence,

$$\sqrt{\frac{p(1 - p)}{n}} \leq \frac{1}{2\sqrt{n}}.$$

Thus, if we want the $(100\gamma)\% = 100(1 - \alpha)\%$ confidence interval for p to be no longer than that given by $y/n \pm \varepsilon$, then the equation

$$\varepsilon = \frac{z_0}{2\sqrt{n}}, \qquad \text{where} \quad \Phi(z_0) = 1 - \frac{\alpha}{2},$$

has a solution for n that provides this protection, namely,

$$n = \frac{z_0^2}{4\varepsilon^2}.$$

Example 6.4–3. A possible gubernatorial candidate wants to assess initial support among the voters before announcing his or her candidacy. If the fraction p of voters who are favorable, without any advance publicity, is around 0.15, the candidate will enter the race. From a poll of n voters selected at random, the candidate would like the estimate y/n to be within 0.05 of p. That is, the decision will be based on a 95.44% confidence interval of the form $y/n \pm 0.05$. Since the candidate has no idea about the magnitude of p, a consulting statistician formulates the equation

$$0.05 = \frac{1.96}{2\sqrt{n}}$$

or, equivalently, $\sqrt{n} = 19.6$. That is, the sample size should be around 385 to achieve the desired reliability and accuracy. Suppose, then, that $n = 400$ voters around the state were selected at random and interviewed and that $y/n = 0.20$ was observed. Accordingly, an approximate 95.44% confidence interval for p is

$$0.20 \pm 1.96 \sqrt{\frac{(0.20)(0.80)}{400}}$$

or, equivalently, $[0.16, 0.24]$. On the basis of this sample, the candidate decides to run for office.

═══ *Exercises* ═══

6.4–1. Let $\overline{X}$ be the mean of a random sample of size n taken from a distribution with a standard deviation $\sigma = 7$. Find n such that $P(\overline{X} - 0.8 \leq \mu \leq \overline{X} + 0.8) = 0.90$, approximately.

6.4–2. Let X equal the amount of "tar" per cigarette for a new brand. Find the number of cigarettes n that must be tested so that $\bar{x} \pm 0.10$ is a 99% confidence interval for $\mu = E(X)$. Assume that $\sigma \approx 0.3$.

6.4–3. Let Y/n be the relative frequency of success in n independent trials, each with probability p of success. If p is unknown, but in the neighborhood of 0.30, determine n such that $P(Y/n - 0.03 \leq p < Y/n + 0.03) = 0.80$, approximately.

6.4–4. Let y/n be the observed relative frequency of success in n Bernoulli trials with unknown probability p of success. What should n equal so that $y/n \pm 0.02$ is an approximate confidence interval for p, with a confidence coefficient equal to at least 90%.

6.4–5. Some dentists were interested in studying the fusion of embryonic rat palates by using a standard transplantation technique. When no treatment is used, the probability of fusion approximately equals 0.89. They would like to estimate p, the probability of fusion, when vitamin A is lacking.
   (a) How large a sample n of rat embryos is needed for $y/n \pm 0.10$ to be a 95% confidence interval for p?
   (b) If $y = 44$ out of $n = 60$ palates showed fusion, give a 95% confidence interval for p.

6.4–6. If Y_1/n and Y_2/n are the respective relative frequencies of successes associated with the two independent binomial distributions $b(n, p_1)$ and $b(n, p_2)$, compute n such that the approximate probability that the random interval $Y_1/n - Y_2/n \pm 0.05$ covers $p_1 - p_2$ is at least 0.80.

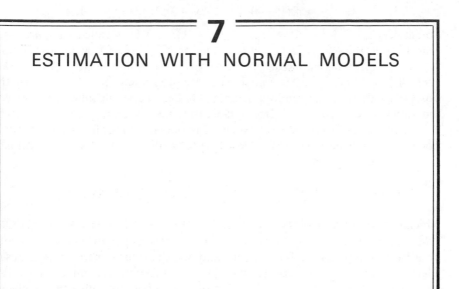

7
ESTIMATION WITH NORMAL MODELS

7.1 Maximum Likelihood Estimation

In the last chapter, we found point and interval estimates for certain distributional characteristics using intuitive concepts and assuming very little about the underlying distribution. In fact, some statisticians would classify many of the statistical inferences in Chapter 6 as distribution free, that is, free of strong distributional assumptions. In this chapter we will assume that we know more about the underlying model; in particular, we know the form of the probability density functions except for one or more unknown parameters. In most, but not all cases, the underlying models will be normal distributions.

Let us begin with an illustration actually considered in Chapter 6, namely, the estimate of the probability of success in a sequence of Bernoulli trials. Recall that the probability of success can be thought of as that percentage of the population with a certain characteristic and, intuitively, we estimated this probability (or percentage) using the relative frequency of the corresponding event in n independent trials. This example provides a more formal reason why the relative frequency is a good estimator.

Suppose that X is $b(1, p)$ so that the p.d.f. of X is

$$f(x; p) = p^x(1 - p)^{1 - x}, \qquad x = 0, 1, \quad 0 \le p \le 1.$$

Sometimes we write $p \in \Omega = \{p: \ 0 \le p \le 1\}$, and Ω is called the *parameter space*, that is, the space of all possible values of the parameter. A random sample $X_1, X_2, \ldots, X_n$ is taken, and the problem is to find an estimator $u(X_1, X_2, \ldots, X_n)$ such that $u(x_1, x_2, \ldots, x_n)$ is a good estimate of p, where $x_1, x_2, \ldots, x_n$ are the observed values of the random sample. That is, we would like the observed function value $u(x_1, x_2, \ldots, x_n)$ to be close to p, and $u(x_1, x_2, \ldots, x_n)$ is said to be a *point estimate* of p. Thus, we are suggesting only one value for the unknown p, not an interval of possible values. Now the probability that $X_1, X_2, \ldots, X_n$ takes these particular values is

$$P(X_1 = x_1, \ldots, X_n = x_n) = \prod_{i=1}^{n} p^{x_i}(1 - p)^{1 - x_i} = p^{\Sigma x_i}(1 - p)^{n - \Sigma x_i},$$

which is the joint p.d.f. of $X_1, X_2, \ldots, X_n$ evaluated at the observed values. One reasonable way to proceed toward a good estimate of p is to regard this probability (or joint p.d.f.) as a function of p and find the value of p that maximizes it. That is, find the p value most likely to have produced these sample values. The joint p.d.f., when regarded as a function of p, is frequently called the *likelihood function* $L(p; x_1, \ldots, x_n)$. Thus, here

$$L(p; x_1, x_2, \ldots, x_n) = f(x_1; p)f(x_2; p) \cdots f(x_n; p)$$
$$= p^{\Sigma x_i}(1 - p)^{n - \Sigma x_i}, \qquad 0 \le p \le 1.$$

Instead of finding the value of p that maximizes $L(p; x_1, x_2, \ldots, x_n) = L(p)$, it is easier to find the p value that maximizes the natural logarithm of $L(p)$. Of course, the p that maximizes $\ln L(p)$ also maximizes $L(p)$. In this case, we have

$$\ln L(p) = \left(\sum_{i=1}^{n} x_i \right) \ln p + \left(n - \sum_{i=1}^{n} x_i \right) \ln(1 - p).$$

To find the maximum, when $0 < p < 1$, consider the first derivative

$$\frac{d[\ln L(p)]}{dp} = \left(\sum_{i=1}^{n} x_i \right)\left(\frac{1}{p} \right) + \left(n - \sum_{i=1}^{n} x_i \right)\left(\frac{-1}{1 - p} \right) = 0.$$

The solution for p is found by solving the equivalent equation

$$(1 - p)\left(\sum_{i=1}^{n} x_i \right) - p\left(n - \sum_{i=1}^{n} x_i \right) = 0$$

to obtain $p = (1/n) \sum x_i = \bar{x}$. By considering the second derivative of $\ln L(p)$, it is easy to see that $\bar{x}$ actually maximizes $\ln L(p)$. The corresponding statistic, namely $\sum X_i/n = \bar{X}$, is called the *maximum likelihood estimator* and is denoted by $\hat{p}$; that is,

$$\hat{p} = \frac{1}{n} \sum_{i=1}^{n} X_i = \bar{X}.$$

Recall that $X_i, i = 1, 2, \ldots, n$, is either zero or one, corresponding to "failure" or "success," respectively. Accordingly, the summation $\sum X_i$ equals the number of successes in n independent trials and $\hat{p} = (1/n) \sum X_i = \bar{X}$ is then the relative frequency of success in these n trials. Hopefully, the observed relative frequency of success, namely $(1/n) \sum x_i = \bar{x}$, would be close to p, which is the probability of success on each single trial, particularly if n is large. Thus, from our understanding of the relative frequency approach to probability, it appears that $\hat{p}$ is an appropriate estimator in this case.

Motivated by the preceding illustration, we present the formal definition of maximum likelihood estimators. This definition is used in the discrete and continuous cases.

Let $X_1, X_2, \ldots, X_n$ be a random sample from a distribution with p.d.f. $f(x; \theta_1, \theta_2, \ldots, \theta_m)$, which depends upon one or more unknown parameters $\theta_1, \theta_2, \ldots, \theta_m$. Suppose $(\theta_1, \theta_2, \ldots, \theta_m)$ is restricted to a given parameter space Ω. Then the joint p.d.f. of $X_1, X_2, \ldots, X_n$, namely,

$$L(\theta_1, \theta_2, \ldots, \theta_m) = f(x_1; \theta_1, \ldots, \theta_m) f(x_2; \theta_1, \ldots, \theta_m)$$

$$\cdots f(x_n; \theta_1, \ldots, \theta_m), \qquad (\theta_1, \theta_2, \ldots, \theta_m) \in \Omega,$$

when regarded as a function of $\theta_1, \theta_2, \ldots, \theta_m$, is called the *likelihood function*. Say

$$[u_1(x_1, \ldots, x_n), u_2(x_1, \ldots, x_n), \ldots, u_m(x_1, \ldots, x_n)]$$

is that m-tuple in Ω that maximizes $L(\theta_1, \theta_2, \ldots, \theta_m)$. Then

$$\hat{\theta}_1 = u_1(X_1, \ldots, X_n),$$
$$\hat{\theta}_2 = u_2(X_1, \ldots, X_n),$$
$$\vdots$$
$$\hat{\theta}_m = u_m(X_1, \ldots, X_n)$$

are *maximum likelihood estimators* of $\theta_1, \theta_2, \ldots, \theta_m$, respectively; and the corresponding observed values of these statistics, namely,

$$u_1(x_1, \ldots, x_n), u_2(x_1, \ldots, x_n), \ldots, u_m(x_1, \ldots, x_n),$$

are called *maximum likelihood estimates*. In many practical cases, these estimators (and estimates) are unique.

Some additional examples will help clarify these definitions.

Example 7.1–1. Let $X_1, X_2, \ldots, X_n$ be a random sample from the exponential distribution with p.d.f.

$$f(x; \theta) = \frac{1}{\theta} e^{-x/\theta}, \qquad 0 < x < \infty, \quad \theta \in \Omega = \{\theta: \ 0 < \theta < \infty\}.$$

The likelihood function is given by

$$L(\theta) = L(\theta; x_1, x_2, \ldots, x_n) = \prod_{i=1}^{n} f(x_i; \theta)$$

$$= \frac{1}{\theta^n} \exp\left(\frac{-\sum_1^n x_i}{\theta}\right), \qquad 0 < \theta < \infty.$$

The natural logarithm of $L(\theta)$ is

$$\ln L(\theta) = -(n) \ln(\theta) - \frac{1}{\theta} \sum_1^n x_i, \qquad 0 < \theta < \infty.$$

Thus,

$$\frac{d[\ln L(\theta)]}{d\theta} = \frac{-n}{\theta} + \frac{\sum_1^n x_i}{\theta^2} = 0.$$

The solution of this equation for θ is

$$\theta = \frac{1}{n} \sum_1^n x_i = \bar{x}.$$

Note that

$$\frac{d[\ln L(\theta)]}{d\theta} = \frac{1}{\theta}\left(-n + \frac{n\bar{x}}{\theta}\right) \begin{cases} > 0, & \theta < \bar{x}, \\ = 0, & \theta = \bar{x}, \\ < 0, & \theta > \bar{x}. \end{cases}$$

Hence, $\ln L(\theta)$ does have a maximum at $\bar{x}$ and thus the maximum likelihood estimator for θ is

$$\hat{\theta} = \bar{X} = \frac{1}{n} \sum_{i=1}^{n} X_i.$$

In the following important example we find the maximum likelihood estimators of the parameters associated with the normal distribution.

Example 7.1–2. Let $X_1, X_2, \ldots, X_n$ be a random sample from $N(\theta_1, \theta_2)$, where

$$\Omega = \{(\theta_1, \theta_2): \quad -\infty < \theta_1 < \infty, \quad 0 < \theta_2 < \infty\}.$$

That is, here we let $\theta_1 = \mu$ and $\theta_2 = \sigma^2$. Then

$$L(\theta_1, \theta_2) = \prod_{i=1}^{n} \frac{1}{\sqrt{2\pi\theta_2}} \exp\left[-\frac{(x_i - \theta_1)^2}{2\theta_2}\right]$$

or, equivalently,

$$L(\theta_1, \theta_2) = \left(\frac{1}{\sqrt{2\pi\theta_2}}\right)^n \exp\left[\frac{-\sum_1^n(x_i - \theta_1)^2}{2\theta_2}\right], \qquad (\theta_1, \theta_2) \in \Omega.$$

The natural logarithm of the likelihood function is

$$\ln L(\theta_1, \theta_2) = -\frac{n}{2}\ln(2\pi\theta_2) - \frac{\sum_1^n (x_i - \theta_1)^2}{2\theta_2}.$$

The partial derivatives with respect to θ_1 and θ_2 are

$$\frac{\partial(\ln L)}{\partial\theta_1} = \frac{1}{\theta_2}\sum_1^n (x_i - \theta_1)$$

and

$$\frac{\partial(\ln L)}{\partial\theta_2} = \frac{-n}{2\theta_2} + \frac{1}{2\theta_2^2}\sum_1^n (x_i - \theta_1)^2.$$

The equation $\partial(\ln L)/\partial\theta_1 = 0$ has the solution $\theta_1 = \bar{x}$. Setting $\partial(\ln L)/\partial\theta_2 = 0$ and replacing θ_1 by $\bar{x}$ yields

$$\theta_2 = \frac{1}{n}\sum_1^n (x_i - \bar{x})^2.$$

By considering the usual condition on the second partial derivatives, these solutions do provide a maximum. Thus, the maximum likelihood estimators are

$$\hat{\theta}_1 = \bar{X} \qquad \text{and} \qquad \hat{\theta}_2 = \frac{1}{n}\sum_1^n (X_i - \bar{X})^2 = S^2.$$

That is, the maximum likelihood estimators of the mean $\theta_1 = \mu$ and the variance $\theta_2 = \sigma^2$ of the normal distribution are the corresponding characteristics of the sample.

═══ Exercises ═══

7.1–1.   Let $X_1, X_2, \ldots, X_n$ be a random sample from $N(\mu, \sigma^2)$ where the mean $\theta = \mu$ is such that $-\infty < \theta < \infty$ and σ^2 is a known positive number. Show that the maximum likelihood estimator for θ is $\hat{\theta} = \bar{X}$.

7.1–2.   A random sample $X_1, X_2, \ldots, X_n$ of size n is taken from $N(\mu, \sigma^2)$, where the variance $\theta = \sigma^2$ is such that $0 < \theta < \infty$ and μ is a known real number. Show that the maximum likelihood estimator for θ is $\hat{\theta} = (1/n)\sum (X_i - \mu)^2$.

7.1–3.   A random sample $X_1, X_2, \ldots, X_n$ of size n is taken from a Poisson distribution with a mean of $\lambda, 0 < \lambda < \infty$. Show that the maximum likelihood estimator for λ is $\hat{\lambda} = \bar{X}$.

7.1–4 Let $X_1, X_2, \ldots, X_n$ be a random sample from distributions with the following probability density functions. In each case find the maximum likelihood estimator $\hat{\theta}$.

(a) $f(x; \theta) = \theta x^{\theta-1}, \qquad 0 < x < 1, \; 0 < \theta < \infty;$

(b) $f(x; \theta) = (1/\theta^2)xe^{-x/\theta}, \qquad 0 < x < \infty, \; 0 < \theta < \infty;$

(c) $f(x; \theta) = (1/2)e^{-|x-\theta|}, \qquad -\infty < x < \infty, \; -\infty < \theta < \infty.$

7.1–5. Find the maximum likelihood estimates for $\theta_1 = \mu$ and $\theta_2 = \sigma^2$ if a random sample of size 15 from $N(\mu, \sigma^2)$ yielded the following values.

31.5	36.9	33.8	30.1	33.9
35.2	29.6	34.4	30.5	34.2
31.6	36.7	35.8	34.5	32.7

7.2 Confidence Intervals for Means Given Known Variances

If a random sample $X_1, X_2, \ldots, X_n$ arises from a normal distribution with unknown mean $\theta = \mu$ and variance σ^2, we found in Example 7.1–2 (and again in Exercise 7.1–1 when σ^2 is known) that $\hat{\mu} = \overline{X}$ is the maximum likelihood estimator of that unknown mean. That is, $\overline{X}$ seems to be an appropriate point estimator of μ. We will use the error structure of the maximum likelihood estimator $\overline{X}$, namely that $\overline{X}$ is $N(\mu, \sigma^2/n)$, to construct a confidence interval for the unknown parameter μ. This development is very similar to the discussion in Section 6.2 when we found an approximate confidence interval for an unknown mean μ of a distribution. But recall that there $\overline{X}$ had only an approximate normal distribution, since the underlying distribution was not necessarily normal.

In our discussion here, we assume that the variance σ^2 is known but the mean μ is unknown. We will consider the case when σ^2 is unknown in Section 7.4. Since the distribution of $\overline{X}$ is $N(\mu, \sigma^2/n)$, we know that $(\overline{X} - \mu)/(\sigma/\sqrt{n})$ is $N(0, 1)$. Let z_0 be a positive number such that

$$\gamma = P\left(-z_0 \leq \frac{\overline{X} - \mu}{\sigma/\sqrt{n}} \leq z_0\right).$$

But $-z_0\sigma/\sqrt{n} \leq \overline{X} - \mu \leq z_0\sigma/\sqrt{n}$ is equivalent to $\overline{X} - z_0\sigma/\sqrt{n} \leq \mu \leq \overline{X} + z_0\sigma/\sqrt{n}$; thus,

$$\gamma = P\left[\overline{X} - z_0\left(\frac{\sigma}{\sqrt{n}}\right) \leq \mu \leq \overline{X} + z_0\left(\frac{\sigma}{\sqrt{n}}\right)\right].$$

That is, the probability that the random interval with endpoints $\overline{X} \pm z_0\sigma/\sqrt{n}$ contains the unknown mean μ is γ.

When the experiment has been run and $x_1, x_2, \ldots, x_n$ are the observed values of the items of the random sample, then

$$\left[\bar{x} - z_0\left(\frac{\sigma}{\sqrt{n}}\right), \bar{x} + z_0\left(\frac{\sigma}{\sqrt{n}}\right)\right]$$

is a $(100\gamma)\%$ confidence interval for μ. This interval is centered at the point estimate $\bar{x}$ and completed by subtracting and adding the quantity $z_0 \sigma/\sqrt{n}$.

Note that as n increases, the quantity $z_0 \sigma/\sqrt{n}$ decreases, resulting in a shorter confidence interval with the same confidence coefficient γ. A shorter confidence interval indicates that we have more reliance in $\bar{x}$ as an estimate of μ. Statisticians who are not restricted by time, money, effort, or availability of items can obviously make the confidence interval as short as they like by increasing the sample size n.

Example 7.2–1. Let $x_1, x_2, \ldots, x_{10}$ be the values of the items of a random sample of size 10 from $N(\mu, 12.34)$. A 90% confidence interval for the unknown mean μ is

$$\left[\bar{x} - 1.645\sqrt{\frac{12.34}{10}}, \bar{x} + 1.645\sqrt{\frac{12.34}{10}}\right].$$

For a particular sample this interval either does or does not contain the mean μ. However, if many such intervals were calculated, it should be true that about 90% of them contain the mean μ. Fifty-two random samples of size 10 from $N(40, 12.34)$ were simulated on the computer. A 90% confidence interval was calculated for each random sample, as if the mean were unknown. Table 7.2–1 lists these 52 intervals. For the 52 calculated intervals, 47 (or 90.3846%) of them contain the mean. We should note that we were extremely fortunate that 47/52 was that close to 90%.

[39.2, 42.8]	[37.5, 41.2]	[38.2, 41.8]	[37.2, 40.9]
[39.3, 42.9]	[37.3, 40.9]	[39.0, 42.6]	[39.5, 43.1]
[35.9, 39.5]	[39.5, 43.2]	[36.6, 40.3]	[38.2, 41.8]
[38.1, 41.8]	[37.5, 41.2]	[37.8, 41.5]	[38.6, 42.3]
[37.2, 40.8]	[40.8, 44.4]	[39.3, 43.0]	[35.9, 39.6]
[37.3, 40.9]	[36.3, 39.9]	[37.3, 41.0]	[36.7, 40.4]
[39.8, 43.5]	[37.8, 41.5]	[38.1, 41.8]	[40.0, 43.7]
[39.2, 42.9]	[38.3, 42.0]	[39.3, 42.9]	[37.0, 40.6]
[37.9, 41.6]	[37.0, 40.7]	[36.9, 40.6]	[36.7, 40.4]
[39.1, 42.8]	[37.1, 40.8]	[37.1, 40.7]	[38.6, 42.2]
[39.6, 43.2]	[36.5, 40.2]	[40.7, 44.3]	[38.0, 41.6]
[37.2, 40.9]	[37.0, 40.6]	[39.4, 43.0]	[39.8, 43.5]
[37.3, 41.0]	[39.2, 42.9]	[39.1, 42.8]	[37.5, 41.1]

TABLE 7.2–1

Let $X_1, X_2, \ldots, X_n$ and $Y_1, Y_2, \ldots, Y_m$ be, respectively, two random samples of sizes n and m from the two independent normal distributions $N(\mu_X, \sigma_X^2)$ and $N(\mu_Y, \sigma_Y^2)$. Suppose, for now, that σ_X^2 and σ_Y^2 are known. If $\overline{X}$ and $\overline{Y}$ denote the respective sample means, what then is the distribution of the difference $\overline{X} - \overline{Y}$? The moment-generating function of $\overline{X} - \overline{Y}$ is

$$E[e^{t(\overline{X} - \overline{Y})}] = E(e^{t\overline{X}} e^{-t\overline{Y}}) = E(e^{t\overline{X}})E(e^{-t\overline{Y}})$$

since $\overline{X}$ and $\overline{Y}$ are means of samples from *independent* normal distributions. Thus, since $\overline{X}$ is $N(\mu_X, \sigma_X^2/n)$ and $\overline{Y}$ is $N(\mu_Y, \sigma_Y^2/m)$, we have

$$E[e^{t(\overline{X} - \overline{Y})}] = \exp\left[t\mu_X + \frac{t^2(\sigma_X^2/n)}{2} \right] \exp\left[-t\mu_Y + \frac{t^2(\sigma_Y^2/m)}{2} \right]$$

$$= \exp\left[t(\mu_X - \mu_Y) + \frac{t^2(\sigma_X^2/n + \sigma_Y^2/m)}{2} \right].$$

However, this is the moment-generating function of a distribution that is

$$N\left(\mu_X - \mu_Y, \frac{\sigma_X^2}{n} + \frac{\sigma_Y^2}{m} \right);$$

thus, $\overline{X} - \overline{Y}$ has this normal distribution. Then the random variable

$$Z = \frac{(\overline{X} - \overline{Y}) - (\mu_X - \mu_Y)}{\sqrt{\sigma_X^2/n + \sigma_Y^2/m}}$$

is $N(0, 1)$. Thus, with $\Phi(z_0) = 1 - \alpha/2$, we have

$$P\left(-z_0 \leq \frac{(\overline{X} - \overline{Y}) - (\mu_X - \mu_Y)}{\sqrt{\sigma_X^2/n + \sigma_Y^2/m}} \leq z_0 \right) = 1 - \alpha = \gamma,$$

which can be rewritten as

$$P[(\overline{X} - \overline{Y}) - z_0 R \leq \mu_X - \mu_Y \leq (\overline{X} - \overline{Y}) + z_0 R] = \gamma,$$

where $R = \sqrt{\sigma_X^2/n + \sigma_Y^2/m}$ is the standard deviation of $\overline{X} - \overline{Y}$. Once the experiments have been performed and the means $\bar{x}$ and $\bar{y}$ computed, then

$$[\bar{x} - \bar{y} - z_0 R, \bar{x} - \bar{y} + z_0 R]$$

or, equivalently, $\bar{x} - \bar{y} \pm z_0 R$ provides a $(100\gamma)\%$ confidence interval for $\mu_X - \mu_Y$. Note that this interval is centered at the point estimate $\bar{x} - \bar{y}$ of $\mu_X - \mu_Y$ and completed by subtracting and adding the quantity $z_0 R$.

Example 7.2–2. In the preceding discussion, let $n = 15$, $m = 8$, $\bar{x} = 70.1$, $\bar{y} = 75.3$, $\sigma_X^2 = 60$, $\sigma_Y^2 = 40$, and $\gamma = 0.90$. Thus, $\alpha = 1 - \gamma = 0.10$ and $1 - \alpha/2 = 0.95 = \Phi(1.645)$. Hence,

$$1.645R = 1.645\sqrt{\frac{60}{15} + \frac{40}{8}} = 4.935,$$

and, since $\bar{x} - \bar{y} = -5.2$, we have that

$$[-5.2 - 4.935, -5.2 + 4.935] = [-10.135, -0.265]$$

is a 90% confidence interval for $\mu_X - \mu_Y$.

=== *Exercises* ===

7.2–1. A random sample of size 16 from $N(\mu, 25)$ yielded $\bar{x} = 73.8$. Find a 95% confidence interval for μ.

7.2–2. A random sample of size 8 from $N(\mu, 72)$ yielded $\bar{x} = 85$. Find **(a)** 99% **(b)** 95%, **(c)** 90%, **(d)** 80% confidence intervals for μ.

7.2–3. The length of life of brand X light bulbs is assumed to be $N(\mu_X, 784)$. The length of life of brand Y light bulbs is assumed to be $N(\mu_Y, 627)$ and independent of that of X. If a random sample of $n = 56$ brand X light bulbs yielded a mean of $\bar{x} = 937.4$ hours and a random sample of size $m = 57$ brand Y light bulbs yielded a mean of $\bar{y} = 988.9$ hours, find a 90% confidence interval for $\mu_X - \mu_Y$.

7.2–4. Let $X_1, X_2, \ldots, X_9$ and $Y_1, Y_2, \ldots, Y_{15}$ be random samples from the independent normal distributions $N(\mu_X, 21)$ and $N(\mu_Y, 25)$, respectively. If the following are the observed values, form a 95% confidence interval for $\mu_X - \mu_Y$:

$$
\begin{array}{lllll}
x_1 = 125.7 & x_2 = 121.0 & x_3 = 118.8 & x_4 = 119.2 & x_5 = 117.6 \\
x_6 = 114.0 & x_7 = 122.2 & x_8 = 110.0 & x_9 = 114.1
\end{array}
$$

and

$$
\begin{array}{lllll}
y_1 = 119.2 & y_2 = 125.6 & y_3 = 126.0 & y_4 = 124.3 & y_5 = 110.3 \\
y_6 = 117.9 & y_7 = 125.9 & y_8 = 125.4 & y_9 = 122.8 & y_{10} = 117.9 \\
y_{11} = 120.8 & y_{12} = 126.6 & y_{13} = 115.2 & y_{14} = 121.6 & y_{15} = 130.8
\end{array}
$$

7.2–5. Suppose that bowling scores for a three game series for Jones and Smith are $N(\mu_X, 30^2)$ and $N(\mu_Y, 38^2)$, respectively. Use the following series scores bowled by Jones and Smith on nine successive weeks to form a 95% confidence interval for $\mu_X - \mu_Y$:

$$
\begin{array}{lllll}
x_1 = 510 & x_2 = 569 & x_3 = 504 & x_4 = 515 & x_5 = 541 \\
x_6 = 546 & x_7 = 480 & x_8 = 526 & x_9 = 597
\end{array}
$$

and

$$
\begin{array}{lllll}
y_1 = 538 & y_2 = 519 & y_3 = 492 & y_4 = 569 & y_5 = 484 \\
y_6 = 474 & y_7 = 527 & y_8 = 517 & y_9 = 590
\end{array}
$$

7.3 Confidence Intervals for Variances

In this section we find confidence intervals for the variance of a normal distribution and for the ratio of the variances of two independent normal distributions.

Just as the confidence interval for the mean μ was based upon the maximum likelihood estimator of that parameter, the confidence interval for the variance σ^2 is based upon the maximum likelihood estimator of σ^2, which, according to Example 7.1–2, is the sample variance S^2. Before discussing this confidence interval, we need the information about the distribution of S^2 contained in the following theorem.

THEOREM 7.3–1. *Let $X_1, X_2, \ldots, X_n$ be a random sample of size n from a distribution $N(\mu, \sigma^2)$. Let*

$$\bar{X} = \frac{1}{n} \sum_{i=1}^{n} X_i$$

and

$$S^2 = \frac{1}{n} \sum_{i=1}^{n} (X_i - \bar{X})^2.$$

Then

(i) *$\bar{X}$ and S^2 are independent.*

(ii) *$\dfrac{nS^2}{\sigma^2} = \dfrac{\sum_{i=1}^{n} (X_i - \bar{X})^2}{\sigma^2}$ is $\chi^2(n-1)$.*

Proof: We are not prepared to prove (i) at this time; so we accept it without proof here. To prove (ii), note that

$$\sum_{i=1}^{n} \left(\frac{X_i - \mu}{\sigma} \right)^2 = \sum_{i=1}^{n} \left[\frac{(X_i - \bar{X}) + (\bar{X} - \mu)}{\sigma} \right]^2$$

$$= \sum_{i=1}^{n} \left(\frac{X_i - \bar{X}}{\sigma} \right)^2 + \frac{n(\bar{X} - \mu)^2}{\sigma^2} \tag{1}$$

because the cross-product term is equal to

$$2 \sum_{i=1}^{n} \frac{(\bar{X} - \mu)(X_i - \bar{X})}{\sigma^2} = \frac{2(\bar{X} - \mu)}{\sigma^2} \sum_{i=1}^{n} (X_i - \bar{X}) = 0.$$

But $Y_i = (X_i - \mu)/\sigma$, $i = 1, 2, \ldots, n$, are mutually independent standardized normal variables; so Y_i^2, $i = 1, 2, \ldots, n$, are mutually independent $\chi^2(1)$ variables by Theorem 4.5–2. Hence, $W = \sum_{i=1}^{n} Y_i^2$ is the sum of n mutually independent $\chi^2(1)$ variables and thus is $\chi^2(n)$. Moreover, since $\bar{X}$ is $N(\mu, \sigma^2/n)$, then $Z = (\bar{X} - \mu)/(\sigma/\sqrt{n})$ is $N(0, 1)$ and $Z^2 = n(\bar{X} - \mu)^2/\sigma^2$ is $\chi^2(1)$. In this notation, equation (1) becomes

$$W = \frac{nS^2}{\sigma^2} + Z^2.$$

However, from (i), $\overline{X}$ and S^2 are independent; thus, Z^2 and S^2 are also independent. In the moment-generating function of W, this independence permits us to write

$$E(e^{tW}) = E[e^{t(nS^2/\sigma^2 + Z^2)}] = E[e^{t(nS^2/\sigma^2)}e^{tZ^2}]$$
$$= E[e^{t(nS^2/\sigma^2)}]E(e^{tZ^2}).$$

Since W and Z^2 have chi-square distributions, we can substitute their moment-generating functions to obtain

$$(1 - 2t)^{-n/2} = E[e^{t(nS^2/\sigma^2)}](1 - 2t)^{-1/2}.$$

Equivalently, we have

$$E[e^{t(nS^2/\sigma^2)}] = (1 - 2t)^{-(n-1)/2}, \qquad t < 1/2.$$

This, of course, is the moment-generating function of a $\chi^2(n - 1)$ variable, and accordingly nS^2/σ^2 has this distribution.

We shall use the fact that nS^2/σ^2 is $\chi^2(n - 1)$ to find a confidence interval for σ^2. Select constants a and b from Appendix Table III with $n - 1$ degrees of freedom such that

$$P\left(a \le \frac{nS^2}{\sigma^2} \le b\right) = \gamma.$$

One way to do this is by selecting a and b so that $P(nS^2/\sigma^2 \le a) = \alpha/2$ and $P(nS^2/\sigma^2 \le b) = 1 - \alpha/2$, where $\alpha = 1 - \gamma$. Then, solving the inequalities, we have

$$P\left(\frac{nS^2}{b} \le \sigma^2 \le \frac{nS^2}{a}\right) = \gamma.$$

That is, the probability that the random interval $[nS^2/b, nS^2/a]$ contains the unknown σ^2 is γ. Once the values of $X_1, X_2, \ldots, X_n$ are observed to be $x_1, x_2, \ldots, x_n$ with variance s^2, then the interval $[ns^2/b, ns^2/a]$ is a $(100\gamma)\%$ confidence interval for σ^2.

Example 7.3–1. Assume that the time in days required for maturation of seeds of a species of *Guardiola*, a flowering plant found in Mexico, is $N(\mu, \sigma^2)$. A random sample of $n = 13$ seeds, both parents having narrow leaves, yielded $\bar{x} = 18.97$ and

$$13s^2 = \sum_{i=1}^{13}(x_i - \bar{x})^2 = 128.41.$$

A 90% confidence interval for σ^2 is

$$\left[\frac{128.41}{21.03}, \frac{128.41}{5.226}\right] = [6.11, 24.57]$$

because $P(13S^2/\sigma^2 \leq 5.226) = 0.05$ and $P(13S^2/\sigma^2 \leq 21.03) = 0.95$ from Appendix Table III with $n - 1 = 12$ degrees of freedom. The corresponding 90% confidence interval for σ is

$$[\sqrt{6.11}, \sqrt{24.57}] = [2.47, 4.96].$$

Although a and b are generally selected so that the probabilities in the two tails are equal, the resulting $(100\gamma)\%$ confidence interval is not the shortest that can be formed using the available data. In the next example, we shall choose a and b so that the length of the confidence interval for the standard deviation σ is minimized.

Example 7.3–2. Let $X_1, X_2, \ldots, X_n$ be a random sample of size n from $N(\mu, \sigma^2)$, where μ and σ^2 are unknown. We shall find the shortest confidence interval for σ using nS^2/σ^2. Select a and b so that

$$\gamma = 1 - \alpha = P\left(a \leq \frac{nS^2}{\sigma^2} \leq b\right) = P\left(\frac{nS^2}{b} \leq \sigma^2 \leq \frac{nS^2}{a}\right).$$

The corresponding observed $(100\gamma)\%$ confidence interval for σ has length

$$k = \sqrt{ns}\left(\frac{1}{\sqrt{a}} - \frac{1}{\sqrt{b}}\right). \tag{2}$$

We shall minimize k, subject to the condition

$$G(b) - G(a) = \int_a^b g(u)\,du = 1 - \alpha, \tag{3}$$

where $G(u)$ and $g(u)$ are the distribution function and the p.d.f., respectively, of a chi-square random variable with $r = n - 1$ degrees of freedom. From (2), since b is a function of a, we obtain

$$\frac{dk}{da} = \sqrt{ns}\left(-\frac{1}{2}a^{-3/2} + \frac{1}{2}b^{-3/2}\frac{db}{da}\right).$$

From (3), we see that

$$\frac{dG(b)}{da} - \frac{dG(a)}{da} = g(b)\frac{db}{da} - g(a) = 0$$

and, hence,

$$\frac{db}{da} = \frac{g(a)}{g(b)}.$$

Accordingly,

$$\frac{dk}{da} = \sqrt{ns}\left\{-\frac{1}{2}a^{-3/2} + \frac{1}{2}b^{-3/2}\left[\frac{g(a)}{g(b)}\right]\right\}.$$

Setting $dk/da = 0$, we obtain

$$a^{3/2}g(a) - b^{3/2}g(b) = 0,$$

or, equivalently,

$$a^{n/2}e^{-a/2} - b^{n/2}e^{-b/2} = 0, \tag{4}$$

which along with equation (3) provides a solution for a and b and thus the interval of the shortest length. For illustration, using Appendix Table IX and the data given in Example 7.3–1, we find that a 90% confidence interval for σ of minimum length is given by

$$\left[\sqrt{\frac{128.41}{24.202}}, \sqrt{\frac{128.41}{5.940}} \right] = [2.30, 4.65].$$

The length of this interval is 2.35, whereas the length of the interval given in Example 7.3–1 is 2.49. Although there seems to be an advantage in using the shortest length confidence interval when n is small, there is not a great difference between the two types of intervals when n is large. Hence, in practice the equal-tail case is usually used since tables for selecting equal probabilities in the two tails are readily available.

There are occasions when it is of interest to compare the variances of two independent normal distributions. We do this by finding a confidence interval for σ_X^2/σ_Y^2 using the ratio of $nS_X^2/[\sigma_X^2(n-1)]$ and $mS_Y^2/[\sigma_Y^2(m-1)]$, where S_X^2 and S_Y^2 are the two variances of the two samples of respective sizes n and m. Since nS_X^2/σ_X^2 and mS_Y^2/σ_Y^2 are independent chi-square variables, we need the following theorem, which we accept without proof.

THEOREM 7.3–2. *Let U and V be independent chi-square variables with r_1 and r_2 degrees of freedom, respectively. Then*

$$F = \frac{U/r_1}{V/r_2}$$

has an F distribution with r_1 and r_2 degrees of freedom. Its p.d.f. is

$$h(w) = \frac{\Gamma[(r_1 + r_2)/2](r_1/r_2)^{r_1/2}w^{r_1/2 - 1}}{\Gamma(r_1/2)\Gamma(r_2/2)(1 + r_1w/r_2)^{(r_1 + r_2)/2}}, \qquad 0 < w < \infty.$$

REMARK. For many years the random variable defined in Theorem 7.3–2 has been called F, a symbol first proposed by George Snedecor to honor R. A. Fisher, who used a modification of this ratio in several statistical applications.

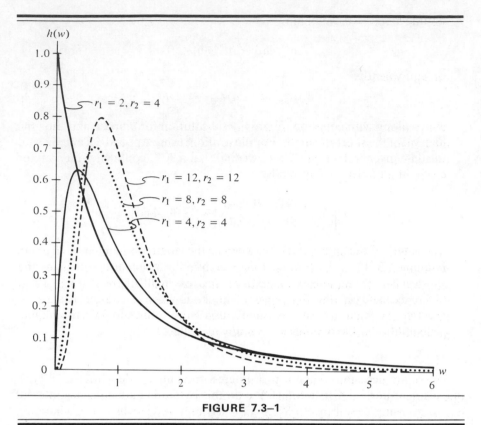

FIGURE 7.3–1

Note that the F distribution depends on two parameters, r_1 and r_2, in that order. The first parameter is the number of degrees of freedom in the numerator and the second is the number of degrees of freedom in the denominator. See Figure 7.3–1 for graphs of the p.d.f. of the F distribution for four pairs of degrees of freedom.

It can be shown that

$$E(F) = \frac{r_2}{r_2 - 2} \quad \text{and} \quad \text{Var}(F) = \frac{2r_2^2(r_1 + r_2 - 2)}{r_1(r_2 - 2)^2(r_2 - 4)}.$$

To verify these two expressions, we note, using the independence of U and V in the definition of F, that

$$E(F) = E\left(\frac{U}{r_1}\right)E\left(\frac{r_2}{V}\right) \quad \text{and} \quad E(F^2) = E\left[\left(\frac{U}{r_1}\right)^2\right]E\left[\left(\frac{r_2}{V}\right)^2\right].$$

In Exercise 7.3–7, the student is asked to find $E(U)$, $E(1/V)$, $E(U^2)$, and $E(1/V^2)$.

Some values of the distribution function $P(F \leq f)$ of the F distribution are given in Appendix Table V. To use this table to find 0.01, 0.025, and 0.05 cumulative probabilities we need to note the following: since $F = (U/r_1)/(V/r_2)$, where U and V are independent and $\chi^2(r_1)$ and $\chi^2(r_2)$, respectively, then $1/F = (V/r_2)/(U/r_1)$ must have an F distribution with r_2 and r_1 degrees of freedom. Note the change in the order of the parameters in the latter distribution.

Example 7.3–3. Let F have an F distribution with r_1 and r_2 degrees of freedom. From Appendix Table V, we see that

$$P(F \leq 3.50) = 0.95 \qquad \text{when } r_1 = 7, \quad r_2 = 8;$$
$$P(F \leq 14.66) = 0.99 \qquad \text{when } r_1 = 9, \quad r_2 = 4;$$
$$P(F \geq 9.01) = 0.05 \qquad \text{when } r_1 = 5, \quad r_2 = 3;$$

$$P(F \leq 0.244) = P\left(\frac{1}{F} \geq \frac{1}{0.244}\right)$$

$$= P\left(\frac{1}{F} \geq 4.10\right) = 0.05 \qquad \text{when } r_1 = 6, \quad r_2 = 9.$$

The last probability was obtained because $1/F$ has an F distribution with nine and six degrees of freedom.

The F distribution has many important applications in normal sampling theory, one of which is presented here. Let $X_1, X_2, \ldots, X_n$ and $Y_1, Y_2, \ldots, Y_m$ be random samples of sizes n and m from two independent normal distributions $N(\mu_X, \sigma_X^2)$ and $N(\mu_Y, \sigma_Y^2)$, respectively. Then, from Theorem 7.3–1 we know that nS_X^2/σ_X^2 is $\chi^2(n-1)$ and mS_Y^2/σ_Y^2 is $\chi^2(m-1)$. The independence of the distributions implies that S_X^2 and S_Y^2 are independent so that

$$F = \frac{mS_Y^2/\sigma_Y^2(m-1)}{nS_X^2/\sigma_X^2(n-1)}$$

has an F distribution with $r_1 = m - 1$ and $r_2 = n - 1$ degrees of freedom. Select c and d so that $P(F \leq c) = \alpha/2$ and $P(F \leq d) = 1 - \alpha/2$. Thus, with $\gamma = 1 - \alpha$, we have that

$$\gamma = P(c \leq F \leq d) = P\left[c \frac{nS_X^2/(n-1)}{mS_Y^2/(m-1)} \leq \frac{\sigma_X^2}{\sigma_Y^2} \leq d \frac{nS_X^2/(n-1)}{mS_Y^2/(m-1)}\right].$$

If s_X^2 and s_Y^2 are the observed values of the variances of the respective samples, then

$$\left[c \frac{ns_X^2/(n-1)}{ms_Y^2/(m-1)}, d \frac{ns_X^2/(n-1)}{ms_Y^2/(m-1)}\right]$$

is a $(100\gamma)\%$ confidence interval for σ_X^2/σ_Y^2.

Example 7.3–4. In Example 7.3–1, let $\sigma_X^2 = \sigma^2$ and $ns_X^2 = 13s^2 = 128.41$. Assume that the time in days required for maturation of seeds of a species of *Guardiola*, both parents having broad leaves, is $N(\mu_Y, \sigma_Y^2)$. A random sample of size $m = 9$ seeds yielded $\bar{y} = 23.20$ and

$$9s_Y^2 = \sum_{i=1}^{9} (y_i - \bar{y})^2 = 36.72.$$

A 98 % confidence interval for σ_X^2/σ_Y^2 is given by

$$\left[\left(\frac{1}{5.67} \right) \frac{(128.41)/12}{(36.72)/8}, (4.50) \frac{(128.41)/12}{(36.72)/8} \right] = [0.41, 10.49].$$

===== *Exercises* =====

7.3–1. A random sample of size 19 from $N(\mu, \sigma^2)$ yielded $\bar{x} = 89.7$ and $s^2 = 11.72$.
  (a) Find a 95 % confidence interval for σ^2.
  (b) Find a 95 % confidence interval for σ.

7.3–2. An observed sample of $n = 11$ American College Testing (ACT) mathematics scores, assumed to be $N(\mu, \sigma^2)$, consisted of

$$\begin{array}{llllll}
x_1 = 26 & x_2 = 31 & x_3 = 27 & x_4 = 28 & x_5 = 29 & x_6 = 28 \\
x_7 = 20 & x_8 = 29 & x_9 = 24 & x_{10} = 31 & x_{11} = 23 &
\end{array}$$

  (a) Find a 95 % confidence interval for σ.
  (b) Find a 90 % confidence interval for σ.

7.3–3. Let $X_1, X_2, X_3, \ldots, X_n$ be a random sample from $N(\mu, \sigma^2)$, with known mean μ. Describe how you would construct a confidence interval for the unknown variance σ^2. HINT: Use the fact that $\sum_{i=1}^{n} (X_i - \mu)^2/\sigma^2$ is $\chi^2(n)$.

7.3–4. Let F have an F distribution with r_1 and r_2 degrees of freedom. Find
  (a) $P(F \geq 3.02)$ when $r_1 = 9, r_2 = 10$;
  (b) $P(F \leq 4.14)$ when $r_1 = 7, r_2 = 15$;
  (c) $P(F \leq 0.1508)$ when $r_1 = 8, r_2 = 5$ (HINT: $0.1508 = 1/6.63$);
  (d) $P(0.1323 \leq F \leq 2.79)$ when $r_1 = 6, r_2 = 15$.

7.3–5. Let F have a F distribution with r_1 and r_2 degrees of freedom. Find numbers a and b such that
  (a) $P(a \leq F \leq b) = 0.90$ when $r_1 = 8, r_2 = 6$;
  (b) $P(a \leq F \leq b) = 0.98$ when $r_1 = 8, r_2 = 6$.

7.3–6. Let $X_1, X_2, \ldots, X_9$ be a random sample of size 9 from a normal distribution $N(54, 10)$ and let Y_1, Y_2, Y_3, Y_4 be a random sample of size 4 from a normal distribution $N(54, 12)$. Compute

$$P\left(0.546 \leq \frac{\sum_{i=1}^{9} (X_i - \bar{X})^2}{\sum_{i=1}^{4} (Y_i - \bar{Y})^2} \leq 61.09 \right).$$

7.3–7. Find the mean and the variance of an F random variable with r_1 and r_2 degrees of freedom by first finding $E(U)$, $E(1/V)$, $E(U^2)$, and $E(1/V^2)$ as suggested in the text.

7.3–8. Assume that bowling scores for a three game series are normally distributed. A random sample of $n = 5$ scores from $N(\mu_X, \sigma_X^2)$ yielded $\bar{x} = 525.3$ and $s_X^2 = 1759$. A random sample of $m = 7$ scores from $N(\mu_Y, \sigma_Y^2)$ yielded $\bar{y} = 510.8$ and $s_Y^2 = 2273$. Find a 98% confidence interval for σ_X^2/σ_Y^2.

7.3–9. Let X and Y have independent normal distributions with known means μ_X and μ_Y. If random samples are taken from each, describe how a confidence interval for σ_X^2/σ_Y^2 can be formed, where σ_X^2 and σ_Y^2 are unknown. HINT: Consider the distribution of the ratio of

$$\frac{\sum_{i=1}^{n}(X_i - \mu_X)^2}{n\sigma_X^2} \quad \text{and} \quad \frac{\sum_{i=1}^{m}(Y_i - \mu_Y)^2}{m\sigma_Y^2}.$$

7.3–10. Assume that high school grade point averages of students at a large university and a small private college are $N(\mu_X, \sigma_X^2)$ and $N(\mu_Y, \sigma_Y^2)$, respectively. Find a 98% confidence interval for σ_X^2/σ_Y^2 if observed random samples of sizes $n = 10$ and $m = 7$ consisted of

$$\begin{array}{lllll}
x_1 = 3.0 & x_2 = 3.2 & x_3 = 3.1 & x_4 = 2.9 & x_5 = 3.4 \\
x_6 = 2.5 & x_7 = 2.8 & x_8 = 3.0 & x_9 = 3.3 & x_{10} = 2.7
\end{array}$$

and

$$\begin{array}{llll}
y_1 = 3.0 & y_2 = 2.6 & y_3 = 3.1 & y_4 = 3.5 \quad y_5 = 3.2 \\
y_6 = 3.9 & y_7 = 2.7
\end{array}$$

7.4 Confidence Intervals for Means

We are again interested in finding a confidence interval for the mean μ of a distribution $N(\mu, \sigma^2)$, assuming now that σ^2 is unknown. To construct the interval it is necessary to define another distribution called the t distribution (or Student's t distribution), which was first discovered by W. S. Gosset. Since the Irish brewery for which he was working did not want other breweries to know they were using statistical methods, Gosset published under the pseudonym of Student.

The necessary sampling distribution theory is summarized in this theorem, which we accept without proof.

THEOREM 7.4–1. *Let Z be a random variable that is $N(0, 1)$, let U be a random variable that is $\chi^2(r)$, and let Z and U be independent. Then*

$$T = \frac{Z}{\sqrt{U/r}}$$

has a t distribution with r degrees of freedom. Its p.d.f. is

$$g(t) = \frac{\Gamma[(r+1)/2]}{\sqrt{\pi r}\,\Gamma(r/2)(1 + t^2/r)^{(r+1)/2}}, \qquad -\infty < t < \infty.$$

Note that the distribution of T is completely determined by the number r. Since it is, in general, difficult to evaluate the distribution function of T, some values of $P(T \leq t)$ are found in Appendix Table VI for $r = 1, 2, 3, \ldots, 30$. We can observe, however, that the graph of the p.d.f. of T is symmetrical with respect to the vertical axis $t = 0$ and is very similar to the graph of the p.d.f. of a distribution $N(0, 1)$. Figure 7.4–1 shows the graphs of these two probability density functions, namely, those of T when $r = 4$ and of $N(0, 1)$. In this figure we see that the tails of the t distribution are heavier than those of a normal one; that is, there is more extreme probability in the t distribution than in the standardized normal one.

Because of the symmetry of the t distribution about $t = 0$, the mean (if it exists) must equal zero. That is, it can be shown that $E(T) = 0$ when $r \geq 2$. No moments exist for this distribution when $r = 1$. The variance of T is

$$E(T^2) = \text{Var}(T) = \frac{r}{r-2}, \qquad \text{when } r \geq 3.$$

The variance does not exist when $r = 1$ or 2. Although it is fairly difficult to compute these moments from the p.d.f. of T, they can be found (Exercise 7.4–3) using the definition of T and the independence of Z and U, namely,

$$E(T) = E(Z)E(\sqrt{r/U}) \qquad \text{and} \qquad E(T^2) = E(Z^2)E\left(\frac{r}{U}\right).$$

Let us consider some illustrations of the use of the table and these formulas.

Example 7.4–1. Let T have a t distribution with seven degrees of freedom. Then, from Appendix Table VI, we have

$$P(T \leq 1.415) = 0.90,$$

$$P(T \leq -1.415) = 1 - P(T \leq 1.415) = 0.10,$$

and

$$P(-1.895 < T < 1.415) = 0.90 - 0.05 = 0.85.$$

Example 7.4–2. Let T have a t distribution with a variance of 5/4. Thus $r/(r-2) = 5/4$ and $r = 10$. Then

$$P(-1.812 \leq T \leq 1.812) = 0.90.$$

Example 7.4–3. Let T have a t distribution with 14 degrees of freedom. Find a constant c such that $P(|T| < c) = 0.90$. From Appendix Table VI we see that $P(T \leq 1.761) = 0.95$ and therefore $c = 1.761$.

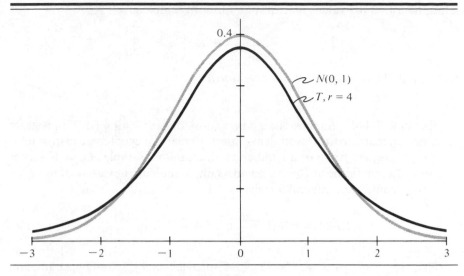

0.4

$N(0, 1)$

$T, r = 4$

−3   −2   −1   0   1   2   3

FIGURE 7.4–1

Recall that if $X_1, X_2, \ldots, X_n$ is a random sample from a distribution $N(\mu, \sigma^2)$, then $(\bar{X} - \mu)/(\sigma/\sqrt{n})$ is $N(0, 1)$, nS^2/σ^2 is $\chi^2(n - 1)$, and the two are independent. Thus, we have that

$$T = \frac{(\bar{X} - \mu)/(\sigma/\sqrt{n})}{\sqrt{nS^2/\sigma^2(n - 1)}} = \frac{\bar{X} - \mu}{S/\sqrt{n - 1}}$$

has a t distribution with $r = n - 1$ degrees of freedom. It was a random variable like this one that motivated Gosset's search for the distribution of T.

Let $\bar{X}$ and S^2 be the mean and the variance of a random sample of size n from a distribution $N(\mu, \sigma^2)$, where μ and σ^2 are both unknown. To construct a confidence interval for μ, we use the fact that

$$T = \frac{\bar{X} - \mu}{S/\sqrt{n - 1}}$$

has a t distribution with $r = n - 1$ degrees of freedom. Let $\gamma = 1 - \alpha$ and select t_0 so that $P(T \leq t_0) = 1 - \alpha/2$. We have that

$$\gamma = 1 - \alpha = P\left(-t_0 \leq \frac{\bar{X} - \mu}{S/\sqrt{n - 1}} \leq t_0\right)$$

$$= P\left[\bar{X} - t_0\left(\frac{S}{\sqrt{n - 1}}\right) \leq u \leq \bar{X} + t_0\left(\frac{S}{\sqrt{n - 1}}\right)\right].$$

Thus, the observed values of items of a random sample give

$$\left[\bar{x} - t_0\left(\frac{s}{\sqrt{n-1}}\right), \bar{x} + t_0\left(\frac{s}{\sqrt{n-1}}\right)\right],$$

which is a $(100\gamma)\%$ confidence interval for μ.

Example 7.4–4. Suppose that American College Testing (ACT) program scores in mathematics for students entering small colleges have a distribution $N(\mu, \sigma^2)$, where μ and σ^2 are unknown. If a random sample of $n = 15$ scores yielded a sample mean $\bar{x} = 27.38$ and a sample standard deviation of $s = 2.26$, a 90% confidence interval for μ is

$$\left[27.38 - 1.761\left(\frac{2.26}{\sqrt{14}}\right), 27.38 + 1.761\left(\frac{2.26}{\sqrt{14}}\right)\right].$$

That is, 27.38 ± 1.06 or $[26.32, 28.44]$ provides a 90% confidence interval for μ.

Let T have a t distribution with $n - 1$ degrees of freedom. If $P(T \le t_0) = 1 - \alpha/2$ and if $\Phi(z_0) = 1 - \alpha/2$, then $t_0 > z_0$. Consequently, we would expect the interval $\bar{X} \pm z_0\sigma/\sqrt{n}$ to be shorter than the interval $\bar{X} \pm t_0 S/\sqrt{n-1}$. After all, we have more information, namely, the value of σ, in constructing the first interval. However, the length of the second interval is very much dependent on the value of the sample standard deviation S. If the observed s is smaller than σ, a shorter confidence interval could result by the second scheme. But on the average, $\bar{X} \pm z_0\sigma/\sqrt{n}$ is the shorter of the two confidence intervals (Exercise 7.4–7).

If we are not able to assume that the underlying distribution is normal but μ and σ are both unknown, approximate confidence intervals for μ can be constructed using

$$T = \frac{\bar{X} - \mu}{S/\sqrt{n-1}},$$

which now only has an approximate t distribution. Generally, if n is greater than 25, this approximation is quite good for many nonnormal distributions (that is, it is robust). However, if the underlying distribution is symmetric, unimodal, and of the continuous type, a much smaller value of n provides a very good approximation. The use of the t approximation is the modification that we mentioned while finding an approximate confidence interval for μ in Chapter 6.

We shall now consider the problem of constructing confidence intervals for the difference of the means of two normal distributions when the variances are unknown. Let $X_1, X_2, \ldots, X_n$ and $Y_1, Y_2, \ldots, Y_m$ be random samples from independent distributions $N(\mu_X, \sigma_X^2)$ and $N(\mu_Y, \sigma_Y^2)$, respectively. If the sample sizes are not large (say considerably smaller than 30), this problem can be a difficult one. However, even in these cases, if we can assume common, but unknown, variances, say $\sigma_X^2 = \sigma_Y^2 = \sigma^2$, there is a way out of our difficulty using the following development. We know that

$$U = \frac{\bar{X} - \bar{Y} - (\mu_X - \mu_Y)}{\sqrt{\sigma^2/n + \sigma^2/m}}$$

is $N(0, 1)$. Also, if S_X^2 and S_Y^2 are the respective sample variances, then, from the independence,

$$V = \frac{nS_X^2}{\sigma^2} + \frac{mS_Y^2}{\sigma^2}$$

has the chi-square distribution with $r = (n - 1) + (m - 1) = n + m - 2$ degrees of freedom. Furthermore U and V are independent. Thus, a random variable T with the t distribution having $r = n + m - 2$ degrees of freedom is given by

$$T = \frac{[\bar{X} - \bar{Y} - (\mu_X - \mu_Y)]/\sqrt{\sigma^2/n + \sigma^2/m}}{\sqrt{(nS_X^2 + mS_Y^2)/\sigma^2(n + m - 2)}}$$

$$= \frac{\bar{X} - \bar{Y} - (\mu_X - \mu_Y)}{\sqrt{[(nS_X^2 + mS_Y^2)/(n + m - 2)](1/n + 1/m)}}.$$

If, using $r = n + m - 2$ degrees of freedom, we have $P(T \leq t_0) = 1 - \alpha/2$, where $\alpha = 1 - \gamma$, then

$$P(-t_0 \leq T \leq t_0) = 1 - \alpha = \gamma.$$

Solving the inequality for $\mu_X - \mu_Y$ yields

$$P(\bar{X} - \bar{Y} - t_0 R \leq \mu_X - \mu_Y \leq \bar{X} - \bar{Y} + t_0 R) = \gamma,$$

where

$$R = \sqrt{\frac{nS_X^2 + mS_Y^2}{n + m - 2}\left(\frac{1}{n} + \frac{1}{m}\right)}.$$

If $\bar{x}$, $\bar{y}$, and r are the observed values of $\bar{X}$, $\bar{Y}$, and R, then

$$[\bar{x} - \bar{y} - t_0 r, \bar{x} - \bar{y} + t_0 r]$$

is a $(100\gamma)\%$ confidence interval for $\mu_X - \mu_Y$.

Example 7.4–5. Suppose that scores on a standardized test in mathematics taken by students from large and small high schools are $N(\mu_X, \sigma^2)$ and $N(\mu_Y, \sigma^2)$, respectively, where σ^2 is unknown. If a random sample of $n = 9$ students from large high schools yielded $\bar{x} = 81.31$, $s_X^2 = 60.76$ and a random sample of $m = 15$ students from small high schools yielded $\bar{y} = 78.61$, $s_Y^2 = 48.24$, the endpoints for a 95% confidence interval for $\mu_X - \mu_Y$ are given by

$$81.31 - 78.61 \pm 2.074 \sqrt{\frac{9(60.76) + 15(48.24)}{22}\left(\frac{1}{9} + \frac{1}{15}\right)},$$

because $P(T \le 2.074) = 0.975$. That is, the 95% confidence interval is $[-3.95, 9.35]$.

The assumption of equal variances, namely, $\sigma_X^2 = \sigma_Y^2$, can be modified somewhat so that we are still able to find a confidence interval for $\mu_X - \mu_Y$. That is, if we know the ratio σ_X^2/σ_Y^2 of the variances, we can still make this type of statistical inference using a random variable with a t distribution (see Exercise 7.4–10).

If the sample sizes are large, we can replace σ_X^2/n and σ_Y^2/m with $s_X^2/(n-1)$ and $s_Y^2/(m-1)$, where s_X^2 and s_Y^2 are the values of the respective sample variances. This means, with $\Phi(z_0) = 1 - \alpha/2$ and $\gamma = 1 - \alpha$, that

$$\bar{x} - \bar{y} \pm z_0 \sqrt{\frac{s_X^2}{n-1} + \frac{s_Y^2}{m-1}}$$

serves as an approximate $(100\gamma)\%$ confidence interval for $\mu_X - \mu_Y$.

=== *Exercises* ===

7.4–1. Let T have a t distribution with r degrees of freedom. Find
(a) $P(T \ge 2.228)$ when $r = 10$, (b) $P(T \le 2.228)$ when $r = 10$,
(c) $P(|T| \ge 2.228)$ when $r = 10$, (d) $P(-1.753 \le T \le 2.602)$ when $r = 15$,
(e) $P(1.330 \le T \le 2.552)$ when $r = 18$.

7.4–2. Let T have a t distribution with $r = 19$. Find c such that
(a) $P(|T| \ge c) = 0.05$, (b) $P(|T| \ge c) = 0.01$,
(c) $P(T \ge c) = 0.025$, (d) $P(|T| \le c) = 0.95$.

7.4–3. Let T have a t distribution with r degrees of freedom. Show that $E(T) = 0, r \ge 2$, and $\mathrm{Var}(T) = r/(r-2)$, provided $r > 3$, by first finding $E(Z)$, $E(1/\sqrt{U})$, $E(Z^2)$, and $E(1/U)$.

7.4–4. Let $\bar{X}$ and S^2 be the mean and the variance of a random sample of size $n = 17$ from a distribution $N(\mu, \sigma^2)$. Find the constant c such that

$$P\left(-c < \frac{4(\bar{X} - \mu)}{S} < c\right) = 0.95.$$

7.4–5. A random sample of $n = 9$ IQ scores of college students, assumed to be $N(\mu, \sigma^2)$, yielded a sample mean of $\bar{x} = 112.72$ and a sample standard deviation of 9.38. Find a 90% confidence interval for μ.

7.4–6. Find an 80% confidence interval for μ if the observed values of a random sample of eight Scholastic Aptitude Test (SAT) scores in mathematics, assumed to be $N(\mu, \sigma^2)$, are

$$x_1 = 624 \qquad x_2 = 532 \qquad x_3 = 565 \qquad x_4 = 492 \qquad x_5 = 407$$
$$x_6 = 591 \qquad x_7 = 611 \qquad x_8 = 558$$

7.4-7. Let $X_1, X_2, \ldots, X_5$ be a random sample of size 5 from the normal distribution $N(\mu, \sigma^2)$. Calculate the expected length of a 95% confidence interval for μ assuming that the variance is   **(a)** known,   **(b)** unknown. HINT: To find $E(S)$ determine $E(\sqrt{nS^2/\sigma^2})$, recalling nS^2/σ^2 is $\chi^2(n-1)$.

7.4-8. Let T have a t distribution with r degrees of freedom. Show that T^2 has an F distribution with 1 and r degrees of freedom. HINT: Consider $T^2 = U^2/(V/r)$.

7.4-9. Let $X_1, X_2, \ldots, X_5$ be a random sample of SAT mathematics scores, assumed to be $N(\mu_X, \sigma^2)$, and let $Y_1, Y_2, \ldots, Y_8$ be an independent random sample of SAT verbal scores, assumed to be $N(\mu_Y, \sigma^2)$. If the following data are observed, find a 90% confidence interval for $\mu_X - \mu_Y$.

$$x_1 = 644 \qquad x_2 = 493 \qquad x_3 = 532 \qquad x_4 = 462 \qquad x_5 = 565$$

$$y_1 = 623 \qquad y_2 = 472 \qquad y_3 = 492 \qquad y_4 = 661 \qquad y_5 = 540$$
$$y_6 = 502 \qquad y_7 = 549 \qquad y_8 = 518$$

7.4-10. Let $\overline{X}, \overline{Y}, S_X^2$, and S_Y^2 be the respective means and variances of random samples of sizes n and m from the independent normal distributions $N(\mu_X, \sigma_X^2)$ and $N(\mu_Y, \sigma_Y^2)$, where μ_X, μ_Y, σ_X^2, and σ_Y^2 are unknown. If, however, $\sigma_X^2/\sigma_Y^2 = d$, a known constant, argue that

(a) $\dfrac{(\overline{X} - \overline{Y}) - (\mu_X - \mu_Y)}{\sqrt{d\sigma_Y^2/n + \sigma_Y^2/m}}$ is $N(0, 1)$;   **(b)** $\dfrac{nS_X^2}{d\sigma_Y^2} + \dfrac{mS_Y^2}{\sigma_Y^2}$ is $\chi^2(n + m - 2)$;

(c) the two random variables in (a) and (b) are independent.

(d) With these results, construct a random variable (not depending upon σ_Y^2) that has a t distribution and can be used to construct a confidence interval for $\mu_X - \mu_Y$.

7.4-11. Independent random samples of the heights of adult males living in two countries yielded the following results: $n = 50$, $\bar{x} = 65.7$ inches, $s_X = 4$ inches and $m = 50$, $\bar{y} = 68.2$ inches, $s_Y = 3$ inches. Find an approximate 98% confidence interval for the difference $\mu_X - \mu_Y$ of the means of the populations of heights. (Do not assume equal variances. The sample sizes are large enough to use approximations.)

7.4-12. A biologist who studies spiders was interested in comparing the lengths of female and male green lynx spiders. Assume that the length X of the female spider is approximately $N(\mu_X, \sigma_X^2)$ and the length Y of the male spider is approximately $N(\mu_Y, \sigma_Y^2)$. Find an approximate 95% confidence interval for $\mu_X - \mu_Y$ using $n = 30$ observations of X:

5.20	4.70	5.75	7.50	6.45	6.55
4.70	4.80	5.95	5.20	6.35	6.95
5.70	6.20	5.40	6.20	5.85	6.80
5.65	5.50	5.65	5.85	5.75	6.35
5.75	5.95	5.90	7.00	6.10	5.80

and $m = 30$ observations of Y:

8.25	9.95	5.90	6.55	8.45	7.55
9.80	10.90	6.60	7.55	8.10	9.10
6.10	9.30	8.75	7.00	7.80	8.00
9.00	6.30	8.35	8.70	8.00	7.50
9.50	8.30	7.05	8.30	7.95	9.60

where the measurements are in millimeters.

7.5 Point Estimation

We continue to consider random variables for which the functional form of the p.d.f. is known but the distribution depends upon an unknown parameter, say θ, that may have any value in a set Ω. Interval estimation of a parameter θ, which has already been considered, can be viewed as an effort by the statistician to select a subset of possible probability density functions by taking a subset of Ω for a confidence interval for the parameter θ. However, in certain instances, it might be necessary for the experimenter to select precisely one member of the family as the p.d.f. of the random variable. That is, the experimenter needs a *point* estimate of the parameter θ, as opposed to an interval estimate of θ. In Section 7.1, the maximum likelihood estimator $\hat{\theta}$ provided a point estimate of θ.

In maximum likelihood estimation, to elicit some information about the unknown parameter θ, we take a random sample from the distribution. That is, we repeat the experiment n independent times, observe the sample items $X_1, X_2, \ldots, X_n$, and try to guess the value of θ using the observations x_1, $x_2, \ldots, x_n$ by maximizing the likelihood function $L(\theta)$. Thus, we select a statistic $u(X_1, X_2, \ldots, X_n)$, called an *estimator*, such that if $x_1, x_2, \ldots, x_n$ are the observed values of $X_1, X_2, \ldots, X_n$, then the number $\hat{\theta} = u(x_1, x_2, \ldots, x_n)$ maximizes $L(\theta)$. We have that $u(x_1, x_2, \ldots, x_n)$, in most situations, is reasonably close to the unknown point θ. In this section we consider some properties of estimators that make the estimates (not only maximum likelihood estimates) tend to be close to θ.

If $Y = u(X_1, X_2, \ldots, X_n)$ is to be a good estimator of θ, what properties should Y enjoy? Note that Y is a random variable and thus has a probability distribution. If Y is to be a good estimator of θ, a fairly desirable property is that its mean be equal to θ, namely, $E(Y) = \theta$. That is, in practice, we would want $E(Y)$ to be reasonably close to θ.

DEFINITION 7.5–1. *If* $E[u(X_1, X_2, \ldots, X_n)] = \theta, u(X_1, X_2, \ldots, X_n)$ *is called an* unbiased *estimator of* θ. *Otherwise it is said to be* biased.

Example 7.5–1. Let $X_1, X_2, \ldots, X_n$ be a random sample of size n from a normal distribution $N(\mu, \sigma^2)$. Since the maximum likelihood estimator of μ is $\overline{X}$, which is $N(\mu, \sigma^2/n)$, then $E(\overline{X}) = \mu$ and thus $\overline{X}$ is an unbiased estimator of μ. However, the maximum likelihood estimator of σ^2 is

$$S^2 = \frac{1}{n} \sum_{i=1}^{n} (X_i - \overline{X})^2,$$

and

$$E(S^2) = E\left[\frac{\sigma^2}{n}\left(\frac{nS^2}{\sigma^2}\right)\right] = \frac{\sigma^2}{n}(n - 1)$$

because nS^2/σ^2 is $\chi^2(n - 1)$. Thus, S^2 is a biased estimator of σ^2. Note that the bias is not great for large n, however; and, in fact, $nS^2/(n - 1)$ is an unbiased estimator of σ^2.

Certainly, most of us would place more reliance on the estimator $\overline{X}$ of μ if it were based on $n = 100$ observations rather than on $n = 10$. Why is this? To present a mathematical answer, let us consider the variance of $\overline{X}$, namely, $\operatorname{Var}(\overline{X}) = \sigma^2/n$. Accordingly, as n increases, the variance and, hence, the spread of the distribution decreases. Since the center of the distribution of $\overline{X}$ is at μ for every n, the probability that $\overline{X}$ will be close to μ increases as n increases. That is,

$$P(|\overline{X} - \mu| < \varepsilon)$$

increases with increasing n. This concept leads to the definition of a consistent estimator.

DEFINITION 7.5–2. *The statistic $Y = u(X_1, X_2, \ldots, X_n)$ is a consistent estimator of θ if, for each positive number ε,*

$$\lim_{n \to \infty} P(|Y - \theta| \geq \varepsilon) = 0$$

or, equivalently,

$$\lim_{n \to \infty} P(|Y - \theta| < \varepsilon) = 1.$$

Chebyshev's inequality, Theorem 2.3–1, helps us prove a theorem that provides sufficient conditions for an estimator to be consistent.

THEOREM 7.5–1. *If Y is an unbiased estimator of θ and $\operatorname{Var}(Y) \to 0$ as $n \to \infty$, then Y is a consistent estimator of θ.*

Proof: By Chebyshev's inequality, given $\varepsilon > 0$,

$$P(|Y - \theta| \geq \varepsilon) \leq \frac{\operatorname{Var}(Y)}{\varepsilon^2} \to 0 \qquad \text{as } n \to \infty.$$

Thus, Y is a consistent estimator of θ.

Example 7.5–2. If the sample arises from $N(\mu, \sigma^2)$, we have noted that $nS^2/(n-1)$ is an unbiased estimator of σ^2. Since nS^2/σ^2 is $\chi^2(n-1)$,

$$\mathrm{Var}\left(\frac{nS^2}{n-1}\right) = \mathrm{Var}\left[\frac{\sigma^2}{n-1}\left(\frac{nS^2}{\sigma^2}\right)\right] = \frac{2(n-1)\sigma^4}{(n-1)^2},$$

which has a limit of zero as $n \to \infty$. Thus, $nS^2/(n-1)$ is both an unbiased and a consistent estimator of σ^2.

From the preceding discussion, we gather that it is desirable for an estimator of an unknown parameter to have an expected value at least close to the parameter and as small a variance as possible. We have seen that the latter can be achieved by increasing the sample size. Most of the time in applications, however, the size of the sample cannot be made too large due to limitations of finances, time, effort, and so on. Thus, with a random sample $X_1, X_2, \ldots, X_n$ of a fixed sample size n, a statistician might like to find that estimator $Y = u(X_1, X_2, \ldots, X_n)$ of an unknown parameter θ which minimizes the mean (expected) value of the square of the error (difference) $Y - \theta$; that is, minimizes

$$E[(Y - \theta)^2] = E\{[u(X_1, X_2, \ldots, X_n) - \theta]^2\}.$$

DEFINITION 7.5–3. *The statistic Y that minimizes $E[(Y - \theta)^2]$ is the one with* minimum mean square error. *If we restrict our attention to only unbiased estimators, then $\mathrm{Var}(Y) = E[(Y - \theta)^2]$, and the unbiased statistic Y that minimizes this expression is said to be the* minimum variance unbiased estimator *of θ.*

Two illustrations will help reinforce these new ideas.

Example 7.5–3. Say that our observations are restricted to a random sample X_1, X_2, X_3 of size $n = 3$ from $N(\mu, \sigma^2)$, $-\infty < \mu < \infty$, and σ^2 is a known positive number. Of course, $\overline{X}$ is unbiased and has variance $\sigma^2/3$. Let us, however, consider another possible estimator, namely,

$$Y = \frac{X_1 + 2X_2 + 3X_3}{6},$$

which is the weighted average of the three sample items. It is possible that some statistician might like a statistic such as Y because it places the most weight on the last observation, namely, X_3. Note that

$$E(Y) = \frac{1}{6} E(X_1 + 2X_2 + 3X_3) = \frac{1}{6}(\mu + 2\mu + 3\mu) = \mu$$

and

$$\text{Var}(Y) = \left(\frac{1}{6}\right)^2 \sigma^2 + \left(\frac{2}{6}\right)^2 \sigma^2 + \left(\frac{3}{6}\right)^2 \sigma^2 = \frac{7\sigma^2}{18}.$$

Thus, we see that Y is also unbiased but that $\overline{X}$ is a better unbiased estimator of μ when the sample arises from a normal distribution since $\text{Var}(\overline{X}) < \text{Var}(Y)$.

Example 7.5-4. We have seen in Examples 7.5-1 and 7.5-2 that the variance S^2 of the sample that arises from $N(\mu, \sigma^2)$, $-\infty < \mu < \infty, 0 < \sigma^2 < \infty$, is a biased estimator of σ^2 and that

$$\frac{nS^2}{n-1} = \frac{1}{n-1}\sum_{i=1}^{n}(X_i - \overline{X})^2$$

is an unbiased estimator of σ^2. This, incidentally, is the reason some authors use the factor $1/(n-1)$ rather than $1/n$ in the definition of the sample variance. Suppose, however, that we wish to find a constant c such that $E[(cS^2 - \sigma^2)^2]$ is minimized. That is, we want to find the minimum mean square error estimator of σ^2, which is of the form cS^2. If we recall that $E(W^2) = \mu_w^2 + \sigma_w^2$ for every random variable (provided these expectations exist), we have, with $W = cS^2 - \sigma^2$, that

$$E[(cS^2 - \sigma^2)^2] = [E(cS^2 - \sigma^2)]^2 + \text{Var}(cS^2 - \sigma^2).$$

Since $E(S^2) = (n-1)\sigma^2/n$ and $\text{Var}(S^2) = 2(n-1)\sigma^4/n^2$, we have that

$$E[(cS^2 - \sigma^2)^2] = \left[c\frac{(n-1)\sigma^2}{n} - \sigma^2\right]^2 + \frac{2c^2(n-1)\sigma^4}{n^2}.$$

The derivative of this expression with respect to c equated to zero provides the equation

$$2\left[c\frac{(n-1)\sigma^2}{n} - \sigma^2\right]\frac{(n-1)\sigma^2}{n} + 4c\frac{(n-1)\sigma^4}{n^2} = 0,$$

the solution of which is $c = n/(n+1)$. It is an easy exercise to verify that this value of c actually minimizes $E[(cS^2 - \sigma^2)^2]$. That is, the minimum mean square error estimator of σ^2 that is of the form cS^2 is

$$\frac{n}{n+1}S^2 = \frac{1}{n+1}\sum_{i=1}^{n}(X_i - \overline{X})^2.$$

This answer also presents another interesting twist to the problem of which coefficient to associate with $\sum(X_i - \overline{X})^2$: $1/(n-1)$ creates the unbiased estimator of σ^2; $1/(n+1)$ yields the minimum mean square error estimator of σ^2; and $1/n$ gives the variance of the empirical distribution and, thus, that of the sample.

===== *Exercises* =====

7.5–1. Let $X_1, X_2, \ldots, X_n$ be a random sample of size n from the exponential distribution whose p.d.f. is $f(x; \theta) = (1/\theta)e^{-x/\theta}, 0 < x < \infty$.
   (a) Show that $\overline{X}$ is an unbiased estimator of θ.
   (b) Show that the variance of $\overline{X}$ is θ^2/n.
   (c) Show that $\overline{X}$ is a consistent estimator of θ.
   (d) What is a good estimate of θ if a random sample of size 5 yielded the sample values: 3.5, 8.1, 0.9, 4.4, 0.5?

7.5–2. Let $X_1, X_2, \ldots, X_n$ be a random sample from $N(\mu, \sigma^2)$.
   (a) Show that $Y = (X_1 + X_2)/2$ is an unbiased estimator of μ.
   (b) Is Y a consistent estimator of μ? Why?
   (c) Show that S^2 is a consistent estimator of σ^2.

7.5–3. Let $X_1, X_2, \ldots, X_n$ be a random sample from a Poisson distribution whose mean is λ.
   (a) Show that $\overline{X}$ is an unbiased estimator of λ.
   (b) Show that $\overline{X}$ is a consistent estimator of λ.

7.5–4. Show that the mean $\overline{X}$ of a random sample of size n from the uniform distribution on the interval $(\theta - 1, \theta + 1)$ is both an unbiased and consistent estimator of θ.

7.5–5. Let $X_1, X_2, \ldots, X_n$ be a random sample from the normal distribution $N(\mu, \sigma^2)$. Find the constant c such that

$$cS = c\sqrt{\frac{\sum (X_i - \overline{X})^2}{n}}$$

is an unbiased estimator of σ. HINT: Find $E(\sqrt{Y})$, where Y is the chi-square variable nS^2/σ^2.

7.6 Functions of Parameters

Sometimes it is necessary to estimate certain functions of unknown parameters. For example, although we know that the relative frequency Y/n of success in n independent trials is the maximum likelihood estimator of the probability p of success, we might also want an estimate of the variance $h(p) = p(1 - p)/n$ of Y/n. It seems reasonable to suggest the estimator $Z = h(Y/n) = (Y/n)(1 - Y/n)/n$. The expectation of $Z = h(Y/n)$ is

$$E(Z) = \frac{1}{n^3} E(nY - Y^2) = \frac{1}{n^3} [nE(Y) - E(Y^2)]$$

$$= \frac{1}{n^3} [n^2 p - (np)^2 - np(1 - p)]$$

$$= \frac{(n - 1)p(1 - p)}{n^2},$$

because $np(1 - p) = E(Y^2) - (np)^2$. Accordingly, $h(Y/n)$ is a biased estimator of $h(p) = p(1 - p)/n$, but

$$\frac{n}{n-1} h\left(\frac{Y}{n}\right) = \frac{(Y/n)(1 - Y/n)}{n - 1}$$

is an unbiased estimator of $h(p)$.

From the preceding discussion, we see that if $\hat{\theta}$ is a maximum likelihood estimator of θ, which is unbiased, then $h(\hat{\theta})$ is not necessarily an unbiased estimator of a function $h(\theta)$ of θ. However, if $h(\theta)$ is a continuous function of θ and if $\hat{\theta}$ is a consistent estimator of θ, then it is true that $h(\hat{\theta})$ is a consistent estimator of $h(\theta)$. To see this, we note from continuity, that for each given $\varepsilon > 0$ there exists a $\delta > 0$ such that if

$$|\hat{\theta} - \theta| < \delta,$$

then

$$|h(\hat{\theta}) - h(\theta)| < \varepsilon.$$

That is, the former of these two preceding events is a subset of the latter; thus

$$P[|\hat{\theta} - \theta| < \delta] \leq P[|h(\hat{\theta}) - h(\theta)| < \varepsilon].$$

However, consistency of $\hat{\theta}$ means that

$$\lim_{n \to \infty} P[|\hat{\theta} - \theta| < \delta] = 1.$$

Since probability is less than or equal to one, it must follow that

$$\lim_{n \to \infty} P[|h(\hat{\theta}) - h(\theta)| < \varepsilon] = 1,$$

and, hence, $h(\hat{\theta})$ is a consistent estimator of $h(\theta)$.

In a more advanced course, it can be shown, in most situations, that maximum likelihood estimators are consistent estimators of the corresponding parameters. Hence, we will usually use the following procedure for estimating a function of one or more unknown parameters. Suppose, for illustration, we wish to estimate the function $h(\theta_1, \theta_2)$ of two parameters θ_1 and θ_2 with the respective maximum likelihood estimators $\hat{\theta}_1$ and $\hat{\theta}_2$. We will find that the estimator $h(\hat{\theta}_1, \hat{\theta}_2)$ is usually adequate for our purposes, although on certain occasions we might make minor modifications in $h(\hat{\theta}_1, \hat{\theta}_2)$ to eliminate bias.

Example 7.6–1. Suppose that we are interested in the percentage of a normal population that is greater than some given constant c. For illustration, we might want to know (assuming a normal distribution) the percentage of students who will earn a grade point average greater than 2.0. More formally, if X is $N(\mu, \sigma^2)$, we wish to estimate

$$P(c < X) = P\left(\frac{c - \mu}{\sigma} < \frac{X - \mu}{\sigma}\right) = 1 - \Phi\left(\frac{c - \mu}{\sigma}\right)$$

when μ and σ^2 are unknown parameters. Of course, the mean $\overline{X}$ and the variance S^2 of a random sample from this distribution are the maximum likelihood estimators of μ and σ^2, respectively. Hence,

$$Z = 1 - \Phi\left(\frac{c - \overline{X}}{S}\right)$$

is a consistent estimator of the probability $P(c < X)$. Now Z is a strictly increasing function of $(\overline{X} - c)/S$; thus, these statistics are equivalent in the sense that if we know either Z or $(\overline{X} - c)/S$, we can compute the other. This suggests that the ratio

$$\frac{\overline{X} - c}{S/\sqrt{n - 1}} = \frac{\sqrt{n}(\overline{X} - c)/\sigma}{\sqrt{\sum_1^n (X_i - \overline{X})^2/(n - 1)\sigma^2}} = T$$

could play a major role in estimation problems. Please note that in Section 7.4 we observed that T has a t distribution with $n - 1$ degrees of freedom provided $c = \mu$; if $c \neq \mu$, then it has what is called a *noncentral t* distribution.

Example 7.6–2. Say that, in a medical situation, the independent random variables X and Y represent the responses using treatment A and using treatment B, respectively. We are interested in $P(X < Y)$ because if this probability is greater than $1/2$ we most likely would use treatment B (provided a bigger response is judged to be a better response). More formally, let us assume that the independent random variables X and Y are $N(\mu_X, \sigma^2)$ and $N(\mu_Y, \sigma^2)$, respectively, where the means μ_X and μ_Y and common variance σ^2 are unknown. We wish to estimate

$$P(X < Y) = P(X - Y < 0) = P\left(\frac{X - Y - (\mu_X - \mu_Y)}{\sqrt{2}\sigma} < \frac{\mu_Y - \mu_X}{\sqrt{2}\sigma}\right).$$

Since it is easy to show (see Exercise 7.6–2) that $X - Y$ is $N(\mu_X - \mu_Y, 2\sigma^2)$, we have that

$$P(X < Y) = \Phi\left(\frac{\mu_Y - \mu_X}{\sqrt{2}\sigma}\right).$$

The likelihood function associated with the two independent random samples $X_1, X_2, \ldots, X_n$ and $Y_1, Y_2, \ldots, Y_m$ of sizes n and m from these respective distributions is

$$L(\mu_X, \mu_Y, \sigma^2) = \left(\frac{1}{2\pi\sigma^2}\right)^{(n+m)/2} \exp\left[-\frac{\sum_1^n (x_i - \mu_X)^2 + \sum_1^m (y_i - \mu_Y)^2}{2\sigma^2}\right].$$

From Exercise 7.6–3, we see that the maximum likelihood estimators are

$$\hat{\mu}_X = \overline{X}, \qquad \hat{\mu}_Y = \overline{Y}, \qquad \hat{\sigma}^2 = \frac{nS_X^2 + mS_Y^2}{n + m},$$

where $\overline{X}$, $\overline{Y}$, S_X^2, and S_Y^2 are the respective sample means and variances. Hence, a consistent estimator of $P(X < Y)$ is $\Phi(cT)$, where

$$c = \sqrt{\frac{(m + n)^2}{2mn(m + n - 2)}}$$

and

$$T = \frac{\overline{Y} - \overline{X}}{\sqrt{[(nS_X^2 + mS_Y^2)/(n + m - 2)](1/n + 1/m)}}.$$

We see (Exercise 7.6–4) that if $\mu_X = \mu_Y$ then T has a t distribution with $m + n - 2$ degrees of freedom and this T is an extremely important random variable in statistical inference.

Example 7.6–3. Let us again consider the two independent normal distributions of Example 7.6–2, only here allow the possibility of different variances σ_X^2 and σ_Y^2, respectively. We might want to estimate the probability

$$P(|Y - \mu_Y| \le |X - \mu_X|),$$

which compares the deviations of the responses from their respective means. In a medical application, if this probability is high, the doctor may prefer the treatment associated with Y, if harmful side effects result from abnormal responses (large deviations from the mean). We first note that the probability is equal to

$$P\left(\frac{(Y - \mu_Y)^2/\sigma_Y^2}{(X - \mu_X)^2/\sigma_X^2} \le \frac{\sigma_X^2}{\sigma_Y^2}\right) = H\left(\frac{\sigma_X^2}{\sigma_Y^2}\right),$$

where H is the distribution function of the ratio of $(Y - \mu_Y)^2/\sigma_Y^2$ to $(X - \mu_X)^2/\sigma_X^2$. However, each of these latter two expressions is the square of a standardized normal variable, and hence each is $\chi^2(1)$. Accordingly, from the independence of X and Y, that ratio has an F distribution with $r_1 = 1$ and $r_2 = 1$ degrees of freedom; so H is the corresponding distribution function. A consistent estimator of $H(\sigma_X^2/\sigma_Y^2)$ is $H(S_X^2/S_Y^2)$. This estimator is a strictly increasing function of

$$F = \frac{nS_X^2/(n - 1)}{mS_Y^2/(m - 1)},$$

when $F > 0$. The statistic F has, when $\sigma_X^2 = \sigma_Y^2$, an F distribution with $n - 1$ and $m - 1$ degrees of freedom (see Section 7.3). This F is important in making certain statistical inferences about the ratio σ_X^2/σ_Y^2 of the variances.

═══ *Exercises* ═══

7.6–1. Let $\bar{X}$ be the mean of a random sample of size n from a distribution with mean μ and variance σ^2. Show that $\bar{X}^2$ is a biased estimator of μ^2.

7.6–2. Show that $X - Y$ is $N(\mu_X - \mu_Y, 2\sigma^2)$, where the independent random variables X and Y are $N(\mu_X, \sigma^2)$ and $N(\mu_Y, \sigma^2)$, respectively. HINT: Find the moment-generating function of $X - Y$.

7.6–3. Let $X_1, X_2, \ldots, X_n$ and $Y_1, Y_2, \ldots, Y_m$ be random samples from the respective independent normal distributions $N(\mu_X, \sigma^2)$ and $N(\mu_Y, \sigma^2)$. By equating to zero each of the three first partial derivatives of the likelihood function $L(\mu_X, \mu_Y, \sigma^2)$, as given in Example 7.6–2, find the maximum likelihood estimators of μ_X, μ_Y, and σ^2.

7.6–4. Use the background given in Example 7.6–2.
    (a) If $\mu_X = \mu_Y$, show that $\bar{Y} - \bar{X}$ is $N(0, \sigma^2/n + \sigma^2/m)$ and hence $Z = (\bar{Y} - \bar{X})/\sqrt{\sigma^2/n + \sigma^2/m}$ is $N(0, 1)$.
    (b) Argue that $W = (nS_X^2 + mS_Y^2)/\sigma^2$ is $\chi^2(m + n - 2)$.
    (c) Why are Z and W independent?
    (d) Show that $Z/\sqrt{W/(m + n - 2)}$ is that T given in Example 7.6–2, and explain why it has a t distribution with $m + n - 2$ degrees of freedom.

7.7 Regression

There is often interest in the relation between two variables, for example, a student's scholastic aptitude test score in mathematics and this same student's grade in calculus. Frequently one of these variables, say x, is known in advance of the other and hence there is interest in predicting a future random variable Y. Since Y is a random variable, we cannot predict its future observed value $Y = y$ with certainty. Thus, let us first concentrate on the problem of estimating the mean of Y, that is, $E(Y)$. Now $E(Y)$ is usually a function of x; for example, in our illustration with the calculus grade, say Y, we would expect $E(Y)$ to increase with increasing mathematics aptitude score x. Sometimes $E(Y) = \mu(x)$ is assumed to be of a given form, such as linear or quadratic or exponential; that is. $\mu(x)$ could be assumed to be equal to $\alpha + \beta x$ or $\alpha + \beta x + \gamma x^2$ or $\alpha e^{\beta x}$. To estimate $E(Y) = \mu(x)$, or equivalently the parameters α, β, and γ, we observe the random variable Y for each of n different values of x, say $x_1, x_2, \ldots, x_n$. Once the n independent experiments have been performed, we have n pairs of known numbers (x_1, y_1), $(x_2, y_2), \ldots, (x_n, y_n)$. These pairs are then used to estimate the mean $E(Y)$. Problems like this are often classified under *regression* because $E(Y) = \mu(x)$ is frequently called a regression curve.

    Let us begin with the case in which $E(Y) = \mu(x)$ is a linear functior. The data points are $(x_1, y_1), (x_2, y_2), \ldots, (x_n, y_n)$; so the first problem is that of fitting a straight line to the set of data (see Figure 7.7–1).

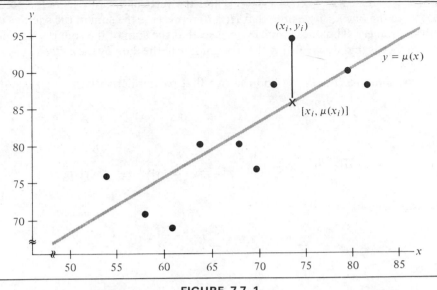

FIGURE 7.7–1

Suppose that in addition to assuming that the mean of Y is a linear function, we can say that Y is normally distributed with unknown variance σ^2. For convenience, rather than taking the mean line to be equal to $\alpha + \beta x$, we let

$$\mu(x) = \alpha + \beta(x - \bar{x}), \qquad \text{where } \bar{x} = \sum_{i=1}^{n} \frac{x_i}{n}.$$

Hence $Y_1, Y_2, \ldots, Y_n$ are mutually independent normal variables with respective means $\alpha + \beta(x_i - \bar{x}), i = 1, 2, \ldots, n$, and unknown variance σ^2. Their joint p.d.f. is therefore the product of the individual probability density functions; that is, the likelihood function equals

$$L(\alpha, \beta, \sigma^2) = \prod_{i=1}^{n} \frac{1}{\sqrt{2\pi\sigma^2}} \exp\left\{ -\frac{[y_i - \alpha - \beta(x_i - \bar{x})]^2}{2\sigma^2} \right\}$$

$$= \left(\frac{1}{2\pi\sigma^2} \right)^{n/2} \exp\left\{ -\frac{\sum_{i=1}^{n} [y_i - \alpha - \beta(x_i - \bar{x})]^2}{2\sigma^2} \right\}.$$

To maximize $L(\alpha, \beta, \sigma^2)$, or, equivalently, to minimize

$$- \ln L(\alpha, \beta, \sigma^2) = \frac{n}{2} \ln(2\pi\sigma^2) + \frac{\sum_{i=1}^{n} [y_i - \alpha - \beta(x_i - \bar{x})]^2}{2\sigma^2},$$

it is obvious that we must select α and β to minimize

$$H(\alpha, \beta) = \sum_{i=1}^{n} [y_i - \alpha - \beta(x_i - \bar{x})]^2.$$

Since $|y_i - \alpha - \beta(x_i - \bar{x})| = |y_i - \mu(x_i)|$ is the vertical distance from the point (x_i, y_i) to the line $\mu(x)$, we note that $H(\alpha, \beta)$ represents the sum of the squares of those distances. Thus, selecting α and β so that the sum of the squares is minimized means that we are fitting the straight line to the data by the *method of least squares*.

To minimize $H(\alpha, \beta)$, we find the two first partial derivatives

$$\frac{\partial H(\alpha, \beta)}{\partial \alpha} = 2 \sum_1^n [y_i - \alpha - \beta(x_i - \bar{x})](-1)$$

and

$$\frac{\partial H(\alpha, \beta)}{\partial \beta} = 2 \sum_1^n [y_i - \alpha - \beta(x_i - \bar{x})][-(x_i - \bar{x})].$$

Setting $\partial H(\alpha, \beta)/\partial \alpha = 0$, we obtain

$$\sum_1^n y_i - n\alpha - \beta \sum_1^n (x_i - \bar{x}) = 0.$$

Since

$$\sum_1^n (x_i - \bar{x}) = 0,$$

we have that

$$\sum_1^n y_i - n\alpha = 0$$

and thus

$$\hat{\alpha} = \bar{Y}.$$

The equation $\partial H(\alpha, \beta)/\partial \beta = 0$ yields, with α replaced by $\bar{y}$,

$$\sum_1^n (y_i - \bar{y})(x_i - \bar{x}) - \beta \sum_1^n (x_i - \bar{x})^2 = 0$$

or, equivalently,

$$\hat{\beta} = \frac{\sum_1^n (Y_i - \bar{Y})(x_i - \bar{x})}{\sum_1^n (x_i - \bar{x})^2} = \frac{\sum_1^n Y_i(x_i - \bar{x})}{\sum_1^n (x_i - \bar{x})^2}.$$

To find the maximum likelihood estimator of σ^2, consider the partial derivative

$$\frac{\partial[-\ln L(\alpha, \beta, \sigma^2)]}{\partial(\sigma^2)} = \frac{n}{2\sigma^2} - \frac{\sum_1^n [y_i - \alpha - \beta(x_i - \bar{x})]^2}{2(\sigma^2)^2}.$$

Setting this equal to zero and replacing α and β by their solutions $\hat{\alpha}$ and $\hat{\beta}$, we obtain

$$\widehat{\sigma^2} = \frac{1}{n} \sum_{i=1}^n [Y_i - \hat{\alpha} - \hat{\beta}(x_i - \bar{x})]^2.$$

Example 7.7–1. The data plotted in Figure 7.7–1 are 10 pairs of test scores of 10 students in a psychology class, x being the score on a preliminary test and y the score on the final examination. The values of x and y are shown in Table 7.7–1.

x	y	x	y
70	77	54	76
74	94	82	88
72	88	64	80
68	80	80	90
58	71	61	69

TABLE 7.7–1

For these data, $\bar{x} = 68.3$, $\bar{y} = 81.3$, $\sum_1^{10} (x_i - \bar{x})^2 = 756.1$, and

$$\sum_1^{10} y_i(x_i - \bar{x}) = 561.1.$$

Thus, $\hat{\alpha} = 81.3$ and $\hat{\beta} = 561.1/756.1 = 0.74$. Hence, the least squares regression line is

$$\mu(x) = 81.3 + 0.74(x - 68.3).$$

During the preceding discussion we have treated $x_1, x_2, \ldots, x_n$ as constants. Of course, many times they can be set by the experimenter; for example, a chemist in experimentation might produce a compound at many different temperatures. But it is also true that these numbers might be observations on an earlier random variable, such as an SAT score or preliminary test grade, but we consider the problem on the *condition* that these x values are given in either case. Thus, in finding the distributions of $\hat{\alpha}$, $\hat{\beta}$, and $\widehat{\sigma^2}$, the only random variables are $Y_1, Y_2, \ldots, Y_n$.

Since $\hat{\alpha}$ is a linear function of independent and normally distributed random variables, $\hat{\alpha}$ has a normal distribution with mean

$$E(\hat{\alpha}) = E\left(\frac{1}{n} \sum_1^n Y_i\right) = \frac{1}{n} \sum_1^n E(Y_i)$$

$$= \frac{1}{n} \sum_1^n [\alpha + \beta(x_i - \bar{x})] = \alpha,$$

and variance

$$\text{Var}(\hat{\alpha}) = \sum_1^n \left(\frac{1}{n}\right)^2 \text{Var}(Y_i) = \frac{\sigma^2}{n}.$$

The estimator $\hat{\beta}$ is also a linear function of $Y_1, Y_2, \ldots, Y_n$ and, hence, has a normal distribution with mean

$$E(\hat{\beta}) = \frac{\sum_1^n (x_i - \bar{x})E(Y_i)}{\sum_1^n (x_i - \bar{x})^2}$$

$$= \frac{\sum_1^n (x_i - \bar{x})[\alpha + \beta(x_i - \bar{x})]}{\sum_1^n (x_i - \bar{x})^2}$$

$$= \frac{\alpha \sum_1^n (x_i - \bar{x}) + \beta \sum_1^n (x_i - \bar{x})^2}{\sum_1^n (x_i - \bar{x})^2} = \beta,$$

and variance

$$\mathrm{Var}(\hat{\beta}) = \sum_1^n \left(\frac{x_i - \bar{x}}{\sum_1^n (x_i - \bar{x})^2} \right)^2 \mathrm{Var}(Y_i)$$

$$= \frac{\sum_1^n (x_i - \bar{x})^2}{[\sum_1^n (x_i - \bar{x})^2]^2} \sigma^2 = \frac{\sigma^2}{\sum_1^n (x_i - \bar{x})^2}.$$

It can be shown that (Exercise 7.7–4)

$$\sum_1^n [Y_i - \alpha - \beta(x_i - \bar{x})]^2 = \sum_1^n \{(\hat{\alpha} - \alpha) + (\hat{\beta} - \beta)(x_i - \bar{x})$$

$$+ [Y_i - \hat{\alpha} - \hat{\beta}(x_i - \bar{x})]\}^2$$

$$= n(\hat{\alpha} - \alpha)^2 + (\hat{\beta} - \beta)^2 \sum_1^n (x_i - \bar{x})^2$$

$$+ \sum_1^n [Y_i - \hat{\alpha} - \hat{\beta}(x_i - \bar{x})]^2. \qquad (1)$$

From the fact that Y_i, $\hat{\alpha}$, and $\hat{\beta}$ have normal distributions, we know that each of

$$\frac{[Y_i - \alpha - \beta(x_i - \bar{x})]^2}{\sigma^2}, \quad \frac{(\hat{\alpha} - \alpha)^2}{\sigma^2/n}, \quad \frac{(\hat{\beta} - \beta)^2}{\sigma^2/\sum_1^n (x_i - \bar{x})^2}$$

has a chi-square distribution with one degree of freedom. Since $Y_1, Y_2, \ldots, Y_n$ are mutually independent, then

$$\frac{\sum_1^n [Y_i - \alpha - \beta(x_i - \bar{x})]^2}{\sigma^2}$$

is $\chi^2(n)$. That is, the left-hand member of equation (1) divided by σ^2 is $\chi^2(n)$ and is equal to the sum of two $\chi^2(1)$ variables and

$$\frac{\sum_1^n [Y_i - \hat{\alpha} - \hat{\beta}(x_i - \bar{x})]^2}{\sigma^2} = \frac{n\widehat{\sigma^2}}{\sigma^2}.$$

Thus, we might then guess that $n\widehat{\sigma^2}/\sigma^2$ is $\chi^2(n - 2)$. This is true, and, moreover, $\hat{\alpha}$, $\hat{\beta}$, and $\widehat{\sigma^2}$ are mutually independent. Since we are not now prepared to prove these facts, we accept them here without proof, but consider them further in Exercise 11.1–5.

Suppose now that we are interested in forming a confidence interval for α. We can use the fact that

$$T_1 = \frac{[\sqrt{n}(\hat{\alpha} - \alpha)]/\sigma}{\sqrt{n\widehat{\sigma^2}/[\sigma^2(n - 2)]}} = \frac{\hat{\alpha} - \alpha}{\sqrt{\widehat{\sigma^2}/(n - 2)}}$$

is a Student's t with $n - 2$ degrees of freedom. Thus, we can select a constant c from the t table so that

$$P\left(-c \leq \frac{\hat{\alpha} - \alpha}{\sqrt{\widehat{\sigma^2}/(n - 2)}} \leq c\right) = \gamma.$$

It follows that $[\hat{\alpha} - c\sqrt{\widehat{\sigma^2}/(n - 2)}, \hat{\alpha} \div c\sqrt{\widehat{\sigma^2}/(n - 2)}]$ is a $(100\gamma)\%$ confidence interval for α.

Similarly,

$$T_2 = \frac{\sqrt{\sum_1^n (x_i - \bar{x})^2}(\hat{\beta} - \beta)/\sigma}{\sqrt{n\widehat{\sigma^2}/[\sigma^2(n - 2)]}}$$

$$= \frac{\hat{\beta} - \beta}{\sqrt{n\widehat{\sigma^2}/[(n - 2)\sum_1^n (x_i - \bar{x})^2]}}$$

has a t distribution with $n - 2$ degrees of freedom. Thus, T_2 can be used to make inferences about β. The fact that $n\widehat{\sigma^2}/\sigma^2$ has a chi-square distribution with $n - 2$ degrees of freedom can be used to make inferences about the variance σ^2.

═══ *Exercises* ═══

7.7–1. The midterm and final exam scores of 10 students in a statistics course are tabulated below.

(a) Calculate the least squares regression line for these data.

(b) Plot the points and the least squares regression line on the same graph.

MIDTERM	FINAL	MIDTERM	FINAL
70	87	67	73
74	79	70	83
80	88	64	79
84	98	74	91
80	96	82	94

7.7–2. The final course grade in calculus was predicted on the basis of the student's high school grade point average in mathematics, Scholastic Aptitude Test (SAT) score in mathematics, and score on a mathematics entrance examination. The predicted grades X and the earned grades Y for 10 students are given (2.0 represents a C, 2.3 a C+, 2.7 a B−, etc.).
   (a) Calculate the least squares regression line for these data.
   (b) Plot the points and the least squares regression line on the same graph.

X	Y	X	Y
2.0	1.3	2.7	3.0
3.3	3.3	4.0	4.0
3.7	3.3	3.7	3.0
2.0	2.0	3.0	2.7
2.3	1.7	2.3	3.0

7.7–3. Students' scores on the mathematics portion of the ACT examination X and on the final examination in first semester calculus (200 points possible) Y are given.
   (a) Calculate the least squares regression line for these data.
   (b) Plot the points and the least squares regression line on the same graph.

X	Y	X	Y
25	138	20	100
20	84	25	143
26	104	26	141
26	112	28	161
28	88	25	124
28	132	31	118
29	90	30	168
32	183		

7.7–4. Show that

$$\sum_1^n [Y_i - \alpha - \beta(x_i - \bar{x})]^2 = n(\hat{\alpha} - \alpha)^2 + (\hat{\beta} - \beta)^2 \sum_1^n (x_i - \bar{x})^2$$

$$+ \sum_1^n [Y_i - \hat{\alpha} - \hat{\beta}(x_i - \bar{x})]^2.$$

7.7–5. A random sample of size $n = 10$ yielded $\hat{\alpha} = 67$, $\hat{\beta} = 2.1$, and $\widehat{\sigma^2} = 288$. Find 95% confidence intervals for α, β, and σ^2, given that

$$\sum_1^{10} (x_i - \bar{x})^2 = 241.$$

7.7–6. Let $\hat{\alpha}$, $\hat{\beta}$, and $\widehat{\sigma^2}$ be the estimators given in this section based upon n random variables. These can be used in the prediction of another normal independent random variable, say Y_{n+1}, having mean $\alpha + \beta(x_{n+1} - \bar{x})$ and unknown variance σ^2, where $\bar{x} = \sum_1^n x_i/n$.

(a)   Recalling that Y_{n+1}, $\hat{\alpha}$, and $\hat{\beta}$ are mutually independent, show that the distribution of $Y_{n+1} - \hat{\alpha} - \hat{\beta}(x_{n+1} - \bar{x})$ is normal with mean zero and variance

$$\sigma^2 \left[\frac{n+1}{n} + \frac{(x_{n+1} - \bar{x})^2}{\sum_1^n (x_i - \bar{x})^2} \right].$$

(b)   Standardize $Y_{n+1} - \hat{\alpha} - \hat{\beta}(x_{n+1} - \bar{x})$ and then divide the resulting $N(0, 1)$ variable by the square root of $n\hat{\sigma}^2/[\sigma^2(n-2)]$; call this ratio T. Argue that T is a Student's t with $n - 2$ degrees of freedom.

(c)   A constant c can be found so that $P(-c < T < c) = \gamma$, where T is that of part (b). Show that the inequalities $-c < T < c$ are equivalent to

$$\hat{\alpha} + \hat{\beta}(x_{n+1} - \bar{x}) - cd < Y_{n+1} < \hat{\alpha} + \hat{\beta}(x_{n+1} - \bar{x}) + cd,$$

where

$$d = \hat{\sigma} \left[\frac{n+1}{n} + \frac{(x_{n+1} - \bar{x})^2}{\sum_1^n (x_i - \bar{x})^2} \right]^{1/2} \left(\frac{n}{n-2} \right)^{1/2}.$$

These latter inequalities provide a $(100\gamma)\%$ *prediction interval* for Y_{n+1}.

8

TESTS OF STATISTICAL HYPOTHESES

8.1 Some Examples and Definitions

A first major area of statistical inference is estimation of parameters and we have introduced some of these basic procedures in Chapters 6 and 7. A second major area is testing statistical hypotheses and some of these ideas are presented in this chapter. We begin by illustrating a relationship between the areas of estimating and testing.

Suppose we know that the breaking strength of a certain type of steel bar has a normal distribution with mean μ and variance $\sigma^2 = 25$. That is, if X is that breaking strength, then X is $N(\mu, 25)$. The manufacturing process has been changed as a result of research, and hence the mean breaking strength μ is now unknown. However, persons who were involved in the decision to change the process have conjectured that μ is about 80. Such an assertion about a distribution of a random variable is called a *statistical hypothesis*; that is, the conjecture that $\mu = 80$ is a statistical hypothesis, here denoted by $H_0: \mu = 80$.

To test the validity of H_0, we take a random sample $X_1, X_2, \ldots, X_n$ of breaking strengths and determine whether the observed sample mean $\bar{x}$ is "close enough" to 80 for us to accept H_0. To help us decide when $\bar{x}$ is "close enough" to 80, we could decide to accept H_0 if a confidence interval for μ contains the hypothesized value $\mu = 80$. That is, since $\overline{X}$ is $N(\mu, 25/n)$, $\bar{x} \pm z_0(5/\sqrt{n})$, where $\Phi(z_0) = 1 - \alpha/2$ and $\bar{x}$ is the observed mean of the sample, is a $100(1 - \alpha)\%$ confidence interval for μ. If the value 80 is in this interval, we certainly would not want to reject $H_0 : \mu = 80$, at least not until we gain more information about the new process. That is, $\mu = 80$ seems plausible enough when we have that

$$\bar{x} - z_0\left(\frac{5}{\sqrt{n}}\right) \leq 80 \leq \bar{x} + z_0\left(\frac{5}{\sqrt{n}}\right)$$

or, equivalently,

$$\frac{|\bar{x} - 80|}{5/\sqrt{n}} \leq z_0.$$

In such a case, we do not reject (that is, we accept) the hypothesis $H_0 : \mu = 80$. If, on the other hand, 80 is not in the interval or, equivalently,

$$\frac{|\bar{x} - 80|}{5/\sqrt{n}} > z_0,$$

we reject the hypothesis that $\mu = 80$. Please note that this *test* of the hypothesis or rule agrees with one's intuition: if the mean $\bar{x}$ of the sample deviates too much from the hypothesized mean 80, we reject the hypothesis that $\mu = 80$. This deviation is measured in terms of the standard deviation $5/\sqrt{n}$ of $\overline{X}$. Of course, it is possible for $H_0 : \mu = 80$ to be true, and yet have the hypothesis rejected by this rule. What then is the probability of such an error? It is, when $\mu = 80$, given by

$$P\left(\frac{|\overline{X} - 80|}{5/\sqrt{n}} > z_0\right) = 1 - P\left(-z_0 \leq \frac{|\overline{X} - 80|}{5/\sqrt{n}} \leq z_0\right).$$

But here $\overline{X}$ is $N(80, 25/n)$ when $\mu = 80$, and thus $(\overline{X} - 80)/(5/\sqrt{n})$ is $N(0, 1)$. Accordingly,

$$P\left(\frac{|\overline{X} - 80|}{5/\sqrt{n}} > z_0\right) = 1 - [\Phi(z_0) - \Phi(-z_0)] = 1 - (1 - \alpha) = \alpha.$$

That is, α is the probability of rejecting $H_0 : \mu = 80$ by this rule when that hypothesis is true. This probability is given a very special name; it is called the *significance level* of the test. In most instances α is selected to be a small value, such as 0.10, 0.05, 0.01, or less.

Example 8.1–1.   In the preceding discussion, take $n = 100$ and $\alpha = 0.05$. Then $z_0 = 1.96$ since $\Phi(1.96) = 0.975 = 1 - \alpha/2$. Suppose the observed sample of breaking strengths had the mean of $\bar{x} = 77.8$. We have

$$\frac{|77.8 - 80|}{5/\sqrt{100}} = 4.4 > 1.96,$$

and thus $H_0 : \mu = 80$ is rejected at the 5% significance level.

Let us summarize the preceding discussion by formally defining a statistical hypothesis, a test of a hypothesis, and the significance level of a test.

DEFINITION 8.1–1.   *A statistical hypothesis is an assertion about a distribution of one or more random variables. It is frequently denoted by symbols such as* H_0 *or* H_1.

DEFINITION 8.1–2.   *A* test *of a statistical hypothesis* H_0 *is a procedure, based upon the observed values of the random variables, that leads to the acceptance or rejection of the hypothesis* H_0.

DEFINITION 8.1–3.   *The* significance level *of the test of a statistical hypothesis* H_0 *is the probability of rejecting the hypothesis* H_0 *when it is true.*

REMARK.   In discussing the test of a statistical hypothesis, the word "accept" might better be replaced by "do not reject." That is, if, in Example 8.1–1, $\bar{x}$ is close enough to 80 so that we accept $\mu = 80$, we do not want that acceptance to imply that μ is actually equal to 80. We want to say that the data do not deviate enough from $\mu = 80$ for us to reject that hypothesis; that is, we do not reject $\mu = 80$ with these observed data. However, with this understanding, we usually use "accept" since it appears so often in the literature.

In the first illustration, we assumed that the variance σ^2 was known. We now take a more realistic position and assume that the variance is unknown. Suppose our hypothesis is $H_0 : \mu = \mu_0$. If a random sample $X_1, X_2, \ldots, X_n$ is taken from a distribution $N(\mu, \sigma^2)$, we recall from Section 7.4 that a confidence interval for μ was based upon

$$T = \frac{\bar{X} - \mu}{\sqrt{S^2/(n-1)}} = \frac{\bar{X} - \mu}{S/\sqrt{n-1}}.$$

This suggests that T might be a good statistic to use for the test of $H_0 : \mu = \mu_0$ with μ replaced by μ_0. In addition, it is the natural statistic to use if we replace σ^2/n by its unbiased estimator $S^2/(n-1)$ in $(\bar{X} - \mu_0)/\sqrt{\sigma^2/n}$. This latter statistic was used in our first illustration with $\mu_0 = 80$ and $\sigma^2 = 25$.

If $\mu = \mu_0$, we know that T has a t distribution with $n - 1$ degrees of freedom. Thus, if we select t_0 so that $P(T \le t_0) = 1 - \alpha/2$, then, with $\mu = \mu_0$,

$$P\left(\frac{|\bar{X} - \mu_0|}{S/\sqrt{n-1}} > t_0\right) = \alpha.$$

Accordingly, if $\bar{x}$ and s are the observed mean and standard deviation of the sample, the rule that rejects $H_0 : \mu = \mu_0$ if and only if

$$\frac{|\bar{x} - \mu_0|}{s/\sqrt{n-1}} > t_0$$

provides a test of this hypothesis with significance level α. Again it should be noted that this rule is equivalent to rejecting $H_0 : \mu = \mu_0$ if μ_0 is not in the $100(1 - \alpha)\%$ confidence interval

$$\left[\bar{x} - t_0\left(\frac{s}{\sqrt{n-1}}\right), \bar{x} + t_0\left(\frac{s}{\sqrt{n-1}}\right)\right].$$

Example 8.1–2. Suppose that the growths of tumors placed in 10 rats provided the following summary of the data: $n = 10$, $\bar{x} = 4.3$, $s = 1.2$. Let us assume normal distributions for these growths and test $H_0 : \mu = \mu_0 = 4.0$, at the $\alpha = 0.10$ significance level. With nine degrees of freedom, we find that $P(T \le 1.833) = 0.95$. Since

$$\frac{|4.3 - 4.0|}{1.2/\sqrt{9}} = \frac{0.3}{0.4} = 0.75 \le 1.833,$$

we do not reject $H_0 : \mu = 4.0$ at the 10% significance level.

A general rule can now be formulated for the cases in which the point estimators of the parameters under consideration have normal or approximate normal distributions. Let $\hat{\theta}$ be the point estimator of the parameter θ and assume that $\hat{\theta}$ has a distribution (approximately, for large n) $N(\theta, \tau^2)$. Frequently, in applications, the standard deviation $\sqrt{\tau^2} = \tau$ of $\hat{\theta}$ is called the standard error of the estimator. In our first illustration $\tau = \sigma/\sqrt{n} = 5/\sqrt{n}$ was known; in the second it was approximated by $s/\sqrt{n-1}$. If the hypothesis is $H_0 : \theta = \theta_0$, then the observed deviation $|\hat{\theta} - \theta_0|$ is compared to the standard deviation τ of $\hat{\theta}$ through the ratio $|\hat{\theta} - \theta_0|/\tau$. If $H_0 : \theta = \theta_0$ is true, the ratio is (approximately) a standardized normal variable. Hence, the rule that rejects H_0 if and only if

$$\frac{|\hat{\theta} - \theta_0|}{\tau} > z_0, \quad \text{where} \quad \Phi(z_0) = 1 - \frac{\alpha}{2},$$

provides a test with (approximate) significance level α. If τ is unknown and we must estimate τ by $\hat{\tau}$, then the rejection rule $|\hat{\theta} - \theta_0|/\hat{\tau} > z_0$ still provides a test with an approximate significance level α.

Example 8.1–3. Suppose we interview 100 persons selected at random, and let Y be the number who favor the President's domestic policy. Assume Y to be $b(100, p)$, where p is the unknown fraction of the population that favors this policy. Of course, $\hat{p} = Y/100$ is approximately $N[p, p(1-p)/100]$. Say $H_0: p = 0.25$ and we actually observed Y is equal to $y = 20$. Hence, the observed relative frequency of persons favoring the policy is $\hat{p} = 0.2$. A biased estimate of $p(1-p)/100$ is $\hat{p}(1-\hat{p})/100$, which in this case equals $(0.2)(0.8)/100 = 0.0016$. Thus,

$$\frac{|0.20 - 0.25|}{\sqrt{0.0016}} = 1.25 \leq 1.96, \qquad \text{where} \quad \Phi(1.96) = 0.975.$$

Accordingly, we do not reject $H_0: p = 0.25$ using the given data at a significance level of $\alpha = 0.05$. We also note that 0.25 is contained in the approximate 95% confidence interval for p,

$$\left[0.20 - 1.96\sqrt{\frac{(0.2)(0.8)}{100}}, 0.20 + 1.96\sqrt{\frac{(0.2)(0.8)}{100}} \right] = [0.122, 0.278].$$

REMARK. In testing the hypothesis $H_0: p = p_0$, many statisticians use $p_0(1-p_0)$ instead of $\hat{p}(1-\hat{p})$ in the denominator of

$$\frac{|\hat{p} - p_0|}{\sqrt{\hat{p}(1-\hat{p})/n}}.$$

We do not have a strong preference one way or the other, since the two methods provide about the same numerical result. Certainly, if H_0 is true, $p_0(1-p_0)/n$ is the variance of Y/n and $\hat{p}(1-\hat{p})/n$ is only an approximation to it. If, however, H_0 is not true, the latter probably provides the better estimate of this variance. Moreover, the form $\hat{p}(1-\hat{p})/n$ seems more consistent with our usual procedure of obtaining a confidence interval for p; hence, we have used it in this text.

Example 8.1–4. Let us assume that the measures of job performances of n persons are $Y_1, Y_2, \ldots, Y_n$, n mutually independent normal variables with respective means $\alpha + \beta(x_i - \bar{x})$, $i = 1, 2, \ldots, n$, and common variance σ^2, where x_i is the college grade point average of the ith person. To test $H_0: \beta = 0$, we observe $n = 12$ such pairs (x_i, y_i), which provide the estimates $\hat{\alpha} = 62.2$, $\hat{\beta} = 3.6$, and $\hat{\sigma}^2 = 123.3$, in addition to $\sum_1^{12}(x_i - \bar{x})^2 = 5.1$. The Student's t suggested in Section 7.7 is, with $\beta = \beta_0 = 0$,

$$\frac{|3.6 - 0|}{\sqrt{[(12)(123.3)]/[(10)(5.1)]}} = 0.6684 < 1.812$$

and, thus, we do not reject $H_0: \beta = 0$ at $\alpha = 0.10$. That is, the college grade point average does not seem to influence job performance too much, in this case, because we accept the hypothesis that the slope β is equal to zero.

══ *Exercises* ══

8.1–1. Let X be $N(\mu, 4)$. To test $H_0 : \mu = 0$ we take a random sample of size $n = 25$ from this distribution and observe $\bar{x} = 0.28$. Do we accept or reject H_0 at the 10% significance level? Is $\mu = 0$ contained in a 90% confidence interval for μ?

8.1–2. Let X be $N(\mu, \sigma^2)$. To test $H_0 : \mu = 75$ we take a random sample of size $n = 65$ and observe that $\bar{x} = 78.8$ and $s = 12.8$. Do we accept or reject H_0 at the 5% significance level?

8.1–3. Work Exercise 8.1–2 with $n = 65$ replaced by $n = 17$.

8.1–4. The claim is that a certain binomial parameter p is equal to 0.43. To test this, $n = 400$ independent observations are made, exactly one half of which are successes. Would you reject that claim at the 10% significance level?

8.1–5. Explain how you would test the equality of the parameters p_1 and p_2 of two independent binomial distributions if the sample sizes n_1 and n_2 are large. (HINT: Let $\theta = p_1 - p_2$ and find a point estimator for θ.) If $\alpha = 0.05$ and if we observed 50 and 60 successes, respectively, out of 100 independent trials for each experiment, do you accept $H_0 : p_1 = p_2$ (or, equivalently, $H_0 : \theta = 0$)? Is $\theta = p_1 - p_2 = 0$ contained in the 95% confidence interval for $p_1 - p_2$?

8.1–6. In the usual regression model in which $\mu(x)$ is a straight line, test $H_0 : \beta = 0$ at the $\alpha = 0.05$ level if $n = 16$, $\hat{\alpha} = 1.2$, $\hat{\beta} = 0.4$, $\widehat{\sigma^2} = 0.11$, and $\sum_1^{16} (x_i - \bar{x})^2 = 7.3$.

8.2 Alternative Hypotheses

Let μ denote the unknown mean of a distribution. The fact that we sometimes reject a statistical hypothesis, such as $H_0 : \mu = \mu_0$, implies that we are willing to accept the possibility that the parameter μ can assume a value other than μ_0; that is, at least one other alternative value of the parameter seems possible. Moreover, in Section 8.1, we were willing to reject $H_0 : \mu = \mu_0$ if the observed sample mean $\bar{x}$ was significantly smaller than μ_0 or significantly larger than μ_0. Such a willingness on our part actually means that μ has possible alternative values both below and above μ_0. As a matter of fact, it might be possible that all other values of μ (not equal to μ_0) are alternatives to the hypothesis $H_0 : \mu = \mu_0$. Such a statement is itself a statistical hypothesis, which may be denoted $H_1 : \mu \neq \mu_0$ and appropriately called the *alternative hypothesis*. To distinguish between the two hypotheses, $H_0 : \mu = \mu_0$ and $H_1 : \mu \neq \mu_0$, the former is frequently called the *null hypothesis*. In this terminology we tested in Section 8.1 a null hypothesis $H_0 : \mu = \mu_0$ against the *two-sided alternative* hypothesis $H_1 : \mu \neq \mu_0$. The corresponding test is usually called a *two-sided test*. Also in Examples 8.1–3 and 8.1–4, we considered two-sided tests of $H_0 : p = p_0$ and $H_0 : \beta = \beta_0$, respectively.

Suppose we have a two-sided alternative; thus, we use a two-sided test. For illustration, we reject $H_0: \mu = \mu_0$ at $\alpha = 0.05$ if

$$\left| \frac{\bar{x} - \mu_0}{\sigma/\sqrt{n}} \right| > 1.96$$

where $\bar{x}$ is the observed value of the mean $\bar{X}$ of a random sample of size n and σ is the known standard deviation of the underlying distribution. With this test, rejection can occur in one of two ways, namely,

$$\bar{x} < \mu_0 - 1.96\left(\frac{\sigma}{\sqrt{n}}\right) \quad \text{or} \quad \mu_0 + 1.96\left(\frac{\sigma}{\sqrt{n}}\right) < \bar{x}.$$

When $H_0: \mu = \mu_0$ is true, the probabilities, under the assumption that the distribution is $N(\mu, \sigma^2)$, of these two events are

$$P\left(\bar{X} < \mu_0 - 1.96\left(\frac{\sigma}{\sqrt{n}}\right)\right) = P\left(\frac{\bar{X} - \mu_0}{\sigma/\sqrt{n}} < -1.96\right) = \Phi(-1.96) = 0.025$$

and

$$P\left(\mu_0 + 1.96\left(\frac{\sigma}{\sqrt{n}}\right) < \bar{X}\right) = P\left(1.96 < \frac{\bar{X} - \mu_0}{\sigma/\sqrt{n}}\right) = 1 - \Phi(1.96) = 0.025,$$

because $\bar{X}$ is $N(\mu_0, \sigma^2/n)$. That is, the significance level $\alpha = 0.05$ is divided into two parts in a two-sided test. Frequently, the two parts are equal as in this case.

Often, however, in testing a hypothesis such as $\mu = \mu_0$, only alternatives on one side of μ_0 are of interest. For illustration, suppose it is known that a certain educational process produces students with a mean score of 70 on a certain type of test. Research involving this process indicates that if certain changes are made, the mean should increase. Thus, plans would be made to test the null ("no change") hypothesis $H_0: \mu = 70$ against the alternative ("research worker's") hypothesis $H_1: \mu > 70$. If the test is to be based upon a random sample $X_1, X_2, \ldots, X_n$ from this new distribution (after changes have been made), which we assume is $N(\mu, \sigma^2)$ with known variance σ^2, we would reject $H_0: \mu = 70$ and accept $H_1: \mu > 70$ if the observed mean $\bar{x}$ of the sample is significantly larger than 70; that is, if $\bar{x} > c$, where c is a constant greater than 70. Of course, we would never want to reject $H_0: \mu = 70$ and accept $H_1: \mu > 70$ with a small value of $\bar{x}$, even though it was substantially smaller than 70.

With the test that rejects H_0 if $\bar{x} > c$, to achieve the significance level α, c must equal $70 + z_0 \sigma/\sqrt{n}$, where $\Phi(z_0) = 1 - \alpha$, because, when $\mu = 70$,

$$P(\bar{X} > c) = P\left(\frac{\bar{X} - 70}{\sigma/\sqrt{n}} > \frac{c - 70}{\sigma/\sqrt{n}}\right) = 1 - \Phi\left(\frac{c - 70}{\sigma/\sqrt{n}}\right) = \alpha$$

only when $(c - 70)/(\sigma/\sqrt{n}) = z_0$. Accordingly, for such a *one-sided* alternative and corresponding *one-sided* test, we note that the significance level α is not divided into two parts but is found at one end of the "null" distribution of the appropriate statistic.

These remarks are illustrated by four examples.

Example 8.2–1. Consider an individual's golf score distribution assumed to be $N(\mu, 9)$. After taking some lessons, this golfer believes that she has improved and should now have an average that is less than her old one of 85. To test $H_0: \mu = 85$ against the one-sided alternative hypothesis $H_1: \mu < 85$, she plans to play $n = 16$ rounds and reject H_0 if her average $\bar{x} < c$, an appropriately selected constant. If the test is to have significance level $\alpha = 0.05$, then

$$c = 85 - 1.645\left(\frac{3}{\sqrt{16}}\right) = 83.77$$

because, for $\mu = 85$,

$$P(\bar{X} < c) = P\left(\frac{\bar{X} - 85}{3/4} < \frac{c - 85}{3/4}\right) = \Phi\left(\frac{c - 85}{3/4}\right) = 0.05$$

if $(c - 85)/(3/4) = 1.645$. Since her average for those 16 rounds equaled 82.94, she rejected H_0 and accepted the fact that $H_1: \mu < 85$; that is, the lessons had helped.

Example 8.2–2. In attempting to control the strength of the wastes discharged into a nearby river, a paper firm has taken a number of measures. They believe that they have reduced the oxygen-consuming power of their wastes from a previous mean μ of 500 (measured from permanganate in parts per million). To test $H_0: \mu = 500$ against $H_1: \mu < 500$, readings on $n = 50$ consecutive working days were taken, resulting in the observed statistics $\bar{x} = 308.8$ and $s = 115.15$. Since

$$\frac{\bar{x} - 500}{s/\sqrt{n - 1}} = \frac{308.8 - 500}{115.15/\sqrt{49}} = -11.62$$

is much smaller than any of the typical critical values found in the normal table, we reject $H_0: \mu = 500$ and accept $H_1: \mu < 500$. It should be noted, however, that although an improvement has been made, there still might exist the question of whether the improvement is adequate. The 95.44% confidence interval $308.8 \pm 2(115.15/7)$ or $[275.9, 341.7]$ for μ might help the company answer that question.

Example 8.2–3. It is claimed that a new treatment of a very serious disease will substantially reduce the mortality of that disease from its present rate of

$p = 1/4$. To test $H_0: p = 1/4$ against $H_1: p < 1/4$ we observe $n = 100$ patients who have submitted to the new treatment and note that only $y = 10$ of them die from this disease. Since

$$\frac{y/n - 1/4}{\sqrt{(y/n)(1 - y/n)/n}} = \frac{1/10 - 1/4}{\sqrt{(1/10)(9/10)/100}} = -5 < -2.326,$$

we reject $H_0: p = 1/4$ and accept $H_1: p < 1/4$ at $\alpha = 0.01$ (that is, the new treatment seems to have made a significant improvement).

Example 8.2–4. The claim is made that students with higher American College Testing (ACT) scores earn higher college grade point averages (GPA's). Let us assume the usual normal regression model with $x = $ ACT score and $y = $ GPA so that the regression is linear with slope β. The null hypothesis is taken to be $H_0: \beta = 0$, so the alternative hypothesis $H_1: \beta > 0$ supports the claim. Say $n = 25$ students were selected at random and it was found that $\hat{\beta} = 0.09$ and $\widehat{\sigma^2} = 0.13$ for $x_1, x_2, \ldots, x_{25}$ such that

$$\sum_1^{25} (x_i - \bar{x})^2 = 9.8.$$

We note that

$$\frac{0.09 - 0}{\sqrt{[(25)(0.13)]/[(23)(9.8)]}} = 0.750 < 1.714,$$

where $t_0 = 1.714$ comes from the t distribution with 23 degrees of freedom and $\alpha = 0.05$. Thus, with only these $n = 25$ observations, we are reluctant to reject $H_0: \beta = 0$ and hence accept it until we have more evidence. That is, we do not believe that these data substantially support the original claim.

=== *Exercises* ===

8.2–1. Assume that IQ scores for a certain population are approximately $N(\mu, 100)$. To test $H_0: \mu = 110$ against the one-sided alternative hypothesis $H_1: \mu > 110$, we take a random sample of size $n = 16$ from this population and observe $\bar{x} = 113.5$. Do we accept or reject H_0 at the   **(a)**   5% significance level?   **(b)**   10% significance level?

8.2–2. Assume that the weight of cereal in a "10-ounce box" is $N(\mu, \sigma^2)$. To test $H_0: \mu = 10$ against $H_1: \mu > 10$, we take a random sample of size $n = 17$ and observe that $\bar{x} = 10.3$ and $s = 0.4$. Do we accept or reject H_0 at the 5% significance level?

8.2–3. A die, claimed to be fair by the manufacturer, is suspected of being biased in favor of sixes. Let p denote the probability of rolling a six with this die. To test $H_0: p = 1/6$ against $H_1: p > 1/6$, this die was cast 50 times. If 10 sixes were observed, would H_0 be rejected at the 5% significance level?

8.2–4. Let p_1 be the proportion of voters in a precinct in the inner city who voted yes on the question of increasing taxes for education and let p_2 be the proportion of yes voters in a precinct containing expensive homes. Would $H_0 : p_1 = p_2$ be rejected in favor of $H_1 : p_1 > p_2$ at the 10% significance level if the vote was 200 yes and 180 no in the inner city precinct and 220 yes and 265 no in the other precinct? HINT: Let $\theta = p_1 - p_2$ so that $H_0 : \theta = 0$ and $H_1 : \theta > 0$; estimate θ with $\hat{\theta} = \hat{p}_1 - \hat{p}_2$ and compare the difference $\hat{\theta} - 0$ with a standard error of $\hat{\theta}$.

8.2–5. In testing the life of motors, we are willing to assume that the life Y of a motor will be a linear function of the temperature x of the room in which it is operated. The manufacturer claims that this temperature will not influence the life too much; that is, $H_0 : \beta = 0$ is true. Others believe, however, that the life is shorter at higher temperatures; that is, $H_1 : \beta < 0$. Twenty motors were run continuously in rooms with different temperatures. This experiment yielded $n = 20$, $\hat{\beta} = -0.009$, $\widehat{\sigma^2} = 0.038$, and $\sum_1^{20} (x_i - \bar{x})^2 = 19{,}162$. Test H_0 against H_1 at the 5% significance level.

8.3　Tests of Variances and Differences of Means

A psychology professor claims that the variance of IQ scores for college students is equal to $\sigma^2 = 100$. To test this claim it was decided to test the hypothesis $H_0 : \sigma^2 = 100$ against a two-sided alternative hypothesis $H_1 : \sigma^2 \neq 100$. A random sample of n students will be selected and the test will be based upon the observed sample variance s^2 of their IQ scores. The hypothesis H_0 will be rejected if s^2 differs "too much" from $\sigma^2 = 100$. That is, H_0 will be rejected if $s^2 < c_1$ or $s^2 > c_2$ for some constants $c_1 < 100$ and $c_2 > 100$. If we assume that IQ scores are normally distributed, the distribution of $nS^2/100$ is $\chi^2(n-1)$ when H_0 is true. Thus, the chi-square probability table can be used in selecting c_1 and c_2 to yield the desired significance level α.

Example 8.3–1. Suppose that in the above discussion $n = 23$ and $\alpha = 0.05$. Then $P(23S^2/100 < 10.98; \sigma^2 = 100) = 0.025$ and $P(23S^2/100 > 36.78; \sigma^2 = 100) = 0.025$. Thus, H_0 will be rejected if $s^2 < (10.98)(100)/23 = 47.74$ or $s^2 > (36.78)(100)/23 = 159.91$. Since the observed value of the sample variance was $s^2 = 147.82$, the hypothesis $H_0 : \sigma^2 = 100$ was not rejected. Note that the 95% confidence interval for σ^2,

$$\left[\frac{(23)(147.82)}{36.78}, \frac{(23)(147.82)}{10.98} \right] = [92.44, 309.64],$$

contains $\sigma^2 = 100$.

If $H_1: \sigma^2 > 100$ had been the alternative hypothesis, $H_0: \sigma^2 = 100$ would have been rejected if $s^2 > c_3$, where $c_3 > 100$ is selected to yield desired significance levels. Using the data in Example 8.3–1, we have

$$P(23S^2/100 > 33.92; \sigma^2 = 100) = 0.05.$$

Thus, $c_3 = [(33.92)(100)]/23 = 147.48$, and H_0 would be rejected in favor of this one-sided alternative hypothesis for $\alpha = 0.05$.

A biologist who studies spiders realizes that female green lynx spiders tend to be longer than male spiders. He also notes that the lengths of the female spiders seem to vary more than those of the male spiders and is interested in testing whether this is true. Suppose that the distribution of the length X of female spiders is $N(\mu_X, \sigma_X^2)$ and that of the length Y of male spiders is $N(\mu_Y, \sigma_Y^2)$. The hypothesis $H_0: \sigma_X^2/\sigma_Y^2 = 1$ will be tested against the alternative hypothesis $H_1: \sigma_X^2/\sigma_Y^2 > 1$. Given random samples of n female spiders and m male spiders, H_0 will be rejected in favor of H_1 if $s_X^2/s_Y^2 > c$, where s_X^2 and s_Y^2 are the observed values of the respective sample variances of the lengths and c is selected to yield the desired significance level, using the F probability table. Recall that nS_X^2/σ_X^2 and mS_Y^2/σ_Y^2 have independent distributions $\chi^2(n-1)$ and $\chi^2(m-1)$, respectively. Thus, when H_0 is true,

$$F = \frac{nS_X^2/(n-1)}{mS_Y^2/(m-1)}$$

has an F distribution with $r_1 = n - 1$ and $r_2 = m - 1$ degrees of freedom.

Example 8.3–2. In the preceding discussion, take $n = 16$, $m = 11$, and $\alpha = 0.05$. Since

$$P\left(\frac{16S_X^2/15}{11S_Y^2/10} > 2.85; \frac{\sigma_X^2}{\sigma_Y^2} = 1\right) = 0.05,$$

H_0 will be rejected if

$$\frac{s_X^2}{s_Y^2} > \left(\frac{11}{10}\right)\left(\frac{15}{16}\right)(2.85) = 2.94.$$

The random samples yielded $s_X^2 = 2.26$ and $s_Y^2 = 0.25$. Since $2.26/0.25 = 9.04$, H_0 is clearly rejected.

A botanist is interested in comparing the growth response of dwarf pea stems to two different levels of the hormone indoleacetic acid (IAA). Using 16-day-old pea plants, the botanist obtains 5 millimeter sections and floats these sections on solutions with different hormone concentrations to observe the effect of the hormone on the growth of the pea stem. Let X and Y denote, respectively, the growth that can be attributed to the hormone during the first 26 hours after sectioning for $(0.5)10^{-4}$ and 10^{-4} levels of concentration of IAA.

Assuming that the distributions of X and Y are $N(\mu_X, \sigma^2)$ and $N(\mu_Y, \sigma^2)$, respectively, the botanist would like to test the hypothesis $H_0: \mu_X - \mu_Y = 0$ against the alternative hypothesis $H_1: \mu_X - \mu_Y < 0$. If random samples of sizes n and m are taken, the test can be based upon the statistic

$$T = \frac{\bar{X} - \bar{Y}}{\sqrt{[(nS_X^2 + mS_Y^2)/(n + m - 2)](1/n + 1/m)}}$$

which has a t distribution with $r = n + m - 2$ degrees of freedom when H_0 is true. The hypothesis H_0 will be rejected in favor of H_1 if the observed value of T is less than t_0, where t_0 is selected from the t table so that $P(T < t_0) = \alpha$.

Example 8.3–3. In the preceding discussion, the botanist measured the growths of pea stem segments, in millimeters, for $n = 11$ observations of X:

$$0.8, \quad 1.8, \quad 1.0, \quad 0.1, \quad 0.9, \quad 1.7, \quad 1.0, \quad 1.4, \quad 0.9, \quad 1.2, \quad 0.5,$$

and $m = 13$ observations of Y:

$$1.0, \quad 0.8, \quad 1.6, \quad 2.6, \quad 1.3, \quad 1.1, \quad 2.4, \quad 1.8, \quad 2.5, \quad 1.4, \quad 1.9, \quad 2.0, \quad 1.2.$$

For these data $\bar{x} = 1.03$, $s_X^2 = 0.22$, $\bar{y} = 1.66$, and $s_Y^2 = 0.33$. Since

$$t = \frac{1.03 - 1.66}{\sqrt{\{[11(0.22) + 13(0.33)]/(11 + 13 - 2)\}(1/11 + 1/13)}}$$

$$= -2.78 < -2.508,$$

H_0 is rejected at an $\alpha = 0.01$ significance level.

If the assumption of equal variances cannot be made and if the sample sizes are large, a test of the hypothesis $H_0: \mu_X - \mu_Y = 0$ can be based upon the statistic

$$Z = \frac{\bar{X} - \bar{Y}}{\sqrt{S_X^2/(n - 1) + S_Y^2/(m - 1)}},$$

which has a distribution that is approximately $N(0, 1)$ when the hypothesis H_0 is true.

══ *Exercises* ══

8.3–1. Let X be $N(\mu, \sigma^2)$. To test the hypothesis $H_0: \sigma^2 = 25.0$ against the alternative hypothesis $H_1: \sigma^2 < 25.0$, a random sample of size $n = 11$ was taken and for this sample $s^2 = 9.88$. Is H_0 rejected for $\alpha = 0.05$?

8.3–2. Let X be $N(\mu, \sigma^2)$. Is the hypothesis $H_0: \sigma^2 = 0.04$ rejected in favor of the two-sided alternative hypothesis $H_1: \sigma^2 \neq 0.04$, if a random sample of size $n = 13$ yielded the sample variance $s^2 = 0.058$? Let $\alpha = 0.05$.

8.3–3. Let X be $N(\mu_X, \sigma_X^2)$ and Y be $N(\mu_Y, \sigma_Y^2)$. Is the hypothesis $H_0 : \sigma_X^2/\sigma_Y^2 = 1$ rejected in favor of $H_1 : \sigma_X^2/\sigma_Y^2 > 1$ if a random sample of size $n = 9$ yielded $\bar{x} = 84.34$, $s_X^2 = 125.13/9 = 13.90$ and a random sample of size $m = 16$ yielded $\bar{y} = 50.29$, $s_Y^2 = 199.71/16 = 12.48$? Let $\alpha = 0.05$.

8.3–4. Let X and Y equal the times in days required for maturation of *Guardiola* seeds from narrow-leaved and broad-leaved parents, respectively. Assume that X is $N(\mu_X, \sigma_X^2)$ and Y is $N(\mu_Y, \sigma_Y^2)$. Test the hypothesis $H_0 : \sigma_X^2/\sigma_Y^2 = 1$ against the alternative hypothesis $H_1 : \sigma_X^2/\sigma_Y^2 > 1$ if a sample of size $n = 13$ yielded $\bar{x} = 18.97$, $s_X^2 = 9.88$ and a sample of size $m = 9$ yielded $\bar{y} = 23.20$, $s_Y^2 = 4.08$. Let $\alpha = 0.05$.

8.3–5. Suppose that scores on a standardized test in mathematics taken by students from large and small high schools are $N(\mu_X, \sigma^2)$ and $N(\mu_Y, \sigma^2)$, respectively. Test the hypothesis $H_0 : \mu_X - \mu_Y = 0$ against the alternative hypothesis $H_1 : \mu_X - \mu_Y > 0$ if a random sample of $n = 9$ students from large high schools yielded $\bar{x} = 81.31$, $s_X^2 = 60.76$ and a random sample of $m = 15$ students from small high schools yielded $\bar{y} = 78.61$, $s_Y^2 = 48.24$. Let $\alpha = 0.05$.

8.3–6. A mathematics professor bowls in a men's league on Tuesday and in a mixed league on Thursday. Assume that the distribution of his scores in the men's league is $N(\mu_X, \sigma^2)$ and in the mixed league is $N(\mu_Y, \sigma^2)$. Test $H_0 : \mu_X - \mu_Y = 0$ against $H_1 : \mu_X - \mu_Y > 0$ at a significance level $\alpha = 0.05$, given that his respective averages are $\bar{x} = 182$ and $\bar{y} = 173$, if the sample standard deviation in each league is $s = 24$. Say $n = 81$ games were rolled in each league.

8.3–7. The botanist in Example 8.3–3 is really interested in testing for synergistic interaction. That is, given two hormones, gibberellin (GA_3) and indoleacetic acid (IAA), let X_1 and X_2 equal the growth responses of dwarf pea stem segments to GA_3 and IAA, respectively. Let $X = X_1 + X_2$ and let Y equal the growth response when both hormones are present. Assuming that X is $N(\mu_X, \sigma^2)$ and Y is $N(\mu_Y, \sigma^2)$, the botanist is interested in testing the hypothesis $H_0 : \mu_X - \mu_Y = 0$ against the alternative hypothesis of synergistic interaction $H_1 : \mu_X - \mu_Y < 0$. Let $\alpha = 0.05$ and perform this test using $n = 10$ observations of X:

$$2.1, \quad 2.6, \quad 2.6, \quad 3.4, \quad 2.1, \quad 1.7, \quad 2.6, \quad 2.6, \quad 2.2, \quad 1.2,$$

and $m = 10$ observations of Y:

$$3.5, \quad 3.9, \quad 3.0, \quad 2.3, \quad 2.1, \quad 3.1, \quad 3.6, \quad 1.8, \quad 2.9, \quad 3.3.$$

8.4 The Power of a Statistical Test

To introduce some new terminology, we begin with an example.

Example 8.4–1. Let X equal the breaking strength of a steel bar. If the steel bar is manufactured by Process I, X is $N(50, 36)$. It is hoped that if Process II (a new process) is used, X will be $N(55, 36)$. Given a large number of steel bars manufactured by Process II, how could we test whether the increase in the mean breaking strength was realized?

That is, we are assuming X is $N(\mu, 36)$ and μ is equal to 50 or 55. We desire to test the null hypothesis $H_0: \mu = 50$ against the alternative hypothesis $H_1: \mu = 55$. Note that each of these hypotheses completely specifies the distribution of X. That is, H_0 states that X is $N(50, 36)$ and H_1 states that X is $N(55, 36)$. An hypothesis that completely specifies the distribution of X is called a *simple* hypothesis; otherwise it is called a *composite* hypothesis (composed of at least two simple hypotheses). For illustration, $H_1: \mu > 50$ would be a composite hypothesis because it is composed of all normal distributions with $\sigma^2 = 36$ and means greater than 50. In order to test which of the two hypotheses, H_0 or H_1, is true, we shall set up a rule based on the breaking strengths $x_1, x_2, \ldots, x_n$ of n bars (the observed values of a random sample of size n from this new normal distribution). The rule leads to a decision to accept or reject H_0; so it is necessary to partition the sample space into two parts, say C and C', so that if $(x_1, x_2, \ldots, x_n) \in C$, H_0 is rejected, and if $(x_1, x_2, \ldots, x_n) \in C'$, H_0 is accepted (not rejected). The rejection region C for H_0 is called the *critical region* for the test. Often the partitioning of the sample space is specified in terms of the values of a statistic called the *test statistic*. In this illustration we could let $\bar{X}$ be the test statistic and, for example, take $C = \{(x_1, x_2, \ldots, x_n): \bar{x} > 53\}$. We could then define the *critical region* as those values of the test statistic for which H_0 is rejected. That is, the given critical region is equivalent to defining $C = \{\bar{x}: \bar{x} > 53\}$ in the $\bar{x}$ space. If $(x_1, x_2, \ldots, x_n) \in C$ when H_0 is true, H_0 would be rejected when it is true. Such an error is called a *Type I error*. If $(x_1, x_2, \ldots, x_n) \in C'$ when H_1 is true, H_0 would be accepted when in fact H_1 is true. Such an error is called a *Type II error*. The probability of a Type I error has been called the significance level of the test and was denoted by α. That is,

$$\alpha = P[(X_1, X_2, \ldots, X_n) \in C; H_0]$$

is the probability that $(X_1, X_2, \ldots, X_n)$ falls in C when H_0 is true. The probability of a Type II error will be denoted by β; that is,

$$\beta = P[(X_1, X_2, \ldots, X_n) \in C'; H_1]$$

is the probability of accepting H_0 when it is false. For illustration, suppose $n = 16$ bars were tested and $C = \{\bar{x}: \bar{x} > 53\}$. Then $\bar{X}$ is $N(50, 36/16)$ when H_0 is true and is $N(55, 36/16)$ when H_1 is true. Thus,

$$\alpha = P(\bar{X} > 53; H_0) = P\left(\frac{\bar{X} - 50}{6/4} > \frac{53 - 50}{6/4}; H_0\right)$$

$$= 1 - \Phi(2) = 0.0228$$

and

$$\beta = P(\bar{X} \leq 53; H_1) = P\left(\frac{\bar{X} - 55}{6/4} \leq \frac{53 - 55}{6/4}; H_1\right)$$

$$= \Phi(-4/3) = 1 - 0.9087 = 0.0913.$$

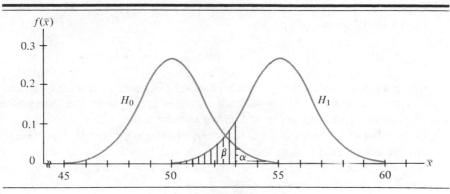

FIGURE 8.4–1

See Figure 8.4–1 for the graphs of the probability density functions of $\overline{X}$ when H_0 and H_1 are true, respectively. Note that a decrease in the size of α leads to an increase in the size of β, and vice versa. Both α and β can be decreased if the sample size n is increased.

We now state formally the new definitions given in Example 8.4–1.

DEFINITION 8.4–1. *A* simple hypothesis *is one that completely specifies the distribution of the random variable under consideration; a* composite hypothesis *does not.*

DEFINITION 8.4–2. *The* critical region C *is the set of points in the sample space which leads to the rejection of the null hypothesis* H_0.

DEFINITION 8.4–3. *A statistic used to define the critical region is called a* test statistic. *The critical region C is often defined as a set of values of the test statistic that leads to the rejection of the null hypothesis* H_0.

DEFINITION 8.4–4. Type I error *is the error made when rejecting H_0 when H_0 is true.* Type II error *is the error made when accepting H_0 when H_1 is true.*

DEFINITION 8.4–5. *The probability of a Type I error is called the* significance level *of the test or the* size *of the critical region C. The significance level is denoted by α and the probability of the Type II error is denoted by β.*

Another example will help to reinforce these definitions and give us the opportunity to introduce a new one.

Example 8.4–2. Suppose a random sample $X_1, X_2, \ldots, X_{10}$ of size 10 comes from a distribution with p.d.f.

$$f(x; p) = p^x (1 - p)^{1-x}, \qquad x = 0, 1,$$

where it is known that $0 < p \leq 1/2$. That is, 10 Bernoulli trials are observed. To test the simple null hypothesis $H_0: p = 1/2$ against the composite alternative hypothesis $H_1: p < 1/2$, we use the test statistic $Y = \sum_{i=1}^{10} X_i$, the number of successes in the 10 independent trials. Of course, Y is $b(10, p)$. Say the critical region is defined by $\{y: \quad y = 0, 1, \text{ or } 2\}$ or, equivalently, by $\{(x_1, x_2, \ldots, x_{10}): \quad \sum_{i=1}^{10} x_i = 0, 1, \text{ or } 2\}$. Since Y is $b(10, 1/2)$ if $p = 1/2$, the significance level of the corresponding test is

$$\alpha = P\left(Y = 0, 1, \text{ or } 2; p = \frac{1}{2}\right) = \sum_{y=0}^{2} \binom{10}{y} \left(\frac{1}{2}\right)^{10} = 0.0547.$$

Of course, the probability β of the Type II error has different values with different values of p selected from the composite alternative hypothesis $H_1: p < 1/2$. For illustration, with $p = 1/4$,

$$\beta = P\left(3 \leq Y \leq 10; p = \frac{1}{4}\right) = \sum_{y=3}^{10} \binom{10}{y} \left(\frac{1}{4}\right)^y \left(\frac{3}{4}\right)^{10-y} = 0.4744,$$

whereas with $p = 1/10$,

$$\beta = P\left(3 \leq Y \leq 10; p = \frac{1}{10}\right) = \sum_{y=3}^{10} \binom{10}{y} \left(\frac{1}{10}\right)^y \left(\frac{9}{10}\right)^{10-y} = 0.0702.$$

Of course, instead of considering the probability β of accepting H_0 when H_1 is true, we could compute the probability K of rejecting H_0 when H_1 is true. After all, β and $K = 1 - \beta$ provide the same information. Since K is a function of p, we denote this explicitly by writing $K(p)$. The probability

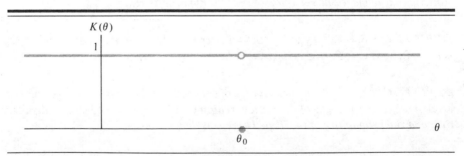

FIGURE 8.4–2

$$K(p) = \sum_{y=0}^{2} \binom{10}{y} p^y (1-p)^{10-y} = (1-p)^8(1 + 8p + 36p^2), \qquad 0 < p \le \frac{1}{2},$$

is called the *power function* of the test. Of course, $\alpha = K(1/2) = 0.0547$, $1 - K(1/4) = 0.4744$, and $1 - K(1/10) = 0.0702$. The value of the power function at a specified p is called the *power* of the test at that point. For illustration, $K(1/4) = 0.5256$ and $K(1/10) = 0.9298$ are the powers at $p = 1/4$ and $p = 1/10$, respectively. An acceptable power function is one that assumes small values when H_0 is true and larger values when p differs much from $p = 1/2$. Since $K(1/4) = 0.5256$ is not very large and yet $p = 1/4$ is not particularly close to $p = 1/2$, the power function in this example is not highly desirable. That is, owing to the small sample size, we do not have a very powerful test.

DEFINITION 8.4–6. *The power function of a test of a null hypothesis H_0 against the alternative hypothesis H_1 is the function that gives the probability of rejecting H_0 for each parameter point in H_0 and H_1. The value of the power function at a parameter point is the* power *of the test at that point.*

When testing $H_0: \theta = \theta_0$ against the two-sided alternative $H_1: \theta \ne \theta_0$, the ideal situation is that in which $\alpha = K(\theta_0) = 0$ and

$$K(\theta) = P[(X_1, X_2, \ldots, X_n) \in C; \theta] = 1,$$

when $\theta \ne \theta_0$. This ideal power function is graphed in Figure 8.4–2.

In general this ideal situation is not possible. If $\theta \ne \theta_0$ but θ is "close" to θ_0, we might feel that it would not be too serious an error to accept H_0 because it is essentially true for most practical purposes. However, if θ is not "close" to θ_0, say $\theta < \theta_1 < \theta_0$ or $\theta > \theta_2 > \theta_0$, we would like to have a high probability that H_0 is rejected. A power function for this situation is graphed in Figure 8.4–3; this is fairly common and the power, when $\theta \ne \theta_0$, can be increased by

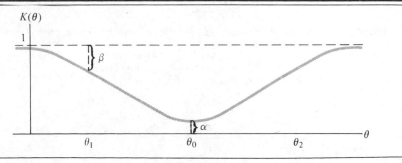

FIGURE 8.4–3

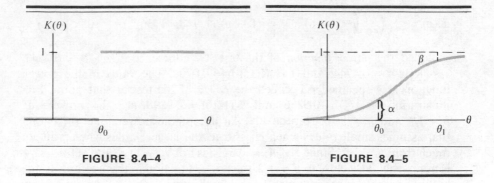

FIGURE 8.4-4   FIGURE 8.4-5

increasing the sample size n. Incidentally, the β depicted in Figure 8.4–3 is that associated with θ_1; that is, here $\beta = 1 - K(\theta_1)$.

Suppose that we are now interested in testing the null hypothesis $H_0: \theta \leq \theta_0$ against the alternative hypothesis $H_1: \theta > \theta_0$. Note that both hypotheses are now composite. An ideal power function for this situation would be like the one graphed in Figure 8.4–4. However, in most instances, we have to settle for a power function like the one graphed in Figure 8.4–5.

The next example indicates how a desirable power function can determine the sample size n needed in a given test procedure.

Example 8.4–3. Let X have a distribution $N(\mu, 36)$. If we have the random sample $X_1, X_2, \ldots, X_n$ when testing $H_0: \mu = 50$ against $H_1: \mu > 50$, we use the critical region defined by $C = \{(x_1, x_2, \ldots, x_n): \bar{x} > c\}$. Determine c and n such that $K(50) = \alpha = 0.05$ and $K(55) = 0.90$, where $K(\mu) = P(\bar{X} > c; \mu)$ is the power function of the test. To find c and n we must solve simultaneously the following two equations:

$$0.05 = P(\bar{X} > c; H_0) \tag{1}$$

and

$$0.90 = P(\bar{X} > c; H_1) \quad \text{or} \quad 0.10 = P(\bar{X} \leq c; H_1). \tag{2}$$

Because $\bar{X}$ is $N(50, 36/n)$ when H_0 is true, equation (1) becomes

$$P\left(\frac{\bar{X} - 50}{6/\sqrt{n}} > \frac{c - 50}{6/\sqrt{n}}\right) = 0.05$$

and, thus,

$$\frac{c - 50}{6/\sqrt{n}} = 1.645. \tag{3}$$

Because $\bar{X}$ is $N(55, 36/n)$ when H_1 is true, equation (2) implies that

$$P\left(\frac{\overline{X} - 55}{6/\sqrt{n}} \le \frac{c - 55}{6/\sqrt{n}}\right) = 0.10$$

and, thus,

$$\frac{c - 55}{6/\sqrt{n}} = -1.282. \tag{4}$$

Solving equations (3) and (4) simultaneously yields the approximate solution $c = 52.8$ and $n = 12$ or 13. Note that since n must be an integer, the probabilities in equations (1) and (2) are only approximated by the final solution.

Example 8.4–4. Let Y be $b(n, p)$. Reject $H_0: p = 1/2$ and accept $H_1: p < 1/2$ if the observed value y of Y is less than the constant c. To find n and c such that $K(1/2) = 0.05$ and $K(1/4) = 0.90$, where $K(p) = P(Y < c: p)$, we proceed as follows:

$$0.05 = P\left(Y < c; p = \frac{1}{2}\right) = P\left(\frac{Y - n/2}{\sqrt{n(1/2)(1/2)}} < \frac{c - n/2}{\sqrt{n(1/2)(1/2)}}\right)$$

implies that

$$\frac{c - n/2}{\sqrt{n/4}} \approx -1.645$$

and

$$0.90 = P\left(Y < c; p = \frac{1}{4}\right) = P\left(\frac{Y - n/4}{\sqrt{n(1/4)(3/4)}} < \frac{c - n/4}{\sqrt{n(1/4)(3/4)}}\right)$$

implies that

$$\frac{c - n/4}{\sqrt{3n/16}} \approx 1.282.$$

That is, we must solve simultaneously the two approximate equalities:

$$c - \frac{n}{2} \approx \frac{-1.645}{2}\sqrt{n}, \qquad c - \frac{n}{4} \approx \frac{1.282\sqrt{3}}{4}\sqrt{n}.$$

Subtracting the first from the second, we obtain

$$\frac{n}{4} \approx \left(\frac{1.645}{2} + \frac{1.282\sqrt{3}}{4}\right)\sqrt{n}$$

or, equivalently,

$$\sqrt{n} \approx 3.29 + 1.282\sqrt{3} = 5.510.$$

Thus, n is about equal to 31 and from either of the first two approximate equalities we find that c is about equal to 10.9. Using $n = 31$ and $c = 10.9$ means that $K(1/2) = 0.05$ and $K(1/4) = 0.90$ are only approximate.

Exercises

8.4–1. Bowl A contains 100 red balls and 200 white balls, bowl B contains 200 red balls and 100 white balls. Let p denote the probability of drawing a red ball from a bowl, but say p is unknown since it is unknown whether bowl A or bowl B is being used. We shall test the simple null hypothesis $H_0: p = 1/3$ against the simple alternative hypothesis $H_1: p = 2/3$. Draw three balls at random, one at a time and with replacement from the selected bowl. Let X equal the number of red balls drawn.

    (a) Define a critical region C for this test in terms of X.

    (b) What are the values of α and β, the probabilities of Type I and Type II errors, respectively, for the critical region defined in part (a)?

8.4–2. A bowl contains two red balls, two white balls, and a fifth ball that is either red or white. Let p denote the probability of drawing a red ball from the bowl. We shall test the simple null hypothesis $H_0: p = 3/5$ against the simple alternative hypothesis $H_1: p = 2/5$. Draw four balls at random from the bowl, one at a time and with replacement. Let X equal the number of red balls drawn.

    (a) Define a critical region C for this test in terms of X.

    (b) For the critical region C defined in part (a), find the values of α and β.

8.4–3. A certain size bag is designed to hold 25 pounds of potatoes. A farmer fills such bags in the field. Assume that the weight X of potatoes in a bag is $N(\mu, 9)$. We shall test the simple null hypothesis $H_0: \mu = 25$ against the composite alternative hypothesis $H_1: \mu < 25$.

    (a) Let X_1, X_2, X_3, X_4 be a random sample of size 4 from this distribution. Define a critical region C for this test using $\overline{X}$ as the test statistic.

    (b) What is the power function $K(\mu)$ of this test? In particular, what is the significance level $\alpha = K(25)$ for your test?

    (c) If the random sample of four bags of potatoes yielded the values $x_1 = 21.24$, $x_2 = 24.81$, $x_3 = 23.62$, $x_4 = 26.82$, would you accept or reject H_0 using your test?

8.4–4. Let X be $N(\mu, 4)$. We would like to test the simple hypothesis, $H_0: \mu = 10$ against the composite alternative hypothesis $H_1: \mu > 10$. Let the critical region be defined by $C = \{(x_1, x_2, \ldots, x_9): \overline{x} > 11.5\}$, where $\overline{x}$ is the sample mean of a random sample of size 9 from this distribution.

    (a) What is the significance level of this test?

    (b) How is the power function $K(\mu)$ defined for this test?

    (c) What are the values of $K(11)$, $K(11.5)$, and $K(12)$?

    (d) Sketch the graph of the power function.

8.4–5. Let X equal the yield of alfalfa in tons per acre. Assume that X is $N(1.5, 0.09)$. It is hoped that new fertilizer will increase the average yield. We shall test the simple null hypothesis $H_0: \mu = 1.5$ against the composite alternative hypothesis $H_1: \mu > 1.5$. Assume that the variance continues to equal $\sigma^2 = 0.09$ with the new fertilizer. Using $\bar{X}$, the mean of a random sample of size n, as the test statistic, reject H_0 if $\bar{X} > c$. Find n and c so that the power function $K(\mu) = P(\bar{X} > c; \mu)$ is such that $\alpha = K(1.5) = 0.05$ and $K(1.7) = 0.95$.

8.4–6. Let X have a Bernoulli distribution with p.d.f.

$$f(x; p) = p^x(1 - p)^{1-x}, \qquad x = 0, 1, \quad 0 \le p \le 1.$$

We would like to test the null hypothesis $H_0: p \le 0.4$ against the alternative hypothesis $H_1: p > 0.4$. For the test statistic use $Y = \sum_{i=1}^{n} X_i$, where $X_1, X_2, \ldots, X_n$ is a random sample of size n from this Bernoulli distribution. Let the critical region be of the form $C = \{y: \ y > c\}$.

(a) Let $n = 100$. On the same set of axes, sketch the graphs of the power functions corresponding to the three critical regions, $C_1 = \{y: \ y > 40\}$, $C_2 = \{y: \ y > 50\}$, and $C_3 = \{y: \ y > 60\}$.

(b) Let $C = \{y: \ y > 0.45n\}$. On the same set of axes, sketch the graphs of the power functions corresponding to the three samples of sizes 10, 100, and 1000.

8.4–7. Let p equal the fraction defective of a certain manufactured item. To test $H_0: p = 1/26$ against $H_1: p > 1/26$, we inspect n items selected at random and let Y be the number of defective items in this sample. We reject H_0 if the observed $Y > c$. Find n and c so that $\alpha = K(1/26) = 0.10$ and $K(1/10) = 0.80$ approximately, where $K(p) = P(Y > c; p)$.

8.5 Binomial Tests for Percentiles

In Chapter 6, we found ways of constructing interval estimates for certain characteristics of the distribution that did not depend too much upon distributional assumptions. For examples, we found distribution-free interval estimates for the $(100p)$th percentile π_p of a distribution of the continuous type. We now take up some tests of statistical hypotheses that are *distribution free*, provided the null hypotheses are true. In this section, in particular, we test the hypothesis $H_0: \pi_p = \pi_0$, where π_0 is given, at some significance level of that test for *every* continuous-type distribution with $(100p)$th percentile $\pi_p = \pi_0$. It should be noted here, however, that the powers of these tests are not distribution free when the null hypotheses are false. Moreover, the resulting nonnull distribution theory is frequently quite complicated, so little will be said here about the power functions.

We begin by considering a hypothesis about the median $\pi_{1/2}$, or, for simplicity, π, of a distribution of the continuous type. Further, we assume that the median π is unique. Suppose it is known from past experience that the median length of sunfish in a particular polluted lake was $\pi = 3.7$ inches. During the past two years the lake was "cleaned up" and the conjecture is made that now $\pi > 3.7$ inches. We consider a procedure for testing this conjecture. In particular, we describe a procedure for testing the null hypothesis $H_0: \pi = 3.7$ against the alternative hypothesis $H_1: \pi > 3.7$.

Let X denote the length of a sunfish selected at random from the lake. If the null hypothesis $H_0: \pi = 3.7$ is true, $P(X \leq 3.7; H_0) = 0.5$. However, if the alternative hypothesis $H_1: \pi > 3.7$ is true, $P(X \leq 3.7; H_1) < 0.5$. If we take a random sample of $n = 10$ fish, we would expect about half the fish to be shorter than 3.7 inches if H_0 is true. However, if H_1 is true, we would expect less than half the fish to be shorter than 3.7 inches. We will decide on the basis of the number, say Y, of fish shorter than 3.7 inches whether to accept H_0 or H_1. Thus, Y can be thought of as the number of "successes" in 10 Bernoulli trials with probability of success given by $p = P(X \leq 3.7)$. If H_0 is true, $p = 1/2$ and Y is $b(10, 1/2)$; whereas if H_1 is true, $p < 1/2$ and Y is $b(10, p)$. We reject H_0 and accept H_1 if and only if the observed value y of Y is sufficiently small, say $y < c$. From Appendix Table I, we find that $P(Y < 3; H_0) = 0.0547$. Thus, if we let the critical region be defined by $C = \{y: y < 3\}$ or, equivalently, $C = \{y: y \leq 2\}$, then $\alpha = 0.0547$.

Suppose that the lengths of 10 sunfish selected at random from this lake were 5.0, 3.9, 5.2, 5.5, 2.8, 6.1, 6.4, 2.6, 1.7, and 4.3 inches. Since $y = 3$ of these lengths are shorter than 3.7, we would not reject $H_0: \pi = 3.7$ at the $\alpha = 0.0547$ significance level by considering only these few data.

Example 8.5–1. Let X denote the length of time in seconds between two calls entering a college switchboard. Let π be the unique median of this continuous-type distribution. We test the null hypothesis $H_0: \pi = 6.2$ against the alternative hypothesis $H_1: \pi < 6.2$. If Y is the number of lengths of time in a random sample of size 20 that are less than 6.2, the critical region $C = \{y: y \geq 14\} = \{y: y > 13\}$ has an approximate size $\alpha = 0.0588$, using the normal approximation. A random sample of size 20 yielded the following data:

6.8	5.7	6.9	5.3	4.1	9.8	1.7	7.0
2.1	19.0	18.9	16.9	10.4	44.1	2.9	2.4
4.8	18.9	4.8	7.9				

Since $y = 9$, the null hypothesis is not rejected.

In many places in the literature the test we have just described is called the *sign test*. The reason for this terminology is that the test is based upon a statistic Y that is equal to the number of negative signs among

$$X_1 - \pi_0, X_2 - \pi_0, \ldots, X_n - \pi_0.$$

The sign test can also be used to test the hypothesis that two continuous-type random variables X and Y are such that $p = P(X > Y) = 1/2$. To test the hypothesis $H_0: p = 1/2$ against an appropriate alternative hypothesis, consider the independent pairs $(X_1, Y_1), (X_2, Y_2), \ldots, (X_n, Y_n)$. Let W denote the number of pairs for which $X_i - Y_i > 0$. When H_0 is true, W is $b(n, 1/2)$, and the test can be based upon the statistic W. It is important to recognize that X_i and Y_i do not need to be independent and are frequently not independent in practice. If they were independent, then there are many ways in which $X_1, X_2, \ldots, X_n$ and $Y_1, Y_2, \ldots, Y_n$ can be paired up. If, however, X is the length of the right foot of a person and Y the length of the corresponding left foot, there is a natural pairing. Here $H_0 : p = P(X > Y) = 1/2$ suggests that it is equally likely that either foot is the longest.

Example 8.5–2. Table 8.5–1 gives 12 pairs of values of two random variables. Also recorded is a plus sign when $x_i - y_i > 0$ and a minus sign when $x_i - y_i < 0$. We test the hypothesis $H_0 : p = P(X > Y) = 1/2$ against a two-sided alternative $H_1 : p \neq 1/2$. Let the critical region be defined by $C = \{w: \; w \leq 3 \quad \text{or} \quad w \geq 9\}$, where w is the number of "plus signs." Since W is $b(12, 1/2)$ when H_0 is true we see that $\alpha = 0.1460$ from the binomial tables. Since $w = 7$ the null hypothesis is not rejected.

x	y	SIGN	x	y	SIGN
37.3	30.0	+	40.5	33.2	+
44.8	35.6	+	34.8	39.0	−
34.5	27.7	+	32.4	35.7	−
31.9	31.8	+	33.5	35.7	−
36.9	34.9	+	33.8	26.5	+
28.6	31.1	−	24.9	37.7	−

TABLE 8.5–1

The sign test can also be used in comparisons when the outcomes are not numerical. For example, each of a number of coffee drinkers selected at random is asked if some freshly perked coffee is preferred over some instant coffee.

There are some objections to the sign test, among them being that the data must be paired and that the test does not take into account the magnitude of differences. How do we decide when to use the sign test? In the situation in which we need a test about a location parameter of a distribution, would we use the sign test to test $\pi = \pi_0$ or a test based upon $\overline{X}$ to test $\mu = \mu_0$? In particular, if the distribution is symmetric, then $\pi = \mu$ and hence each procedure would be testing the hypothesis that the common value of π and μ is equal to a known constant. In more advanced theory and practice, we discover that if there are outliers (that is, extreme x values that deviate greatly from most of the others), the sign test is usually better than that based upon $\overline{X}$. This certainly appeals to our intuition because a few extreme x values can influence the average $\bar{x}$ a great deal; but, in the sign test, each extreme value is associated with only one sign no matter how far it is from π_0. Hence, in cases with highly skewed data or data with outliers, the distribution-free sign test would probably be preferred to that based upon $\overline{X}$. In the next section, we consider another distribution-free test, first proposed by Wilcoxon, which does account somewhat for the magnitudes of the deviations $X_1 - \pi_0, X_2 - \pi_0, \ldots, X_n - \pi_0$ and is an excellent test in many situations.

Another big advantage of the sign test is the fact that it can easily be generalized to percentiles other than the median. To test the hypothesis that the $(100p)$th percentile π_p of a continuous-type distribution is equal to a specified value π_0, let Y be the number of the items of the random sample of size n that are less than π_0. If $H_0: \pi_p = \pi_0$ is true, then Y is $b(n, p)$; where here, of course, p is a known fraction. If, for example, the alternative hypothesis is $H_1: \pi_p > \pi_0$, then the critical region would be of the form $C = \{y: \quad y < c\}$.

Example 8.5–3. Suppose, from past testing, we know that the 25th percentile of ninth-grade general mathematics students taking a standardized examination is 62.4. Statisticians believe that by introducing more statistics involving real problems into such a course the students will see the usefulness of mathematics. It is hoped that this approach will motivate the poorer students in these classes and increase some of the lower percentiles. In particular, they believe that if these courses were changed as suggested then $H_1: \pi_{0.25} > 62.4$. To test $H_0: \pi_{0.25} = 62.4$ against H_1, 192 general mathematics students were selected at random and given this new type of course using more statistics. Let Y be the number of students scoring less than 62.4. Of course, if H_0 is true, then Y is $b(192, 1/4)$. If H_1 is true, we would expect smaller values of Y and thus the critical region is of the form $y < c$. To determine c, we can use the fact that Y, under H_0, has an approximate normal distribution with mean $192(1/4) = 48$ and variance $192(1/4)(3/4) = 36$. Thus, for the significance level to be about 0.05, we want

$$c \approx 48 - (1.645)(6) = 38.13.$$

Accordingly, if we take $c = 39$, then the significance level is

$$P(Y < 39) = P(Y < 38.5) = P\left(\frac{Y - 48}{6} < \frac{-9.5}{6}\right)$$

$$\approx \Phi(-1.583) = 1 - 0.9433 = 0.0567.$$

Suppose, after the course is over, these 192 students take the standardized examination and only 31 have grades less than 62.4. We would reject $H_0: \pi_{0.25} = 62.4$ and accept $H_1: \pi_{0.25} > 62.4$; that is, it seems as if the 25th percentile is greater in the new type of general mathematics course than that associated with the old one.

Exercises

8.5–1. For the following set of data, test the null hypothesis H_0: $\pi = 4.8$ against a two-sided alternative hypothesis H_1: $\pi \neq 4.8$. Use an approximate 10% significance level.

1.0	10.3	16.7	38.4	2.4
2.6	8.9	36.3	27.1	3.8
1.9	0.9	0.4	9.2	3.0

8.5–2. Test the null hypothesis that $P(X > Y) = 1/2$ against a two-sided alternative at an approximate 10% significance level using the following set of paired data (x, y).

(46.8, 49.5)	(48.8, 51.1)	(52.4, 49.1)	(49.5, 47.8)
(46.1, 47.7)	(48.1, 45.6)	(45.7, 46.7)	(44.8, 51.7)
(45.4, 49.5)	(47.5, 51.1)	(47.0, 49.4)	(48.1, 47.8)

8.5–3. A course in economics was taught to two groups of students, one in a classroom situation and the other by TV. There were 24 students in each group. These students were first paired according to cumulative grade point averages and background in economics, and then assigned to the courses by a flip of a coin (this was repeated 24 times). At the end of the course each class was given the same final examination. Use the sign test to test the hypothesis that the two methods of teaching are equally effective against a two-sided alternative. The differences in final scores for each pair of students, the TV student's score having been subtracted from the corresponding classroom student's score, were as follows.

14	-4	-6	-2	-1	18
6	12	8	-4	13	7
2	6	21	7	-2	11
-3	-14	-2	17	-4	-5

8.5–4. The administration claims that Professor A is more popular than Professor B. We shall test the hypothesis H_0 of no difference in popularity against the administration's claim H_1 at an approximate 0.05 significance level. If 12 out of 17 students prefer Professor A to Professor B, is H_0 rejected and the administration's claim supported?

8.5–5. In a certain region, it was known that about 20% of the drivers exceeded the speed limit of 55 by more than 5 miles per hour. The highway patrol then established a new enforcement procedure and claimed that this percentage had been reduced substantially. How would you test the null hypothesis H_0: $\pi_{0.8} = 60$ against the patrol's claim H_1: $\pi_{0.8} < 60$ at $\alpha = 0.05$ if you could observe the speeds of 100 drivers selected at random? If, in fact, out of the 100 drivers, you found that the speeds of 89 were under 60 miles per hour, what decision would you make?

8.6 The Wilcoxon Test

Let X be a continuous-type random variable. Let π denote the median of X. To test the hypothesis H_0: $\pi = \pi_0$ against an appropriate alternative hypothesis, the sign test, discussed in the previous section, can be used. That is, if X_1, X_2,

..., X_n denote the items of a random sample from this distribution and if we count the number of negative differences among $X_1 - \pi_0$, $X_2 - \pi_0$, ..., $X_n - \pi_0$, this number has the binomial distribution $b(n, 1/2)$ under H_0 and is the test statistic for the sign test. One major objection to this test is that it does not take into account the magnitude of the differences $X_1 - \pi_0, \ldots, X_n - \pi_0$.

In this section we shall discuss a test that does take into account the magnitude of the differences $|X_i - \pi_0|$, $i = 1, 2, \ldots, n$. However, in addition to the assumption that the random variable X is of the continuous type, we must also assume that the p.d.f. of X is symmetric about the median in order to find the distribution of this new statistic. Because of the continuity assumption, we assume, in the following discussion, that no two observations are equal and that no observation is equal to the median.

We are interested in testing the hypothesis $H_0 : \pi = \pi_0$ where π_0 is some given constant. With our random sample $X_1, X_2, \ldots, X_n$, we rank the differences $X_1 - \pi_0, X_2 - \pi_0, \ldots, X_n - \pi_0$ in ascending order according to magnitude. That is, for $i = 1, 2, \ldots, n$, we let R_i denote the rank of $|X_i - \pi_0|$ among $|X_1 - \pi_0|, |X_2 - \pi_0|, \ldots, |X_n - \pi_0|$. Note that $R_1, R_2, \ldots, R_n$ is a permutation of the first n positive integers, $1, 2, \ldots, n$. Now with each R_i we associate the sign of the difference $X_i - \pi_0$; that is, if $X_i - \pi_0 > 0$, we use R_i, but if $X_i - \pi_0 < 0$, we use $-R_i$. The Wilcoxon statistic Z is the sum of these n signed ranks.

Example 8.6–1. Consider an education testing situation. Let $\pi_0 = 63$ and let $n = 10$ randomly selected scores on the test be

$$x_i: \quad 42, \quad 61, \quad 58, \quad 93, \quad 77, \quad 55, \quad 69, \quad 41, \quad 60, \quad 38.$$

Thus we have

$x_i - \pi_0$:	$-21, \; -2, \; -5, \; 30, \; 14, \; -8, \; 6, \; -22, \; -3, \; -25;$
$\lvert x_i - \pi_0 \rvert$:	$21, \quad 2, \quad 5, \; 30, \; 14, \quad 8, \; 6, \quad 22, \quad 3, \quad 25;$
Ranks:	$7, \quad 1, \quad 3, \; 10, \quad 6, \quad 5, \; 4, \quad 8, \quad 2, \quad 9;$
Signed ranks:	$-7, \; -1, \; -3, \; 10, \quad 6, \; -5, \; 4, \quad -8, \; -2, \; -9.$

Therefore, the Wilcoxon statistic is equal to

$$Z = -7 - 1 - 3 + 10 + 6 - 5 + 4 - 8 - 2 - 9 = -15.$$

Incidentally, the negative answer seems reasonable because the number of the ten values less than 63 is seven, which is the statistic used in the sign test.

It seems that the hypothesis $H_0 : \pi = \pi_0$ is supported if the observed value of Z is close to zero. If the alternative hypothesis is $H_1 : \pi > \pi_0$, we would reject H_0 if the observed $Z = z$ is too large since, in this case, the larger deviations $|X_i - \pi_0|$ would usually be associated with observations for which $x_i - \pi_0 > 0$. That is, the critical region would be of the form $\{z : \; z > c_1\}$. If the alternative hypothesis is $H_1 : \pi < \pi_0$, the critical region would be of the form $\{z : \; z < c_2\}$.

Of course, the critical region would be of the form $\{z: \ z < c_3 \ \text{ or } \ z > c_4\}$, for a two-sided alternative hypothesis $H_1: \pi \neq \pi_0$. In order to find the values of c_1, c_2, c_3, c_4 that yield desired significance levels, it is necessary to determine the distribution of Z under H_0. We now consider certain characteristics of this distribution.

When $H_0: \pi = \pi_0$ is true,

$$P(X_i < \pi_0) = P(X_i > \pi_0) = \frac{1}{2}, \qquad i = 1, 2, \ldots, n.$$

Hence, the probability that the sign associated with the rank R_i of $|X_i - \pi_0|$ is negative is 1/2. Moreover the assignment of these n signs are independent because $X_1, X_2, \ldots, X_n$ are mutually independent. In addition, Z is a sum that contains the integers $1, 2, \ldots, n$, each integer with a positive or negative sign. Thus, intuitively, it seems that Z has the same distribution as the random variable

$$V = \sum_{1}^{n} V_i, \qquad \text{where } V_1, V_2, \ldots, V_n \text{ are independent}$$

and

$$P(V_i = i) = P(V_i = -i) = \frac{1}{2}, \qquad i = 1, 2, \ldots, n.$$

That is, V is a sum that contains the integers $1, 2, \ldots, n$, and these integers receive their algebraic signs by independent assignments.

Since Z and V have the same distribution, their means and variances are equal, and we can easily find those of V. Now the mean of V_i is

$$E(V_i) = -i\left(\frac{1}{2}\right) + i\left(\frac{1}{2}\right) = 0$$

and, thus,

$$E(Z) = E(V) = \sum_{1}^{n} E(V_i) = 0.$$

The variance of V_i is

$$\text{Var}(V_i) = E(V_i^2) = (-i)^2\left(\frac{1}{2}\right) + (i)^2\left(\frac{1}{2}\right) = i^2.$$

Thus

$$\text{Var}(Z) = \text{Var}(V) = \sum_{1}^{n} \text{Var}(V_i) = \sum_{1}^{n} i^2 = \frac{n(n+1)(2n+1)}{6}.$$

We will not try to find the distribution of Z in general since that p.d.f. does not have a convenient expression. However, we demonstrate how we could tabulate the distribution of one probability of Z (or V) with enough patience and computer support. For these comments, recall that the moment-generating function of V_k is

$$M_k(t) = e^{t(-k)}\left(\frac{1}{2}\right) + e^{t(+k)}\left(\frac{1}{2}\right) = \frac{e^{-kt} + e^{kt}}{2}, \qquad k = 1, 2, \ldots, n.$$

Let $n = 2$, so the moment-generating function of $V_1 + V_2$ is

$$M(t) = E[e^{t(V_1 + V_2)}].$$

From the independence of V_1 and V_2, we obtain

$$M(t) = E(e^{tV_1})E(e^{tV_2})$$

$$= \left(\frac{e^{-t} + e^t}{2}\right)\left(\frac{e^{-2t} + e^{2t}}{2}\right)$$

$$= \frac{e^{-3t} + e^{-t} + e^t + e^{3t}}{4}.$$

This means that each of the points $-3, -1, 1, 3$ in the support of $V_1 + V_2$ has probability 1/4.

Next let $n = 3$, so the moment-generating function of $V_1 + V_2 + V_3$ is

$$M(t) = E[e^{t(V_1 + V_2 + V_3)}]$$

$$= E[e^{t(V_1 + V_2)}]E(e^{tV_3})$$

$$= \left(\frac{e^{-3t} + e^{-t} + e^t + e^{3t}}{4}\right)\left(\frac{e^{-3t} + e^{3t}}{2}\right)$$

$$= \frac{e^{-6t} + e^{-4t} + e^{-2t} + 2e^0 + e^{2t} + e^{4t} + e^{6t}}{8}.$$

Thus, the points $-6, -4, -2, 0, 2, 4, 6$ in the support of $V_1 + V_2 + V_3$ have the respective probabilities 1/8, 1/8, 1/8, 2/8, 1/8, 1/8, 1/8. Obviously this procedure can be continued for $n = 4, 5, 6, \ldots$, but it is rather tedious. Fortunately, however, even though $V_1, V_2, \ldots, V_n$ are not identically distributed random variables, the sum V of them still has an approximate normal distribution. To obtain this normal approximation for V (or Z), a more general form of the Central Limit Theorem, due to Liapounov, can be used that allows us to say that the standardized random variable

$$\frac{Z - 0}{\sqrt{n(n + 1)(2n + 1)/6}}$$

is approximately $N(0, 1)$ when H_0 is true. We accept this without proof and thus we can approximate probabilities such as $P(Z > c; H_0)$ when the sample size n is sufficiently large using this normal distribution.

Example 8.6–2. Let π be the median of a symmetric distribution of the continuous type. To test the hypothesis $H_0: \pi = 160$ against the alternative hypothesis $H_1: \pi > 160$, we take a random sample of size $n = 16$. For an approximate significance level of $\alpha = 0.05$, H_0 is rejected if the computed $Z = z$ is such that

$$\frac{z}{\sqrt{16(17)(33)/6}} > 1.645,$$

or

$$z > 1.645 \sqrt{\frac{16(17)(33)}{6}} = 63.626.$$

Say the observed values of a random sample are 176.9, 158.3, 152.1, 158.8, 172.4, 169.8, 159.7, 162.7, 156.6, 174.5, 184.4, 165.2, 147.8, 177.8, 160.1, 160.5. In Table 8.6–1 the magnitudes of the differences $|x_i - 160|$ have been

0.1	0.3	0.5	1.2	1.7	2.7	3.4	5.2
1	2	3	4	5	6	7	8
7.9	9.8	12.2	12.4	14.5	16.9	17.8	24.4
9	10	11	12	13	14	15	16

TABLE 8.6–1

ordered and ranked. Those differences $x_i - 160$ that were negative have been underlined. For this set of data

$$z = 1 - 2 + 3 - 4 - 5 + 6 + \cdots + 16 = 60.$$

Since $60 < 63.626$, H_0 is not rejected at the 0.05 significance level. It is interesting to note that H_0 would have been rejected at $\alpha = 0.10$. This would indicate to us that the data are too few to reject H_0, but if the pattern continues, we will most certainly reject with a larger sample size.

Although theoretically we could ignore the possibilities that $x_i = \pi_0$, for some i, and that $|x_i - \pi_0| = |x_j - \pi_0|$ for some $i \neq j$, these situations do occur in applications. Usually, in practice, if $x_i = \pi_0$ for some i, that observation is deleted and the test is performed with a reduced sample size. If the absolute value of the differences from π_0 of two or more observations are equal, each observation is assigned the average of the corresponding ranks. The change this causes in the distribution of Z is not very great, and thus we continue using the same normal approximation.

In modern statistics, generalizations of tests such as the sign and Wilcoxon are becoming so popular that we should at least mention them. In this discussion, we continue to assume an underlying distribution that is, under H_0, symmetric about π_0. Again let R_i be the rank of $|X_i - \pi_0|$. Instead of using the sum of the signed ranks, we replace each rank by a function of that rank and proceed as before. That is, let $a(R_i)$ be a function, called a *score*, such that

$$a(1) \le a(2) \le \cdots \le a(n).$$

We sign each score $a(R_i)$ as before and add these signed scores to obtain a generalization of the Wilcoxon test statistic. Obviously, this sum is distributed exactly like $\sum_1^n V_i$, where $V_1, V_2, \ldots, V_n$ are mutually independent with

$$P[V_i = a(i)] = P[V_i = -a(i)] = \frac{1}{2}, \qquad i = 1, 2, \ldots, n.$$

It is interesting to note that if

$$a(1) = a(2) = \cdots = a(n) = 1,$$

then the sum of signed scores is the difference between the numbers of positive and negative signs. But this statistic is equivalent to knowing the number of negative signs; that is, if we know the number of negative signs, we can easily compute that difference. Thus, the test is essentially the same as the sign test.

Of course, the question of which score function to use in a particular case always arises. Usually, $a(i) = 1, i = 1, 2, \ldots, n$ (that is, the sign test), is used if we suspect some large outliers. The score function of the Wilcoxon statistic namely, $a(i) = i, i = 1, 2, \ldots, n$, is quite *robust*; that is, good over a wide range of distributions, particularly for distribution with somewhat heavier tails than the normal, such as the logistic distribution. The *normal* scores defined by

$$\frac{i}{n+1} = \int_0^{a(i)} \sqrt{\frac{2}{\pi}} e^{-x^2/2} \, dx, \qquad i = 1, 2, \ldots, n,$$

are quite good for distributions with normal (or somewhat lighter) tails. When there is doubt, the Wilcoxon statistic is usually a reasonable selection.

=== *Exercises* ===

8.6–1. The observed values of a random sample of size 16 from a continuous distribution that is symmetric about the median π are

2.1	8.0	3.1	4.0
8.5	1.1	6.7	1.8
8.3	4.9	0.3	6.9
1.3	9.9	4.3	2.2

Use the Wilcoxon statistic to test the hypothesis $H_0: \pi = 5.0$ against the two-sided alternative hypothesis $H_1: \pi \ne 5.0$. Let $\alpha = 0.10$. How does this decision compare to that obtained if the sign test was used?

8.6–2. Twenty observations of a random variable X are

68.54	70.32	78.73	80.06	82.62
59.59	53.02	67.11	98.85	67.19
86.20	71.51	81.64	71.96	87.07
80.47	72.90	65.02	57.28	75.80

Use the Wilcoxon statistic to test the hypothesis $H_0: \pi = 70.0$ against the alternative hypothesis $H_1: \pi > 70.0$, where π is the median of the symmetric distribution. Let $\alpha = 0.05$. Also use the sign test with these data.

8.6–3. Let $n = 20$ and consider the generalized statistic associated with the scores given by $a(i) = 1, i = 1, 2, \ldots, 10$, and $a(i) = i - 10, i = 11, 12, \ldots, 20$. Find the mean and the variance of Z. With the data in Exercise 8.6–2, use this statistic (and its normal approximation) and compare the decision with those of the Wilcoxon and sign tests. Incidentally, the type of test associated with these scores would, in general, be better with light-tailed distributions (in which outliers would be very unlikely).

8.6–4. With the data in Exercise 8.5–3, use the Wilcoxon statistic to test the hypothesis that the median of the differences in final scores equals zero.

8.7 Two Sample Distribution-Free Tests

In Sections 8.5 and 8.6, we considered some distribution-free tests based upon a sample from one distribution. In this section, we consider corresponding tests associated with the characteristics of two independent distributions.

The first test corresponds to the sign test and is called the *median test*. Let π_X and π_Y be the respective medians of two independent distributions of the continuous type. By taking random samples $X_1, X_2, \ldots, X_n$ and $Y_1, Y_2, \ldots, Y_m$ from these two independent distributions, respectively, we wish to test the hypothesis $H_0: \pi_X = \pi_Y$. To do this, combine the two samples and count the number, say V, of X values in the lower half of this combined sample. If $H_0: \pi_X = \pi_Y$ is true, then we would expect V to equal some number around $n/2$. If, as an alternative, $\pi_X < \pi_Y$, we would expect V to be somewhat larger; and, of course, the alternative $\pi_X > \pi_Y$ would suggest a smaller value of V.

Let us see what this means in terms of the distribution functions $F(x)$ and $G(y)$ of the respective distributions. Of course, if $F(z) = G(z)$, then $H_0: \pi_X = \pi_Y$ is true. Since we cannot find the distribution of V knowing only that $\pi_X = \pi_Y$, we will find its distribution assuming that $F(z) = G(z)$. If $F(z) \geq G(z)$, $\pi_X \leq \pi_Y$ as depicted in Figure 8.7–1. If the observed value of V is quite large—that is, if the number of values of X falling below the median of the combined sample is large—we would suspect that $\pi_X < \pi_Y$. Thus, the critical region for testing $H_0: \pi_X = \pi_Y$ against $H_1: \pi_X < \pi_Y$ is of the form $v > c$, where c is to be determined to yield the desired significance level [when $F(z) = G(z)$]. Similarly,

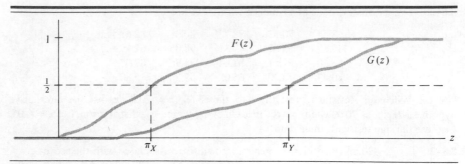

FIGURE 8.7–1

the critical region for testing $H_0: \pi_X = \pi_Y$ against $H_1: \pi_X > \pi_Y$ is of the form $v < c$.

When $F(z) = G(z)$ is true and still assuming continuous-type distributions, we shall argue that V has a hypergeometric distribution. To simplify the discussion, say that $n + m = 2k$, where k is a positive integer. To compute $P(V = v)$, we need the probability that exactly v of $X_1, X_2, \ldots, X_n$ are in the lower half of the ordered combined sample. (Under our assumptions, the probability is zero that any two of the $2k$ random variables are equal.) The smallest k of the $n + m = 2k$ items can be selected in any one of $\binom{2k}{k}$ ways, each having the same probability provided $F(z) = G(z)$. Of these $\binom{2k}{k}$ ways, the number in which exactly v of the n values of X and $k - v$ of the m values of Y appear in the lower k items is $\binom{n}{v}\binom{m}{k - v}$. Hence,

$$h(v) = P(V = v) = \frac{\binom{n}{v}\binom{m}{k - v}}{\binom{n + m}{k}}, \quad v = 0, 1, \ldots, n,$$

with the understanding that $\binom{j}{i} = 0$ if $i > j$.

Example 8.7–1. Let X and Y denote the weights of ground cinnamon in "120 gram" tins packaged by Companies A and B, respectively. We shall test the hypothesis $H_0: \pi_X = \pi_Y$ against the one-sided alternative hypothesis $H_1: \pi_X < \pi_Y$. The weights of $n = 8$ and $m = 8$ tins of cinnamon packaged by Companies A and B, respectively, selected at random, yielded the following observations of X:

117.2,   121.3,   127.8,   121.9,   117.4,   124.5,   119.5,   115.1,

and the following observations of Y:

123.5,   125.3,   126.5,   127.9,   122.1,   125.6,   129.8,   117.2.

The critical region is of the form $v > c$. To determine the value of c when $F(z) = G(z)$, we compute $P(V = v)$ for $v = 6, 7,$ and 8. Now

$$\binom{n+m}{k} = \binom{8+8}{8} = 12{,}870.$$

Thus,

$$h(8) = \frac{\binom{8}{8}\binom{8}{0}}{12{,}870} = \frac{1}{12{,}870};$$

$$h(7) = \frac{\binom{8}{7}\binom{8}{1}}{12{,}870} = \frac{64}{12{,}870};$$

$$h(6) = \frac{\binom{8}{6}\binom{8}{2}}{12{,}870} = \frac{784}{12{,}870}.$$

Since

$$h(8) + h(7) + h(6) = \frac{849}{12{,}870} = 0.066,$$

we shall reject H_0 if $v > 5$ at an $\alpha = 0.066$ significance level. The combined ordered sample of weights is listed below with the observations of X underlined.

| 115.1 | 117.1 | 117.2 | 117.4 | 119.5 | 121.3 | 121.9 | 122.1 |
| 123.5 | 124.5 | 125.3 | 125.6 | 126.5 | 127.8 | 127.9 | 129.8 |

We see that $v = 6$ and, thus, H_0 is rejected with $\alpha = 0.066$.

Of course, the median test can be generalized as easily as the sign test. Instead of letting V be the number of X values in the lower half of the combined sample of size $n + m$, let V be the number of X values in the lower i values of the combined sample. Thus, if $i/(n + m) = p$, we would use V to test the equality of the $(100p)$th percentiles of the two distributions. If V is much larger than np, we would suspect that the $(100p)$th percentile of the X distribution is smaller than that of the Y distribution. If V is much smaller than np, we would guess that it would be the other way around. Since, under $F(z) = G(z)$, all orderings of the n values of X and m values of Y have the same probability, the p.d.f. of V is

$$h(v) = P(V = v) = \frac{\binom{n}{v}\binom{m}{i - v}}{\binom{n + m}{i}}, \qquad v = 0, 1, 2, \ldots, n.$$

There is another method (due to Wilcoxon) for testing the equality of the two locations of two distributions of the continuous type. Order the combined sample of $X_1, X_2, \ldots, X_n$ and $Y_1, Y_2, \ldots, Y_m$ in increasing order of magnitude. Assign to the ordered values the ranks $1, 2, 3, \ldots, n + m$. In case of ties, assign the average of the ranks associated with the tied values. Let T equal the sum of the ranks of $Y_1, Y_2, \ldots, Y_m$. If the distribution of Y is shifted to the right of that of X, the values of Y would tend to be larger than the values of X and T would usually be larger than expected when $F(z) = G(z)$. Thus, the critical region for testing $H_0: \pi_X = \pi_Y$ against $H_1: \pi_X < \pi_Y$ would be of the form $T > t$. Similarly, if the alternative hypothesis is $\pi_X > \pi_Y$, the critical region would be of the form $T < t$.

We shall not derive the distribution of T. However if n and m are both greater than 7, a normal approximation can be used. With $F(z) = G(z)$, the mean and variance of T are

$$\mu_T = \frac{m(n + m + 1)}{2}$$

and

$$\text{Var}(T) = \frac{nm(n + m + 1)}{12};$$

thus, the statistic

$$Z = \frac{T - m(n + m + 1)/2}{\sqrt{nm(n + m + 1)/12}}$$

is approximately $N(0, 1)$.

Example 8.7–2.   We illustrate the two-sample Wilcoxon test using the data given in Example 8.7–1. The critical region for testing $H_0: \pi_X = \pi_Y$ against $H_1: \pi_X < \pi_Y$ is of the form $T > t$. Since $n = m = 8$, at an approximate $\alpha = 0.05$ significance level, H_0 is rejected if

$$z = \frac{t - 8(8 + 8 + 1)/2}{\sqrt{[(8)(8)(8 + 8 + 1)]/12}} > 1.645;$$

that is, if

$$t > 1.645 \sqrt{\frac{(8)(8)(17)}{12}} + 4(17) = 83.66.$$

From the data we see that the computed T is

$$t = 3 + 8 + 9 + 11 + 12 + 13 + 15 + 16 = 87 > 83.66.$$

Thus, H_0 is rejected, an action consistent with that of the median test.

It should be mentioned that, shortly after Wilcoxon proposed his test, Mann and Whitney suggested a test based upon the estimate of the probability $P(X < Y)$. Namely, they let U equal the number of times that $X_i < Y_j$, $i = 1, 2, \ldots, n$ and $j = 1, 2, \ldots, m$. Using the data in Example 8.7–1, we find that the computed U is $u = 51$ among all $nm = (8)(8) = 64$ pairs of (X, Y). Thus, the estimate of $P(X < Y)$ is $51/64$ or, in general, u/nm. At the time of the Mann–Whitney suggestion, it was noted that U was just a linear function of Wilcoxon's T and hence really provided the same test. That relationship is

$$U = T - \frac{m(m+1)}{2},$$

which in our special case is

$$51 = 87 - \frac{8(9)}{2} = 87 - 36.$$

Thus, we often read about the test of Mann, Whitney, and Wilcoxon.

It should be noted here that the median and Wilcoxon tests are much less sensitive to extreme values than is Student's test based upon $\bar{X} - \bar{Y}$. Therefore, if there is much skewness or contamination, these proposed distribution-free tests are much safer. In particular, that of Wilcoxon is quite good and does not lose too much in case the distributions are close to normal ones.

═══ *Exercises* ═══

8.7–1. Let us compare the failure times of a certain type of light bulb produced by two different manufacturers, A and B, by testing ten bulbs selected at random from each of the outputs. The data are (in hundreds of hours used before failure)

A: 5.6,   4.6,   6.8,   4.9,   6.1,   5.3,   4.5,   5.8,   5.4,   4.7.
B: 7.2,   8.1,   5.1,   7.3,   6.9,   7.8,   5.9,   6.7,   6.5,   7.1.

Use one-sided median and Wilcoxon tests to test the equality of medians of the two processes at the approximate 5% significance level. Then modify the median test to test the equality of the 25th percentiles of the two processes. HINT: Let V equal the number of A scores in the lower five values of the combined sample.

8.7–2. Use the median and Wilcoxon tests to test $H_0: \pi_X = \pi_Y$ against $H_1: \pi_X < \pi_Y$ if $n = m = 7$ and we observe that the values of X and Y are

$$X: \quad 4.213, \quad 0.325, \quad 8.725, \quad 6.094, \quad 9.932, \quad 5.019, \quad 11.486.$$
$$Y: \quad 19.881, \quad 14.288, \quad 21.280, \quad 7.376, \quad 15.827, \quad 4.490, \quad 23.554.$$

8.7–3. Let X and Y denote the heights of blue spruce trees, measured in centimeters, in two large fields. We shall compare these heights by measuring 12 trees selected at random from each of the fields. Use the statistic T, the sum of the ranks of the observations of Y in the combined sample, to test at $\alpha = 0.05$, approximately, the hypothesis $H_0: \pi_X = \pi_Y$, against the alternative hypothesis $H_1: \pi_X < \pi_Y$, based on $n = 12$ observations of X:

90.4	77.2	75.9	83.2	84.0	90.2
87.6	67.4	77.6	69.3	83.3	72.7

and $m = 12$ observations of Y:

92.7	78.9	82.5	88.6	95.0	94.4
73.1	88.3	90.4	86.5	84.7	87.5

8.8 Run Test and Test for Randomness

Under the assumption that the random variables X and Y are of the continuous type and have distribution functions $F(x)$ and $G(y)$, respectively, we describe another test of the hypothesis $H_0: F(z) = G(z)$. This new test can also be used to test for randomness. For these tests we need the concept of runs, which we now define.

Suppose that we have n observations of the random variable X and m observations of the random variable Y. The combination of two sets of observations into one collection of $n + m$ observations, placed in ascending order of magnitude, might yield an arrangement

$$\underline{y\,y\,y}\;\underline{x\,x}\;\underline{y}\;\underline{x}\;\underline{y}\;\underline{x\,x}\;\underline{y\,y},$$

where x denotes an observation of X and y an observation of Y in the ordered arrangement. We have underlined groups of successive values of X and of Y. Each underlined group is called a *run*. Thus, we have a run of three values of Y, followed by a run of two values of X, followed by a run of one value of Y, and so on. In this example there are seven runs.

We give two more examples to show what might be indicated by the number of runs. If the five x's and seven y's had the orderings

$$\underline{x\,x\,x\,x}\;\underline{y}\;\underline{x}\;\underline{y\,y\,y\,y\,y\,y},$$

we might suspect that $F(z) \geq G(z)$ (see Figure 8.7–1). Note that there are four runs in this ordering. The ordered arrangement

$$\underline{y\,y\,y}\;\underline{x\,x}\;\underline{y}\;\underline{x\,x\,x}\;\underline{y\,y\,y}$$

might suggest that the medians of the two distributions are equal but that the spread of the Y distribution is greater than the spread of the X distribution, for example, that $\sigma_Y > \sigma_X$. These examples suggest that the hypothesis $F(z) = G(z)$ should be rejected if the number of runs is too small, where a small number of runs could be caused by differences in the location or in the spread of the two distributions.

Let the random variable R equal the number of runs in the combined ordered sample of n observations of X and m observations of Y. We shall find the distribution of R when $F(z) = G(z)$ and then describe a test of the hypothesis $H_0: F(z) = G(z)$.

Under H_0, all permutations of the n observations of X and m observations of Y have equal probabilities. We can select the n positions for the n values of X in $\binom{n+m}{n}$ ways, the probability of each arrangement being $1 \Big/ \binom{m+n}{n}$. To find $P(R = r)$ we must determine the number of permutations which yield r runs.

First suppose that $r = 2k$, where k is a positive integer. In this case the n ordered values of X and the m ordered values of Y must each be separated into k runs. We can form k runs of the n values of X by inserting $k - 1$ dividers into the $n - 1$ spaces between the values of X, with no more than one divider per space. This can be done in $\binom{n-1}{k-1}$ ways (see Exercise 8.8–4). Similarly k runs of the m values of Y can be formed in $\binom{m-1}{k-1}$ ways. These two sets of runs can be placed together to form $r = 2k$ runs, of which $\binom{n-1}{k-1}\binom{m-1}{k-1}$ begin with a run of x's and $\binom{m-1}{k-1}\binom{n-1}{k-1}$ begin with a run of y's. Thus,

$$P(R = 2k) = \frac{2\binom{n-1}{k-1}\binom{m-1}{k-1}}{\binom{n+m}{n}},$$

where $2k$ is an element of the space of R.

When $r = 2k + 1$, it is possible to have $k + 1$ runs of the ordered values of X and k runs of the ordered values of Y or k runs of x's and $k + 1$ runs of y's. We can form $k + 1$ runs of the n values of X by inserting k dividers into the $n - 1$ spaces between the values of X with no more than one divider per space in $\binom{n-1}{k}$ ways. Similarly k runs of m values of Y can be done in $\binom{m-1}{k-1}$ ways. These two sets of runs can be placed together to form $2k + 1$ runs in $\binom{n-1}{k}\binom{m-1}{k-1}$ ways. In addition, $k + 1$ runs of the m values of Y and k runs of the n values of X can be placed together to form $\binom{m-1}{k}\binom{n-1}{k-1}$ sets of $2k+1$ runs. Hence,

$$P(R = 2k+1) = \frac{\binom{n-1}{k}\binom{m-1}{k-1} + \binom{n-1}{k-1}\binom{m-1}{k}}{\binom{n+m}{k}}$$

for $2k+1$ in the space of R.

A test based upon the number of runs can be used for testing the hypothesis $H_0: F(z) = G(z)$. The hypothesis is rejected if the observed number of runs r is too small. That is, the critical region is of the form $r < c$, where the constant c is determined by using the p.d.f. of R to yield the desired significance level. The run test is sensitive both to differences in location and differences in spread of the two distributions.

Example 8.8–1. We shall test the hypothesis $H_0: F(z) = G(z)$ based upon five observations of X: 79.4, 79.6, 81.4, 80.6, 81.2, and seven observations of Y: 85.2, 81.8, 76.8, 82.1, 78.9, 84.4, 82.7. Using the p.d.f. of R with $n = 5$ and $m = 7$, we see that

$$P(R = 2) = \frac{2}{792}, \qquad P(R = 3) = \frac{10}{792}, \qquad P(R = 4) = \frac{48}{792}.$$

Thus, we can take $\{r: \ r < 5\}$ as the critical region at a significance level $\alpha = 60/792 = 0.076$. Ordering the combined sample yields the arrangement

$$\underline{76.8 \ 78.9} \ \underline{79.4 \ 79.6 \ 80.6 \ 81.2 \ 81.4} \ \underline{81.8 \ 82.1 \ 82.7 \ 84.4 \ 85.2}$$

in which we have a run of y's, a run of x's, and then a run of y's. Since $r = 3$ the hypothesis is rejected.

When m and n are large, say each is greater than 10, R can be approximated with a normally distributed random variable. That is, it can be shown that

$$\mu_R = E(R) = \frac{2mn}{m+n} + 1,$$

$$\text{Var}(R) = \frac{(\mu_R - 1)(\mu_R - 2)}{m+n-1} = \frac{2mn(2mn - m - n)}{(m+n)^2(m+n-1)},$$

and

$$Z = \frac{R - \mu_R}{\sqrt{\text{Var}(R)}}$$

is approximately $N(0, 1)$. The critical region for testing the hypothesis $H_0: F(z) = G(z)$ is of the form $z < c$, where $c < 0$ and is selected from the normal table (Appendix Table IV) to yield the desired significance level.

Applications of the run test include tests for randomness. Analysis of runs can also be useful in quality-control studies. To illustrate these applications let $x_1, x_2, \ldots, x_k$ be the observed values of a random variable X where the subscripts now designate the order in which the outcomes were observed and the observations are *not* arranged in order of magnitude. It is possible, in a quality-control situation, that the observations are made systematically every hour, for example. Assume that k is even. The median divides the k numbers into a lower and an upper half. Replace each observation by L if it falls below the median and by U if it falls above the median. Then, for example, a sequence such as

$$U\ U\ U\ L\ U\ L\ L\ L$$

might suggest a trend toward decreasing values of X. If trend is the alternative hypothesis to randomness, the critical region would be of the form $r < c$. On the other hand, if we have a sequence such as

$$U\ L\ U\ L\ U\ L\ U\ L,$$

we would suspect a cyclic effect and would reject the hypothesis of randomness if r were too large. To test both for trend and cyclic effect, the critical region for testing the hypothesis of randomness is of the form $r < c_1$ or $r > c_2$.

If the sample size k is odd, the number of observations in the "upper half" and "lower half" will differ by one. That is, include the sample median in one of the two halves.

Example 8.8–2. We shall use a sample of size $k = 14$ to test for both trend and cyclic effect. To determine the critical region for rejecting the hypothesis of randomness, we use the p.d.f. of R with $m = n = 7$. Since

$$P(R = 2) = P(R = 14) = \frac{2}{3432},$$

$$P(R = 3) = P(R = 13) = \frac{12}{3432},$$

$$P(R = 4) = P(R = 12) = \frac{72}{3432},$$

the critical region $\{r: \quad r < 5 \quad$ or $\quad r > 11\}$ would yield a test at a significance level of $\alpha = 172/3432 = 0.05$. The 14 observations are

$$81.4, \quad 76.3, \quad 85.6, \quad 76.4, \quad 88.4, \quad 80.2, \quad 85.6,$$
$$84.6, \quad 78.3, \quad 82.8, \quad 88.1, \quad 85.4, \quad 87.7, \quad 86.6.$$

The median of these outcomes is $(84.6 + 85.4)/2 = 85.0$. Replacing each outcome with L if it falls below 85.0 and U if it falls above 85.0 yields the sequence

$$\underline{L}\,\underline{L}\,\underline{U}\,\underline{L}\,\underline{U}\,\underline{L}\,\underline{U}\,\underline{L}\,\underline{L}\,\underline{L}\,\underline{U}\,\underline{U}\,\underline{U}\,\underline{U}.$$

Since $r = 8$, the hypothesis of randomness is not rejected.

=== *Exercises* ===

8.8–1. Let the total lengths of the male and female trident lynx spider be denoted by X and Y, respectively, with corresponding distribution functions $F(x)$ and $G(y)$. Measurement of the lengths, in millimeters, of eight male and eight female spiders yielded the following observations of X:

$$5.40, \quad 5.55, \quad 6.00, \quad 5.00, \quad 5.70, \quad 5.20, \quad 5.45, \quad 4.95,$$

and of Y:

$$6.20, \quad 6.25, \quad 5.75, \quad 5.85, \quad 6.55, \quad 6.05, \quad 5.50, \quad 6.65.$$

Use these data to test the hypothesis $H_0 : F(z) = G(z)$. Let $\alpha = 0.10$, approximately.

8.8–2. Let X and Y denote the times in hours per week that students in two different schools watch television. Let $F(x)$ and $G(y)$ denote the respective distribution functions. To test the hypothesis $H_0 : F(z) = G(z)$, a random sample of eight students was selected from each school. Test this hypothesis using the following observations.

X:	16.75	19.25	22.00	20.50	22.50	15.50	17.25	20.75,
Y:	24.75	21.50	19.75	17.50	22.75	23.50	13.00	19.00.

8.8–3. Test the hypothesis $H_0 : F(z) = G(z)$ based upon 12 observations of X:

$$113.7, \quad 103.0, \quad 105.7, \quad 105.4, \quad 111.2, \quad 107.6, \quad 101.3, \quad 99.8, \quad 114.2, \quad 110.6,$$
$$115.6, \quad 116.5,$$

and 12 observations of Y:

$$122.7, \quad 118.7, \quad 109.8, \quad 117.9, \quad 125.3, \quad 101.6, \quad 112.1, \quad 116.6, \quad 101.5, \quad 102.5,$$
$$121.3, \quad 118.7,$$

by using the normal approximation.

8.8–4. Given six values of x, list the $\binom{5}{3} = 10$ ways in which three dividers can be inserted between the six values of x, no more than one divider per space.

8.8–5. Use the following sample of size $k = 14$ to test the hypothesis of randomness against the alternative hypothesis of a trend effect.

$$12.4, \quad 14.2, \quad 11.7, \quad 14.0, \quad 12.7, \quad 15.7, \quad 12.8,$$
$$14.1, \quad 17.9, \quad 18.4, \quad 17.5, \quad 20.2, \quad 20.8, \quad 20.3,$$

8.8–6. Use the following sample of size $k = 16$ to test the hypothesis of randomness against the alternative hypothesis of a cyclic effect.

$$12.4, \quad 31.8, \quad 22.2, \quad 24.5, \quad 17.9, \quad 24.6, \quad 15.7, \quad 27.3,$$
$$22.7, \quad 26.0, \quad 14.5, \quad 22.0, \quad 21.8, \quad 31.9, \quad 11.5, \quad 28.3.$$

8.9 Kolmogorov–Smirnov Goodness of Fit Test

In this section we discuss a test that considers the goodness of fit between a hypothesized distribution function and an empirical distribution function. The definition of the empirical distribution function, which was given in Section 3.1, is repeated here in terms of the order statistics. Let $y_1 < y_2 < \cdots < y_n$ be the observed values of the order statistics of a random sample $x_1, x_2, \ldots, x_n$ of size n. When no two observations are equal, the empirical distribution function is defined by

$$F_n(x) = \begin{cases} 0, & x < y_1, \\ \dfrac{k}{n}, & y_k \leq x < y_{k+1}, \quad k = 1, 2, \ldots, n - 1, \\ 1, & y_n \leq x. \end{cases}$$

In this case the empirical distribution function has a jump of magnitude $1/n$ occurring at each observation. If n_k observations are equal to x_k, a jump of magnitude n_k/n occurs at x_k.

Suppose that a random sample of size n is taken from a distribution of the continuous type that has the distribution function $F(x)$. How can we measure the "closeness" of $F(x)$ and the empirical distribution function $F_n(x)$? How does the sample size affect this closeness? We give some theoretical results to help answer these questions and then give a test for goodness of fit.

Let $X_1, X_2, \ldots, X_n$ denote a random sample of size n from a distribution of the continuous type with the distribution function $F(x)$. Consider a fixed value of x. Then $W_n = F_n(x)$, the value of the empirical distribution function at x, can be thought of as a random variable that takes on the values $0, 1/n, 2/n, \ldots, 1$. Now $nW_n = k$ if, and only if, exactly k observations are less than or equal to x (say success) and $n - k$ observations are greater than x. The probability that an observation is less than or equal to x is given by $F(x)$. That is, the probability of success is $F(x)$. Because of the independence of the random variables

$X_1, X_2, \ldots, X_n$, the probability of k successes is given by the binomial distribution, namely,

$$P(nW_n = k) = P\left(W_n = \frac{k}{n}\right)$$

$$= \binom{n}{k}[F(x)]^k[1 - F(x)]^{n-k}, \qquad k = 0, 1, 2, \ldots, n.$$

Since nW_n has a binomial distribution with $p = F(x)$, the mean and variance of nW_n are given by

$$E(nW_n) = nF(x) \qquad \text{and} \qquad \text{Var}(nW_n) = n[F(x)][1 - F(x)].$$

Hence, since $W_n = F_n(x)$,

$$E[F_n(x)] = E(W_n) = F(x)$$

and

$$\text{Var}[F_n(x)] = \text{Var}(W_n) = \frac{F(x)[1 - F(x)]}{n}.$$

Thus, we see that pointwise the relative frequency $F_n(x)$ converges in probability to $F(x)$ and $F_n(x)$ is a consistent estimator of $F(x)$. A theorem by Glivenko, the proof of which is beyond the level of this book, states that with probability one, $F_n(x)$ converges to $F(x)$ uniformly in x as $n \to \infty$.

Because of the convergence of the empirical distribution function to the theoretical distribution function, it makes sense to construct a goodness of fit test based upon the closeness of the empirical and a hypothesized distribution function, say $F_n(x)$ and $F_0(x)$, respectively. We shall use the Kolmogorov–Smirnov statistic defined by

$$D_n = \sup_x [\,|F_n(x) - F_0(x)|\,].$$

That is, D_n is the least upper bound of all pointwise differences $|F_n(x) - F_0(x)|$.

The exact distribution of the statistic D_n can be derived. We shall not derive this distribution but do give some values of the distribution function of D_n, namely, $P(D_n \le d)$, in Appendix Table VII that will be used for goodness of fit tests. We would like to point out that the distribution of D_n does not depend upon the particular function $F_0(x)$ of the continuous type. [This is essentially due to the fact that $Y = F_0(X)$ has a uniform distribution $U(0, 1)$.] Thus, D_n can be thought of as a distribution-free statistic.

We are interested in using the Kolmogorov–Smirnov statistic D_n to test the hypothesis $H_0: F(x) = F_0(x)$ against all alternatives, $H_1: F(x) \ne F_0(x)$, where $F_0(x)$ is some specified distribution function. Intuitively we accept H_0 if the empirical distribution function $F_n(x)$ is sufficiently close to $F_0(x)$, that is, if the value of D_n is sufficiently small. The hypothesis H_0 is rejected if the observed value of D_n is greater than the critical value selected from Appendix Table VII,

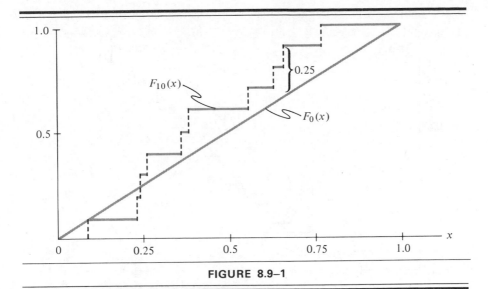

FIGURE 8.9–1

this critical value depending upon the desired significance level and sample size.

The use of the Kolmogorov–Smirnov statistic is illustrated by two examples.

Example 8.9–1. We shall test the hypothesis $H_0: F(x) = F_0(x)$ against $H_1: F(x) \neq F_0(x)$, where

$$F_0(x) = \begin{cases} 0, & x < 0, \\ x, & 0 \leq x < 1, \\ 1, & 1 \leq x. \end{cases}$$

That is, the null hypothesis is that X is $U(0, 1)$. If the test is based upon a sample of size $n = 10$ and if $\alpha = 0.10$, the critical region is $C = \{d_{10}: d_{10} > 0.37\}$ where d_{10} is the observed value of the Kolmogorov–Smirnov statistic D_{10}. Suppose that the observed values of the random sample are 0.62, 0.36, 0.23, 0.76, 0.65, 0.09, 0.55, 0.26, 0.38, and 0.24. In Figure 8.9–1 we have plotted the empirical and hypothesized distribution functions for $0 \leq x \leq 1$. We see that $d_{10} = F_{10}(0.65) - F_0(0.65) = 0.25$ and hence H_0 is not rejected.

Example 8.9–2. When observing a Poisson process with a mean rate of arrivals $\lambda = 1/\theta$, the random variable W, which denotes the waiting time until the hth arrival, has a gamma distribution. The p.d.f. of W is

$$f(w) = \frac{w^{h-1}e^{-w/\theta}}{(h-1)!\theta^h}, \qquad 0 \leq w < \infty.$$

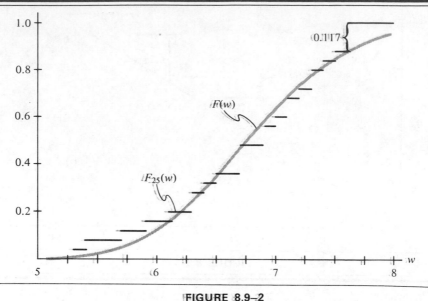

FIGURE 8.9–2

A Geiger counter was set up to record the waiting time W in seconds to observe $h = 100$ alpha particle emissions of barium 133. It is claimed that the number of counts per second has a Poisson distribution with $\lambda = 14.7$ and hence $\theta = 0.068$. We shall test the hypothesis

$$H_0 : F(w) = \int_{-\infty}^{w} f(t)\, dt,$$

where $f(t)$ is the gamma p.d.f. with $\theta = 0.068$ and $h = 100$. Based upon 25 observations, H_0 is rejected if $d_{25} > 0.24$ for $\alpha = 0.10$. For 25 observations, the empirical and theoretical distribution functions are depicted in Figure 8.9–2. For these data, $d_{25} = 0.117$ and hence H_0 is not rejected.

You will note that we have been assuming that $F(x)$ is a continuous function. That is, we have only considered random variables of the continuous type. This procedure may also be applied in the discrete case. However, in the discrete case, the true significance level will be at most α. That is, the resulting test will be conservative.

Another application of the Kolmogorov–Smirnov statistic is in forming a confidence band for an unknown distribution function $F(x)$. To form a confidence band based upon a sample of size n, select a number d such that

$$P(D_n > d) = \alpha.$$

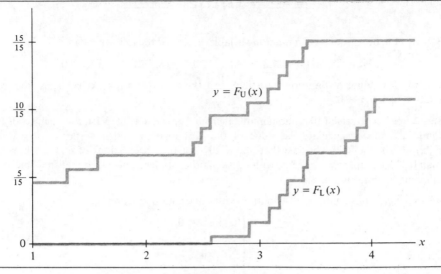

FIGURE 8.9–3

Then

$$1 - \alpha = P\left[\sup_x |F_n(x) - F(x)| \le d\right]$$

$$= P[|F_n(x) - F(x)| \le d \text{ for all } x]$$

$$= P[F_n(x) - d \le F(x) \le F_n(x) + d \text{ for all } x].$$

Let

$$F_L(x) = \begin{cases} 0, & F_n(x) - d \le 0 \\ F_n(x) - d, & F_n(x) - d > 0, \end{cases}$$

and

$$F_U(x) = \begin{cases} F_n(x) + d, & F_n(x) + d < 1, \\ 1, & F_n(x) + d \ge 1. \end{cases}$$

The two-step functions $F_L(x)$ and $F_U(x)$ yield a $100(1 - \alpha)\%$ confidence band for the unknown distribution function $F(x)$.

Example 8.9–3. A random sample of size $n = 15$ from an unknown distribution yielded the sample values 3.88, 3.97, 4.03, 2.49, 3.18, 3.08, 2.91, 3.43, 2.41, 1.57, 3.78, 3.25, 1.29, 2.57, and 3.40. Now

$$P(D_{15} > 0.30) = 0.10.$$

A 90% confidence band for the unknown distribution function $F(x)$ is depicted in Figure 8.9–3.

Exercises

8.9–1. Ten observations of a random variable X yielded the following data:

$$32.4, \quad 6.2, \quad 11.4, \quad 27.3, \quad 29.2, \quad 17.0, \quad 30.6, \quad 21.6, \quad 18.7, \quad 8.0.$$

Use the Kolmogorov–Smirnov statistic to test the hypothesis that X has a distribution $N(20, 100)$. Let $\alpha = 0.20$.

8.9–2. Select 10 sets of 10 random numbers from Appendix Table VIII. For each set of 10 random numbers calculate the value of the Kolmogorov–Smirnov statistic d_{10} with $F_0(x) = x, 0 \leq x \leq 1$. Is it true that about 20% of the observations of D_{10} are greater than 0.32? (Compare your results with those of other students, provided that they selected different random numbers.)

8.9–3. Let X have a mixed distribution with distribution function

$$F(x) = \begin{cases} 0, & x < 0 \\ \dfrac{x}{2}, & 0 \leq x < 1 \\ \dfrac{1}{2}, & 1 \leq x < 2 \\ 1, & 2 \leq x. \end{cases}$$

[For example, flip a coin. If the outcome is heads, $X = 2$. If the outcome is tails, let X equal a number selected at random from the interval $(0, 1)$.] Note that $E(X) = 5/4$ and $\mathrm{Var}(X) = 29/48$. Let $X_1 X_2, \ldots, X_{13}$ denote 13 observations of X. Let W_{13} be defined by, if $n = 13$,

$$W_{13} = \frac{\bar{X} - 5/4}{\sqrt{29/48n}}.$$

We shall use the Kolmogorov–Smirnov statistic to test the hypothesis that W_{13} has a distribution $N(0, 1)$.

  (a) If 25 observations of W_{13} are taken, what is a critical region of size $\alpha = 0.10$?

  (b) Use the following 25 observations of W_{13}, which have been ordered,

$$-2.11, \quad -1.55, \quad -1.24, \quad -1.06, \quad -0.99, \quad -0.98, \quad -0.89, \quad -0.44, \quad -0.24,$$
$$-0.23, \quad -0.21, \quad -0.05, \quad -0.01, \quad 0.09, \quad 0.22, \quad 0.31, \quad 0.36, \quad 0.41,$$
$$0.71, \quad 1.00, \quad 1.05, \quad 1.29, \quad 1.47, \quad 2.05, \quad 2.10,$$

to sketch the empirical distribution function of W_{13}. Also sketch the distribution function for the standard normal distribution.

  (c) Is the hypothesis rejected?

8.9–4. Construct a 90% confidence band for the unknown distribution function $F(x)$ using the following 15 observations of X:

$$20.2, \quad 85.4, \quad 59.9, \quad 72.7, \quad 88.0, \quad 33.7, \quad 87.1, \quad 99.5,$$
$$93.8, \quad 18.4, \quad 60.6, \quad 98.9, \quad 90.9, \quad 86.9, \quad 74.2.$$

8.9–5. Use the Kolmogorov–Smirnov statistic to test the hypothesis that X has a Poisson distribution with a mean of $\lambda = 5.4$ given the following observations of X:

$$7, \quad 4, \quad 4, \quad 11, \quad 0, \quad 6, \quad 8, \quad 4, \quad 5, \quad 4, \quad 7, \quad 3, \quad 2, \quad 8, \quad 6, \quad 7, \quad 4, \quad 2, \quad 9, \quad 8.$$

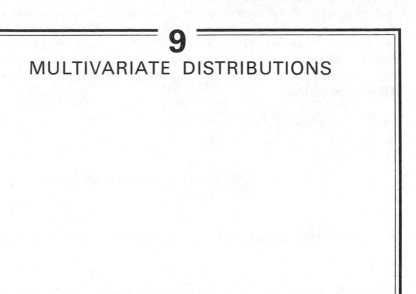

9
MULTIVARIATE DISTRIBUTIONS

9.1 Multivariate Distributions of the Discrete Type

We have already considered situations that involve more than one random variable. However, the variables in each case have been independent. We now turn to situations that involve two or more dependent random variables. For example, suppose a bowl contains three red, four white, and five blue balls. Draw two balls from the bowl at random and without replacement. We could let X denote the number of red balls drawn, and let Y denote the number of white balls drawn. Of course, $2 - X - Y$ would then denote the number of blue balls drawn. And we see that, in some sense, X and Y must be dependent because if $X = 2$ we know that $Y = 0$.

Other illustrations might involve the height X and the weight Y of individuals in a given population. Or manufactured items might be classified into three or more categories: here X might represent the number of good items among n items, Y would be the number of "seconds," and the number of defectives would then be $n - X - Y$. In order to deal with situations such as these, it will be necessary to extend certain definitions as well as give new ones.

DEFINITION 9.1–1. Let *X* and *Y* be two functions defined on a discrete probability space. Let *R* denote the corresponding two-dimensional space of *X* and *Y*, the two random variables of the discrete type. The probability that $X = x$ and $Y = y$ is denoted by $f(x, y) = P(X = x, Y = y)$, and it is induced from the discrete probability space through the functions *X* and *Y*. The function $f(x, y)$ is called the joint probability density function (*joint p.d.f.*) of *X* and *Y* and has the following properties:

(i) $0 \le f(x, y) \le 1.$

(ii) $\sum_{(x, y) \in R} \sum f(x, y) = 1.$

(iii) $P[(X, Y) \in A] = \sum_{(x, y) \in A} \sum f(x, y)$, where *A* is a subset of the space *R*.

The following example will make this definition more meaningful.

Example 9.1–1. Roll a pair of unbiased dice. For each of the 36 sample points with probability 1/36, let *X* denote the smaller and *Y* the larger outcome on the dice. For example, if the outcome is (3, 2) then the observed values are $X = 2$, $Y = 3$; if the outcome is (2, 2), then the observed values are $X = Y = 2$. The joint p.d.f. of *X* and *Y* is given by the induced probabilities

$$f(x, y) = \begin{cases} \dfrac{1}{36}, & 1 \le x = y \le 6, \\[2mm] \dfrac{2}{36}, & 1 \le x < y \le 6, \end{cases}$$

when *x* and *y* are integers. Figure 9.1–1 depicts the probabilities of the various points of the space *R*.

Notice that certain numbers have been recorded in the bottom and left-hand margins of Figure 9.1–1. These numbers are the respective column and row totals of the probabilities. The column totals are the respective probabilities that *X* will assume the values in the *x* space $R_1 = \{1, 2, 3, 4, 5, 6\}$, and the row totals are the respective probabilities that *Y* will assume the values in the *y* space $R_2 = \{1, 2, 3, 4, 5, 6\}$. That is, the totals describe probability density functions of *X* and *Y*, respectively. Since each collection of these probabilities is frequently recorded in the margins and satisfies the properties of a p.d.f. of one random variable, each is called a marginal p.d.f.

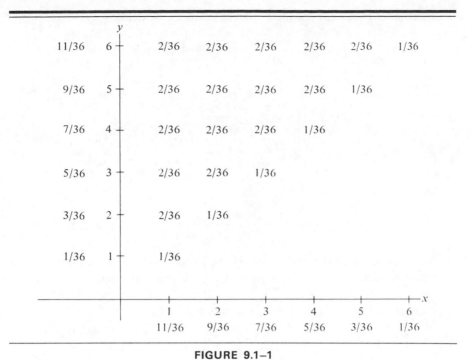

FIGURE 9.1–1

DEFINITION 9.1–2. *Let X and Y have the joint probability density function* $f(x, y)$ *with space R. The probability density function of X alone, called the* marginal probability density function *of X, is defined by*

$$f_1(x) = \sum_y f(x, y), \qquad x \in R_1,$$

where the summation is taken over all possible y values for each given x in the x space R_1. *That is, the summation is over all* (x, y) *in R with a given x value. Similarly, the* marginal probability density function *of Y is defined by*

$$f_2(y) = \sum_x f(x, y), \qquad y \in R_2,$$

where the summation is taken over all possible x values for each given y in the y space R_2. *The random variables X and Y are* independent *if and only if*

$$f(x, y) \equiv f_1(x) f_2(y), \qquad x \in R_1, \quad y \in R_2;$$

otherwise X and Y are said to be dependent.

Example 9.1–2. Let the joint p.d.f. of X and Y be defined by

$$f(x, y) = \frac{x + y}{21}, \qquad x = 1, 2, 3, \quad y = 1, 2.$$

Then

$$f_1(x) = \sum_y f(x, y) = \sum_{y=1}^{2} \frac{x + y}{21}$$

$$= \frac{x + 1}{21} + \frac{x + 2}{21} = \frac{2x + 3}{21}, \qquad x = 1, 2, 3;$$

and

$$f_2(y) = \sum_x f(x, y) = \sum_{x=1}^{3} \frac{x + y}{21} = \frac{6 + 3y}{21}, \qquad y = 1, 2.$$

Note that both $f_1(x)$ and $f_2(y)$ satisfy the properties of a probability density function. Since $f(x, y) \neq f_1(x)f_2(y)$, X and Y are dependent.

Example 9.1–3. Let the joint p.d.f. of X and Y be

$$f(x, y) = \frac{xy^2}{30}, \qquad x = 1, 2, 3, \quad y = 1, 2.$$

The marginal probability density functions are

$$f_1(x) = \sum_{y=1}^{2} \frac{xy^2}{30} = \frac{x}{6}, \qquad x = 1, 2, 3,$$

and

$$f_2(y) = \sum_{x=1}^{3} \frac{xy^2}{30} = \frac{y^2}{5}, \qquad y = 1, 2.$$

Thus, $f(x, y) \equiv f_1(x)f_2(y)$ for $x = 1, 2, 3$ and $y = 1, 2$, and X and Y are independent.

Example 9.1–4. Let the joint p.d.f. of X and Y be

$$f(x, y) = \frac{xy^2}{13}, \qquad (x, y) = (1, 1), (1, 2), (2, 2).$$

Then the p.d.f. of X is

$$f_1(x) = \begin{cases} \dfrac{5}{13}, & x = 1, \\[2mm] \dfrac{8}{13}, & x = 2, \end{cases}$$

and that of Y is

$$f_2(y) = \begin{cases} \dfrac{1}{13}, & y = 1, \\[2mm] \dfrac{12}{13}, & y = 2. \end{cases}$$

Thus, $f(x, y) \neq f_1(x)f_2(y)$ for $x = 1, 2$ and $y = 1, 2$, and X and Y are dependent.

REMARK. Note that in Example 9.1–4 the support R of X and Y is "triangular." Whenever this support R is not "rectangular," the random variables must be dependent because R cannot then equal the product set $\{(x, y): \ x \in R_1, y \in R_2\}$. That is, if we observe that the support R of X and Y is not a product set, then X and Y must be dependent. For illustration, in Example 9.1–4, X and Y are dependent because $R = \{(1, 1), (1, 2), (2, 2)\}$ is not a product set. On the other hand, if R equals the product set $\{(x, y): \ x \in R_1, y \in R_2\}$ and if the formula for $f(x, y)$ is the product of an expression in x alone and an expression in y alone, then X and Y are independent, as illustrated in Example 9.1–3. Example 9.1–2 illustrates the fact that the support can be rectangular but the formula for $f(x, y)$ is not such a product and thus X and Y are dependent.

The notion of a joint p.d.f. of two discrete random variables can be extended to a joint p.d.f. of n random variables of the discrete type. Briefly, the joint p.d.f. of the n random variables $X_1, X_2, \ldots, X_n$ is defined by

$$f(x_1, x_2, \ldots, x_n) = P(X_1 = x_1, X_2 = x_2, \ldots, X_n = x_n)$$

over an appropriate space R. Furthermore, $f(x_1, x_2, \ldots, x_n)$ satisfies properties similar to those given in Definition 9.1–1. In addition, the *marginal probability density function* of one of n discrete random variables, say X_k, is found by summing $f(x_1, x_2, \ldots, x_n)$ over all x_i's except x_k; that is,

$$f_k(x_k) = \sum_{x_1} \cdots \sum_{x_{k-1}} \sum_{x_{k+1}} \cdots \sum_{x_n} f(x_1, x_2, \ldots, x_n), \qquad x_k \in R_k.$$

The random variables $X_1, X_2, \ldots, X_n$ are *mutually independent* if and only if

$$f(x_1, x_2, \ldots, x_n) = f_1(x_1)f_2(x_2) \cdots f_n(x_n), \qquad x_1 \in R_1, x_2 \in R_2, \ldots, x_n \in R_n.$$

If $X_1, X_2, \ldots, X_n$ are not independent, they are said to be *dependent*.

A joint marginal p.d.f. of X_j and X_k is found by summing $f(x_1, x_2, \ldots, x_n)$ over all x_i's except x_j and x_k; that is,

$$f_{j,k}(x_j, x_k) = \sum_{x_1} \cdots \sum_{x_{j-1}} \sum_{x_{j+1}} \cdots \sum_{x_{k-1}} \sum_{x_{k+1}} \cdots \sum_{x_n} f(x_1, x_2, \ldots, x_n).$$

Extensions of these marginal probability density functions to more than two random variables are made in an obvious way.

Example 9.1–5. Consider a population of 200 students who have just finished a first course in calculus. Of these 200, 40 have earned A's, 60 B's, 70 C's, 20 D's, and 10 F's. A sample of size 25 is taken at random and without replacement from this population so that each possible sample has probability $1 / \binom{200}{25}$. Within the sample of 25, let X be the number of A students, Y the number of B students, Z the number of C students, W the number of D students, and $25 - X - Y - Z - W$ the number of F students. The space R of (X, Y, Z, W) is defined by the collection of ordered 4-tuplets of nonnegative integers (x, y, z, w) such that $x + y + z + w \leq 25$. The joint p.d.f. of $X, Y, Z,$ and W is

$$f(x, y, z, w) = \frac{\binom{40}{x}\binom{60}{y}\binom{70}{z}\binom{20}{w}\binom{10}{25 - x - y - z - w}}{\binom{200}{25}},$$

$$(x, y, z, w) \in R.$$

Without actually summing, we know that the marginal p.d.f. of Z is

$$f_3(z) = \frac{\binom{70}{z}\binom{130}{25 - z}}{\binom{200}{25}}, \qquad z = 0, 1, 2, \ldots, 25$$

and the joint marginal p.d.f. of X and Y is

$$f_{12}(x, y) = \frac{\binom{40}{x}\binom{60}{y}\binom{100}{25 - x - y}}{\binom{200}{25}}, \qquad x + y = 0, 1, 2, \ldots, 25.$$

Of course, $f_3(z)$ is a hypergeometric p.d.f. and $f_{12}(x, y)$ and $f(x, y, z, w)$ are extensions of that type of p.d.f. It is easy to see that

$$f(x, y, z, w) \neq f_1(x)f_2(y)f_3(z)f_4(w),$$

and thus $X, Y, Z,$ and W are dependent.

We now consider an extension of the binomial distribution, namely the multinomial distribution. Consider a sequence of repetitions of an experiment for which the following conditions are satisfied:

(a) The experiment has k possible outcomes that are mutually exclusive and exhaustive, say $A_1, A_2, \ldots, A_k$.
(b) n independent trials of this experiment are observed.
(c) $P(A_i) = p_i, i = 1, 2, \ldots, k,$ on each trial with $\sum_{i=1}^k p_i = 1$.

Let the random variable X_i be the number of times A_i occurs in the n trials, $i = 1, 2, \ldots, k$. If $x_1, x_2, \ldots, x_k$ are nonnegative integers such that their sum equals n, then, for such a sequence, the probability that A_i occurs x_i times, $i = 1, 2, \ldots, k$, is given by

$$P(X_1 = x_1, X_2 = x_2, \ldots, X_k = x_k) = \frac{n!}{x_1! \, x_2! \cdots x_k!} \, p_1^{x_1} p_2^{x_2} \cdots p_k^{x_k}.$$

To see that this is correct, note that the number of distinguishable arrangements of $x_1 \, A_1$'s, $x_2 \, A_2$'s, $\ldots$, $x_k \, A_k$'s is

$$\binom{n}{x_1, x_2, \ldots, x_k} = \frac{n!}{x_1! \, x_2! \cdots x_k!}$$

and that the probability of each of these distinguishable arrangements is

$$p_1^{x_1} p_2^{x_2} \cdots p_k^{x_k}.$$

Hence, the product of these two latter expressions gives the correct probability, which is in agreement with the expression for $P(X_1 = x_1, X_2 = x_2, \ldots, X_k = x_k)$.

We say that $X_1, X_2, \ldots, X_k$ have a multinomial distribution. The reason is that

$$\sum \frac{n!}{x_1! \, x_2! \cdots x_k!} \, p_1^{x_1} p_2^{x_2} \cdots p_k^{x_k} = (p_1 + p_2 + \cdots + p_k)^n = 1,$$

where the summation is over the set of all nonnegative integers $x_1, x_2, \ldots, x_k$ whose sum is n. That is, $P(X_1 = x_1, X_2 = x_2, \ldots, X_k = x_k)$ is a typical term in the expansion of the nth power of the multinomial $(p_1 + p_2 + \cdots + p_k)$.

Example 9.1–6. A bowl contains three red, four white, two blue, and five green balls. One ball is drawn at random from the bowl and then replaced. This is repeated 20 independent times. Let X_1, X_2, X_3, and X_4 denote the numbers of red, white, blue, and green balls drawn, respectively. The probability of drawing five balls of each color is

$$P(X_1 = 5, X_2 = 5, X_3 = 5, X_4 = 5) = f(5, 5, 5, 5)$$

$$= \frac{20!}{5! \, 5! \, 5! \, 5!} \left(\frac{3}{14}\right)^5 \left(\frac{4}{14}\right)^5 \left(\frac{2}{14}\right)^5 \left(\frac{5}{14}\right)^5.$$

When $k = 3$, we often let $X = X_1$ and $Y = X_2$; then $n - X - Y = X_3$. We say that X and Y have a *trinomial distribution*. The joint p.d.f. of X and Y is

$$f(x, y) = \frac{n!}{x! \, y! \, (n - x - y)!} \, p_1^x p_2^y (1 - p_1 - p_2)^{n - x - y},$$

where x and y are nonnegative integers such that $x + y \leq n$. Since the marginal distributions of X and Y are, respectively, $b(n, p_1)$ and $b(n, p_2)$, it is obvious that the product of their probability density functions does not equal $f(x, y)$ and hence they are dependent random variables.

═══ Exercises ═══

9.1–1. Let the joint p.d.f. of X and Y be defined by

$$f(x, y) = \frac{x + y}{32}, \qquad x = 1, 2, \quad y = 1, 2, 3, 4.$$

Find
(a) $f_1(x)$, the marginal p.d.f. of X;   (b) $f_2(y)$, the marginal p.d.f. of Y;
(c) $P(X > Y)$;   (d) $P(Y = 2X)$;
(e) $P(X + Y = 3)$;   (f) $P(X \leq 3 - Y)$.
(g) Are X and Y independent or dependent?

9.1–2. Draw 13 cards at random and without replacement from an ordinary deck of playing cards. Among these 13 cards let X be the number of spades, Y the number of hearts, Z the number of diamonds, and $13 - X - Y - Z$ the number of clubs.
(a) Determine $P(X = 5, Y = 4, Z = 3)$.
(b) Among the 13 cards, what is the probability that the numbers of cards in the four suits are 5, 4, 3, and 1? HINT: Part (a) presents one way this could occur, but there are also other ways, for example, $X = 3, Y = 5, Z = 1$.

9.1–3. Toss a fair die 12 independent times. Let X_i denote the number of times i occurs, $i = 1, 2, 3, 4, 5, 6$.
(a) What is the joint p.d.f. of $X_1, X_2, \ldots, X_6$?
(b) Find the probability that each outcome occurs two times.
(c) Find $P(X_i = 2)$.
(d) Are $X_1, X_2, \ldots, X_6$ mutually independent?

9.1–4. A manufactured item is classified as good, a "second," or defective, with probabilities 6/10, 3/10, and 1/10, respectively. Find the probability that of 15 such items, 10 are good and 4 are seconds. State the assumptions needed in your solution.

9.1–5. A particular grass seed mixture contains 50% Kentucky blue grass, 30% red fescue, 10% annual rye, and 10% Dutch clover. Let X_1, X_2, X_3, X_4 denote the number of seeds of each type, respectively, in a sample of size n drawn from a 50 pound barrel.
(a) Give the joint p.d.f. of X_1, X_2, X_3, X_4.
(b) Give the marginal p.d.f. of X_2.
(c) Give the joint marginal p.d.f. of X_1 and X_3.
(d) If $n = 10$, give $P(X_1 = 5, X_2 = 3, X_3 = 1, X_4 = 1)$.

9.1–6. A box contains 100 Christmas tree light bulbs of which 30 are red, 35 are blue, 15 are white, and 20 are green. Fifteen bulbs are to be drawn at random from the box to fill a string with 15 sockets. Let X denote the number of red, Y the number of blue, and Z the number of white bulbs drawn.

   (a)   Define the joint p.d.f. of X, Y, and Z.
   (b)   Define the marginal p.d.f. of X and find $P(X = 10)$.

9.2 The Correlation Coefficient

In Section 5.1 we considered the mathematical expectations of functions of independent random variables. In this section we first extend the definition of expectation to functions of two or more random variables that are possibly dependent.

DEFINITION 9.2–1.   *If* $u(X_1, X_2, \ldots, X_n)$ *is a function of n random variables of the discrete type that have a joint p.d.f.* $f(x_1, x_2, \ldots, x_n)$ *and space R, then*

$$E[u(X_1, X_2, \ldots, X_n)] = \sum \cdots \sum_{(x_1, \ldots, x_n)} u(x_1, x_2, \ldots, x_n) f(x_1, x_2, \ldots, x_n),$$

if it exists, is called the mathematical expectation (*or* expected value) *of*

$$u(X_1, X_2, \ldots, X_n).$$

Several functions u are of particular interest and some of them are listed in the next definition.

DEFINITION 9.2–2.   *Let* $X_1, X_2, \ldots, X_n$ *have a joint distribution. The following mathematical expectations, subject to their existence, have special names.*
   (i)   *If* $u_1(X_1, X_2, \ldots, X_n) = X_i$, *then*

$$E[u_1(X_1, X_2, \ldots, X_n)] = E(X_i) = \mu_i$$

   is called the mean *of* $X_i, i = 1, 2, \ldots, n$.
   (ii)   *If* $u_2(X_1, X_2, \ldots, X_n) = (X_i - \mu_i)^2$, *then*

$$E[u_2(X_1, X_2, \ldots, X_n)] = E[(X_i - \mu_i)^2] = \sigma_i^2 = \text{Var}(X_i)$$

   is called the variance *of* $X_i, i = 1, 2, \ldots, n$.
   (iii)   *If* $u_3(X_1, X_2, \ldots, X_n) = (X_i - \mu_i)(X_j - \mu_j), i \neq j, then$

$$E[u_3(X_1, X_2, \ldots, X_n)] = E[(X_i - \mu_i)(X_j - \mu_j)] = \sigma_{ij} = \text{Cov}(X_i, X_j)$$

   is called the covariance *of* X_i *and* X_j.

(iv) *If the standard deviations σ_i and σ_j are positive, then*

$$\rho_{ij} = \frac{\text{Cov}(X_i, X_j)}{\sigma_i \sigma_j} = \frac{\sigma_{ij}}{\sigma_i \sigma_j}$$

is called the correlation coefficient *of X_i and X_j.*

It is convenient to observe that the mean and the variance of X_i can be computed from either the joint p.d.f. or the marginal p.d.f. of X_i. For example, if $n = 2$,

$$\mu_1 = E(X_1) = \sum_{x_1} \sum_{x_2} x_1 f(x_1, x_2)$$

$$= \sum_{x_1} x_1 \left[\sum_{x_2} f(x_1, x_2) \right] = \sum_{x_1} x_1 f_1(x_1).$$

Before attempting to explain the meaning of the covariance and the correlation coefficient, let us note a few simple facts. With $i \neq j$,

$$E[(X_i - \mu_i)(X_j - \mu_j)] = E(X_i X_j - \mu_i X_j - \mu_j X_i + \mu_i \mu_j)$$
$$= E(X_i X_j) - \mu_i E(X_j) - \mu_j E(X_i) + \mu_i \mu_j$$

because it is true that even in the multivariate situation E is still a linear or distributive operator (see Exercise 9.2–3). Thus,

$$\text{Cov}(X_i X_j) = E(X_i X_j) - \mu_i \mu_j - \mu_j \mu_i + \mu_i \mu_j = E(X_i X_j) - \mu_i \mu_j.$$

However, since $\rho_{ij} = \text{Cov}(X_i X_j)/\sigma_i \sigma_j$, we have

$$E(X_i X_j) = \mu_i \mu_j + \rho_{ij} \sigma_i \sigma_j.$$

That is, the expected value of the product of two random variables is equal to the product $\mu_i \mu_j$ of their expectations plus their covariance $\rho_{ij} \sigma_i \sigma_j$.

A simple example at this point would be helpful.

Example 9.2–1. Let X_1 and X_2 have the joint p.d.f.

$$f(x_1, x_2) = \frac{x_1 + 2x_2}{18}, \qquad x_1 = 1, 2, \quad x_2 = 1, 2.$$

The marginal probability density functions are, respectively,

$$f_1(x_1) = \sum_{x_2 = 1}^{2} \frac{x_1 + 2x_2}{18} = \frac{2x_1 + 6}{18}, \qquad x_1 = 1, 2$$

and

$$f_2(x_2) = \sum_{x_1 = 1}^{2} \frac{x_1 + 2x_2}{18} = \frac{3 + 4x_2}{18}, \qquad x_2 = 1, 2.$$

Since $f(x_1, x_2) \neq f_1(x_1)f_2(x_2)$, X_1 and X_2 are dependent. The mean and the variance of X_1 are

$$\mu_1 = \sum_{x_1=1}^{2} x_1 \frac{2x_1 + 6}{18} = (1)\left(\frac{8}{18}\right) + (2)\left(\frac{10}{18}\right) = \frac{14}{9}$$

and

$$\sigma_1^2 = \sum_{x_1=1}^{2} x_1^2 \frac{2x_1 + 6}{18} - \left(\frac{14}{9}\right)^2 = \frac{24}{9} - \frac{196}{81} = \frac{20}{81}.$$

The mean and the variance of X_2 are

$$\mu_2 = \sum_{x_2=1}^{2} x_2 \frac{3 + 4x_2}{18} = (1)\left(\frac{7}{18}\right) + (2)\left(\frac{11}{18}\right) = \frac{29}{18}$$

and

$$\sigma_2^2 = \sum_{x_2=1}^{2} x_2^2 \frac{3 + 4x_2}{18} - \left(\frac{29}{18}\right)^2 = \frac{51}{18} - \frac{841}{324} = \frac{77}{324}.$$

The covariance of X_1 and X_2 is

$$\text{Cov}(X_1, X_2) = \sum_{x_2=1}^{2} \sum_{x_1=1}^{2} x_1 x_2 \frac{x_1 + 2x_2}{18} - \left(\frac{14}{9}\right)\left(\frac{29}{18}\right)$$

$$= (1)(1)\left(\frac{3}{18}\right) + (2)(1)\left(\frac{4}{18}\right) + (1)(2)\left(\frac{5}{18}\right)$$

$$+ (2)(2)\left(\frac{6}{18}\right) - \left(\frac{14}{9}\right)\left(\frac{29}{18}\right)$$

$$= \frac{45}{18} - \frac{406}{162} = -\frac{1}{162}.$$

Hence, the correlation coefficient is

$$\rho = \frac{-1/162}{\sqrt{(20/81)(77/324)}} = \frac{-1}{\sqrt{1540}} = -0.025.$$

Insight into the correlation coefficient ρ of two discrete random variables X and Y may be gained by thoughtfully examining its definition

$$\rho = \frac{\sum_R (x - \mu_X)(y - \mu_Y)f(x, y)}{\sigma_X \sigma_Y}$$

where μ_X, μ_Y, σ_X, and σ_Y denote the respective means and standard deviations. If positive probability is assigned to pairs (x, y) in which both x and y are either simultaneously above or simultaneously below their respective means, the corresponding terms in the summation that defines ρ are positive because both factors $(x - \mu_X)$ and $(y - \mu_Y)$ will be positive or both will be negative. If pairs (x, y), which yield large positive products $(x - \mu_X)(y - \mu_Y)$, contain most of the probability of the distribution, the correlation coefficient will tend to be positive. If, on the other hand, the points (x, y), in which one component is below its mean and the other above its mean, have most of the probability, then the coefficient of correlation will tend to be negative because the products

$$(x - \mu_X)(y - \mu_Y)$$

are negative. This interpretation of the sign of the correlation coefficient will play an important role in subsequent work.

To gain additional insight into the meaning of the correlation coefficient ρ, consider the following problem. Think of the points (x, y) in the space R and their corresponding probabilities. Let us consider all possible lines in two-dimensional space, each with finite slope, that pass through the point associated with the means, namely (μ_X, μ_Y). These lines are of the form $y - \mu_Y = b(x - \mu_X)$ or, equivalently, $y = \mu_Y + b(x - \mu_X)$. For each point in R, say (x_0, y_0) so that $f(x_0, y_0) > 0$, consider the vertical distance from that point to one of these lines. Since y_0 is the height of the point above the x axis and $\mu_Y + b(x_0 - \mu_X)$ is the height of the point on the line that is directly above or below point (x_0, y_0), then the absolute value of the difference of these two heights is the vertical distance from point (x_0, y_0) to the line $y = \mu_Y + b(x - \mu_X)$. That is, the required distance is $|y_0 - \mu_Y - b(x_0 - \mu_X)|$. Let us now square this distance and take the weighted average of all such squares; that is, let us consider the mathematical expectation

$$E\{[(Y - \mu_Y) - b(X - \mu_X)]^2\} = K(b).$$

The problem is to find that line (or that b) which minimizes this expectation of the square $\{Y - \mu_Y - b(X - \mu_X)\}^2$. This is another application of the principle of *least squares*.

The solution of the problem is very easy since

$$K(b) = E\{(Y - \mu_Y)^2 - 2b(X - \mu_X)(Y - \mu_Y) + b^2(X - \mu_X)^2\}$$
$$= \sigma_Y^2 - 2b\rho\sigma_X\sigma_Y + b^2\sigma_X^2,$$

because E is a linear operator and $E[(X - \mu_X)(Y - \mu_Y)] = \rho\sigma_X\sigma_Y$. Accordingly, the derivative

$$K'(b) = -2\rho\sigma_X\sigma_Y + 2b\sigma_X^2$$

equals zero at $b = \rho\sigma_Y/\sigma_X$; and we see that $K(b)$ obtains its minimum for that b since $K''(b) = 2\sigma_X^2 > 0$. Consequently the line of the given form that is the best fit in the above sense is

$$y = \mu_Y + \rho\frac{\sigma_Y}{\sigma_X}(x - \mu_X).$$

Of course, if $\rho > 0$, the slope of the line is positive; but if $\rho < 0$, the slope is negative.

It is also instructive to note the value of the minimum of

$$K(b) = E\{[(Y - \mu_Y) - b(X - \mu_X)]^2\} = \sigma_Y^2 - 2b\rho\sigma_X\sigma_Y + b^2\sigma_X^2.$$

It is

$$K\left(\rho\frac{\sigma_Y}{\sigma_X}\right) = \sigma_Y^2 - 2\rho\frac{\sigma_Y}{\sigma_X}\rho\sigma_X\sigma_Y + \left(\rho\frac{\sigma_Y}{\sigma_X}\right)^2\sigma_X^2$$

$$= \sigma_Y^2 - 2\rho^2\sigma_Y^2 + \rho^2\sigma_Y^2 = \sigma_Y^2(1 - \rho^2).$$

Since $K(b)$ is the expected value of a square, it must be nonnegative for all b, and we see that $\sigma_Y^2(1 - \rho^2) \geq 0$; that is, $\rho^2 \leq 1$, and hence $-1 \leq \rho \leq 1$, which is an important property of the correlation coefficient ρ. Note that if $\rho = 0$, then $K(\rho\sigma_Y/\sigma_X) = \sigma_Y^2$; on the other hand, $K(\rho\sigma_Y/\sigma_X)$ is relatively small if ρ is close to one or negative one. That is, the vertical deviations of the points with positive density from the line $y = \mu_Y + \rho(\sigma_Y/\sigma_X)(x - \mu_X)$ are small if ρ is close to one or negative one because $K(\rho\sigma_Y/\sigma_X)$ is the expectation of the square of those deviations. Thus, ρ measures, in this sense, the amount of *linearity* in the probability distribution. As a matter of fact, in the discrete case, all the points of positive density lie on this straight line if and only if ρ is equal to one or negative one.

REMARK. More generally, we could have fitted the line $y = a + bx$ by the same application of the principle of least squares. We would have then proved that the "best" line actually passes through the point (μ_X, μ_Y). Recall that in the above discussion we assumed our line to be of that form. Students will find this derivation to be an interesting exercise (see Exercise 9.2–4).

Suppose that X and Y are independent so that $f(x, y) \equiv f_1(x)f_2(y)$, and we want to find the expected value of the product $(X - \mu_X)(Y - \mu_Y)$. Subject to the existence of the expectations, we know that

$$E[u(X)v(Y)] = \sum_R \sum u(x)v(y)f(x, y)$$

$$= \sum_{R_1} \sum_{R_2} u(x)v(y)f_1(x)f_2(y)$$

$$= \sum_{R_1} u(x)f_1(x) \sum_{R_2} v(y)f_2(y)$$

$$= E[u(X)]E[v(Y)].$$

We have used this result in Chapter 5. However, it is interesting that it can also be used to show that the correlation coefficient of two independent variables is zero. For, in a standard notation, we have

$$Cov(X, Y) = E[(X - \mu_X)(Y - \mu_Y)]$$

$$= E(X - \mu_X)E(Y - \mu_Y) = 0.$$

The converse of this fact is not necessarily true, however; zero correlation does not in general imply independence. It is most important to keep this straight: independence implies zero correlation, but zero correlation does not necessarily imply independence. The latter is now illustrated.

Example 9.2–2. Let X and Y have the joint p.d.f.

$$f(x, y) = \frac{1}{3}, \qquad (x, y) = (0, 1), (1, 0), (2, 1).$$

Since the support is not "rectangular," X and Y must be dependent. The means of X and Y are $\mu_X = 1$ and $\mu_Y = 2/3$, respectively. Hence,

$$\text{Cov}(X, Y) = E(XY) - \mu_X \mu_Y$$

$$= (0)(1)\left(\frac{1}{3}\right) + (1)(0)\left(\frac{1}{3}\right) + (2)(1)\left(\frac{1}{3}\right) - (1)\left(\frac{2}{3}\right) = 0.$$

That is, $\rho = 0$ but X and Y are dependent.

Exercises

9.2–1. Let the random variables X and Y have the joint p.d.f.

$$f(x, y) = \frac{x + y}{32}, \qquad x = 1, 2, \quad y = 1, 2, 3, 4.$$

Find the means μ_X and μ_Y, the variances σ_X^2 and σ_Y^2, and the correlation coefficient ρ. Are X and Y independent or dependent?

9.2–2. Let X and Y have the joint p.d.f. described as follows.

(x, y)	$f(x, y)$
(0, 0)	1/6
(1, 0)	2/6
(1, 1)	2/6
(2, 1)	1/6

Find the correlation coefficient ρ and the "best" fitting line. HINT: First depict the points in R and their corresponding probabilities.

9.2–3. In the multivariate situation, show that E is a linear or distributive operator. For convenience, let $n = 2$ and show that

$$E[a_1 u_1(X_1, X_2) + a_2 u_2(X_1, X_2)] = a_1 E[u_1(X_1, X_2)] + a_2 E[u_2(X_1, X_2)].$$

9.2–4. Let X and Y be random variables with respective means μ_X and μ_Y, respective variances σ_X^2 and σ_Y^2, and correlation coefficient ρ. Fit the line $y = a + bx$ by the method of least squares to the probability distribution by minimizing the expectation

$$K(a, b) = E[(Y - a - bX)^2]$$

with respect to a and b.

9.2–5. Let X and Y have a trinomial distribution with parameters $n = 3$, $p_1 = 1/6$, and $p_2 = 1/2$. Find **(a)** $E(X)$, **(b)** $E(Y)$, **(c)** $\text{Var}(X)$, **(d)** $\text{Var}(Y)$, **(e)** $\text{Cov}(X, Y)$, **(f)** ρ. Note that $\rho = -\sqrt{p_1 p_2/(1 - p_1)(1 - p_2)}$.

9.2–6. Let the joint p.d.f. of X and Y be $f(x, y) = 1/4$, $(x, y) \in R = \{(0, 0), (1, 1), (1, -1), (2, 0)\}$.

    (a) Are X and Y independent?
    (b) Calculate $\text{Cov}(X, Y)$ and ρ.
This also illustrates the fact that dependent random variables can have a correlation coefficient of zero.

9.3 Conditional Distributions

Let X and Y have a joint discrete distribution with p.d.f. $f(x, y)$ with space R and marginal probability density functions $f_1(x)$ and $f_2(y)$ with spaces R_1 and R_2, respectively. Let event $A = \{X = x\}$ and event $B = \{Y = y\}$, $(x, y) \in R$. Thus $A \cap B = \{X = x, Y = y\}$. Because

$$P(A \cap B) = P(X = x, Y = y) = f(x, y)$$

and

$$P(B) = P(Y = y) = f_2(y) > 0 \qquad (\text{since } y \in R_2),$$

we see that the conditional probability of event A given event B is

$$P(A \mid B) = \frac{P(A \cap B)}{P(B)} = \frac{f(x, y)}{f_2(y)}.$$

This leads to the following definition.

DEFINITION 9.3–1. *The* conditional probability density function *of X, given that $Y = y$, is defined by*

$$g(x \mid y) = \frac{f(x, y)}{f_2(y)}, \qquad f_2(y) > 0.$$

Similarly, the conditional probability density function of Y, given that $X = x$, is defined by

$$h(y|x) = \frac{f(x, y)}{f_1(x)}, \qquad f_1(x) > 0.$$

Example 9.3–1. Let X and Y have the joint p.d.f.

$$f(x, y) = \frac{x + y}{21}, \qquad x = 1, 2, 3, \quad y = 1, 2.$$

In Example 9.1–2 we showed that

$$f_1(x) = \frac{2x + 3}{21}, \qquad x = 1, 2, 3$$

and

$$f_2(y) = \frac{3y + 6}{21}, \qquad y = 1, 2.$$

Thus, the conditional p.d.f. of X, given $Y = y$, is equal to

$$g(x|y) = \frac{(x + y)/21}{(3y + 6)/21} = \frac{x + y}{3y + 6}, \qquad x = 1, 2, 3, \text{ when } y = 1 \text{ or } 2.$$

For example,

$$P(X = 2|Y = 2) = g(2|2) = \frac{4}{12} = \frac{1}{3}.$$

Similarly, the conditional p.d.f. of Y, given $X = x$, is equal to

$$h(y|x) = \frac{x + y}{2x + 3}, \qquad y = 1, 2, \text{ when } x = 1, 2, \text{ or } 3.$$

Note that $0 \leq g(x|y)$ and

$$\sum_x g(x|y) = \sum_x \frac{f(x, y)}{f_2(y)} = \frac{f_2(y)}{f_2(y)} = 1.$$

Thus, $g(x|y)$ satisfies the conditions of a probability density function, and so we can compute conditional probabilities such as

$$P(a < X < b|Y = y) = \sum_{\{x:\, a < x < b\}} g(x|y)$$

and conditional expectations such as

$$E[u(X)|Y = y] = \sum_x u(x)g(x|y)$$

in a manner similar to those associated with probabilities and expectations. We now define two special conditional expectations.

DEFINITION 9.3–2. *The* conditional mean of X, given $Y = y$, *is defined by*

$$\mu_{X|y} = E(X|y) = \sum_x xg(x|y);$$

the conditional variance of X, given $Y = y$, *is defined by*

$$\sigma^2_{X|y} = E\{[X - E(X|y)]^2|y\} = \sum_x [x - E(X|y)]^2 g(x|y)$$

and can be computed using

$$\sigma^2_{X|y} = E(X^2|y) - [E(X|y)]^2.$$

The conditional mean $\mu_{Y|x}$ and the conditional variance $\sigma^2_{Y|x}$ are given by similar expressions.

Example 9.3–2. We use the background of Example 9.3–1 and compute $\mu_{X|y}$ and $\sigma^2_{X|y}$, when $y = 2$:

$$\mu_{X|2} = E(X|y = 2) = \sum_{x=1}^{3} xg(x|2) = \sum_{x=1}^{3} x\frac{x+2}{12} = \frac{13}{6}$$

and

$$\sigma^2_{X|2} = E\left[\left(X - \frac{13}{6}\right)^2\middle| y = 2\right] = \sum_{x=1}^{3}\left(x - \frac{13}{6}\right)^2\frac{x+2}{12}$$

$$= \sum_{x=1}^{3} x^2\frac{x+2}{12} - \left(\frac{13}{6}\right)^2 = \frac{64}{12} - \frac{169}{36} = \frac{23}{36}.$$

The conditional mean of X, given $Y = y$, is a function of y alone and the conditional mean of Y, given $X = x$, is a function of x alone. Suppose the latter conditional mean is a linear function of x; that is, $E(Y|x) = a + bx$. Let us find the constants a and b in terms of characteristics μ_Y, μ_Y, σ^2_X, σ^2_Y, and ρ. This development will shed additional light on the correlation coefficient ρ; accordingly we assume that the respective standard deviations σ_X and σ_Y are both positive so that the correlation coefficient will exist.

It is given that

$$\sum_y y\frac{f(x, y)}{f_1(x)} = a + bx, \qquad x \in R_1,$$

where R_1 is the space of X. Hence,

$$\sum_y yf(x, y) = (a + bx)f_1(x), \qquad x \in R_1, \tag{1}$$

and

$$\sum_{x \in R_1}\sum_y yf(x, y) = \sum_{x \in R_1}(a + bx)f_1(x);$$

that is, with μ_X and μ_Y representing the respective means, we have

$$\mu_Y = a + b\mu_X. \tag{2}$$

In addition, if we multiply both members of equation (1) by x and sum, we obtain

$$\sum_{x \in R_1} \sum_y xyf(x, y) = \sum_{x \in R_1} (ax + bx^2)f_1(x);$$

that is,

$$E(XY) = aE(X) + bE(X^2)$$

or, equivalently,

$$\mu_X \mu_Y + \rho\sigma_X\sigma_Y = a\mu_X + b(\mu_X^2 + \sigma_X^2). \tag{3}$$

The solution of equations (2) and (3) is

$$a = \mu_Y - \rho\frac{\sigma_Y}{\sigma_X}\mu_X \quad \text{and} \quad b = \rho\frac{\sigma_Y}{\sigma_X},$$

which implies that if $E(Y|x)$ is linear, it is defined by

$$E(Y|x) = \mu_Y + \rho\frac{\sigma_Y}{\sigma_X}(x - \mu_X).$$

That is, if the conditional mean of Y, given $X = x$, is linear, it is exactly the same as the "best" fitting line considered in Section 9.2.

Of course, if the conditional mean of X, given $Y = y$, is linear, it is given by

$$E(X|y) = \mu_X + \rho\frac{\sigma_X}{\sigma_Y}(y - \mu_Y).$$

We see that the point $x = \mu_X$, $E(Y|x) = \mu_Y$ satisfies the expression for $E(Y|x)$; and $E(X|y) = \mu_X$, $y = \mu_Y$ satisfies the expression for $E(X|y)$. That is, the point (μ_X, μ_Y) is on each of the two lines. In addition, we note that the product of the coefficient of x in $E(Y|x)$ and the coefficient of y in $E(X|y)$ equals ρ^2, and the ratio of these two coefficients equals σ_Y^2/σ_X^2. These observations sometimes prove useful in particular problems.

Example 9.3–3. Let X and Y have the trinomial p.d.f. with parameters n, p_1, p_2, and $1 - p_1 - p_2 = p_3$. That is,

$$f(x, y) = \frac{n!}{x!y!(n - x - y)!}p_1^x p_2^y p_3^{n-x-y},$$

where x and y are nonnegative integers such that $x + y \leq n$. From the development of the trinomial distribution, it is obvious that X and Y have the binomial distributions $b(n, p_1)$ and $b(n, p_2)$, respectively. Thus,

$$h(y|x) = \frac{f(x, y)}{f_1(x)} = \frac{(n - x)!}{y!(n - x - y)!}\left(\frac{p_2}{1 - p_1}\right)^y\left(\frac{p_3}{1 - p_1}\right)^{n-x-y},$$

$$y = 0, 1, 2, \ldots, n - x.$$

That is, the conditional p.d.f. of Y, given $X = x$, is binomial

$$b\left[n - x, \frac{p_2}{1 - p_1}\right]$$

and thus has conditional mean

$$E(Y \mid x) = (n - x)\frac{p_2}{1 - p_1}.$$

In a similar manner, we obtain

$$E(X \mid y) = (n - y)\frac{p_1}{1 - p_2}.$$

Since each of the conditional means is linear, the product of the respective coefficients of x and y is

$$\rho^2 = \frac{-p_2}{1 - p_1}\frac{-p_1}{1 - p_2} = \frac{p_1 p_2}{(1 - p_1)(1 - p_2)}.$$

Since both coefficients are negative, ρ must be negative and is equal to

$$\rho = -\sqrt{\frac{p_1 p_2}{(1 - p_1)(1 - p_2)}}.$$

═══ *Exercises* ═══

9.3–1. Let X and Y have the joint p.d.f.

$$f(x, y) = \frac{x + y}{32}, \qquad x = 1, 2, \quad y = 1, 2, 3, 4.$$

Find

(a) $g(x \mid y)$,
(c) $E[Y \mid X = 1]$,
(e) $P(1 \leq Y \leq 3 \mid X = 1)$,

(b) $h(y \mid x)$,
(d) $E[Y \mid X = 2]$,
(f) the conditional variance of Y, given $X = 1$.

9.3–2. Let the joint p.d.f. $f(x, y)$ of X and Y be given by the following:

(x, y)	$f(x, y)$
(1, 1)	3/8
(2, 1)	1/8
(1, 2)	1/8
(2, 2)	3/8

Find the two conditional probability density functions and the corresponding means and variances.

9.3–3. An unbiased die is cast 30 independent times. Let X be the number of one's and Y the number of two's.
   (a)   What is the joint p.d.f. of X and Y?
   (b)   Find the conditional p.d.f. of X, given $Y = y$.
   (c)   Compute $E(X^2 - 4XY + 3Y^2)$.

9.3–4. Let X and Y have a uniform distribution on the set of points with integer coordinates in $R = \{(x, y): \ 0 \le x \le 7, \ x \le y \le x + 2\}$. That is, $f(x, y) = 1/24$, $(x, y) \in R$, and both x and y are integers. Find   **(a)** $f_1(x)$,   **(b)** $h(y|x)$,   **(c)** $E(Y|x)$,   **(d)** $\sigma^2_{Y|x}$,   **(e)** $f_2(y)$.

9.3–5. Let $f_1(x) = 1/10$, $x = 0, 1, 2, \ldots, 9$, and $h(y|x) = 1/(10 - x)$, $y = x, x + 1, \ldots, 9$. Find   **(a)** $f(x, y)$,   **(b)** $f_2(y)$,   **(c)** $E(Y|x)$.

9.4 Multivariate Distributions of the Continuous Type

In this section we extend the idea of the p.d.f. of one random variable of the continuous type to that of two or more random variables of the continuous type. As in the one variable case, the definitions are the same as those in the discrete case except that integrals replace summations. For the most part, we simply accept this substitution, and thus this section consists mainly of examples and exercises.

The *joint probability density function* of n random variables $X_1, X_2, \ldots, X_n$ of the continuous type is an integrable function $f(x_1, x_2, \ldots, x_n)$ with the following properties:

(a) $f(x_1, x_2, \ldots, x_n) \ge 0$.

(b) $\displaystyle\int_{-\infty}^{\infty} \cdots \int_{-\infty}^{\infty} f(x_1, x_2, \ldots, x_n) \, dx_1 \cdots dx_n = 1$.

(c) $\displaystyle P[(X_1, X_2, \ldots, X_n) \in A] = \int \cdots \int_A f(x_1, x_2, \ldots, x_n) \, dx_1 \cdots dx_n$,

    where $(X_1, X_2, \ldots, X_n) \in A$ is an event defined in n-dimensional Euclidean space.

For the special case of a joint distribution of two random variables X and Y note that

$$P[(X, Y) \in A] = \iint_A f(x, y) \, dx \, dy,$$

and thus $P[(X, Y) \in A)]$ is the volume of the solid over the region A in the xy plane and bounded by the surface $z = f(x, y)$.

The *marginal p.d.f.* of any one of these n random variables, say X_k, is given by the $(n-1)$-fold integral

$$f_k(x_k) = \int_{-\infty}^{\infty} \cdots \int_{-\infty}^{\infty} f(x_1, x_2, \ldots, x_n) \, dx_1 \cdots dx_{k-1} dx_{k+1} \cdots dx_n.$$

The definitions associated with mathematical expectations are the same as those associated with the discrete case after replacing the summations by integrations.

Example 9.4–1. Let X and Y have the joint p.d.f.

$$f(x, y) = 2, \qquad 0 \le x \le y \le 1.$$

Then $R = \{(x, y): 0 \le x \le y \le 1\}$ is the support and, for illustration,

$$P\left(0 \le X \le \frac{1}{2}, 0 \le Y \le \frac{1}{2}\right) = P\left(0 \le X \le Y, 0 \le Y \le \frac{1}{2}\right)$$

$$= \int_0^{1/2} \int_0^{y} 2 \, dx \, dy = \int_0^{1/2} 2y \, dy = \frac{1}{4}.$$

The shaded region in Figure 9.4–1 is the region of integration that is a subset of R, and the given probability is the volume above that region under the surface $z = 2$. The marginal p.d.f.'s are given by

$$f_1(x) = \int_x^1 2 \, dy = 2(1 - x), \qquad 0 \le x \le 1,$$

and

$$f_2(y) = \int_0^y 2 \, dx = 2y, \qquad 0 \le y \le 1.$$

Three illustrations of expected values are

$$E(X) = \int_0^1 \int_x^1 2x \, dy \, dx = \int_0^1 2x(1 - x) \, dx = \frac{1}{3},$$

$$E(Y) = \int_0^1 \int_0^y 2y \, dx \, dy = \int_0^1 2y^2 \, dy = \frac{2}{3},$$

and

$$\text{Cov}(X, Y) = E[(X - \mu_X)(Y - \mu_Y)] = E(XY) - \mu_X \mu_Y$$

$$= \int_{-\infty}^{\infty} \int_{-\infty}^{\infty} xyf(x, y) \, dx \, dy - \left(\frac{1}{3}\right)\left(\frac{2}{3}\right)$$

$$= \int_0^1 \int_0^y 2xy \, dx \, dy - \frac{2}{9} = \frac{1}{4} - \frac{2}{9} = \frac{1}{36}.$$

From these calculations it is obvious that $E(X)$ and $E(Y)$ could be calculated using the marginal p.d.f.'s as well as the joint one.

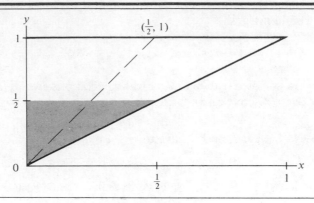

FIGURE 9.4–1

Let X and Y have a distribution of the continuous type with joint p.d.f. $f(x, y)$ and marginal p.d.f.'s $f_1(x)$ and $f_2(y)$, respectively. So, in accord with our policy of transition from the discrete to the continuous case, we have that the conditional p.d.f., mean, and variance of X, given $Y = y$, are, respectively,

$$g(x|y) = \frac{f(x, y)}{f_2(y)}, \quad \text{provided } f_2(y) > 0,$$

$$E(X|y) = \int_{-\infty}^{\infty} xg(x|y)\, dx,$$

and

$$E\{[X - E(X|y)]^2|y\} = \int_{-\infty}^{\infty} [x - E(X|y)]^2 g(x|y)\, dx$$

$$= E[X^2|y] - [E(X|y)]^2.$$

Similar expressions are associated with the conditional distribution of Y, given $X = x$.

Example 9.4–2. Let X and Y be the random variables of Example 9.4–1. Thus,

$$f(x, y) = 2, \quad 0 \leq x \leq y \leq 1,$$
$$f_1(x) = 2(1 - x), \quad 0 \leq x \leq 1,$$

and

$$f_2(y) = 2y, \quad 0 \leq y \leq 1.$$

Before we actually find the conditional p.d.f. of X, given $Y = y$, we shall give an intuitive argument. The joint p.d.f. is constant over the triangular region shown in Figure 9.4–1. If the value of Y is known, say $Y = y$, then the possible values of X are between 0 and y. Furthermore we would expect X to be uniformly distributed on the interval $[0, y]$. That is, we would anticipate that $g(x \mid y) = 1/y, 0 \le x \le y$. More formally now, we have by definition that

$$g(x \mid y) = \frac{f(x, y)}{f_2(y)} = \frac{2}{2y}, \qquad 0 \le x \le y, \ 0 \le y \le 1.$$

The conditional mean of X, given $Y = y$, is

$$E(X \mid y) = \int_0^y x \frac{1}{y} \, dx = \left(\frac{x^2}{2y}\right)_0^y = \frac{y}{2}, \qquad 0 \le y \le 1.$$

Note that for a given y, the conditional mean of X lies on the dotted line in Figure 9.4–1, a result that also agrees with our intuition. Similarly, it could be shown that

$$E(Y \mid x) = \frac{x + 1}{2}, \qquad 0 \le x \le 1.$$

The conditional variance of X, given $Y = y$, is

$$E\{[X - E(X \mid y)]^2 \mid y\} = \int_0^y \left(x - \frac{y}{2}\right)^2 \frac{1}{y} \, dx$$

$$= \left[\frac{1}{3y}\left(x - \frac{y}{2}\right)^3\right]_0^y = \frac{y^2}{12}.$$

An illustration of the computation of a conditional probability is

$$P\left(\frac{1}{8} \le X \le \frac{1}{4} \,\middle|\, Y = \frac{3}{4}\right) = \int_{1/8}^{1/4} g\left(x \,\middle|\, \frac{3}{4}\right) dx = \int_{1/8}^{1/4} \frac{1}{3/4} \, dx = \frac{1}{6}.$$

In general, if $E(X \mid y)$ is linear, it is equal to

$$E(X \mid y) = \mu_X + \rho\left(\frac{\sigma_X}{\sigma_Y}\right)(y - \mu_Y);$$

if $E(Y \mid x)$ is linear, then

$$E(Y \mid x) = \mu_Y + \rho\left(\frac{\sigma_Y}{\sigma_X}\right)(x - \mu_X).$$

Thus, in Example 9.4–2, we see that the product of the coefficients of x in $E(Y \mid x)$ and y in $E(X \mid y)$ is $\rho^2 = 1/4$. Thus, $\rho = 1/2$ since each coefficient is positive. Since the ratio of those coefficients is equal to $\sigma_Y^2/\sigma_X^2 = 1$, we have that $\sigma_X^2 = \sigma_Y^2$.

Of course, the definition of independent random variables of the continuous type carries over naturally from the discrete case. That is, X and Y are *independent* if and only if the joint p.d.f. factors into the product of their marginal p.d.f.'s, namely,

$$f(x, y) = f_1(x)f_2(y), \qquad x \in R_1, \qquad y \in R_2.$$

In addition the rules that allow us to determine easily dependent and independent random variables are also valid here. For illustration, X and Y in Example 9.4–1 are obviously dependent because the support R is not a product space since it is bounded by the diagonal line $y = x$.

═══ *Exercises* ═══

9.4–1. Let $f(x, y) = 2e^{-x-y}$, $0 \le x \le y < \infty$, be the joint p.d.f. of X and Y. Find $f_1(x)$ and $f_2(y)$, the marginal p.d.f.'s of X and Y, respectively. Are X and Y independent?

9.4–2. Let $f(x, y) = 3/2$, $x^2 \le y \le 1, 0 \le x \le 1$, be the joint p.d.f. of X and Y. Find

(a) $P\left(0 \le X \le \dfrac{1}{2}\right)$,

(b) $P\left(\dfrac{1}{2} \le Y \le 1\right)$,

(c) $P\left(\dfrac{1}{2} \le X \le 1, \dfrac{1}{2} \le Y \le 1\right)$,

(d) $P\left(X \ge \dfrac{1}{2}, Y \ge \dfrac{1}{2}\right)$.

Are X and Y independent?

9.4–3. Let $f(x, y) = 1/4$, $0 \le x \le 2$, $0 \le y \le 2$, be the joint p.d.f. of X and Y. Find $f_1(x)$ and $f_2(y)$, the marginal probability density functions. Are the two random variables independent?

9.4–4. Let X and Y have the joint p.d.f. $f(x, y) = x + y$, $0 \le x \le 1, 0 \le y \le 1$. Find the marginal p.d.f.'s $f_1(x)$ and $f_2(y)$ and show that $f(x, y) \ne f_1(x)f_2(y)$. Thus, X and Y are dependent. Compute the correlation coefficient ρ.

9.4–5. Let $f(x, y) = e^{-x-y}$, $0 \le x < \infty$, $0 \le y < \infty$, be the joint p.d.f. of X and Y. Argue that X and Y are independent and compute
(a) $P(X < Y)$,
(b) $P(X > 1, Y > 1)$,
(c) $P(X = Y)$,
(d) $P(X < 2)$.

9.4–6. Let $f(x, y) = 1/20$, $x \le y \le x + 2, 0 \le x \le 10$, be the joint p.d.f. of X and Y.
(a) Sketch the region for which $f(x, y) > 0$, that is, the support.
(b) Find $f_1(x)$, the marginal p.d.f. of X.
(c) Find $h(y|x)$, the conditional p.d.f. of Y, given $X = x$.
(d) Find the conditional mean and variance of Y, given $X = x$.

9.4–7. Let X have a distribution $U(0, 2)$ and let the conditional distribution of Y, given $X = x$, be $U(0, x^2)$.
(a) Define the joint p.d.f. of X and Y, $f(x, y)$.
(b) Calculate $f_2(y)$, the marginal p.d.f. of Y.
(c) Find $E(X|y)$, the conditional mean of X given $Y = y$.

9.4–8. The *joint moment-generating function* of the random variables X and Y of the continuous type with joint p.d.f. $f(x, y)$ is defined by

$$M(t_1, t_2) = \int_{-\infty}^{\infty} \int_{-\infty}^{\infty} e^{t_1 x + t_2 y} f(x, y) \, dx \, dy,$$

if this integral exists for $-h_1 < t_1 < h_1, -h_2 < t_2 < h_2$. If X and Y are independent, show that

$$M(t_1, t_2) = M(t_1, 0) M(0, t_2).$$

From the uniqueness property of the moment-generating function, it can be argued that this is also a sufficient condition for independence. These statements are also true in the discrete case. That is, X and Y are independent if and only if $M(t_1, t_2) = M(t_1, 0) M(0, t_2)$.

9.4–9. Let (X, Y) denote a point selected at random from the rectangle

$$R = \{(x, y): \ 0 \le x \le 1, 0 \le y \le e\}.$$

Compute $P[(X, Y) \in A]$ where $A = \{(x, y): \ y \le e^x\} \cap R$.

9.4–10. Let X_1, X_2 be independent and have distributions $U(0, 1)$. The joint p.d.f. of X_1 and X_2 is $f(x_1, x_2) = 1, 0 \le x_1 \le 1, 0 \le x_2 \le 1$.
   (a) Show that $P(X_1^2 + X_2^2 \le 1) = \pi/4$.
   (b) Using pairs of random numbers find an approximation of $\pi/4$. HINT: For n pairs of random numbers, the relative frequency

$$\frac{N[\{(x_1, x_2): \ x_1^2 + x_2^2 \le 1\}]}{n}$$

is an approximation of $\pi/4$.

9.5 The Bivariate Normal Distribution

Let X and Y be random variables with joint p.d.f. $f(x, y)$ of the continuous type. Many applications are concerned with the conditional distribution of one of the random variables, say Y, given that $X = x$. For example, X and Y might be a student's grade point averages from high school and from his first year in college, respectively. Persons in the field of educational testing and measurement are extremely interested in the conditional distribution of Y, given $X = x$, in such situations.

Suppose that we have an application in which we can make the following three assumptions about the conditional distribution of Y, given $X = x$:

   (a) It is normal for each real x.
   (b) Its mean $E(Y|x)$ is a linear function of x.
   (c) Its variance is constant; that is, it does not depend upon the given value of x.

Of course, assumption (b), along with a result given in Section 9.4, implies

$$E(Y|x) = \mu_Y + \rho \frac{\sigma_Y}{\sigma_X}(x - \mu_X).$$

Let us now consider the implication of assumption (c). The conditional variance is given by

$$\sigma^2_{Y|x} = \int_{-\infty}^{\infty} \left[y - \mu_Y - \rho \frac{\sigma_Y}{\sigma_X}(x - \mu_X) \right]^2 h(y|x)\, dy.$$

Let us multiply each member of this equation by $f_1(x)$ and integrate on x. Since $\sigma^2_{Y|x}$ is a constant, the left-hand member is equal to $\sigma^2_{Y|x}$. Thus, we have

$$\sigma^2_{Y|x} = \int_{-\infty}^{\infty} \int_{-\infty}^{\infty} \left[(y - \mu_Y) - \rho \frac{\sigma_Y}{\sigma_X}(x - \mu_X) \right]^2 h(y|x) f_1(x)\, dy\, dx.$$

However, $h(y|x)f_1(x) = f(x, y)$ and, hence, the right-hand member is just an expectation and the equation can be written as

$$\sigma^2_{Y|x} = E\left\{ (Y - \mu_Y)^2 - 2\rho \frac{\sigma_Y}{\sigma_X}(X - \mu_X)(Y - \mu_Y) + \rho^2 \frac{\sigma^2_Y}{\sigma^2_X}(X - \mu_X)^2 \right\}.$$

But using the fact that the expectation E is a linear operator, we have, recalling $E[(X - \mu_X)(Y - \mu_Y)] = \rho \sigma_X \sigma_Y$, that

$$\sigma^2_{Y|x} = \sigma^2_Y - 2\rho \frac{\sigma_Y}{\sigma_X} \rho \sigma_X \sigma_Y + \rho^2 \frac{\sigma^2_Y}{\sigma^2_X} \sigma^2_X$$

$$= \sigma^2_Y - 2\rho^2 \sigma^2_Y + \rho^2 \sigma^2_Y = \sigma^2_Y(1 - \rho^2).$$

That is, the conditional variance of Y, for each given x, is $\sigma^2_Y(1 - \rho^2)$. These facts about the conditional mean and variance, along with assumption (a), require that the conditional p.d.f. of Y, given $X = x$, be

$$h(y|x) = \frac{1}{\sigma_Y \sqrt{2\pi} \sqrt{1 - \rho^2}} \exp\left[-\frac{[y - \mu_Y - \rho(\sigma_Y/\sigma_X)(x - \mu_X)]^2}{2\sigma^2_Y(1 - \rho^2)} \right],$$

$$-\infty < y < \infty, \text{ for every real } x.$$

Up to this point, nothing has been said about the distribution of X other than that it has mean μ_X and positive variance σ^2_X. Suppose, in addition, we assume that this distribution is also normal; that is, the marginal p.d.f. of X is

$$f_1(x) = \frac{1}{\sigma_X \sqrt{2\pi}} \exp\left[-\frac{(x - \mu_X)^2}{2\sigma^2_X} \right], \qquad -\infty < x < \infty.$$

Hence, the joint p.d.f. of X and Y is given by the product

$$f(x, y) = h(y|x)f_1(x) = \frac{1}{2\pi \sigma_X \sigma_Y \sqrt{1 - \rho^2}} \exp\left[-\frac{q(x, y)}{2} \right], \qquad (1)$$

where it is easy to show (see Exercise 9.5–1) that

$$q(x, y) = \frac{1}{1 - \rho^2}\left[\left(\frac{x - \mu_X}{\sigma_X}\right)^2 - 2\rho\left(\frac{x - \mu_X}{\sigma_X}\right)\left(\frac{y - \mu_Y}{\sigma_Y}\right) + \left(\frac{y - \mu_Y}{\sigma_Y}\right)^2\right].$$

A joint p.d.f. of this form is called a *bivariate normal p.d.f.*

Example 9.5–1. Let us assume that in a certain population of college students, the respective grade point averages, say X and Y, in high school and first year in college have an approximate bivariate normal distribution with parameters $\mu_X = 2.9$, $\mu_Y = 2.4$, $\sigma_X = 0.4$, $\sigma_Y = 0.5$, and $\rho = 0.8$. The conditional p.d.f. of Y, given $X = 3.2$, is normal with mean

$$2.4 + (0.8)\left(\frac{0.5}{0.4}\right)(3.2 - 2.9) = 2.7$$

and standard deviation $(0.5)\sqrt{1 - 0.64} = 0.3$. Thus, for illustration, we have that

$$P(2.1 < Y < 3.3 \,|\, X = 3.2) = \Phi(2) - \Phi(-2) = 0.9544.$$

From a practical point of view, however, the reader should be warned that the correlation coefficient of these grade averages is, in most instances, much smaller than 0.8.

Since x and y enter the bivariate normal p.d.f. in a similar manner, the roles of X and Y could have been interchanged. That is, Y could have been assigned the marginal normal p.d.f. $N(\mu_Y, \sigma_Y^2)$ and the conditional p.d.f. of X, given $Y = y$, would have then been normal with mean $\mu_X + \rho(\sigma_X/\sigma_Y)(y - \mu_Y)$ and variance $\sigma_X^2(1 - \rho^2)$. Although this is fairly obvious, we do want to make special note of it.

We close this section by observing another important property of the correlation coefficient ρ if X and Y have a bivariate normal distribution. In equation (1) of the product $h(y|x)f_1(x)$, let us consider the factor $h(y|x)$ if $\rho = 0$. We see that this product, which is the joint p.d.f. of X and Y, equals $f_1(x)f_2(y)$ because $h(y|x)$ is, when $\rho = 0$, a normal p.d.f. with mean μ_Y and variance σ_Y^2. That is, if $\rho = 0$, the joint p.d.f. factors into the product of the two marginal probability density functions, and, hence, X and Y are independent random variables. Of course, if X and Y are any independent random variables (not necessarily normal), we know that ρ, if it exists, is always equal to zero. Thus, we have proved the following.

THEOREM 9.5–1. *If X and Y have a bivariate normal distribution with correlation coefficient ρ, then X and Y are independent if and only if $\rho = 0$.*

Thus, in the bivariate normal case, $\rho = 0$ does imply independence of X and Y.

Exercises

9.5-1. Show that the expression in the exponent of equation (1) is equal to the function $q(x, y)$ given in the text.

9.5-2. Let X and Y have a bivariate normal distribution with parameters $\mu_X = -3$, $\mu_Y = 10$, $\sigma_X^2 = 25$, $\sigma_Y^2 = 9$, and $\rho = 3/5$. Compute
  (a) $P(-5 < X < 5)$,   (b) $P(-5 < X < 5 | Y = 13)$,
  (c) $P(7 < Y < 16)$,   (d) $P(7 < Y < 16 | X = 2)$.

9.5-3. Let X and Y have a bivariate normal distribution. Find two different lines, $a(x)$ and $b(x)$, parallel to and equidistant from $E(Y|x)$, such that

$$P[a(x) < Y < b(x) | X = x] = 0.9544$$

for all real x. Plot $a(x)$, $b(x)$, and $E(Y|x)$ when $\mu_X = 2$, $\mu_Y = -1$, $\sigma_X = 3$, $\sigma_Y = 5$, and $\rho = 3/5$.

9.5-4. Let X and Y have a bivariate normal distribution with parameters $\mu_X = 2.8$, $\mu_Y = 110$, $\sigma_X^2 = 0.16$, $\sigma_Y^2 = 100$, and $\rho = 0.6$. Compute
  (a) $P(106 < Y < 124)$,   (b) $P(106 < Y < 124 | X = 3.2)$.

9.5-5. Let X and Y have a bivariate normal distribution with parameters $\mu_X = 50$, $\mu_Y = 70$, $\sigma_X^2 = 64$, $\sigma_Y^2 = 100$, and $\rho = 4/5$. Calculate
  (a) $P(65.8 < Y \le 85.3)$,   (b) $P(65.8 < Y \le 85.3 | X = 55.7)$.

9.5-6. Let $f(x, y)$ be the general expression for the joint bivariate normal p.d.f. given in equation (1). Show that $f(x, y) = c$, where the constant c is such that

$$0 < c < 1/(2\pi\sigma_X\sigma_Y\sqrt{1 - \rho^2})$$

defines an ellipse. That is, each plane parallel to the xy plane that intersects the surface $f(x, y)$ creates an elliptical figure.

9.6 Sampling from Bivariate Distributions

In Chapter 3 we found that we could compute characteristics of a sample in exactly the same way that we computed the corresponding characteristics of a distribution of one random variable. Of course, this can be done because the sample characteristics are actually those of a distribution, namely the empirical distribution. And since the empirical distribution approximates the actual distribution when the sample size is large, the sample characteristics can be used as estimates of the corresponding characteristics of the actual distribution. These notions can be extended to samples from distributions of two or more variables, and the purpose of this section is to illustrate this extension.

Let (X_1, Y_1), (X_2, Y_2), ..., (X_n, Y_n) be n independent observations of the pair of random variables (X, Y) with a fixed, but possibly unknown, distribution.

We say that the collection of pairs $(X_1, Y_1), (X_2, Y_2), \ldots, (X_n, Y_n)$ is a *random sample* from this distribution involving two random variables. The definition of a random sample can obviously be extended to samples from distributions with more than two random variables and hence will not be given formally. If we now assign the weight $1/n$ to each pair, we create a discrete-type distribution of probability in two-dimensional space. Using this distribution, we can compute the means, the variances, the covariances, the correlation coefficient, and the "best-fitting" line. As before, those are called characteristics of the sample because knowledge of the sample items and the empirical distribution are equivalent. In addition, these characteristics can be used as estimates of the corresponding ones of the distribution (or the population) from which the sample arose.

The means, variances, and correlation coefficient are given by

$$\overline{X} = \frac{1}{n} \sum_{i=1}^{n} X_i; \qquad \overline{Y} = \frac{1}{n} \sum_{i=1}^{n} Y_i;$$

$$S_X^2 = \frac{1}{n} \sum_{i=1}^{n} (X_i - \overline{X})^2 = \frac{1}{n} \sum_{i=1}^{n} X_i^2 - \overline{X}^2;$$

$$S_Y^2 = \frac{1}{n} \sum_{i=1}^{n} (Y_i - \overline{Y})^2 = \frac{1}{n} \sum_{i=1}^{n} Y_i^2 - \overline{Y}^2;$$

and

$$R = \frac{(1/n) \sum_{i=1}^{n} (X_i - \overline{X})(Y_i - \overline{Y})}{S_X S_Y} = \frac{(1/n) \sum_{i=1}^{n} X_i Y_i - \overline{X}\,\overline{Y}}{S_X S_Y}.$$

For brevity, we write the last equation as

$$R = \frac{S_{XY}}{S_X S_Y};$$

thus,

$$S_{XY} = R S_X S_Y.$$

The symbols $\bar{x}, \bar{y}, s_x, s_y, s_{xy}$, and r will denote the respective observed values of the means, the standard deviations, the covariance, and the correlation coefficient of the sample. Hence, the "best-fitting" line is

$$y = \bar{y} + r \frac{s_y}{s_x} (x - \bar{x}) = \bar{y} + \frac{s_{xy}}{s_x^2} (x - \bar{x})$$

since $r s_x s_y = s_{xy}$.

Recall that in Section 9.2 we fit that straight line by minimizing a certain expectation, which in the notation of this section would be

$$\frac{1}{n} \sum_{i=1}^{n} (y_i - a - bx_i)^2.$$

That is, we find a and b by minimizing a sum of squares or more simply, by least squares. The reader will observe that this is exactly the same problem we considered in Section 7.7, but there the x values were treated as observations of a nonrandom variable. Obviously, the resulting expressions must be exactly the same.

Of course, we do not expect that each observed pair (x_i, y_i) will lie on this line. That is, y_i does not usually equal $\bar{y} + r(s_y/s_x)(x_i - \bar{x})$, but we would expect this line to fit the collection of points $(x_1, y_1), (x_2, y_2), \ldots, (x_n, y_n)$ in some best way, namely, best by the principle of least squares.

Example 9.6–1.   To simplify the calculations, we take a small sample size, namely $n = 5$. Let the five observed points be

$$(3, 2), \quad (6, 0), \quad (5, 2), \quad (1, 6), \quad (3, 5).$$

Then

$$\bar{x} = \frac{3 + 6 + 5 + 1 + 3}{5} = 3.6, \qquad \bar{y} = \frac{2 + 0 + 2 + 6 + 5}{5} = 3,$$

$$s_x^2 = \frac{3^2 + 6^2 + 5^2 + 1^2 + 3^2}{5} - (3.6)^2 = 16 - 12.96 = 3.04,$$

$$s_y^2 = \frac{2^2 + 0^2 + 2^2 + 6^2 + 5^2}{5} - 3^2 = 13.8 - 9 = 4.8,$$

$$s_{xy} = \frac{6 + 0 + 10 + 6 + 15}{5} - (3.6)(3) = -3.4,$$

and

$$r = \frac{-3.4}{\sqrt{(3.04)(4.8)}} = -0.9,$$

approximately. Thus, the "best-fitting" line is

$$y = 3 - \frac{3.4}{3.04}(x - 3.6) = (-1.12)x + 7.03,$$

approximately. We plot the five observed points and the "best-fitting" line on the same graph (Figure 9.6–1) in order to compare the fit of this line to the collection of points.

Example 9.6–2.   Let the joint p.d.f. of the random variables X and Y of the discrete type be $f(x, y) = 1/15$, $1 \leq y \leq x \leq 5$, where x and y are integers. Then the marginal p.d.f. of X is $f_1(x) = x/15$, $x = 1, 2, 3, 4, 5$, and

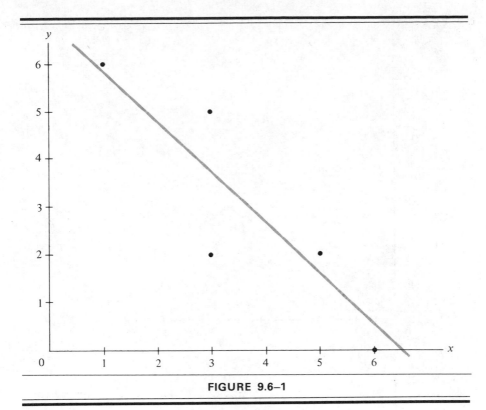

FIGURE 9.6–1

the marginal p.d.f. of Y is $f_2(y) = (6 - y)/15$, $y = 1, 2, 3, 4, 5$. It is easy to show that $\mu_X = 11/3$, $\sigma_X^2 = 14/9$, $\mu_Y = 7/3$, $\sigma_Y^2 = 14/9$, and $\rho = 1/2$. Thus, the "best-fitting" line associated with this distribution is

$$y = \mu_Y + \rho\left(\frac{\sigma_Y}{\sigma_X}\right)(x - \mu_X)$$

$$= \frac{7}{3} + \frac{1}{2}\sqrt{\frac{(14/9)}{(14/9)}}\left(x - \frac{11}{3}\right) = \frac{x}{2} + \frac{1}{2}.$$

A random sample of $n = 30$ observations from this joint distribution was generated on the computer and yielded the following data:

(2, 2)	(5, 2)	(5, 2)	(4, 2)	(4, 3)
(4, 1)	(3, 2)	(4, 3)	(4, 2)	(5, 2)
(5, 4)	(1, 1)	(5, 1)	(2, 1)	(4, 4)
(3, 1)	(5, 1)	(4, 3)	(4, 3)	(5, 5)
(4, 1)	(3, 2)	(4, 2)	(4, 1)	(3, 1)
(2, 1)	(5, 4)	(4, 2)	(3, 3)	(3, 2)

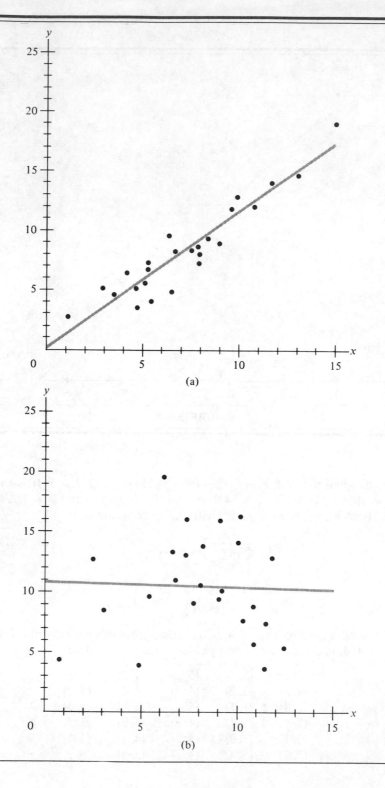

(a)

(b)

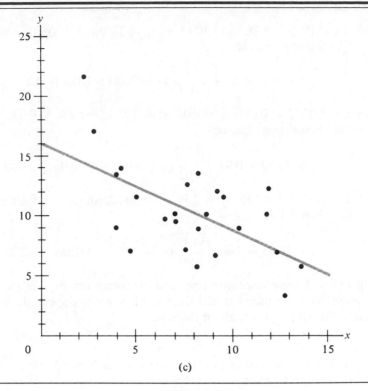

FIGURE 9.6–2

For these 30 observations, in which some points were observed more than once, we have $\bar{x} = 3.77$, $\bar{y} = 2.13$, $s_X^2 = 1.11$, $s_Y^2 = 1.18$, and $r = 0.41$. Thus, the observed "best fitting" line for these 30 points is

$$y = 0.42x + 0.56;$$

this should be compared to the "best-fitting" line of the distribution.

To visualize better the relationship between r and a plot of n observed points $(x_1, y_1), \ldots, (x_n, y_n)$, we have generated three different sets of 25 pairs of observations from three bivariate normal distributions. In the next example we list the corresponding values of $\bar{x}, \bar{y}, s_X^2, s_Y^2, r$, and the observed "best-fitting" line. Each set of points and corresponding line are plotted on the same graph.

Example 9.6–3. Three random samples, each of size $n = 25$, were taken from three different bivariate normal distributions. The corresponding sample characteristics are

(a) $\bar{x} = 7.19$, $\bar{y} = 8.28$, $s_X^2 = 10.13$, $s_Y^2 = 14.37$, and $r = 0.94$, so that the "best-fitting" line is

$$y = 8.28 + 0.94 \sqrt{\frac{14.37}{10.13}} (x - 7.19) = 1.12x + 0.23.$$

(b) $\bar{x} = 8.02$, $\bar{y} = 10.50$, $s_X^2 = 9.02$, $s_Y^2 = 17.23$, and $r = -0.03$, so that the "best-fitting" line is

$$y = 10.50 - 0.03 \sqrt{\frac{17.23}{9.02}} (x - 8.02) = -0.04x + 10.83.$$

(c) $\bar{x} = 7.84$, $\bar{y} = 10.35$, $s_X^2 = 9.80$, $s_Y^2 = 14.32$, and $r = -0.60$, so that the "best-fitting" line is

$$y = 10.35 - 0.60 \sqrt{\frac{14.32}{9.80}} (x - 7.84) = -0.88x + 17.22.$$

In Figure 9.6–2, these respective lines and the corresponding sample points are plotted. Note the effect that the value of r has on the slope of the line and the variability of the points about that line.

The next example shows that two random variables X and Y may be clearly related (dependent) but yet have a correlation coefficient ρ close to zero. This, however, is not unexpected since we recall that ρ does, in a sense, measure the *linear* relationship between two random variables. That is, the linear relationship between X and Y could be zero, while higher order ones could be quite strong.

Example 9.6–4. Let a random number X be selected uniformly from the interval $(1, 9)$. For each observed value of $X = x$, let a random number Y be selected uniformly from the interval $(-x^2 + 10x - 4, \ -x^2 + 10x)$. Twenty-five observations of X and Y, generated on a computer, are

(6.91, 17.52)	(4.32, 22.69)	(2.38, 17.61)	(7.98, 14.29)
(8.26, 10.77)	(2.00, 12.87)	(3.10, 18.63)	(7.69, 16.77)
(2.21, 14.97)	(3.42, 19.16)	(8.18, 11.15)	(5.39, 22.41)
(1.19, 7.50)	(3.21, 19.06)	(5.47, 23.89)	(7.35, 16.63)
(2.32, 15.09)	(7.54, 14.75)	(1.72, 10.75)	(7.33, 17.42)
(8.41, 9.40)	(8.72, 9.83)	(6.09, 22.33)	(5.30, 21.37)
(7.30, 17.36)			

For this set of observations $\bar{x} = 5.33$, $\bar{y} = 16.17$, $s_X^2 = 6.26$, $s_Y^2 = 20.03$, and $r = -0.06$. Note that r is very close to zero even though X and Y seem very dependent; that is, it seems that a quadratic expression would fit the data

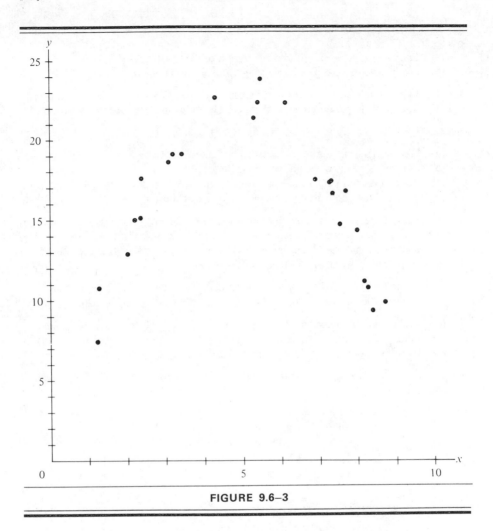

FIGURE 9.6–3

very well. In Exercise 9.6–6, the reader is asked to fit $y = a + bx + cx^2$ to these 25 points by the method of least squares. See Figure 9.6–3 for a plot of the 25 points.

════ *Exercises* ════

9.6–1. Three observed values of the pair of random variables (X, Y) yielded the three points $(2, 3)$, $(4, 7)$, and $(6, 5)$.

(a) Calculate $\bar{x}$, $\bar{y}$, s_X^2, s_Y^2, r, and the equation of the "best-fitting" line.

(b) Plot the points and the "best-fitting" line on the same graph.

9.6–2. Three observed values of the pair of random variables (X, Y) yielded the three points $(1, 2), (3, 1)$, and $(2, 3)$.
   (a)   Calculate $\bar{x}, \bar{y}, s_X^2, s_Y^2, r$, and the equation of the "best-fitting" line.
   (b)   Plot the points and the "best-fitting" line on the same graph.

9.6–3. A pair of unbiased dice was rolled six independent times. Let X denote the smaller outcome and Y the larger outcome on the dice. The following outcomes were observed:

$$(2, 5), \quad (3, 5), \quad (3, 6), \quad (2, 3), \quad (5, 5), \quad (1, 3).$$

   (a)   Find $\bar{x}, \bar{y}, s_X^2, s_Y^2, r$, and the "best-fitting" line of the sample.
   (b)   Plot the points and the line on the same graph.
   (c)   Define the joint p.d.f. of X and Y (see Example 9.1–1) and then calculate μ_X, μ_Y, $\sigma_X^2, \sigma_Y^2, \rho$, and the "best-fitting" line of this joint distribution.

9.6–4. Ten college students took the Undergraduate Record Exam (URE) when they were juniors and the Graduate Record Exam (GRE) when they were seniors. The Quantitative URE score (x) and the Quantitative GRE score (y) for each of these 10 students is given in the following list of ordered pairs (x, y):

   (550, 570)   (670, 730)   (490, 450)   (410, 510)   (570, 560)
   (490, 400)   (450, 420)   (490, 520)   (780, 710)   (520, 620)

   (a)   Verify that $\bar{x} = 542.0, \bar{y} = 553.0, s_X^2 = 10{,}836.0, s_Y^2 = 11{,}081.0$, and $r = 0.80$.
   (b)   Find the equation of the "best-fitting" line.
   (c)   Plot the 10 points and the line on the same graph.

9.6–5. The respective high school and college grade point averages for 20 college seniors as ordered pairs (x, y) are

   (3.75, 3.19)   (3.45, 3.34)   (2.87, 2.23)   (3.60, 3.46)
   (3.42, 2.97)   (4.00, 3.79)   (2.65, 2.55)   (3.10, 2.50)
   (3.47, 3.15)   (2.60, 2.26)   (4.00, 3.76)   (2.30, 2.11)
   (2.47, 2.11)   (3.36, 3.01)   (3.60, 2.92)   (3.65, 3.09)
   (3.30, 3.05)   (2.58, 2.63)   (3.80, 3.22)   (3.79, 3.27)

   (a)   Verify that $\bar{x} = 3.29, \bar{y} = 2.93, s_X^2 = 0.27, s_Y^2 = 0.25$, and $r = 0.92$.
   (b)   Find the equation of the "best-fitting" line.
   (c)   Plot the 20 points and the line on the same graph.

9.6–6. The respective high school grade point average and the SAT mathematics score for 25 college students as ordered pairs (x, y) are

   (4.00, 577)   (2.53, 453)   (3.45, 407)   (2.48, 539)
   (2.69, 534)   (2.82, 584)   (2.33, 464)   (2.21, 525)
   (2.59, 545)   (3.37, 499)   (3.00, 446)   (2.93, 446)
   (3.25, 491)   (2.90, 433)   (3.64, 556)   (3.23, 394)
   (2.46, 497)   (2.62, 460)   (2.75, 413)   (2.82, 440)
   (3.51, 608)   (4.00, 657)   (3.72, 449)   (2.78, 323)
   (3.33, 413)

   (a)   Verify that $\bar{x} = 3.02, \bar{y} = 486.12, s_X^2 = 0.25, s_Y^2 = 5640.57$, and $r = 0.28$.
   (b)   Find the equation of the "best-fitting" line.
   (c)   Plot the 25 points and the line on the same graph.

9.6–7. Let the set $R = \{(x, y): \ x/2 < y < x/2 + 2, 0 < x < 10\}$. Let (X, Y) denote a random point selected uniformly from R.

(a) Sketch the set R in the xy plane. Does it seem intuitive to you that $E(Y|x) = x/2 + 1$?

(b) Twenty-five points selected at random from R by the computer are

(6.58, 3.58)	(4.73, 2.36)	(1.52, 1.96)	(7.35, 4.68)
(9.17, 5.50)	(9.17, 5.50)	(9.96, 5.68)	(6.43, 3.61)
(9.06, 4.81)	(2.05, 1.92)	(3.39, 2.70)	(4.78, 4.05)
(1.70, 0.97)	(3.37, 3.62)	(2.75, 2.25)	(6.57, 4.26)
(5.29, 3.17)	(3.13, 1.60)	(8.06, 4.34)	(1.79, 1.26)
(9.81, 6.39)	(1.32, 1.87)	(9.30, 5.95)	(0.46, 2.04)
(5.27, 3.65)			

Use the 25 observed values to obtain the "best-fitting" line of this sample. Note that it is close to $E(Y|x) = x/2 + 1$.

9.6–8. We would like to fit the quadratic curve $y = a + bx + cx^2$ to a set of points $(x_1, y_1), (x_2, y_2), \ldots, (x_n, y_n)$ by the method of least squares. To do this, let

$$h(a, b, c) = \sum_{i=1}^{n} (y_i - a - bx_i - cx_i^2)^2.$$

(a) By setting the three first partial derivatives of h with respect to a, b, and c equal to zero, show that a, b, and c satisfy the following set of equations, all sums going from 1 to n:

$$an + b \sum x_i + c \sum x_i^2 = \sum y_i;$$
$$a \sum x_i + b \sum x_i^2 + c \sum x_i^3 = \sum x_i y_i;$$
$$a \sum x_i^2 + b \sum x_i^3 + c \sum x_i^4 = \sum x_i^2 y_i.$$

(b) For the data given in Example 9.6–4, $\sum x_i = 133.34$, $\sum x_i^2 = 867.75$, $\sum x_i^3 = 6{,}197.21$, $\sum x_i^4 = 46{,}318.88$, $\sum y_i = 404.22$, $\sum x_i y_i = 2{,}138.38$, $\sum x_i^2 y_i = 13{,}380.30$. Show that $a = -1.88$, $b = 9.89$, and $c = -0.995$.

(c) Plot the points and this least squares quadratic regression curve on the same graph.

9.6–9. Let a random number X be selected uniformly from the interval $(1, 9)$. For each observed value of $X = x$, let a random number Y be selected uniformly from the interval $(x^2 - 10x + 26, x^2 - 10x + 30)$. Twenty-five observations of X and Y generated on a computer are

(4.16, 2.66)	(2.88, 8.60)	(4.97, 3.76)	(2.02, 12.81)
(2.69, 10.14)	(2.54, 8.69)	(1.49, 14.07)	(2.13, 10.36)
(2.44, 8.04)	(3.20, 4.41)	(4.20, 3.01)	(8.74, 17.15)
(3.17, 6.79)	(5.39, 1.63)	(8.43, 14.23)	(6.10, 4.75)
(5.47, 1.82)	(8.17, 14.55)	(2.18, 12.76)	(3.18, 6.56)
(8.26, 12.89)	(6.62, 6.72)	(2.68, 9.53)	(8.06, 11.63)
(6.87, 5.96)			

(a) For these data, $\sum x_i = 116.04$, $\sum x_i^2 = 675.35$, $\sum x_i^3 = 4{,}551.52$, $\sum x_i^4 = 33{,}331.38$, $\sum y_i = 213.52$, $\sum x_i y_i = 1{,}036.97$, $\sum x_i^2 y_i = 6{,}661.79$. Show that the equation of the least squares quadratic regression curve is equal to $1.026x^2 - 10.296x + 28.612$.

(b) Plot the points and the least squares quadratic regression curve on the same graph.

9.7 The Sample Correlation Coefficient

Let X and Y have a bivariate normal distribution. We know that if the correlation coefficient ρ is zero, then X and Y are independent random variables (see Section 9.5). Furthermore the value of ρ gives a measure of the linear relationship between X and Y. In this section we shall give methods for using the sample correlation coefficient to test the hypothesis H_0: $\rho = 0$ and also to form a confidence interval for ρ.

Let $(X_1, Y_1), (X_2, Y_2), \ldots, (X_n, Y_n)$ denote a random sample from a bivariate normal distribution with parameters $\mu_X, \mu_Y, \sigma_X^2, \sigma_Y^2$, and ρ. Recall that the sample correlation coefficient is

$$R = \frac{(1/n) \sum_1^n (X_i - \overline{X})(Y_i - \overline{Y})}{\sqrt{(1/n) \sum_1^n (X_i - \overline{X})^2} \sqrt{(1/n) \sum_1^n (Y_i - \overline{Y})^2}} = \frac{S_{XY}}{S_X S_Y}.$$

We have also noted in Section 9.6 that

$$R \frac{S_Y}{S_X} = \frac{S_{XY}}{S_X^2}$$

is exactly the solution that we obtained for $\hat{\beta}$ in Section 7.7 when the X values were fixed at $X_1 = x_1, X_2 = x_2, \ldots, X_n = x_n$. Let us consider these values fixed temporarily so that we are considering conditional distributions, given $X_1 = x_1, \ldots, X_n = x_n$. Moreover, if H_0: $\rho = 0$ is true, then the distributions of $Y_1, Y_2, \ldots, Y_n$ are independent of $x_1, x_2, \ldots, x_n$ and thus $\beta = 0$. Under these conditions, the conditional distribution of

$$\hat{\beta} = R \frac{S_Y}{S_X} = \frac{S_{XY}}{S_X^2}$$

is $N(0, \sigma_Y^2/nS_X^2)$ when $S_X^2 > 0$. Moreover, recall from Section 7.7 that the conditional distribution, given $X_1 = x_1, \ldots, X_n = x_n$, of

$$\frac{\sum_{i=1}^n [Y_i - \overline{Y} - (S_{XY}/S_X^2)(x_i - \bar{x})]^2}{\sigma_Y^2} = \frac{nS_Y^2(1 - R^2)}{\sigma_Y^2}$$

is $\chi^2(n - 2)$ and is independent of $\hat{\beta}$. Thus, when $\rho = 0$, the conditional distribution of

$$T = \frac{(R S_Y/S_X)/(\sigma_Y/\sqrt{nS_X})}{\sqrt{[nS_Y^2(1 - R^2)/\sigma_Y^2][1/(n - 2)]}} = \frac{\sqrt{n - 2R}}{\sqrt{1 - R^2}}$$

is Student's t with $n - 2$ degrees of freedom. However, since the conditional distribution of T, given $X_1 = x_1, \ldots, X_n = x_n$, does not depend upon x_1,

$x_2, \ldots, x_n$, the unconditional distribution of T must be Student's t with $n - 2$ degrees of freedom and T and $(X_1, X_2, \ldots, X_n)$ are independent when $\rho = 0$.

REMARK.   It is interesting to note that in the discussion about the distribution of T, nothing was said about the distribution of $X_1, X_2, \ldots, X_n$. This means that if X and Y are independent and Y has a normal distribution, then T has a Student's t distribution whatever the distribution of X. Obviously, the roles of X and Y can be reversed in all of this development. In particular, if X and Y are independent, then T and $Y_1, Y_2, \ldots, Y_n$ are also independent.

Now T can be used to test $H_0: \rho = 0$. If the alternative hypothesis is $H_1: \rho > 0$, we would use the critical region defined by $T > c$ since large T implies large R. Obvious modifications would be made for the alternative hypotheses $H_1: \rho < 0$ and $H_1: \rho \neq 0$, the latter leading to a two-sided test.

Using the p.d.f. $h(t)$ of T, we can find the distribution function of R when $-1 < r < 1$:

$$G(r) = P(R \leq r) = P\left(T \leq \frac{\sqrt{n - 2}\, r}{\sqrt{1 - r^2}} \right)$$

$$= \int_{-\infty}^{\sqrt{n-2}\, r/\sqrt{1-r^2}} h(t)\, dt$$

$$= \int_{-\infty}^{\sqrt{n-2}\, r/\sqrt{1-r^2}} \frac{\Gamma[(n-1)/2]}{\Gamma(1/2)\Gamma[(n-2)/2]} \frac{1}{\sqrt{n-2}} \left(1 + \frac{t^2}{n-2} \right)^{-(n-1)/2} dt.$$

The derivative of $G(r)$, with respect to r, is

$$g(r) = h\left(\frac{\sqrt{n-2}\, r}{\sqrt{1-r^2}} \right) \frac{d(\sqrt{n-2}\, r/\sqrt{1-r^2})}{dr},$$

which equals

$$g(r) = \frac{\Gamma[(n-1)/2]}{\Gamma(1/2)\Gamma[(n-2)/2]} (1 - r^2)^{(n-4)/2}, \qquad -1 < r < 1.$$

Thus, for example, to test the hypothesis $H_0: \rho = 0$ against the alternative hypothesis $H_1: \rho \neq 0$ at a significance level α, select either a constant c_1 or a constant c_2 so that

$$\alpha = P(|R| \geq c_1; H_0) = P(|T| \geq c_2; H_0),$$

depending upon the availability of R or T tables.

It is also possible to obtain an approximate test of size α by using the fact that

$$W = \frac{1}{2} \ln \frac{1 + R}{1 - R}$$

has an approximate normal distribution with mean $(1/2)\ln[(1 + \rho)/(1 - \rho)]$ and variance $1/\sqrt{n - 3}$. We accept this statement without proof now. Thus, a test of $H_0: \rho = 0$ can be based on the statistic

$$Z = \frac{(1/2)\ln[(1 + R)/(1 - R)] - (1/2)\ln[(1 + \rho)/(1 - \rho)]}{\sqrt{1/(n - 3)}}$$

with $\rho = 0$ so that $(1/2)\ln[(1 + \rho)/(1 - \rho)] = 0$. However, with Z, we can also test hypotheses such as $H_0: \rho = \rho_0$ against $H_1: \rho \neq \rho_0$, where ρ_0 is not necessarily zero. In that case, the hypothesized mean of Z is $(1/2)\ln[(1 + \rho_0)/(1 - \rho_0)]$.

Example 9.7–1. We would like to test the hypothesis $H_0: \rho = 0$ against $H_1: \rho \neq 0$ at an $\alpha = 0.05$ significance level. A random sample of size 18 from a bivariate normal distribution yielded a sample correlation coefficient of $r = 0.35$. Using the t distribution, we would reject H_0 if $|t| \geq 2.120$. Since

$$t = \frac{0.35\sqrt{16}}{\sqrt{1 - (0.35)^2}} = 1.495,$$

H_0 is not rejected. If we had used the normal approximation, H_0 would be rejected if $|z| \geq 1.96$. Since

$$z = \frac{(1/2)\ln[(1 + 0.35)/(1 - 0.35)] - 0}{\sqrt{1/(18 - 3)}} = 1.415,$$

H_0 is not rejected. If the sample size had been $n = 32$, and it was still true that $r = 0.35$, then $t = 2.046$ and $z = 1.968$. Since $t > 2.042$ and $z > 1.96$, H_0 would be rejected.

To develop an approximate $(100\gamma)\%$ confidence interval for ρ, we shall use the normal approximation for the distribution of Z. Thus we select a constant c from Appendix Table IV so that

$$P\left(-c \leq \frac{(1/2)\ln[(1 + R)/(1 - R)] - (1/2)\ln[(1 + \rho)/(1 - \rho)]}{\sqrt{1/(n - 3)}} \leq c\right) = \gamma.$$

After several steps, this becomes

$$P\left(\frac{1 + R - (1 - R)\exp(2c/\sqrt{n - 3})}{1 + R + (1 - R)\exp(2c/\sqrt{n - 3})}\right.$$

$$\left. \leq \rho \leq \frac{1 + R - (1 - R)\exp(-2c/\sqrt{n - 3})}{1 + R + (1 - R)\exp(-2c/\sqrt{n - 3})}\right) = \gamma.$$

Example 9.7–2. Suppose that a random sample of size 12 from a bivariate normal distribution yielded a correlation coefficient of $r = 0.6$. An approximate 95% confidence interval for ρ would be

$$\left[\frac{1 + 0.6 - (1 - 0.6)\exp[2(1.96)/3]}{1 + 0.6 + (1 - 0.6)\exp[2(1.96)/3]}, \frac{1 + 0.6 - (1 - 0.6)\exp[-2(1.96)/3]}{1 + 0.6 + (1 - 0.6)\exp[-2(1.96)/3]}\right]$$

$$= [0.040, 0.873].$$

If the sample size had been $n = 39$, the approximate 95% confidence interval would have been $[0.351, 0.770]$.

═══ *Exercises* ═══

9.7–1. A random sample of size $n = 27$ from a bivariate normal distribution yielded a sample correlation coefficient of $r = -0.45$. Would the hypothesis $H_0 : \rho = 0$ be rejected in favor of $H_1 : \rho \neq 0$ at an $\alpha = 0.05$ significance level?

9.7–2. When bowling it is often possible to score well on the first game and then bowl poorly on the second game, or vice versa. The following six pairs of numbers give the scores of the first and second games bowled by the same person on six consecutive Tuesday evenings. Assume a bivariate normal distribution and use these scores to test the hypothesis $H_0 : \rho = 0$ against $H_1 : \rho \neq 0$ at $\alpha = 0.10$.

Game 1: 170	190	200	183	187	178
Game 2: 197	178	150	176	205	153

9.7–3. A random sample of size 28 from a bivariate normal distribution yielded a sample correlation coefficient of $r = 0.65$. Find an approximate 90% confidence interval for ρ.

9.7–4. For each part of Example 9.6–3, find an approximate 95% confidence interval for ρ.

10

CHI–SQUARE TESTS OF MODELS

10.1 The Basic Chi-Square Statistic

In this chapter we consider applications of the very important chi-square statistic first proposed by Karl Pearson in 1900. As the reader will see, it is a very adaptable test statistic and can be used for many different types of tests. In particular, one application of it allows us to test the appropriateness of different probabilistic models.

We begin our study by considering the basic chi-square statistic, which, interestingly enough, has only an approximate chi-square distribution. It is based upon some random variables $Y_1, Y_2, \ldots, Y_k$ that have a multinomial distribution with parameters n and $p_1, p_2, \ldots, p_k$. So that the reader can get some idea why Pearson first proposed it, we start with the binomial case.

Let Y_1 be $b(n, p_1)$, where $0 < p_1 < 1$. According to the Central Limit Theorem,

$$Z = \frac{Y_1 - np_1}{\sqrt{np_1(1 - p_1)}}$$

has a distribution that is approximately $N(0, 1)$ for large n, particularly when $np_1 \geq 5$ and $n(1 - p_1) \geq 5$. Thus, it is not surprising that $Q_1 = Z^2$ is approximately $\chi^2(1)$. If we let $Y_2 = n - Y_1$ and $p_2 = 1 - p_1$, we see that Q_1 may be written as

$$Q_1 = \frac{(Y_1 - np_1)^2}{np_1(1 - p_1)} = \frac{(Y_1 - np_1)^2}{np_1} + \frac{(Y_1 - np_1)^2}{n(1 - p_1)};$$

or, equivalently, since

$$(Y_1 - np_1)^2 = (n - Y_2 - n + np_2)^2 = (Y_2 - np_2)^2,$$

$$Q_1 = \frac{(Y_1 - np_1)^2}{np_1} + \frac{(Y_2 - np_2)^2}{np_2}.$$

Let us now carefully consider each term in this last expression for Q_1. Of course, Y_1 is the number of "successes" and np_1 is the expected number of "successes"; that is, $E(Y_1) = np_1$. Likewise, Y_2 and np_2 are, respectively, the number and the expected number of "failures." So each consists of the square of the difference of the number and expected number. In this case, Q_1 can be written as

$$Q_1 = \sum_{i=1}^{2} \frac{(Y_i - np_i)^2}{np_i},$$

and we have seen intuitively that it has an approximate chi-square distribution with one degree of freedom. In a sense, Q_1 measures the "closeness" of the observed numbers to the corresponding expected numbers. For example, if the observed values of Y_1 and Y_2 equal their expected values, then the computed Q_1 is equal to $q_1 = 0$; but if they differ much from them, then the computed $Q_1 = q_1$ is relatively large.

To generalize, we let an experiment have k (instead of only two) mutually exclusive and exhaustive outcomes, say $A_1, A_2, \ldots, A_k$. Let $p_i = P(A_i)$ and thus $\sum_{i=1}^{k} p_i = 1$. The experiment is repeated n independent times, and we let Y_i represent the number of times the experiment results in A_i, $i = 1, 2, \ldots, k$. Then $Y_1, Y_2, \ldots, Y_k$ have a multinomial distribution with parameters $n, p_1, p_2, \ldots, p_k$. Say Y_i is observed to be equal to y_i. A statistic that measures the "closeness" of the observed number y_i of occurrences A_i to the expected number np_i, $i = 1, 2, \ldots, k$, is

$$q_{k-1} = \sum_{i=1}^{k} \frac{(y_i - np_i)^2}{np_i}.$$

The corresponding random variable, namely,

$$Q_{k-1} = \sum_{i=1}^{k} \frac{(Y_i - np_i)^2}{np_i},$$

can be shown to have, for large n, an approximate chi-square distribution with $k - 1$ degrees of freedom. The proof of this is beyond the level of this text; however, it is based upon the fact that $Y_1, Y_2, \ldots, Y_{k-1}$ have an approximate multivariate normal distribution. That is, it is similar to the argument given in the case $k = 2$ in which Y_1 has an approximate normal distribution with mean np_1 and variance $np_1(1 - p_1)$.

Some writers suggest that n should be large enough so that $np_i \geq 5$, $i = 1, 2, \ldots, k$, to be certain that the approximating distribution is adequate. This is probably good advice for the beginner to follow, although we have seen the approximation work very well when $np_i \geq 1, i = 1, 2, \ldots, k$. The important thing to guard against is allowing some particular np_i to become so small that the corresponding term in q_{k-1}, namely, $(y_i - np_i)^2/np_i$, tends to dominate the others because of its small denominator. In any case it is important to realize that Q_{k-1} has only an approximate chi-square distribution.

We shall now show how we can use the fact that Q_{k-1} is approximately $\chi^2(k - 1)$ to test hypotheses about probabilities of various outcomes. Let an experiment have k mutually exclusive and exhaustive outcomes, $A_1, A_2, \ldots, A_k$. We would like to test whether $p_i = P(A_i)$ is equal to a known number π_i, $i = 1, 2, \ldots, k$. That is, we shall test the hypothesis

$$H_0: \quad p_i = \pi_i, \qquad i = 1, 2, \ldots, k.$$

In order to test such a hypothesis, we shall take a sample of size n, that is, repeat the experiment n independent times. We tend to favor H_0 if the observed number of times that A_i occurred, say y_i, and the number of times A_i was expected to occur if H_0 were true, namely $n\pi_i$, are approximately equal. That is, if

$$q_{k-1} = \sum_{i=1}^{k} \frac{(y_i - n\pi_i)^2}{n\pi_i}$$

is "small," we tend to favor H_0. What is needed is a table that helps us decide what values of q_{k-1} are "small." This is where we make use of the fact that the corresponding Q_{k-1} is approximately $\chi^2(k - 1)$ when H_0 is true. A number c is selected from the chi-square table with $k - 1$ degrees of freedom so that $P(Q_{k-1} > c) = \alpha$. We shall accept H_0 if $q_{k-1} \leq c$; otherwise we reject H_0. Note that it is possible to make the incorrect decision. That is, if H_0 is true, it is possible for $q_{k-1} > c$. However, since $P(Q_{k-1} > c) = \alpha$, this probability α is the significance level of the test. The significance level can be decreased by decreasing α. However, as α decreases, c increases. If c is too large, it is possible that the test would not detect when H_0 is false; that is, the probability of the Type II error would increase.

Example 10.1–1. Say we wish to test whether a particular die is unbiased. That is, if $A_i = \{i\}, i = 1, 2, \ldots, 6$, we wish to test $H_0: p_i = 1/6, i = 1, 2, \ldots, 6$, against all possible alternatives. Here, in the notation of our discussion, each given π_i is equal to $1/6$. To perform this test, we decide to cast the die 120 independent times and record the numbers of ones, twos, ..., sixes, say $Y_1, Y_2, \ldots, Y_6$. Then $Q_5 = \sum_1^6 (Y_i - 20)^2/20$ because $n\pi_i = 120(1/6) = 20$, $i = 1, 2, \ldots, 6$. If H_0 is true, $P(Q_5 > 11.07) = 0.05$, approximately. Suppose the experimental Y values are $y_1 = 26$, $y_2 = 13$, $y_3 = 21$, $y_4 = 29$, $y_5 = 20$, $y_6 = 11$. Then the observed value of Q_5 is

$$q_5 = \frac{(26-20)^2}{20} + \frac{(13-20)^2}{20} + \cdots + \frac{(11-20)^2}{20}$$

$$= \frac{36+49+1+81+0+81}{20} = \frac{248}{20} = 12.4.$$

Since $12.4 > 11.07$, the hypothesis that the die is unbiased, namely, $H_0 : p_i = 1/6$, is rejected at $\alpha = 0.05$.

Example 10.1–2. If persons are asked to record a string of random digits, such as

3, 7, 2, 4, 1, 9, 7, 2, 1, 5, 0, 8, 4, 6, 1, 9, 6, ...,

we usually find that they are reluctant to record the same or even the two closest numbers in adjacent positions. And yet, in true random number generation, the probability of the next digit being the same as the preceding one is 1/10 and the probability of the next being only one away from the preceding is 2/10 (assuming 0 is one away from 9). To test one person's concept of a random sequence, we asked him to record a string of 51 digits that seems to represent a random generation. This resulted in

5, 8, 3, 1, 9, 4, 6, 7, 9, 2, 6, 3, 0, 8, 7, 5, 1, 3,
6, 2, 1, 9, 5, 4, 8, 0, 3, 7, 1, 4, 6, 0, 4, 3, 8, 2,
1, 3, 9, 8, 5, 6, 1, 8, 7, 0, 3, 5, 2, 5, 2.

We went through this listing and observed how many times the next digit was the same as or was one away from the preceding one.

	FREQUENCY	EXPECTED NUMBER
Same	0	50(1/10) = 5
One away	8	50(2/10) = 10
Other	42	50(7/10) = 35
Total	50	50

The computed chi-square statistic is

$$\frac{(0-5)^2}{5} + \frac{(8-10)^2}{10} + \frac{(42-35)^2}{35} = 6.8 > 5.991,$$

where 5.991 is the rejection number associated with $k - 1 = 3 - 1 = 2$ degrees of freedom and $\alpha = 0.05$. Thus, we would say that this string of 51 digits does not seem to be random.

One major disadvantage in the use of the chi-square test is that it is a many-sided test. That is, the alternative hypothesis is very general, and it would be difficult to restrict alternatives to situations such as, with $k = 3$, $H_1: p_1 > \pi_1$, $p_2 > \pi_2, p_3 < \pi_3$. As a matter of fact, some statisticians would probably test H_0 against this particular alternative H_1 by using a linear function of Y_1, Y_2, and Y_3. However, this sort of discussion is beyond the scope of this text because it involves knowing more about the distributions of linear functions of the dependent random variables Y_1, Y_2, and Y_3. In any case, the student who truly recognizes that this chi-square statistic tests $H_0: p_i = \pi_i, i = 1, 2, \ldots, k$ against *all* alternatives can usually appreciate the fact that it is more difficult to reject H_0 at a given significance level α using the chi-square statistic than it would be if some appropriate "one-sided" test statistic were available.

═══ *Exercises* ═══

10.1–1. It is suspected that a certain die is not fair. Let p_i denote the probability that the outcome, when rolling this die, is equal to $i, i = 1, 2, 3, \ldots, 6$. We shall test the hypothesis

$$H_0: \quad p_1 = p_2 = p_3 = p_4 = p_5 = p_6 = \frac{1}{6}$$

against all alternatives. If the experimental frequencies of the six possible outcomes in 60 independent trials are 4, 14, 10, 13, 5, and 14, respectively, is H_0 rejected at the approximate 5% significance level?

10.1–2. A bowl contains red, white, blue, green, and yellow chips. Let p_R denote the probability of drawing a red chip from the bowl, etc. We shall test the hypothesis

$$H_0: \quad p_R = p_W = 0.05, p_B = p_G = 0.25, p_Y = 0.40$$

at the approximate 5% significance level. Is H_0 rejected if a random sample of size 200, sampling with replacement, yielded 15 red chips, 9 white chips, 46 blue chips, 44 green chips, and 86 yellow chips?

10.1–3. In a biology laboratory the mating of two red-eyed fruit flies yielded $n = 432$ offspring, among which 254 were red-eyed, 69 were brown-eyed, 87 were scarlet-eyed, and 22 were white-eyed. Use these data to test, with $\alpha = 0.05$, the hypothesis that the ratio among the offspring would be $9 : 3 : 3 : 1$, respectively.

10.1–4. In a large bin of crocus bulbs it is claimed that 1/4 will produce yellow crocuses, 1/4 will produce white crocuses, and 1/2 will produce purple crocuses. If 40 bulbs produced 6 yellow, 7 white, and 27 purple crocuses, would the claim be rejected?

10.1–5. The college registrar was interested in determining whether class offerings were uniformly distributed over the first eight periods of the day on Monday, Wednesday, and Friday. Test the null hypothesis that class offerings are uniformly distributed at an $\alpha = 0.05$ significance level using the following data.

PERIOD	NUMBER OF CLASSES	PERIOD	NUMBER OF CLASSES
1	35	5	31
2	46	6	54
3	47	7	45
4	33	8	29

10.2 Testing Probabilistic Models

Many experiments yield a set of data, say $x_1, x_2, \ldots, x_n$; and the experimenter is often interested in determining whether these data can be treated as the observed values of a random sample $X_1, X_2, \ldots, X_n$ from a given distribution. That is, would this proposed distribution be a reasonable probabilistic model for these sample items? To see how the chi-square test can help us answer questions of this sort, consider a very simple example.

Example 10.2–1. Let X denote the number of heads that occur when four coins are tossed at random. Under the assumptions that the four coins are independent and the probability of heads on each coin is 1/2, X is $b(4, 1/2)$. One hundred repetitions of this experiment resulted in 0, 1, 2, 3, and 4 heads being observed on 7, 18, 40, 31, and 4 trials, respectively. Do these results support the assumptions? That is, is $b(4, 1/2)$ a reasonable model for the distribution of X? To answer this, we begin by letting $A_1 = \{0\}$, $A_2 = \{1\}$, $A_3 = \{2\}$, $A_4 = \{3\}$, $A_5 = \{4\}$. If $\pi_i = P(X \in A_i)$ when X is $b(4, 1/2)$,

$$\pi_1 = \pi_5 = 0.0625 = \binom{4}{0}\left(\frac{1}{2}\right)^4,$$

$$\pi_2 = \pi_4 = 0.25 = \binom{4}{1}\left(\frac{1}{2}\right)^4,$$

$$\pi_3 = 0.375 = \binom{4}{2}\left(\frac{1}{2}\right)^4.$$

At an approximate $\alpha = 0.05$ significance level, the null hypothesis

$$H_0: \quad p_1 = \pi_1, p_2 = \pi_2, p_3 = \pi_3, p_4 = \pi_4, p_5 = \pi_5$$

is rejected if the observed value of Q_4 is greater than $c = 9.488$. Using the 100 repetitions of this experiment that resulted in the observed values of $Y_1, Y_2, \ldots, Y_5$ of $y_1 = 7$, $y_2 = 18$, $y_3 = 40$, $y_4 = 31$, and $y_5 = 4$, the computed value of Q_4 is

$$q_4 = \frac{(7 - 6.25)^2}{6.25} + \frac{(18 - 25)^2}{25} + \frac{(40 - 37.5)^2}{37.5} + \frac{(31 - 25)^2}{25} + \frac{(4 - 6.25)^2}{6.25}$$

$$= 4.47.$$

Since $4.47 < 9.488$, the hypothesis is not rejected. That is, the data support the hypothesis that $b(4, 1/2)$ is a reasonable probabilistic model for X.

In Section 10.1 and in Example 10.2–1, all the hypotheses H_0 tested with the chi-square statistic Q_{k-1} have been simple ones (that is, completely specified, namely, in $H_0: p_i = \pi_i$, $i = 1, 2, \ldots, k$, each π_i has been a known fraction). This is not always the case, and it frequently happens that $\pi_1, \pi_2, \ldots, \pi_k$ are functions of one or more unknown parameters. For example, suppose that the hypothesized model for X in Example 10.2–1 was $H_0: X$ is $b(4, p)$, $0 < p < 1$. Then

$$\pi_i = P(X \in A_i) = \frac{4!}{(i - 1)!(5 - i)!} p^{i-1}(1 - p)^{5-i}, \qquad i = 1, 2, \ldots, 5,$$

which is a function of the unknown parameter p. Of course, if $H_0: p_i = \pi_i$, $i = 1, 2, \ldots, 5$, is true, for large n,

$$Q_4 = \sum_{i=1}^{5} \frac{(Y_i - n\pi_i)^2}{n\pi_i}$$

still has an approximate chi-square distribution with four degrees of freedom. The difficulty is that when $Y_1, Y_2, \ldots, Y_5$ are observed to be equal to $y_1, y_2, \ldots, y_5$, Q_4 cannot be computed since $\pi_1, \pi_2, \ldots, \pi_5$ (and hence Q_4) are functions of the unknown parameter p.

One way out of the difficulty would be to estimate p from the data and then carry out the computations using this estimate. It is interesting to note the following. Say the estimation of p is carried out by minimizing Q_4 with respect to p yielding $\tilde{p}$. This $\tilde{p}$ is sometimes called a *minimum chi-square* estimator of p. If then this $\tilde{p}$ is used in Q_4, the statistic Q_4 still has an approximate chi-square distribution but with only $4 - 1 = 3$ degrees of freedom. That is, the number of degrees of freedom of the approximating chi-square distribution is reduced by one for each parameter estimated by the minimum chi-square technique. We accept this result without proof (as it is a rather difficult one). Although we have considered this when π_i, $i = 1, 2, \ldots, k$, is a function of only one parameter, it holds when there is more than one unknown parameter, say r. Hence, in a more general situation, the test would be completed by computing Q_{k-1} using y_i and the estimated π_i, $i = 1, 2, \ldots, k$, to obtain q_{k-1} (that is, q_{k-1} is the minimized chi-square). This value q_{k-1} would then be compared to a critical value c taken from a chi-square table with $k - 1 - r$ degrees of freedom. In our special case, the computed (minimized) chi-square q_4 would be compared to a c from Appendix Table III with three degrees of freedom.

There is still one trouble with all of this: it is usually very difficult to find minimum chi-square estimators. Hence, most statisticians usually use some reasonable method (maximum likelihood is satisfactory) of estimating the parameters. They then compute q_4, recognizing that it is somewhat larger than the minimized chi-square, and compare it with a critical value c from a chi-square distribution with $k - 1 - r$ degrees of freedom. Note that this provides a slightly larger probability of rejecting H_0 than would the scheme in which the minimized chi-square were used.

Example 10.2–2. Let X denote the number of alpha particles emitted by barium-133 in 1/10 of a second. The following 50 observations of X were taken with a Geiger counter in a fixed position:

7	4	3	6	4	4	5	3	5	3
5	5	3	2	5	4	3	3	7	6
6	4	3	11	9	6	7	4	5	4
7	3	2	8	6	7	4	1	9	8
4	8	9	3	9	7	7	9	3	10

The experimenter is interested in determining whether X has a Poisson distribution. To test H_0: X is Poisson, we first estimate the mean of X, say λ, with the sample mean, $\bar{x} = 5.4$, of these 50 observations. We then partition the set of outcomes for this experiment into the sets $A_1 = \{0, 1, 2, 3\}$, $A_2 = \{4\}$, $A_3 = \{5\}$, $A_4 = \{6\}$, $A_5 = \{7\}$, and $A_6 = \{8, 9, 10, \ldots\}$. We combine 0, 1, 2, 3 into one set A_1 and 8, 9, 10, ... into another A_6 so that the expected number of outcomes for each set is at least five when H_0 is true. In Table 10.2–1 the data are grouped, and the estimated probabilities specified by the hypothesis that Y has a Poisson distribution with an estimated $\lambda = 5.4$ are given. Since one parameter was estimated, Q_{6-1} has an approximate chi-square distribution with $r = 5 - 1 = 4$ degrees of freedom.

	OUTCOME					
	A_1	A_2	A_3	A_4	A_5	A_6
Frequency	13	9	6	5	7	10
Probability	0.213	0.160	0.173	0.156	0.120	0.178

TABLE 10.2–1

Since

$$q_5 = \frac{[13 - 50(0.213)]^2}{50(0.213)} + \cdots + \frac{[10 - 50(0.178)]^2}{50(0.178)}$$

$$= 2.763 < 9.488,$$

H_0 is not rejected at the 5% significance level. That is, with only these data, we are quite willing to accept the model that X has a Poisson distribution.

Let us now consider the problem of testing a model for the distribution of a random variable W of the continuous type. That is, if $F(w)$ is the distribution function of W, we wish to test

$$H_0: \quad F(w) = F_0(w),$$

where $F_0(w)$ is some known distribution function of the continuous type. In order to use the chi-square statistic, we must partition the set of possible values of W into k sets. One way this can be done is as follows. Partition the interval $[0, 1]$ into k sets with the points $b_0, b_1, b_2, \ldots, b_k$, where

$$0 = b_0 < b_1 < b_2 < \cdots < b_k = 1.$$

Let $a_i = F_0^{-1}(b_i)$, $i = 1, 2, \ldots, k - 1$; $A_1 = (-\infty, a_1]$, $A_i = (a_{i-1}, a_i]$ for $i = 2, 3, \ldots, k - 1$, and $A_k = (a_{k-1}, \infty)$; $p_i = P(W \in A_i)$, $i = 1, 2, \ldots, k$. Let Y_i denote the number of times the observed value of W belongs to $A_i, i = 1, 2, \ldots, k$, in n independent repetitions of the experiment. Then $Y_1, Y_2, \ldots, Y_k$ have a multinomial distribution with parameters $n, p_1, p_2, \ldots, p_k$. Let $\pi_i = P(W \in A_i)$ when the distribution function of W is $F_0(w)$. The hypothesis that we actually test is a modification of H_0, namely

$$H_0': \quad p_i = \pi_i, \quad i = 1, 2, \ldots, k.$$

This hypothesis is rejected if the observed value of the chi-square statistic

$$Q_{k-1} = \sum_{i=1}^{k} \frac{(Y_i - n\pi_i)^2}{n\pi_i}$$

is at least as great as c, where c is selected to yield the desired significance level. If the hypothesis $H_0': p_i = \pi_i$, $i = 1, 2, \ldots, k$, is not rejected, we do not reject the hypothesis $H_0: F(w) = F_0(w)$.

Example 10.2–3. Let W denote the outcome of a random experiment. Let $F(w)$ denote the distribution function of W and let

$$F_0(w) = \begin{cases} 0, & w < -1, \\ \dfrac{1}{2}(w^3 + 1), & -1 \le w < 1, \\ 1, & 1 \le w. \end{cases}$$

We shall test the hypothesis $H_0 : F(w) = F_0(w)$ against all alternatives. The interval $[0, 1]$ can be partitioned into 10 sets of equal probability with the points $b_i = i/10$, $i = 0, 1, \ldots, 10$. If $a_i = F_0^{-1}(b_i) = (2b_i - 1)^{1/3}$, $i = 1, 2, \ldots, 9$, then the sets $A_1 = [-1, a_1], A_2 = (a_1, a_2], \ldots, A_{10} = (a_9, 1]$ will each have probability $1/10$ when H_0 is true. If a random sample of size $n = 50$ is observed, then $50(1/10) = 5$ is the expected number of outcomes for each set A_i, $i = 1, 2, \ldots, 10$. Letting Y_i denote the number of outcomes in A_i, we see that the hypothesis is rejected if the calculated value of

$$Q_9 = \sum_{i=1}^{10} \frac{(Y_i - 5)^2}{5}$$

is greater than 16.92 at an approximate 5 % significance level. The outcomes of 50 independent repetitions of this experiment simulated on the computer were

−0.75	−0.28	−0.59	0.26	−0.89	−0.60	0.63
−0.68	−0.62	0.75	−0.81	0.99	0.76	−0.88
0.35	−0.87	0.68	0.81	−0.82	−0.90	0.65
0.43	0.56	0.62	0.96	0.72	−0.67	−0.40
−0.97	−0.96	0.83	0.90	0.83	−0.55	0.69
0.75	0.91	0.88	−0.70	0.96	−0.98	−0.93
−0.77	−0.96	−0.29	0.83	−0.93	0.96	0.79
−0.76						

A summary of these data is given in Table 10.2–2. The values of $a_0, a_1, \ldots, a_{10}$ are -1.00, -0.928, -0.843, -0.737, -0.585, 0.00, 0.585, 0.737, 0.843, 0.928, and 1.00, respectively. The calculated value of Q_9 is

$$q_9 = \frac{(6 - 5)^2}{5} + \frac{(4 - 5)^2}{5} + \cdots + \frac{(4 - 5)^2}{5} = 4.0 \le 16.92,$$

and thus the hypothesis is not rejected.

OUTCOME	FREQUENCY	OUTCOME	FREQUENCY
A_1	6	A_6	4
A_2	4	A_7	6
A_3	5	A_8	8
A_4	6	A_9	3
A_5	4	A_{10}	4

TABLE 10.2–2

Let $X_1, X_2, \ldots, X_n$ denote a random sample of size n from a distribution with mean μ and positive variance σ^2. Let $W = (\bar{X} - \mu)/(\sigma/\sqrt{n})$, where $\bar{X}$ is the sample mean. From the Central Limit Theorem, we claim that W has a distribution that is approximately $N(0, 1)$ when n is sufficiently large. For what values of n can we say that W is approximately $N(0, 1)$? The answer will, of course, depend upon the distribution from which the sample is taken. For example, if the underlying distribution is normal, $n = 1$ is sufficient. In the next example, we shall give a partial answer for the U-shaped distribution used in Example 10.2–3.

Example 10.2–4. Let $X_1, X_2, \ldots, X_n$ denote a random sample of size n from the distribution with p.d.f. $f(x) = (3/2)x^2$, $-1 \leq x \leq 1$. For this distribution $\mu = 0$ and $\sigma^2 = 3/5$. We shall test the hypothesis that $W = \overline{X}/\sqrt{3/(5n)}$ is $N(0, 1)$ when $n = 6$. One hundred observations of W were generated by the computer. Table 10.2–3 gives eight sets, their frequencies, and their probabilities when sampling from $N(0, 1)$. For these data,

$$q_7 = \frac{(1 - 3.59)^2}{3.59} + \frac{(13 - 7.92)^2}{7.92} + \cdots + \frac{(2 - 3.59)^2}{3.59} = 11.44.$$

Since $11.44 < 14.07$, the hypothesis is not rejected at the 5% significance level. That is, even though we know that W is not exactly $N(0, 1)$, its distribution is close enough to that of $N(0, 1)$ so that we can accept the approximation suggested by the Central Limit Theorem even though n is as small as 6.

A_i	Y_i	π_i
$(-\infty, -1.80]$	1	0.0359
$(-1.80, -1.20]$	13	0.0792
$(-1.20, -0.60]$	18	0.1592
$(-0.60, \ 0.00]$	13	0.2257
$(\ 0.00, \ 0.60]$	24	0.2257
$(\ 0.60, \ 1.20]$	20	0.1592
$(\ 1.20, \ 1.80]$	9	0.0792
$(\ 1.80, +\infty)$	2	0.0359

TABLE 10.2–3

It is also true, in dealing with models of random variables of the continuous type, that we must frequently estimate unknown parameters. For example, let H_0 be that W is $N(\mu, \sigma^2)$, where μ and σ^2 are unknown. With a random sample $W_1, W_2, \ldots, W_n$, we first can estimate μ and σ^2, possibly with $\overline{W}$ and S_W^2. We partition the space $\{w : -\infty < w < \infty\}$ into k mutually disjoint sets $A_1, A_2, \ldots, A_k$. We then use the estimated μ and σ^2 to estimate

$$\pi_i = \int_{A_i} \frac{1}{\sigma\sqrt{2\pi}} \exp\left[-\frac{(w - \mu)^2}{2\sigma^2}\right] dw,$$

$i = 1, 2, \ldots, k$. Using the observed frequencies $y_1, y_2, \ldots, y_k$ of $A_1, A_2, \ldots, A_k$, respectively, from the observed random sample $w_1, w_2, \ldots, w_n$, and $\pi_1, \pi_2, \ldots, \pi_k$ estimated with $\bar{w}$ and s_W^2, we compare the computed

$$q_{k-1} = \sum_{i=1}^{k} \frac{(y_i - n\pi_i)^2}{n\pi_i}$$

to a critical c found in the chi-square table with $k - 1 - 2$ degrees of freedom. This value q_{k-1} will again be somewhat larger than that which would be found using minimum chi-square estimation and certain caution should be observed. Exercises 10.2–4 and 10.2–6 illustrate the procedure in which one or more parameters must be estimated.

Finally, it should be noted that the methods given in this section frequently are classified under the more general title of *goodness of fit tests*. In particular, then, the tests in this section would be *chi-square goodness of fit tests*.

=== *Exercises* ===

10.2–1. Let X denote the outcome of a random experiment. We shall test the hypothesis that X has a Poisson distribution with mean 4. The experiment was repeated 200 independent times, yielding the frequencies in Table 10.2–4. Do these data support the null hypothesis at the 1% significance level? HINT: You might want to combine certain outcomes to create the extreme sets A_1 and A_k. Why?

OUTCOME	FREQUENCY	OUTCOME	FREQUENCY
0	3	6	14
1	9	7	13
2	36	8	7
3	33	9	0
4	47	10	2
5	36		

TABLE 10.2–4

10.2–2. Test whether the following data represent the values of a random sample from a Poisson distribution. Use a 5% significance level. Use $\bar{x}$ to estimate the mean λ.

6	3	6	4	2	5	4	6	3	5
3	4	5	5	3	9	8	3	5	6
6	7	8	4	5	7	4	3	3	1
8	6	5	7	5	5	5	6	9	2
7	1	3	7	2	9	6	4	3	7

10.2–3. Appendix Table VIII lists a set of numbers between 0 and 1. Use a chi-square goodness of fit test to test the hypothesis that these numbers give the values of a random sample of size 315 from a uniform distribution on $(0, 1)$. Let $A_i = [(i - 1)/10, i/10)$, $i = 1, 2, \ldots, 10$. Let $\alpha = 0.05$.

10.2–4. Test at the 10% significance level the hypothesis that the following data give the values of a random sample of size 50 from an exponential distribution with p.d.f. $f(x; \theta) = (1/\theta)e^{-x/\theta}$, $0 < x < \infty$, where $\theta > 0$. HINT: Estimate θ and use about 6 to 10 mutually disjoint sets.

6.88	6.92	4.80	9.85	7.05
19.06	6.54	3.67	2.94	4.89
69.82	6.92	4.34	13.45	5.74
10.07	16.91	7.47	5.04	7.97
15.74	0.32	4.14	5.19	18.69
2.45	23.69	44.10	1.70	2.14
5.79	3.02	9.87	2.44	18.99
18.90	5.42	1.54	1.55	20.99
7.99	5.38	2.36	9.66	0.97
4.82	10.43	15.06	0.49	2.81

10.2–5. It is claimed that SAT scores in mathematics for a certain population of students have a normal distribution $N(500, 6400)$. Test this claim using the following sample of 100 scores. Let $\alpha = 0.05$.

479	525	597	571	591	504	310	615	361	518
486	413	407	545	520	539	479	519	657	460
406	528	551	572	507	492	639	571	379	452
453	538	683	512	381	499	470	549	591	502
533	479	543	492	504	556	604	545	433	407
424	578	518	499	556	446	551	465	625	400
449	431	492	690	520	432	440	571	525	460
479	525	491	558	335	499	315	512	440	584
445	528	597	492	492	524	663	611	525	464
427	624	403	466	571	446	558	565	525	427

10.2–6. Test whether the following data give the values of a random sample of size 100 from a distribution $N(\mu, \sigma^2)$. Let $\alpha = 0.05$. HINT: Estimate μ and σ^2 with $\bar{w} = 50.24$ and $s_W^2 = 12.52$, respectively.

48.74	51.35	56.21	51.32	48.68	52.29	59.15	58.26
56.62	49.23	52.09	41.20	48.56	54.17	46.03	48.13
52.49	51.10	51.96	47.07	52.43	56.04	53.90	50.01
52.73	56.98	47.84	40.95	48.31	49.92	45.78	51.88
48.24	50.85	47.71	46.82	48.18	51.79	49.65	53.76
52.12	48.73	47.59	52.70	48.06	49.67	53.53	51.63
47.99	50.60	47.46	42.57	47.93	43.54	53.40	53.51
55.87	48.48	47.34	48.85	51.81	57.42	53.28	51.38
51.74	46.35	55.21	50.32	43.68	51.29	49.15	49.26
51.62	52.23	51.09	52.20	43.56	57.17	53.03	51.13
47.49	50.10	46.96	46.07	51.43	51.04	52.90	45.01
51.37	47.98	50.84	51.95	55.31	52.92	48.78	46.88
51.24	49.85	50.71	41.82				

10.3 Additional Models

The binomial, Poisson, gamma, chi-square, and normal models are certainly the most frequently used models in statistics. However, many other interesting and very useful models can be found by modifying one of the postulates of an approximate Poisson process as given in Section 2.7. In that definition, the numbers of changes occurring in nonoverlapping intervals were independent and the probability of at least two changes in a sufficiently small interval is essentially zero. We continue to use these postulates, but now we say that the probability of exactly one change in a sufficiently short interval of length h is approximately λh, *where λ is a nonnegative function of the position of this interval.* To be explicit, say $p(x, w)$ is the probability of x changes in the interval $(0, w)$, $0 \leq w$. Then, the last postulate, in more formal terms, becomes

$$p(x + 1, w + h) - p(x, w) \approx \lambda(w)h,$$

where $\lambda(w)$ is a nonnegative function of w. This means that if we want the approximate probability of zero changes in the interval $(0, w + h)$ we could take, from the independence, the probability of zero changes in the interval $(0, w)$ times that of zero changes in the interval $(w, w + h)$. That is,

$$p(0, w + h) \approx p(0, w)[1 - \lambda(w)h]$$

because the probability of one or more changes in $(w, w + h)$ is about equal to $\lambda(w)h$. Equivalently,

$$\frac{p(0, w + h) - p(0, w)}{h} \approx -\lambda(w)p(0, w).$$

Taking limits as $h \to 0$, we have

$$D_w[p(0, w)] = -\lambda(w)p(0, w).$$

That is, the resulting differential equation is

$$\frac{D_w[p(0, w)]}{p(0, w)} = -\lambda(w);$$

and, thus,

$$\ln p(0, w) = -\int \lambda(w)\, dw + c_1.$$

Therefore,

$$p(0, w) = \exp\left[-\int \lambda(w)\, dw + c_1\right] = c_2 \exp\left[-\int \lambda(w)\, dw\right],$$

where $c_2 = e^{c_1}$. However, the boundary condition of the probability of zero changes in an interval of length zero must be one; that is, $p(0, 0) = 1$. So if we select

$$H(w) = \int \lambda(w)\, dw$$

to be such that $H(0) = 0$, then $c_2 = 1$. That is,

$$p(0, w) = e^{-H(w)},$$

where $H'(w) = \lambda(w)$ and $H(0) = 0$. Hence,

$$H(w) = \int_0^w \lambda(t)\, dt.$$

Suppose that we now let the continuous-type random variable W be the interval necessary to produce the first change, then the distribution function of W is

$$G(w) = P(W \le w) = 1 - P(W > w), \qquad 0 \le w.$$

Because zero changes in the interval $(0, w)$ are the same as $W > w$, then

$$G(w) = 1 - p(0, w) = 1 - e^{-H(w)}, \qquad 0 \le w.$$

The p.d.f. of W is

$$g(w) = G'(w) = H'(w)e^{-H(w)} = \lambda(w) \exp\left[-\int_0^w \lambda(t)\, dt \right], \qquad 0 \le w.$$

In many applications of this result, W can be thought of as a random time interval. For example, if one change means "death" or "failure" of the item under consideration, then W is actually the length of life of the item. Usually $\lambda(w)$, which is commonly called the *failure rate* or *force of mortality*, is an increasing function of w. That is, the larger w (the older the item) is, the better chance of failure within a short interval of length h, namely, $\lambda(w)h$. As we review the exponential distribution of Section 4.3, we note that there $\lambda(w)$ is a constant; that is, the failure rate or force of mortality does not increase as the item gets older. If this were true in human populations, it would mean that a person 80 years old would have as much chance of living another year as would a person 20 years old (sort of a mathematical "fountain of youth"). However, a constant failure rate (force of mortality) is not the case in most human populations nor is it usual in populations of manufactured items. That is, the failure rate $\lambda(w)$ is usually an increasing function of w. We give two important examples of useful probabilistic models.

Example 10.3–1.   Let the failure rate be a power function,

$$H(w) = \left(\frac{w}{\beta}\right)^{\alpha}, \qquad 0 \le w,$$

where $\alpha > 0$, $\beta > 0$. Then the p.d.f. of W is

$$g(w) = \frac{\alpha w^{\alpha-1}}{\beta^{\alpha}} \exp\left[-\left(\frac{w}{\beta}\right)^{\alpha} \right], \qquad 0 \leq w.$$

Frequently, in engineering, this distribution, with appropriate values of α and β, is excellent for describing the life of a manufactured item. Often α is greater than one but less than five. This p.d.f. is frequently called that of the *Weibull distribution* and, in model fitting, is a strong competitor to the gamma p.d.f.

Example 10.3–2.   Persons are often shocked to learn that human mortality increases almost exponentially once a person reaches 25 years of age. Depending upon which mortality table is used, one finds that this increase is about 10% each year, which means that the rate of mortality will double about every 7 years. Although this fact can be shocking, we can be thankful that the force of mortality starts very low. The probability that a person in reasonably good health at age 52 dies within the next year is only about 1%. Now, assuming an exponential force of mortality, we have

$$\lambda(w) = H'(w) = ae^{bw}, \qquad a > 0, \quad b > 0.$$

Thus,

$$H(w) = \frac{a}{b} e^{bw} + c.$$

However, $H(0) = 0$; so $c = -a/b$. Thus,

$$G(w) = 1 - \exp\left[-\frac{a}{b} e^{bw} + \frac{a}{b} \right], \qquad 0 \leq w$$

and

$$g(w) = ae^{bw} \exp\left[-\frac{a}{b} e^{bw} + \frac{a}{b} \right], \qquad 0 \leq w$$

are the distribution function and p.d.f. associated with the famous Gompertz law found in actuarial science.

Both the gamma and Weibull distributions are skewed. In many studies (life testing, response times, incomes, etc.) these are valuable distributions for model selection. Another attractive one is called the *lognormal distribution*.

Example 10.3–3.   Say that X is $N(\mu, \sigma^2)$. If we let $W = e^X$, then the distribution function of W is

$$G(w) = P(W \leq w) = P(e^X \leq w) = P(X \leq \ln w), \qquad 0 < w.$$

That is,

$$G(w) = \int_{-\infty}^{\ln w} \frac{1}{\sigma\sqrt{2\pi}} \exp\left[-\frac{(x-\mu)^2}{2\sigma^2}\right] dx, \qquad 0 < w,$$

and, thus, the p.d.f. of W is

$$g(w) = G'(w) = \frac{1}{\sigma w\sqrt{2\pi}} \exp\left[-\frac{(\ln w - \mu)^2}{2\sigma^2}\right], \qquad 0 < w.$$

It is most important to observe that μ and σ^2 are not the mean and the variance of W, but those of $X = \ln W$, which has a normal distribution. Incidentally, it is easy to see why this distribution has been called the *lognormal*.

Suppose that we have some data for which it is obvious that the support of the distribution is the set of positive numbers. We wish to find the best fit among the gamma, Weibull, and lognormal distributions. None of these are as easy to use as the normal; and, in each case, a computer or an easier scheme of estimation is almost necessary to determine reasonable estimates of the parameters:

(a) *Gamma Model.* Rather than using the method of maximum likelihood to estimate α and β of the gamma distribution, we usually use the *method of moments*. That is, if $w_1, w_2, \ldots, w_n$ are the observed values of the items of the random sample, we would estimate α and β by equating, respectively, the first and the second moments of the distribution and the sample, namely,

$$\alpha\beta = \bar{w}, \qquad \alpha\beta^2 = s^2.$$

Thus, the estimates for the parameters are given by

$$\tilde{\alpha} = \frac{\bar{w}^2}{s^2}, \qquad \tilde{\beta} = \frac{s^2}{\bar{w}}.$$

(b) *Weibull Model.* The likelihood function is

$$L(\alpha, \beta) = \frac{\alpha^n \prod_{i=1}^{n} w_i^{\alpha-1}}{\beta^{n\alpha}} \exp\left[-\frac{\sum_{i=1}^{n} w_i^{\alpha}}{\beta^{\alpha}}\right].$$

Now

$$\ln L = n \ln \alpha + (\alpha - 1)\sum_{i=1}^{n} \ln w_i - (n\alpha) \ln \beta - \frac{\sum_{i=1}^{n} w_i^{\alpha}}{\beta^{\alpha}}$$

and, thus,

$$\frac{\partial \ln L}{\partial \beta} = -\frac{n\alpha}{\beta} + \alpha \frac{\sum_{i=1}^{n} w_i^{\alpha}}{\beta^{\alpha+1}}.$$

Setting $\partial \ln L / \partial \beta = 0$, we have that the solution for β is

$$\beta = \left(\frac{1}{n} \sum_{i=1}^{n} w_i^{\alpha} \right)^{1/\alpha}.$$

That is, if α is known, we would have that

$$\hat{\beta} = \left(\frac{1}{n} \sum W_i^{\alpha} \right)^{1/\alpha}.$$

Unfortunately α is usually not known; so we consider this "solution" along with $\partial \ln L / \partial \alpha = 0$ and solve them simultaneously with the computer using an iterative process.

(c) *Lognormal Model.* Probably the easiest estimates for μ and σ^2 can be obtained by first taking logarithms

$$x_1 = \ln w_1, x_2 = \ln w_2, \ldots, x_n = \ln w_n$$

and by then using estimates given by

$$\tilde{\mu} = \bar{x}, \qquad \tilde{\sigma}^2 = s_X^2.$$

After fitting one or more of these distributions to the data, the goodness of the fits could be checked by a chi-square test.

═══ *Exercises* ═══

10.3–1. Let the life W (in years) of the usual family car have a Weibull distribution with $\alpha = 2$.

    (a) Show that β must equal 10 for $P(W > 5) = e^{-1/4} \approx 0.7788$. HINT: $P(W > 5) = e^{-H(5)}$.

    (b) Take a random sample of size 8 from the uniform distribution $(0, 1)$ and simulate eight values from the Weibull $(\alpha = 2, \beta = 10)$ distribution. HINT: You may use the random number table. Also recall that if U has that uniform distribution, then $F^{-1}(U)$ has a distribution with distribution function F.

10.3–2. Suppose the length W of a human life does follow the Gompertz distribution with $\lambda(w) = a(1.1)^w = ae^{(\ln 1 \cdot 1)w}$, $P(W \le 53 | 52 < W) = 0.01$. Determine the constant a and $P(W \le 71 | 70 < W)$.

10.3–3. Let $Y_1 < Y_2 < Y_3$ be the order statistics of a random sample W_1, W_2, W_3 of size $n = 3$ from a Weibull distribution with parameters α and β. Show that Y_1 has a Weibull distribution. What are the parameters of this latter distribution? HINT:

$$G(y_1) = P(Y_1 \le y_1) = 1 - P(y_1 < W_i, i = 1, 2, 3) = 1 - [P(y_1 < W_1)]^3.$$

10.3–4. A frequent force of mortality used in actuarial science is $\lambda(w) = ae^{bw} + c$. Find the distribution function and p.d.f. associated with this *Makeham's law*.

10.4 Comparisons of Several Distributions

In this section, we try to demonstrate the flexibility of the chi-square test. We begin simply by testing the equality of two multinomial distributions.

Suppose each of two independent experiments can end in one of the k mutually exclusive and exhaustive events $A_1, A_2, \ldots, A_k$. Let

$$p_{ij} = P(A_i), \qquad i = 1, 2, \ldots, k, \quad j = 1, 2.$$

That is, $p_{11}, p_{21}, \ldots, p_{k1}$ are the probabilities of the events in the first experiment and $p_{12}, p_{22}, \ldots, p_{k2}$ are those associated with the second experiment. Let the experiments be repeated n_1 and n_2 independent times, respectively. Also let $Y_{11}, Y_{21}, \ldots, Y_{k1}$ be the frequencies of $A_1, A_2, \ldots, A_k$ associated with the n_1 independent trials of the first experiment. Similarly, let $Y_{12}, Y_{22}, \ldots, Y_{k2}$ be the respective frequencies associated with the n_2 trials of the second experiment. Of course, $\sum_{i=1}^{k} Y_{ij} = n_j$, $j = 1, 2$. From the sampling distribution theory corresponding to the basic chi-square test, we know that each of

$$\sum_{i=1}^{k} \frac{(Y_{ij} - n_j p_{ij})^2}{n_j p_{ij}}$$

has an approximate chi-square distribution with $k - 1$ degrees of freedom. Since the two experiments are independent (and thus the two chi-squares are independent), the sum

$$\sum_{j=1}^{2} \sum_{i=1}^{k} \frac{(Y_{ij} - n_j p_{ij})^2}{n_j p_{ij}}$$

is approximately chi-square with $k - 1 + k - 1 = 2k - 2$ degrees of freedom.

Usually p_{ij}, $i = 1, 2, \ldots, k$, $j = 1, 2$, are unknown but frequently we wish to test the hypothesis

$$H_0: \quad p_{11} = p_{12}, p_{21} = p_{22}, \ldots, p_{k1} = p_{k2};$$

that is, the hypothesis that the corresponding probabilities associated with the two independent experiments are equal. Under H_0, we can estimate the unknown $p_{i1} = p_{i2}$, $i = 1, 2, \ldots, k$, by using the relative frequency $(Y_{i1} + Y_{i2})/(n_1 + n_2)$, $i = 1, 2, \ldots, k$. That is, if H_0 is true, we can say that the two experiments are actually parts of a large one in which $Y_{i1} + Y_{i2}$ is the frequency of the event A_i, $i = 1, 2, \ldots, k$. Note that we only have to estimate the $k - 1$ probabilities

$$p_{i1} = p_{i2}$$

using

$$\frac{Y_{i1} + Y_{i2}}{n_1 + n_2}, \qquad i = 1, 2, \ldots, k - 1,$$

since the sum of the k probabilities must equal one. That is, the estimator of $p_{k1} = p_{k2}$ is

$$1 - \frac{Y_{11} + Y_{12}}{n_1 + n_2} - \cdots - \frac{Y_{k-1,1} + Y_{k-1,2}}{n_1 + n_2} = \frac{Y_{k1} + Y_{k2}}{n_1 + n_2}.$$

Substituting these estimators, we have that

$$Q = \sum_{j=1}^{2} \sum_{i=1}^{k} \frac{[Y_{ij} - n_j(Y_{i1} + Y_{i2})/(n_1 + n_2)]^2}{n_j(Y_{i1} + Y_{i2})/(n_1 + n_2)}$$

has an approximate chi-square distribution with $2k - 2 - (k - 1) = k - 1$ degrees of freedom. Here $k - 1$ is subtracted from $2k - 2$ because that is the number of parameters estimated. The critical region for testing H_0 is of the form $q > c$, where c is selected from Appendix Table III with $k - 1$ degrees of freedom to achieve the correct (but approximate) significance level α.

Example 10.4–1. To test two methods of instruction, 50 students are selected at random for each of two groups. At the end of the instruction period each student is assigned a grade (A, B, C, D, or F) by an evaluating team. The data are recorded as follows.

	GRADE					
	A	B	C	D	F	TOTALS
Group I	8	13	16	10	3	50
Group II	4	9	14	16	7	50

Accordingly, if the hypothesis H_0 that the corresponding probabilities are equal is true, the respective estimates of the probabilities are

$$\frac{8 + 4}{100} = 0.12, \ 0.22, \ 0.30, \ 0.26, \ \frac{3 + 7}{100} = 0.10.$$

Thus, the estimates of $n_1 p_{i1} = n_2 p_{i2}$ are 6, 11, 15, 13, 5, respectively. Hence, the computed value of Q is

$$q = \frac{(8 - 6)^2}{6} + \frac{(13 - 11)^2}{11} + \frac{(16 - 15)^2}{15} + \frac{(10 - 13)^2}{13} + \frac{(3 - 5)^2}{5}$$

$$+ \frac{(4 - 6)^2}{6} + \frac{(9 - 11)^2}{11} + \frac{(14 - 15)^2}{15} + \frac{(16 - 13)^2}{13} + \frac{(7 - 5)^2}{5}$$

$$= \frac{4}{6} + \frac{4}{11} + \frac{1}{15} + \frac{9}{13} + \frac{4}{5} + \frac{4}{6} + \frac{4}{11} + \frac{1}{15} + \frac{9}{13} + \frac{4}{5} = 5.18.$$

Now, under H_0, Q has an approximate chi-square distribution with $k - 1 = 4$ degrees of freedom so the $\alpha = 0.05$ critical region is $q > 9.488$.

Here $q = 5.18 < 9.488$, and hence H_0 is not rejected at the 5% significance level. That is, with these data, we cannot say there is a difference between the two methods of instruction.

It is fairly obvious how this procedure can be extended to testing the equality of h independent multinomial distributions. That is, let

$$p_{ij} = P(A_i), \quad i = 1, 2, \ldots, k, \quad j = 1, 2, \ldots, h$$

and test

$$H_0: \quad p_{i1} = p_{i2} = \cdots = p_{ih} = p_i, \quad i = 1, 2, \ldots, k.$$

Repeat the jth experiment n_j independent times and let $Y_{1j}, Y_{2j}, \ldots, Y_{kj}$ denote the frequencies of the respective events $A_1, A_2, \ldots, A_k$. Now

$$Q = \sum_{j=1}^{h} \sum_{i=1}^{k} \frac{(Y_{ij} - n_j p_{ij})^2}{n_j p_{ij}}$$

has an approximate chi-square distribution with $h(k - 1)$ degrees of freedom. Under H_0, we must estimate $k - 1$ probabilities using

$$\hat{p}_i = \frac{\sum_{j=1}^{h} Y_{ij}}{\sum_{j=1}^{h} n_j}, \quad i = 1, 2, \ldots, k - 1, k,$$

because the estimate of p_k follows from $p_k = 1 - p_1 - p_2 - \cdots - p_{k-1}$. We use these estimates to obtain

$$Q = \sum_{j=1}^{h} \sum_{i=1}^{k} \frac{(Y_{ij} - n_j \hat{p}_i)^2}{n_j \hat{p}_i},$$

which has an approximate chi-square distribution with $h(k - 1) - (k - 1) = (h - 1)(k - 1)$ degrees of freedom.

Let us see how we can use the above procedures to test the equality of two or more independent distributions that are not necessarily multinomial. Suppose first we are given random variables U and V with distribution functions $F(u)$ and $G(v)$ respectively. It is sometimes of interest to test the hypothesis $H_0: F(x) = G(x)$ for all x. We have previously considered tests such as $\mu_u = \mu_v$, $\sigma_u^2 = \sigma_v^2$, and equality of the medians assuming certain properties of the underlying distributions. Now we shall only assume that the distributions are independent and of the continuous type.

We are interested in testing the hypothesis $H_0: F(x) = G(x)$ for all x. This hypothesis will be replaced by another one. Partition the real line into k mutually disjoint sets $A_1, A_2, \ldots, A_k$. Let

$$p_{i1} = P(U \in A_i), \quad i = 1, 2, \ldots, k,$$

and

$$p_{i2} = P(V \in A_i), \quad i = 1, 2, \ldots, k.$$

We observe that if $F(x) = G(x)$ for all x, then $p_{i1} = p_{i2}, i = 1, 2, \ldots, k$. We shall replace the hypothesis $H_0: F(x) = G(x)$ with the less restrictive hypothesis $H_0': p_{i1} = p_{i2}, i = 1, 2, \ldots, k$. That is, we are now essentially interested in testing the equality of two multinomial distributions.

Let n_1 and n_2 denote the number of independent observations of U and V, respectively. For $i = 1, 2, \ldots, k$, let Y_{ij} denote the number of these observations of U and $V, j = 1, 2$, respectively, that fall into a set A_i. At this point, we proceed to make the test of H_0' as described earlier. Of course, if H_0' is rejected at the (approximate) significance level α, then H_0 is rejected with the same probability. However, if H_0' is true, H_0 is not necessarily true. Thus if H_0' is accepted, it is probably better to say that we do not reject H_0 than to say H_0 is accepted. However, it has been pointed out earlier that in most cases this terminology is preferred although we continue to use "is accepted" because it arises more often in the literature.

In applications, the question of how to select $A_1, A_2, \ldots, A_k$ is frequently raised. Obviously, there is not a single choice for k nor the dividing marks of the partition. But it is interesting to observe that the combined sample can be used in this selection without upsetting the approximate distribution of Q. For example, suppose $n_1 = n_2 = 20$, we could easily select the dividing marks of the partition so that $k = 4$ and one fourth of the combined sample falls in each of the four sets.

Example 10.4–2. Select, at random, 20 cars of each of two comparable major-brand models. All 40 cars are submitted to accelerated life testing; that is, they are driven many miles over very poor roads in a short time and their failure times are recorded (in weeks):

Brand U: 25 31 20 42 39 19 35 36 44 26
            38 31 29 41 43 36 28 31 25 38

Brand V: 28 17 33 25 31 21 16 19 31 27
            23 19 25 22 29 32 24 20 34 26

If we use 23.5, 28.5, and 34.5 as dividing marks, we note that exactly one-fourth of the 40 cars fall in each of the resulting four sets. Thus, the data can be summarized as follows:

	A_1	A_2	A_3	A_4	TOTALS
Brand U	2	4	4	10	20
Brand V	8	6	6	0	20

The estimate of each p_i is $10/40 = 1/4$, which multiplied by $n_j = 20$ gives 5.

Hence, the computed Q is

$$q = \frac{(2-5)^2}{5} + \frac{(4-5)^2}{5} + \frac{(4-5)^2}{5} + \frac{(10-5)^2}{5} + \frac{(8-5)^2}{5}$$

$$+ \frac{(6-5)^2}{5} + \frac{(6-5)^2}{5} + \frac{(0-5)^2}{5}$$

$$= \frac{72}{5} = 14.4 > 7.815,$$

the latter number being found in Appendix Table III and associated with $\alpha = 0.05$ and 3 degrees of freedom. Hence, it seems that the two brands of cars have different distributions for the length of life under accelerated life testing.

Again it is clear how this can be extended to more than two distributions and this extension will be illustrated in the exercises.

===== *Exercises* =====

10.4–1. We wish to test to see if two groups of nurses distribute their time in six different categories about the same way. That is, the hypothesis under consideration is $H_0: p_{i1} = p_{i2}, i = 1, 2, \ldots, 6$. To test this, nurses are observed at random throughout several days, each observation resulting in a mark in one of the six categories. The summary is given by the following frequency table.

	CATEGORY						
	1	2	3	4	5	6	TOTALS
Group I	95	36	71	21	45	32	300
Group II	53	26	43	18	32	28	200

Use a chi-square test with $\alpha = 0.05$.

10.4–2. Suppose that a third group of nurses was observed along with Groups I and II of Exercise 10.4–1, resulting in the respective frequencies 130, 75, 136, 33, 61, and 65. Test $H_0: p_{i1} = p_{i2} = p_{i3}, i = 1, 2, \ldots, 6$, at the $\alpha = 0.025$ significance level.

10.4–3. Each of two comparable classes of 15 students responded to two different methods of instructions with the following scores on a standardized test:

$$
\begin{array}{llllllll}
\text{Class U:} & 91 & 42 & 39 & 62 & 55 & 82 & 67 & 44 \\
& 51 & 77 & 61 & 52 & 76 & 41 & 59 \\
\text{Class V:} & 80 & 71 & 55 & 67 & 61 & 93 & 49 & 78 \\
& 57 & 88 & 79 & 81 & 63 & 51 & 75
\end{array}
$$

Use a chi-square test with $\alpha = 0.05$ to test the equality of the distributions of test scores by dividing the combined sample into three equal parts (low, middle, high).

10.4–4. Suppose a third class (W) of 15 students was observed along with Classes U and V of Exercise 10.4–3, resulting in scores of

$$
\begin{array}{llllllll}
91 & 73 & 67 & 83 & 59 & 98 & 87 & 69 \\
78 & 80 & 65 & 94 & 82 & 74 & 85
\end{array}
$$

Again use a chi-square with $\alpha = 0.025$ to test the equality of the three distributions by dividing the combined sample into three equal parts.

10.4–5. Two bowls contain red, white, blue, green, and yellow chips. Samples of size 1000 from each of the bowls, in which the sampling is with replacement, yielded the following data. Test at the 5 % significance level whether the distributions of chips in the two bowls are the same.

	COLOR					
	RED	WHITE	BLUE	GREEN	YELLOW	TOTALS
Bowl I	198	202	201	189	210	1000
Bowl II	160	221	206	209	204	1000

10.4–6. Test the hypothesis that the distribution of letters is the same   (a)   in a psychology textbook and a science textbook,   (b)   on the editorial page and sports page of a newspaper,   (c)   in two types of written material of your choice. Since some letters do not occur very frequently, you might want to group $\{j, k\}$, $\{p, q\}$, and $\{x, y, z\}$. Take a sample of 600 to 1000 letters and let $\alpha = 0.05$.

10.5 Contingency Tables

Let us suppose that a random experiment results in an outcome that can be classified by two different attributes, such as height and weight. Assume that the first attribute is assigned to one and only one of k mutually exclusive and exhaustive events, say $A_1, A_2, \ldots, A_k$, and the second attribute falls in one and only one of h mutually exclusive and exhaustive events, say $B_1, B_2, \ldots, B_h$. Let the probability of $A_i \cap B_j$ be defined by

$$p_{ij} = P(A_i \cap B_j), \qquad i = 1, 2, \ldots, k, \quad j = 1, 2, \ldots, h.$$

The random experiment is to be repeated n independent times and Y_{ij} will denote the frequency of the event $A_i \cap B_j$. Since there are kh such events as $A_i \cap B_j$, the random variable

$$Q_{kh-1} = \sum_{j=1}^{h} \sum_{i=1}^{k} \frac{(Y_{ij} - np_{ij})^2}{np_{ij}}$$

has an approximate chi-square distribution with $kh - 1$ degrees of freedom, provided n is large.

Suppose that we wish to test the hypothesis

$$H_0: \quad P(A_i \cap B_j) = P(A_i)P(B_j), \qquad i = 1, 2, \ldots, k, \quad j = 1, 2, \ldots, h.$$

Let us denote $P(A_i)$ by $p_{i\cdot}$ and $P(B_j)$ by $p_{\cdot j}$; that is,

$$p_{i\cdot} = \sum_{j=1}^{h} p_{ij} = P(A_i), \qquad \text{and} \qquad p_{\cdot j} = \sum_{i=1}^{k} p_{ij} = P(B_j),$$

and, or course,

$$1 = \sum_{j=1}^{h} \sum_{i=1}^{k} p_{ij} = \sum_{j=1}^{h} p_{\cdot j} = \sum_{i=1}^{k} p_{i\cdot}.$$

Then the hypothesis can be formulated as

$$H_0: \quad p_{ij} = p_{i\cdot}p_{\cdot j}, \qquad i = 1, 2, \ldots, k, \quad j = 1, 2, \ldots, h.$$

To test H_0, we can use Q_{kh-1} with p_{ij} replaced by $p_{i\cdot}p_{\cdot j}$. But if $p_{i\cdot}, i = 1, 2, \ldots, k$, and $p_{\cdot j}, j = 1, 2, \ldots, h$, are unknown, as they usually are in the applications, we cannot compute Q_{kh-1} once the frequencies are observed. In such a case we estimate these unknown parameters by

$$\hat{p}_{i\cdot} = \frac{y_{i\cdot}}{n}, \qquad \text{where} \quad y_{i\cdot} = \sum_{j=1}^{h} y_{ij}$$

is the observed frequency of $A_i, i = 1, 2, \ldots, k$; and

$$\hat{p}_{\cdot j} = \frac{y_{\cdot j}}{n}, \qquad \text{where} \quad y_{\cdot j} = \sum_{i=1}^{k} y_{ij}$$

is the observed frequency of $B_j, j = 1, 2, \ldots, h$. Since $\sum_i p_{i\cdot} = \sum_j p_{\cdot j} = 1$, we actually estimate only $k - 1 + h - 1 = k + h - 2$ parameters. So if these estimates are used in Q_{kh-1}, with $p_{ij} = p_{i\cdot}p_{\cdot j}$, then, according to the rule stated earlier, the random variable

$$Q = \sum_{j=1}^{h} \sum_{i=1}^{k} \frac{[Y_{ij} - n(Y_{i\cdot}/n)(Y_{\cdot j}/n)]^2}{n(Y_{i\cdot}/n)(Y_{\cdot j}/n)}$$

has an approximate chi-square distribution with $kh - 1 - (k + h - 2) = (k - 1)(h - 1)$ degrees of freedom, provided H_0 is true. The hypothesis H_0 is then rejected if the computed value of this statistic exceeds the constant c, where c is selected from the chi-square probability table so that the test has the desired significance level α, approximately.

Example 10.5–1. In league bowling three games are bowled by each team member. Table 10.5–1 classifies 63 games according to whether the game was bowled first, second, or third on a particular night and also according to the score. Such a display is usually called a *contingency table*.

GAME	SCORE			TOTALS
	0–162	163–177	178–300	
1	4	4	13	21
2	8	7	6	21
3	6	10	5	21
TOTALS	18	21	24	63

TABLE 10.5–1

In this example, A_i, $i = 1, 2, 3$, is the number of the game bowled; B_1 is the number of scores less than or equal to 162, B_2 is the number of scores from 163 to 177, inclusive, and B_3 is the number of scores greater than or equal to 178. We shall test the null hypothesis that the two attributes are independent against all alternatives.

Now Q has $(k - 1)(h - 1) = (3 - 1)(3 - 1) = 4$ degrees of freedom; thus the null hypothesis will be rejected if the computed value of Q exceeds 9.488 at the 5% significance level. Since $\hat{p}_{1.} = \hat{p}_{2.} = \hat{p}_{3.} = 21/63$ and $\hat{p}_{.1} = 18/63$, $\hat{p}_{.2} = 21/63$, $\hat{p}_{.3} = 24/63$, we have

$$q = \frac{[4 - 63(18/63)(21/63)]^2}{63(18/63)(21/63)} + \cdots + \frac{[5 - 63(24/63)(21/63)]^2}{63(24/63)(21/63)}$$

$$= 32.65 > 9.488.$$

Of course, the null hypothesis is rejected, which means that the score seems to depend upon which of the three games is being bowled. From our data, it looks as if this bowler has a better chance of getting a higher score on the first game. This may or may not be the case for all bowlers; some bowler might want to perform this experiment.

It is fairly obvious how to extend the above testing procedure to more than two attributes. For example, if the third attribute falls in one and only one of m mutually exclusive and exhaustive events, say $C_1, C_2, \ldots, C_m$, then we test the independence of the three attributes by using

$$Q = \sum_{r=1}^{m} \sum_{j=1}^{h} \sum_{i=1}^{k} \frac{[Y_{ijr} - n(Y_{i..}/n)(Y_{.j.}/n)(Y_{..r}/n)]^2}{n(Y_{i..}/n)(Y_{.j.}/n)(Y_{..r}/n)}$$

where Y_{ijr}, $Y_{i..}$, $Y_{.j.}$, and $Y_{..r}$ are the respective observed frequencies of the events $A_i \cap B_j \cap C_r$, A_i, B_j, and C_r in n independent trials of the experiment. If n is large and if the three attributes are independent, then Q has an approximate chi-square distribution with $khm - 1 - (k-1) - (h-1) - (m-1) = khm - k - h - m + 2$ degrees of freedom.

Rather than explore this extension further, it is more instructive to note some interesting uses of contingency tables.

Example 10.5–2. Say we observed 30 values $x_1, x_2, \ldots, x_{30}$ said to represent the items of a random sample. That is, the corresponding random variables $X_1, X_2, \ldots, X_{30}$ were supposed to be mutually independent and to have the same distribution. Say, however, by looking at the 30 values we detect an upward trend that indicates there might have been some dependence and/or the items did not actually have the same distribution. One simple way to test if they could be thought of as being observed items of a random sample is the following. Mark each x high (H) or low (L) depending upon whether it is above or below the sample median. Then divide the x values into three groups: $x_1, \ldots, x_{10}$; $x_{11}, \ldots, x_{20}$; $x_{21}, \ldots, x_{30}$. Certainly if the items are those of a random sample we would expect five H's and five L's in each group. That is, the attribute classified as H or L should be independent of the group number. The summary of these data provides a 3×2 contingency table. For example, say the 30 values are

5.6	8.2	7.8	4.8	5.5	8.1	6.7	7.7	9.3	6.9
8.2	10.1	7.5	6.9	11.1	9.2	8.7	10.3	10.7	10.0
9.2	11.6	10.3	11.7	9.9	10.6	10.0	11.4	10.9	11.1

The median can be taken to be the average of the two middle items in magnitude, namely, 9.2 and 9.3. Marking each item H or L after comparing it with this median, we obtain the following 3×2 contingency table.

GROUP	L	H	TOTALS
1	9	1	10
2	5	5	10
3	1	9	10
TOTALS	15	15	30

Here each $n(y_i./n)(y._j/n) = 30(10/30)(15/30) = 5$ so that the computed value of Q is

$$q = \frac{(9-5)^2}{5} + \frac{(1-5)^2}{5} + \frac{(5-5)^2}{5} + \frac{(5-5)^2}{5} + \frac{(1-5)^2}{5} + \frac{(9-5)^2}{5}$$

$$= 12.8 > 5.991,$$

where 5.991 is the rejection number corresponding to $(k-1)(h-1) = 2$ degrees of freedom and $\alpha = 0.05$. Hence, we reject the conjecture that these 30 values could be the items of a random sample. Obviously, modifications could be made to this scheme: dividing the sample into more (or less) than three groups and rating items differently, such as low (L), middle (M), and high (H).

It cannot be emphasized enough that the chi-square statistic can be used fairly effectively in almost any situation in which there should be independence. For illustration, suppose that we have a group of workers who have essentially *the same qualifications* (training, experience, etc.). Many believe that salary and sex of the workers should be independent attributes; yet there have been several claims in special cases that there is a dependence—or discrimination—in salary for one group or another. The following example illustrates some of the calculations associated with such a problem.

Example 10.5–3. Two groups of workers have the same qualifications for a particular type of work. Their experience in salaries is summarized by the following 2×5 contingency table, in which the upper bound of each salary range is not included in that listing.

	SALARY IN THOUSANDS OF DOLLARS					
GROUP	8–10	10–12	12–14	14–16	16—	TOTALS
1	6	11	16	14	13	60
2	5	9	8	6	2	30
TOTALS	11	20	24	20	15	90

To test if the group assignment and the salaries seem to be independent with these data at the $\alpha = 0.05$ significance level, we compute

$$q = \frac{[6 - 90(60/90)(11/90)]^2}{90(60/90)(11/90)} + \cdots + \frac{[2 - 90(30/90)(15/90)]^2}{90(30/90)(15/90)}$$

$$= 1.769 < 9.488,$$

where 9.488 is the critical value for $(2-1)(5-1) = 4$ degrees of freedom and $\alpha = 0.05$. Hence, group assignment and salaries seem to be independent.

Before turning to the exercises, note that we could have thought of each example in this section as testing the equality of two or more multinomial distributions. In Example 10.5–1, the three games define three trinomial distributions; in Example 10.5–2, the three groups define three binomial distributions; and in Example 10.5–3, the two groups define two multinomial distributions. What would have happened if we had used the computations outlined in Section 10.4? It is interesting to note that we obtain exactly the same value of chi-square and in each case, the number of degrees of freedom is equal to $(k - 1)(h - 1)$. Hence, it makes no difference whether we think of it as a test of independence or a test of the equality of several multinomial distributions. Our advice is to use the terminology that seems most natural for the particular situations.

=== *Exercises* ===

10.5–1. Sixty men and forty women were asked to state their preferences among three colas. Test whether brand preference is independent of sex at an approximate 5% significance level.

SEX	COLA 1	COLA 2	COLA 3	TOTALS
	PRODUCT			
MALE	36	10	14	60
FEMALE	12	9	19	40
TOTALS	48	19	33	100

10.5–2. A five-man bowling team bowled three games on each of 32 nights. The totals of their five scores have been classified according to whether the game was bowled first, second, or third and also according to score. At the 5% significance level test whether the score is independent of the game being bowled.

GAME	720–820	821–860	861–900	901–1000	TOTALS
		SCORE			
1	10	11	6	5	32
2	7	5	13	7	32
3	5	10	7	10	32
TOTALS	22	26	26	22	96

10.5–3.   While high school grades and testing scores, such as SAT or ACT, can be used to predict first-year college grade point average (GPA), many educators claim that a more important factor influencing that GPA is the living conditions of students. In particular, it is claimed that the roommate of the student will have a great influence on his or her grades. To test this, suppose we selected at random 200 students and classified each according to the following two attributes:

(1)   Ranking of the student's roommate from 1 to 5, from a person who was difficult to live with and discouraged scholarship to one who was congenial but encouraged scholarship.

(2)   The student's first-year GPA.

Say this gives the 5 × 4 contingency table.

RANK OF ROOMMATE	GRADE POINT AVERAGE				
	UNDER 2.00	2.00–2.69	2.70–3.19	3.20–4.00	TOTALS
1	8	9	10	4	31
2	5	11	15	11	42
3	6	7	20	14	47
4	3	5	22	23	53
5	1	3	11	12	27
TOTALS	23	35	78	64	200

Compute the chi-square statistic used to test the independence of the two attributes and compare it to the critical value associated with $\alpha = 0.05$.

10.5–4.   For a Sunday afternoon project, the son of a statistician wrote down the color and the three digit number on the license plate of cars passing his house. Test the null hypothesis that the attributes of color and license number are independent. Let $\alpha = 0.05$.

COLOR	NUMBER		TOTALS
	000–499	500–999	
RED	5	8	13
YELLOW	8	7	15
BLUE	6	7	13
BROWN	16	7	23
GREEN	7	6	13
OTHER	6	7	13
TOTALS	48	42	90

11

ANALYSIS OF VARIANCE

11.1 A Chi-Square Decomposition Theorem

We begin this section by reviewing an important test statistic, but in a slightly different notation. Let $X_{11}, X_{12}, X_{13}, \ldots, X_{1n_1}$ and $X_{21}, X_{22}, X_{23}, \ldots, X_{2n_2}$ be random samples from the respective independent normal distribution $N(\mu_1, \sigma^2)$ and $N(\mu_2, \sigma^2)$. Recall that the statistic used to test the hypothesis $H_0 \colon \mu_1 = \mu_2$ is

$$\frac{\overline{X}_1. - \overline{X}_2.}{\sqrt{\{[\sum_1^{n_1} (X_{1j} - \overline{X}_1.)^2 + \sum_1^{n_2}(X_{2j} - \overline{X}_2.)^2]/(n_1 + n_2 - 2)\}(1/n_1 + 1/n_2)}},$$

where $\overline{X}_1.$ and $\overline{X}_2.$ are the respective sample means. Of course, this ratio has a t distribution because

$$\frac{\overline{X}_1. - \overline{X}_2.}{\sqrt{\sigma^2/n_1 + \sigma^2/n_2}} \quad \text{and} \quad \frac{\sum_1^{n_1} (X_{1j} - \overline{X}_1.)^2 + \sum_1^{n_2} (X_{2j} - \overline{X}_2.)^2}{\sigma^2}$$

are independent random variables having, under H_0, distributions $N(0, 1)$ and $\chi^2(n_1 + n_2 - 2)$, respectively.

Let us look at the same situation a little differently by examining the total sum of squares of the deviations of the items X_{ij} from the mean $\bar{X}_{..}$ of the combined samples. That is, with

$$\bar{X}_{..} = \sum_{i=1}^{2} \sum_{j=1}^{n_i} X_{ij}/(n_1 + n_2),$$

we wish to analyze

$$\sum_{i=1}^{2} \sum_{j=1}^{n_i} (X_{ij} - \bar{X}_{..})^2.$$

It is easy to accomplish this as follows; the sum of squares is equal to

$$\sum_{i=1}^{2} \sum_{j=1}^{n_i} [(X_{ij} - \bar{X}_{i.}) + (\bar{X}_{i.} - \bar{X}_{..})]^2$$

$$= \sum_{i=1}^{2} \sum_{j=1}^{n_i} (X_{ij} - \bar{X}_{i.})^2 + 2\sum_{i=1}^{2} \sum_{j=1}^{n_i} (X_{ij} - \bar{X}_{i.})(\bar{X}_{i.} - \bar{X}_{..}) \quad (1)$$

$$+ \sum_{i=1}^{2} \sum_{j=1}^{n_i} (\bar{X}_{i.} - \bar{X}_{..})^2.$$

Three observations help simplify equation (1):

(a) $\displaystyle\sum_{i=1}^{2} \sum_{j=1}^{n_i} (X_{ij} - \bar{X}_{i.})^2 = \sum_{j=1}^{n_1} (X_{1j} - \bar{X}_{1.})^2 + \sum_{j=1}^{n_2} (X_{2j} - \bar{X}_{2.})^2.$

(b) $\displaystyle 2\sum_{i=1}^{2} \sum_{j=1}^{n_i} (X_{ij} - \bar{X}_{i.})(\bar{X}_{i.} - \bar{X}_{..}) = 2\sum_{i=1}^{2} (\bar{X}_{i.} - \bar{X}_{..})\sum_{j=1}^{n_i} (X_{ij} - \bar{X}_{i.})$

$$= 2\sum_{i=1}^{2} (\bar{X}_{i.} - \bar{X}_{..})\left[\sum_{j=1}^{n_i} X_{ij} - n_i \bar{X}_{i.}\right] = 0$$

because $\bar{X}_{i.} = \sum_{j=1}^{n_i} X_{ij}/n_i$.

(c) $\displaystyle\sum_{i=1}^{2} \sum_{j=1}^{n_i} (\bar{X}_{i.} - \bar{X}_{..})^2 = n_1(\bar{X}_{1.} - \bar{X}_{..})^2 + n_2(\bar{X}_{2.} - \bar{X}_{..})^2$

$$= n_1\left[\frac{n_2(\bar{X}_{1.} - \bar{X}_{2.})}{n_1 + n_2}\right]^2 + n_2\left[\frac{n_1(\bar{X}_{2.} - \bar{X}_{1.})}{n_1 + n_2}\right]^2$$

since $\bar{X}_{..} = (n_1\bar{X}_{1.} + n_2\bar{X}_{2.})/(n_1 + n_2)$, and this then equals

$$\frac{(n_1 n_2^2 + n_2 n_1^2)(\bar{X}_{1.} - \bar{X}_{2.})^2}{(n_1 + n_2)^2} = \frac{n_1 n_2(\bar{X}_{1.} - \bar{X}_{2.})^2}{n_1 + n_2}.$$

If we substitute (a), (b), and (c) in equation (1), we have, after dividing by σ^2, that

$$\frac{\sum_{i=1}^{2}\sum_{j=1}^{n_i}(X_{ij}-\bar{X}..)^2}{\sigma^2} = \frac{\sum_{j=1}^{n_1}(X_{1j}-\bar{X}_1.)^2}{\sigma^2} + \frac{\sum_{j=1}^{n_2}(X_{2j}-\bar{X}_2.)^2}{\sigma^2}$$

$$+ \frac{n_1 n_2(\bar{X}_1. - \bar{X}_2.)^2}{(n_1+n_2)\sigma^2},$$

or, for brevity,

$$\frac{Q}{\sigma^2} = \frac{Q_1}{\sigma^2} + \frac{Q_2}{\sigma^2} + \frac{Q_3}{\sigma^2}.$$

Please note that the last term Q_3/σ^2 in the right-hand member is the square of that $N(0,1)$ variable in the numerator of the t variable and measures the variation between the two sample means. On the other hand, the sum $(Q_1 + Q_2)/\sigma^2$ of the first two terms is that $\chi^2(n_1 + n_2 - 2)$ variable in the denominator of the t variable, and it measures the variation among the items in both samples. Of course, the square, namely Q_3/σ^2, of the $N(0, 1)$ variable is $\chi^2(1)$ and, from the independence of this normal variable and the chi-square random variable, $(Q_1 + Q_2)/\sigma^2$ and Q_3/σ^2 are also independent.

The symbols $Q, Q_1, Q_2,$ and Q_3 are used because the corresponding expressions are quadratic forms in $X_{ij}; i = 1, 2; j = 1, 2, \ldots, n_i$. That is, if we remember the definitions of $\bar{X}.., \bar{X}_1.,$ and $\bar{X}_2.,$ we easily note that, by squaring, each term of $Q, Q_1, Q_2,$ and Q_3 is of second degree in X_{ij}. Any homogeneous polynomial of degree two in certain variables is called a *quadratic form* in those variables.

Example 11.1–1. The form $X_1^2 + 2X_1X_2 + X_2^2$ is a quadratic form in X_1 and X_2, but the form

$$(X_1 - \mu_1)^2 + (X_2 - \mu_2)^2 = X_1^2 + X_2^2 - 2\mu_1 X_1 - 2\mu_2 X_2 + \mu_1^2 + \mu_2^2$$

is not a quadratic form in X_1 and X_2, although it is a quadratic form in the variables $X_1 - \mu_1$ and $X_2 - \mu_2$. Let $\bar{X}$ and S^2 denote, respectively, the mean and variance of a random sample $X_1, X_2, \ldots, X_n$ from an arbitrary distribution. Then

$$nS^2 = \sum_1^n (X_i - \bar{X})^2 = \sum_1^n \left[X_i - \frac{X_1 + X_2 + \cdots + X_n}{n}\right]^2$$

$$= \frac{n-1}{n}(X_1^2 + X_2^2 + \cdots + X_n^2)$$

$$- \frac{2}{n}(X_1 X_2 + \cdots + X_1 X_n + \cdots + X_{n-1}X_n)$$

is a quadratic form in the n variables $X_1, X_2, \ldots, X_n$.

In all these illustrations of quadratic forms, the coefficients have been real numbers. If both the variables and the coefficients are real, the form is called a *real quadratic form*. In the statistical tests that will be given later in this chapter, we shall need the distributions of certain real quadratic forms in normally distributed mutually independent random variables. The discussion of the t variable, and particularly the decomposition

$$\frac{Q}{\sigma^2} = \frac{Q_1}{\sigma^2} + \frac{Q_2}{\sigma^2} + \frac{Q_3}{\sigma^2},$$

illustrates the following theorem, which we accept without proof because it is beyond the level of this book.

THEOREM 11.1–1. *Let $Q = Q_1 + Q_2 + \cdots + Q_k$, where $Q, Q_1, \ldots, Q_k$ are $k + 1$ real quadratic forms in n mutually independent random variables normally distributed with the same variance σ^2. Let Q/σ^2, $Q_1/\sigma^2, \ldots, Q_{k-1}/\sigma^2$ have chi-square distributions with $r, r_1, \ldots, r_{k-1}$ degrees of freedom, respectively. If Q_k is nonnegative, then*

 (i) $Q_1, \ldots, Q_k$ *are mutually independent, and hence,*
 (ii) Q_k/σ^2 *has a chi-square distribution with* $r - (r_1 + \cdots + r_{k-1}) = r_k$ *degrees of freedom.*

Please note that in our example Q/σ^2 is $\chi^2(n_1 + n_2 - 1)$ since $Q/(n_1 + n_2)$ would be the variance of a sample of size $n_1 + n_2$ when $H_0: \mu_1 = \mu_2$ is true. Of course, we know that Q_1/σ^2 is $\chi^2(n_1 - 1)$, Q_2/σ^2 is $\chi^2(n_2 - 1)$, and the nonnegative expression Q_3/σ^2 is $\chi^2(1)$, and Q_1, Q_2, and Q_3 are mutually independent. Also note that the number of degrees of freedom in each member is the same, namely,

$$n_1 + n_2 - 1 = (n_1 - 1) + (n_2 - 1) + 1.$$

Example 11.1–2. Let X_1, X_2, X_3 be a random sample of size $n = 3$ from $N(\mu, \sigma^2)$. Note that

$$\frac{\sum_{i=1}^{3} (X_i - \bar{X})^2}{\sigma^2} = \frac{(X_1 - X_2)^2}{2\sigma^2} + \frac{[(X_1 + X_2)/2 - X_3]^2}{3\sigma^2/2}$$

or, for brevity,

$$\frac{Q}{\sigma^2} = \frac{Q_1}{\sigma^2} + \frac{Q_2}{\sigma^2}.$$

This is easy to verify (Exercise 11.1–1) by replacing $\bar{X}$ by $(X_1 + X_2 + X_3)/3$, squaring, and showing that each member is equal to

$$\frac{2}{3\sigma^2} (X_1^2 + X_2^2 + X_3^2 - X_1 X_2 - X_1 X_3 - X_2 X_3).$$

Of course, $nS^2/\sigma^2 = Q/\sigma^2$ is $\chi^2(2)$ since $n = 3$. Also $X_1 - X_2$ is $N(0, 2\sigma^2)$, so $[(X_1 - X_2)/\sqrt{2\sigma}]^2 = Q_1/\sigma^2$ is $\chi^2(1)$. Since $Q_2 \geq 0$, Theorem 11.1–1 implies that Q_1 and Q_2 are independent and Q_2/σ^2 is $\chi^2(2 - 1 = 1)$. Incidentally, it is interesting to note that a constant times the sample variance, namely $3S^2 = Q$, is decomposed into a measure of variation between X_1 and X_2 and a measure of variation between the average $(X_1 + X_2)/2$ and X_3. An analysis of the variance like this plays a major role in statistics and generalizations of it usually come under the broad heading of *analysis of variance* (ANOVA).

Example 11.1–3.   Let $X_1, X_2, \ldots, X_n$ be a random sample of size n from $N(\mu, \sigma^2)$. The quadratic form $Q = \sum_{i=1}^{n} (X_i - \mu)^2$ in $X_1 - \mu, X_2 - \mu, \ldots, X_n - \mu$ is such that we know that Q/σ^2 is $\chi^2(n)$. It is also true that

$$\frac{\sum_{i=1}^{n} (X_i - \mu)^2}{\sigma^2} = \frac{n(\overline{X} - \mu)^2}{\sigma^2} + \frac{\sum_{i=1}^{n} (X_i - \overline{X})^2}{\sigma^2}$$

or, for brevity,

$$\frac{Q}{\sigma^2} = \frac{Q_1}{\sigma^2} + \frac{Q_2}{\sigma^2},$$

where Q_1 and Q_2 are also quadratic forms in $X_1 - \mu, X_2 - \mu, \ldots, X_n - \mu$. But $\overline{X}$ is $N(\mu, \sigma^2/n)$ and hence $(\overline{X} - \mu)/(\sigma/\sqrt{n})$ is $N(0, 1)$. Thus

$$\left(\frac{\overline{X} - \mu}{\sigma/\sqrt{n}}\right)^2 = \frac{n(\overline{X} - \mu)^2}{\sigma^2} = \frac{Q_1}{\sigma^2}$$

is $\chi^2(1)$. But $Q_2 = nS^2 \geq 0$, so Theorem 11.1–1 declares that Q_1 and Q_2 are independent and $Q_2/\sigma^2 = nS^2/\sigma^2$ is $\chi^2(n - 1)$. This is another way of looking at Theorem 7.3–1, part of which we accepted without proof.

═══ *Exercises* ═══

11.1–1.   Show that each member of the display in Example 11.1–2 is equal to the expression given in that example and hence equal to each other.

11.1–2.   Let X_1, X_2, X_3, X_4 be a random sample of size $n = 4$ from $N(\mu, 1)$. Show that

$$\sum_{i=1}^{4} (X_i - \overline{X})^2 = \frac{(X_1 - X_2)^2}{2} + \frac{[(X_1 + X_2)/2 - X_3]^2}{3/2}$$

$$+ \frac{[(X_1 + X_2 + X_3)/3 - X_4]^2}{4/3}.$$

With the aid of Example 11.1–2 and Theorem 11.1–1, argue that the three terms in the right-hand member are mutually independent chi-square variables, each with one degree of freedom.

11.1–3. Let X_1, X_2, X_3, X_4 be a random sample of size $n = 4$ from $N(\mu, 1)$. Find c_1, c_2, and c_3 such that

$$\sum_{i=1}^{4} (X_i - \bar{X})^2 = c_1(X_1 + X_2 - X_3 - X_4)^2 + c_2(X_1 + X_3 - X_2 - X_4)^2$$
$$+ c_3(X_1 + X_4 - X_2 - X_3)^2.$$

Use Theorem 11.1–1 to show that the three terms in the right-hand member are mutually independent chi-square variables, each with one degree of freedom.

11.1–4. Let X_1, X_2, X_3, X_4, X_5 be a random sample of size 5 from $N(\mu, \sigma^2)$. Show that

$$\sum_{i=1}^{5} (X_i - \bar{X})^2 = \frac{[\sum_{i=1}^{5}(i-3)X_i]^2}{10} + \sum_{i=1}^{5}\left[X_i - \bar{X} - \left(\frac{\sum_1^5 (i-3)X_i}{10}\right)(i-3)\right]^2,$$

say, for brevity, $Q = Q_1 + Q_2$. Prove that Q_1/σ^2 is $\chi^2(1)$ by demonstrating that it is the square of a $N(0, 1)$ variable. Use Theorem 11.1–1 to argue that Q_1 and Q_2 are independent and Q_2/σ^2 is $\chi^2(3)$.

11.1–5. In Section 7.7 we noted that

$$\sum_1^n [Y_i - \alpha - \beta(x_i - \bar{x})]^2 = n(\hat{\alpha} - \alpha)^2 + \sum_1^n (x_i - \bar{x})^2(\hat{\beta} - \beta)^2 + n\widehat{\sigma^2},$$

say $Q = Q_1 + Q_2 + Q_3$, all of which are quadratic forms in $y_i - \alpha - \beta(x_i - \bar{x})$. There we proved that Q/σ^2 is $\chi^2(n)$, Q_1/σ^2 is $\chi^2(1)$, and Q_2/σ^2 is $\chi^2(1)$. Show that $Q_3/\sigma^2 = n\widehat{\sigma^2}/\sigma^2$ is $\chi^2(n-2)$ and that $\hat{\alpha}$, $\hat{\beta}$, and $\widehat{\sigma^2}$ are mutually independent.

11.2 Tests of the Equality of Means

Suppose that we have several independent normal distributions and that we wish to make one or more statistical inferences about their means. In particular, let us consider k independent normal distributions with unknown means $\mu_1, \mu_2, \ldots, \mu_k$, respectively, and an unknown but common variance σ^2. Let $X_{i1}, X_{i2}, \ldots, X_{in_i}$ represent a random sample of size n_i from $N(\mu_i, \sigma^2)$, $i = 1, 2, \ldots, k$. One inference that we wish to consider is the test of the equality of the k means, namely $H_0: \mu_1 = \mu_2 = \ldots = \mu_k = \mu$, μ unspecified, against all possible alternative hypotheses H_1.

To determine a critical region for this test we shall decompose the sum of squares associated with the variance of the combined samples into two parts and analyze each of them.

Let

$$TS = \sum_{i=1}^{k} \sum_{j=1}^{n_i} (X_{ij} - \bar{X}..)^2$$

where, with $n = n_1 + n_2 + \cdots + n_k$,

$$\bar{X}_{..} = \frac{1}{n} \sum_{i=1}^{k} \sum_{j=1}^{n_i} X_{ij} \quad \text{and} \quad \bar{X}_{i\cdot} = \frac{1}{n_i} \sum_{j=1}^{n_i} X_{ij}, \quad i = 1, 2, \ldots, k.$$

Then

$$TS = \sum_{i=1}^{k} \sum_{j=1}^{n_i} (X_{ij} - \bar{X}_{i\cdot} + \bar{X}_{i\cdot} - \bar{X}_{..})^2$$

$$= \sum_{i=1}^{k} \sum_{j=1}^{n_i} (X_{ij} - \bar{X}_{i\cdot})^2 + \sum_{i=1}^{k} \sum_{j=1}^{n_i} (\bar{X}_{i\cdot} - \bar{X}_{..})^2$$

$$+ 2 \sum_{i=1}^{k} \sum_{j=1}^{n_i} (X_{ij} - \bar{X}_{i\cdot})(\bar{X}_{i\cdot} - \bar{X}_{..}).$$

The last term of the right-hand member of this identity may be written

$$2 \sum_{i=1}^{k} \left[(\bar{X}_{i\cdot} - \bar{X}_{..}) \sum_{j=1}^{n_i} (X_{ij} - \bar{X}_{i\cdot}) \right] = 2 \sum_{i=1}^{k} (\bar{X}_{i\cdot} - \bar{X}_{..})(n_i \bar{X}_{i\cdot} - n_i \bar{X}_{i\cdot}) = 0,$$

and the preceding term

$$\sum_{i=1}^{k} \sum_{j=1}^{n_i} (\bar{X}_{i\cdot} - \bar{X}_{..})^2 = \sum_{i=1}^{k} n_i (\bar{X}_{i\cdot} - \bar{X}_{..})^2.$$

Thus,

$$TS = \sum_{i=1}^{k} \sum_{j=1}^{n_i} (X_{ij} - \bar{X}_{i\cdot})^2 + \sum_{i=1}^{k} n_i (\bar{X}_{i\cdot} - \bar{X}_{..})^2.$$

For notation let

$$TS = \sum_{i=1}^{k} \sum_{j=1}^{n_i} (X_{ij} - \bar{X}_{..})^2, \quad \text{the total sum of squares;}$$

$$WS = \sum_{i=1}^{k} \sum_{j=1}^{n_i} (X_{ij} - \bar{X}_{i\cdot})^2, \quad \text{the sum of squares within groups or within classes;}$$

$$AS = \sum_{i=1}^{k} n_i (\bar{X}_{i\cdot} - \bar{X}_{..})^2, \quad \text{the sum of squares among groups or among classes.}$$

Thus,

$$TS = WS + AS.$$

When H_0 is true, we may regard X_{ij}, $i = 1, 2, \ldots, k, j = 1, 2, \ldots, n_i$, as a random sample of size $n = n_1 + n_2 + \cdots + n_k$ from $N(\mu, \sigma^2)$. Then $TS/(n-1)$ is an unbiased estimator of σ^2 because TS/σ^2 is $\chi^2(n-1)$. An unbiased estimator of σ^2 based only on the sample from the ith distribution is

$$V_i = \sum_{j=1}^{n_i} \frac{(X_{ij} - \bar{X}_{i\cdot})^2}{n_i - 1} \quad \text{for } i = 1, 2, \ldots, k,$$

because $(n_i - 1)V_i/\sigma^2$ is $\chi^2(n_i - 1)$. Thus,

$$\sum_{i=1}^{k} (n_i - 1)\frac{V_i}{\sigma^2} = \frac{WS}{\sigma^2}$$

is $\chi^2(n - k)$ and, hence, $WS/(n - k)$ is an unbiased estimator of σ^2. Since $TS = WS + AS, TS/\sigma^2$ is $\chi^2(n - 1)$, WS/σ^2 is $\chi^2(n - k)$, and $AS \geq 0$, it follows from Theorem 11.1–1 that WS and AS are independent quadratic forms and AS/σ^2 is $\chi^2(k - 1)$. Moreover, $AS/(k - 1)$ is also an unbiased estimator of σ^2 when $\mu_1 = \mu_2 = \cdots = \mu_k$.

The estimator of σ^2, which is based upon WS, is always unbiased whether H_0 is true or false. However, if the means $\mu_1, \mu_2, \ldots, \mu_k$, are not equal, the expected value of the estimator based upon AS will be greater than σ^2. To make this last statement clear, we have

$$E(AS) = E\left[\sum_{i=1}^{k} n_i(\overline{X}_{i.} - \overline{X}_{..})^2\right] = E\left[\sum_{i=1}^{k} n_i\overline{X}_{i.}^2 - n\overline{X}_{..}^2\right]$$

$$= \sum_{i=1}^{k} n_i\{\text{Var}(\overline{X}_{i.}) + [E(\overline{X}_{i.})]^2\} - n\{\text{Var}(\overline{X}_{..}) + [E(\overline{X}_{..})]^2\}$$

$$= \sum_{i=1}^{k} n_i\left\{\frac{\sigma^2}{n_i} + \mu_i^2\right\} - n\left\{\frac{\sigma^2}{n} + \bar{\mu}^2\right\}$$

$$= (k - 1)\sigma^2 + \sum_{i=1}^{k} n_i(\mu_i - \bar{\mu})^2,$$

where $\bar{\mu} = (1/n)\sum_{i=1}^{k} n_i\mu_i$. If $\mu_1 = \mu_2 = \cdots = \mu_k = \mu$,

$$E\left[\frac{AS}{k - 1}\right] = \sigma^2.$$

If the means are not all equal,

$$E\left[\frac{AS}{k - 1}\right] = \sigma^2 + \sum_{i=1}^{k} n_i\frac{(\mu_i - \bar{\mu})^2}{k - 1} > \sigma^2.$$

Exercise 11.2–4 also illustrates the fact that the estimator using AS is usually bigger than that using WS.

We can base our test of H_0 on the ratio of $AS/(k - 1)$ and $WS/(n - k)$, both of which are unbiased estimators of σ^2 provided $H_0: \mu_1 = \mu_2 = \cdots = \mu_k$ is true so that, under H_0, the ratio would assume values near one. However, as the means $\mu_1, \mu_2, \ldots, \mu_k$ begin to differ, this ratio tends to become large since $E[AS/(k - 1)]$ gets bigger. Under H_0, the ratio

$$\frac{AS/(k - 1)}{WS/(n - k)} = \frac{(AS/\sigma^2)/(k - 1)}{(WS/\sigma^2)/(n - k)} = F$$

has an F distribution with $k - 1$ and $n - k$ degrees of freedom because AS/σ^2 and WS/σ^2 are independent chi-square variables. We would reject H_0 if the observed value of F is too large, because this would indicate that we have a relatively large AS which suggests that the means are unequal. Thus, the critical region is of the form $F > c$, where c is selected from the probability table for the F distribution (Appendix Table V) with $k - 1$ and $n - k$ degrees of freedom such that $P(F > c) = \alpha$, the desired significance level.

Example 11.2–1. Let X_1, X_2, X_3, X_4 be independent random variables that are $N(\mu_i, \sigma^2)$, $i = 1, 2, 3, 4$. We shall test

$$H_0: \quad \mu_1 = \mu_2 = \mu_3 = \mu_4 = \mu$$

against all alternatives based on a random sample of size $n_i = 3$ from each of the four distributions. A critical region of size $\alpha = 0.05$ is given by

$$F = \frac{AS/(4 - 1)}{WS/(12 - 4)} > 4.07.$$

The observed data are given in Table 11.2–1. (Clearly these data are not observations from normal distributions. They were selected to illustrate the calculations.)

				$\bar{X}_{i\cdot}$
X_1	13	8	9	10
X_2	15	11	13	13
X_3	8	12	7	9
X_4	11	15	10	12
$\bar{X}_{\cdot\cdot}$				11

TABLE 11.2–1

For these data, the calculated TS, WS, and AS are

$$ts = (13 - 11)^2 + (8 - 11)^2 + \cdots + (15 - 11)^2 + (10 - 11)^2 = 80;$$
$$ws = (13 - 10)^2 + (8 - 10)^2 + \cdots + (15 - 12)^2 + (10 - 12)^2 = 50;$$
$$as = 3[(10 - 11)^2 + (13 - 11)^2 + (9 - 11)^2 + (12 - 11)^2] = 30.$$

Note that since $ts = ws + as$, only two of the three values need to be calculated from the data directly. Here the computed value of F is

$$\frac{30/3}{50/8} = 1.6 < 4.07,$$

and H_0 is not rejected.

Formulas that sometimes simplify the calculations of *ts*, *as*, and *ws* are

$$ts = \sum_{i=1}^{k} \sum_{j=1}^{n_i} x_{ij}^2 - \frac{1}{n} \left[\sum_{i=1}^{k} \sum_{j=1}^{n_i} x_{ij} \right]^2,$$

$$as = \sum_{i=1}^{k} \frac{1}{n_i} \left[\sum_{j=1}^{n_i} x_{ij} \right]^2 - \frac{1}{n} \left[\sum_{i=1}^{k} \sum_{j=1}^{n_i} x_{ij} \right]^2,$$

and

$$ws = ts - as.$$

It is interesting to note that in these formulas each square is divided by the number of items in the sum being squared: x_{ij}^2 by one, $(\sum_{j=1}^{n_i} x_{ij})^2$ by n_i, and $(\sum_{i=1}^{k} \sum_{j=1}^{n_i} x_{ij})^2$ by n. These formulas are used in Example 11.2–2.

Example 11.2–2. Let $X_1, X_2, \ldots, X_5$ be independent random variables that are $N(\mu_i, \sigma^2)$, $i = 1, 2, \ldots, 5$. We shall test the null hypothesis

$$H_0: \quad \mu_1 = \mu_2 = \mu_3 = \mu_4 = \mu_5 = \mu$$

against all alternatives. The observed values of random samples of size 4 from each of these distributions are given in Table 11.2–2. At a significance level of $\alpha = 0.05$, H_0 is rejected if the computed

$$F = \frac{AS/(5-1)}{WS/(20-5)} > 3.06.$$

For these data

$$ts = 45{,}050.15 - \frac{1}{20}(947.1)^2 = 200.23;$$

$$as = \frac{1}{4}[(183.3)^2 + (190.8)^2 + \cdots + (185.9)^2] - \frac{1}{20}(947.1)^2 = 22.38;$$

$$ws = ts - as = 177.85.$$

					$\sum_{j=1}^{4} x_{ij}$	$\sum_{j=1}^{4} x_{ij}^2$
X_1	52.6	43.3	40.3	47.1	183.3	8484.15
X_2	49.7	51.7	46.1	43.3	190.8	9143.08
X_3	50.3	46.2	48.4	50.0	194.9	9507.09
X_4	44.0	47.7	50.8	49.6	192.2	9562.02
X_5	44.6	45.7	49.6	46.0	185.9	8653.81
TOTALS					947.1	45,050.15

TABLE 11.2–2

Since the computed

$$F = \frac{22.38/4}{177.85/15} = 0.47 < 3.06,$$

the hypothesis is not rejected.

When the sum of squares AS among groups is large so that the F statistic leads to the rejection of $H_0 : \mu_1 = \mu_2 = \cdots = \mu_k$, the experimenter is frequently puzzled about what action to take. That is, the experimenter is willing to say that the means $\mu_1, \mu_2, \ldots, \mu_k$ are unequal, but why are they unequal? Which means seem to differ? Which one is the largest mean? Does μ_1 seem to be unequal to μ_2 or μ_3 or even something like the average $(\mu_2 + \cdots + \mu_k)/(k - 1)$? And so on. Accordingly, it seems that AS should be analyzed more by decomposing it into two or more parts in such a way that these parts, divided by σ^2, are independent chi-square variables. This approach will be illustrated with $k = 4$ in Example 11.2–3. Obviously, there is not just one way to make such a decomposition, but this procedure does allow the experimenter to analyze the total variance.

Example 11.2–3.   With $k = 4$, we have that

$$AS = \sum_{i=1}^{4} n_i(\overline{X}_{i\cdot} - \overline{X}_{\cdot\cdot})^2.$$

Suppose we wish to find a term that reflects the difference between μ_1 and μ_2 and another that measures the difference between the weighted average of μ_1 and μ_2 and that of μ_3 and μ_4. Natural choices for these would be

$$Y_1 = \overline{X}_{1\cdot} - \overline{X}_{2\cdot}.$$

and

$$Y_2 = \frac{n_1 \overline{X}_{1\cdot} + n_2 \overline{X}_{2\cdot}}{n_1 + n_2} - \frac{n_3 \overline{X}_{3\cdot} + n_4 \overline{X}_{4\cdot}}{n_3 + n_4}.$$

Of course, we know, under H_0, that Y_1 is $N[0, \sigma^2(1/n_1 + 1/n_2)]$ and Y_2 is also normal with mean zero and variance

$$\sigma^2\left[\left(\frac{n_1}{n_1 + n_2}\right)^2 \frac{1}{n_1} + \left(\frac{n_2}{n_1 + n_2}\right)^2 \frac{1}{n_2} + \left(\frac{n_3}{n_3 + n_4}\right)^2 \frac{1}{n_3} + \left(\frac{n_4}{n_3 + n_4}\right)^2 \frac{1}{n_4}\right]$$

$$= \sigma^2\left[\frac{1}{n_1 + n_2} + \frac{1}{n_3 + n_4}\right].$$

If we let

$$Q_1 = \frac{Y_1^2}{1/n_1 + 1/n_2} \quad \text{and} \quad Q_2 = \frac{Y_2^2}{1/(n_1 + n_2) + 1/(n_3 + n_4)},$$

then each of Q_1/σ^2 and Q_2/σ^2 is $\chi^2(1)$ and it can be shown (Exercise 11.2–6) that

$$AS = Q_1 + Q_2 + Q_3,$$

where

$$Q_3 = \frac{(\overline{X}_3. - \overline{X}_4.)^2}{1/n_3 + 1/n_4} \geq 0.$$

Moreover, $TS = WS + Q_1 + Q_2 + Q_3$; hence, by Theorem 11.1–1, WS, Q_1, Q_2, and Q_3 are mutually independent and Q_3 is $\chi^2(1)$. By comparing each of Q_r, $r = 1, 2, 3$, to $WS/(n-4)$, we can determine the reason (or reasons) for the large variation in AS. Obviously, AS could also be decomposed in other ways, as illustrated in Exercise 11.2–6.

===== *Exercises* =====

11.2–1. Let μ_1, μ_2, μ_3 be, respectively, the means of three independent normal distributions with a common but unknown variance σ^2. In order to test, at the $\alpha = 5\%$ significance level, the hypothesis $H_0: \mu_1 = \mu_2 = \mu_3$ against all possible alternative hypotheses, we take a random sample of size 4 from each of these distributions. Determine whether we accept or reject H_0 if the observed values from these three distributions are, respectively,

$$\begin{array}{lcccc}
X_1: & 5 & 9 & 6 & 8 \\
X_2: & 11 & 13 & 10 & 12 \\
X_3: & 10 & 6 & 9 & 9
\end{array}$$

11.2–2. Let μ_i be the average yield in bushels per acre of variety i of corn, $i = 1, 2, 3, 4$. In order to test, at the 5% significance level, the hypothesis $H_0: \mu_1 = \mu_2 = \mu_3 = \mu_4$, four test plots for each of the four varieties of corn are planted. Determine whether we accept or reject H_0 if the yield in bushels per acre of the four varieties of corn are, respectively,

$$\begin{array}{lcccc}
X_1: & 68.82 & 76.99 & 74.30 & 78.73 \\
X_2: & 86.84 & 75.69 & 77.87 & 76.18 \\
X_3: & 90.16 & 78.84 & 80.65 & 83.58 \\
X_4: & 61.58 & 73.51 & 74.57 & 70.75
\end{array}$$

11.2–3. Four groups of three pigs each were fed four different feeds for a specified length of time to test the hypothesis, $H_0: \mu_1 = \mu_2 = \mu_3 = \mu_4$, where μ_i is the mean weight gain for each of the feeds, $i = 1, 2, 3, 4$. Determine whether the null hypothesis is accepted or rejected at a 5% significance level if the observed weight gains are, respectively,

$$\begin{array}{lccc}
X_1: & 194.11 & 182.80 & 187.43 \\
X_2: & 216.06 & 203.50 & 216.88 \\
X_3: & 178.10 & 189.20 & 181.33 \\
X_4: & 197.11 & 202.68 & 209.18
\end{array}$$

11.2–4. For the following set of data show that the computed $WS/(n - k) = 1$ and $AS/(k - 1) = 75$. This suggests that the unbiased estimate of σ^2 based upon AS is usually greater than σ^2 when the true means are unequal.

$$
\begin{array}{llll}
X_1: & 4 & 5 & 6 \\
X_2: & 9 & 10 & 11 \\
X_3: & 14 & 15 & 16
\end{array}
$$

11.2–5. In a tulip display area, several varieties of tulips have long stems. The lengths of the flower stems were measured for four varieties of tulips, the colors being white, pink, red trimmed with yellow, and dark red. We shall call them varieties 1, 2, 3, and 4. Let $\mu_i, i = 1, 2, 3, 4$, denote the mean of the lengths of the stems for variety i. Test the hypothesis $H_0: \mu_1 = \mu_2 = \mu_3 = \mu_4$ against all alternatives, making reasonable assumptions. Use a 5% significance level. Clearly specify the critical region for this test.

$$
\begin{array}{llllll}
X_1: & 13.75 & 13.00 & 14.25 & 12.00 & 12.25 \\
X_2: & 13.75 & 14.50 & 12.75 & 14.50 & 13.25 \\
X_3: & 12.75 & 12.50 & 12.00 & 14.00 & 13.00 \\
X_4: & 16.75 & 14.25 & 14.50 & 15.50 & 13.75
\end{array}
$$

11.2–6. In Example 11.2–3, prove that $AS = Q_1 + Q_2 + Q_3$ by showing that the left-hand and right-hand members equal the same quadratic form in $\overline{X}_1., \overline{X}_2., \overline{X}_3., \overline{X}_4.$. Find a decomposition of AS using

$$ Z_1 = \overline{X}_3. - \overline{X}_4., $$

$$ Z_2 = \overline{X}_2. - \frac{n_3 \overline{X}_3. + n_4 \overline{X}_4.}{n_3 + n_4}, $$

$$ Z_3 = \overline{X}_1. - \frac{n_2 \overline{X}_2. + n_3 \overline{X}_3. + n_4 \overline{X}_4.}{n_2 + n_3 + n_4}. $$

HINT: Under H_0, find the distribution of Z_1, Z_2, and Z_3. Find c_i so that $c_i Z_i^2/\sigma^2$ is $\chi^2(1)$, $i = 1, 2, 3$.

11.3 Two-Factor Analysis of Variance

The test of the equality of several means, considered in Section 11.2, is an example of a statistical inference method called the *analysis of variance* (ANOVA). This method derives its name from the fact that the quadratic form $TS = nS^2$, the total sum of squares about the combined sample mean, is decomposed into component parts and analyzed. In this section, other problems in the analysis of variance will be investigated; here we restrict our considerations to the two-factor case, but the reader can see how it can be extended to three-factor and other cases.

Consider a situation in which it is desirable to investigate the effects of two factors that influence an outcome of an experiment. For example, a teaching method (lecture, discussion, computer assisted, television, etc.) and the size of a class might influence a student's score on a standard test; or the type of car and the grade of gasoline used might change the number of miles per gallon. In this latter example, if the number of miles per gallon is not affected by the grade of gasoline, we would no doubt use the least expensive grade.

The first analysis of variance model that we discuss is referred to as a *two-way classification with one observation per cell*. In particular, assume that there are two factors (attributes), one of which has a levels and the other has b levels. There are thus $n = ab$ possible combinations, each of which determines a cell. Let us think of these cells as being arranged in a rows and b columns. Here we take one observation per cell and we denote the observation in the ith row and jth column by X_{ij}. Further assume that X_{ij}, $i = 1, 2, \ldots, a$, and $j = 1, 2, \ldots, b$, are n random variables that are mutually independent and are $N(\mu_{ij}, \sigma^2)$, respectively. [The assumptions of normality and homogeneous (same) variances can be *somewhat* relaxed in applications with little change in the significance levels of the resulting tests.] We shall assume that the means μ_{ij} are composed of a row effect, a column effect, and an overall effect in some additive way, namely, $\mu_{ij} = \mu + \alpha_i + \beta_j$, where $\sum_{i=1}^{a} \alpha_i = 0$ and $\sum_{j=1}^{b} \beta_j = 0$. The parameter α_i represents the ith row effect and the parameter β_j represents the jth column effect.

REMARK.   There is no loss in generality in assuming that

$$\sum_{1}^{a} \alpha_i = \sum_{1}^{b} \beta_j = 0.$$

To see this, let $\mu_{ij} = \mu' + \alpha_i' + \beta_j'$. Write

$$\bar{\alpha}' = \left(\frac{1}{a}\right) \sum_{1}^{a} \alpha_i' \quad \text{and} \quad \bar{\beta}' = \left(\frac{1}{b}\right) \sum_{1}^{b} \beta_j'.$$

We have

$$\mu_{ij} = (\mu' + \bar{\alpha}' + \bar{\beta}') + (\alpha_i' - \bar{\alpha}') + (\beta_j' - \bar{\beta}') = \mu + \alpha_i + \beta_j,$$

where $\sum_{1}^{a} \alpha_i = 0$ and $\sum_{1}^{b} \beta_j = 0$. The reader is asked to find μ, α_i, and β_j for one display of μ_{ij} in Exercise 11.3–2.

To test the hypothesis that there is no row effect, we would test $H_R : \alpha_1 = \alpha_2 = \cdots = \alpha_a = 0$ since $\sum_{1}^{a} \alpha_i = 0$. Similarly, to test that there is no column effect, we would test $H_C : \beta_1 = \beta_2 = \cdots = \beta_b = 0$ since $\sum_{1}^{b} \beta_j = 0$. To test these hypotheses we shall again decompose the total sum of squares into several component parts. Letting

$$\bar{X}_{i\cdot} = \frac{1}{b} \sum_{j=1}^{b} X_{ij}, \qquad \bar{X}_{\cdot j} = \frac{1}{a} \sum_{i=1}^{a} X_{ij}, \qquad \bar{X}_{\cdot\cdot} = \frac{1}{ab} \sum_{j=1}^{b} \sum_{i=1}^{a} X_{ij},$$

we have

$$
TS = \sum_{j=1}^{b} \sum_{i=1}^{a} (X_{ij} - \bar{X}..)^2
$$

$$
= \sum_{j=1}^{b} \sum_{i=1}^{a} [(\bar{X}_{i\cdot} - \bar{X}..) + (\bar{X}_{\cdot j} - \bar{X}..) + (X_{ij} - \bar{X}_{i\cdot} - \bar{X}_{\cdot j} + \bar{X}..)]^2
$$

$$
= b \sum_{i=1}^{a} (\bar{X}_{i\cdot} - \bar{X}..)^2 + a \sum_{j=1}^{b} (\bar{X}_{\cdot j} - \bar{X}..)^2 + \sum_{j=1}^{b} \sum_{i=1}^{a} (X_{ij} - \bar{X}_{i\cdot} - \bar{X}_{\cdot j} + \bar{X}..)^2
$$

$$
= RS + CS + ES,
$$

where RS is the sum of squares among rows, CS is the sum of squares among columns, and ES is the error or residual sum of squares. It is an easy exercise (Exercise 11.3–4) to show that the three "cross-product" terms in the square of the trinomial sum to zero. The distribution of the residual sum of squares does not depend upon the mean μ_{ij}, provided the additive model is correct, and hence whether H_R or H_C is true; thus, ES acts as a "measuring stick" as did WS in Section 11.2. This can be seen more clearly by writing

$$
ES = \sum_{j=1}^{b} \sum_{i=1}^{a} (X_{ij} - \bar{X}_{i\cdot} - \bar{X}_{\cdot j} + \bar{X})^2
$$

$$
= \sum_{j=1}^{b} \sum_{i=1}^{a} [X_{ij} - (\bar{X}_{i\cdot} - \bar{X}) - (\bar{X}_{\cdot j} - \bar{X}) - \bar{X}]^2
$$

and noting the similarity of the summand in the right-hand member with

$$
X_{ij} - \mu_{ij} = X_{ij} - \alpha_i - \beta_j - \mu.
$$

We now show that RS/σ^2, CS/σ^2, and ES/σ^2 are independent chi-square variables provided both H_R and H_C are true, that is, when all the means μ_{ij} have a common value μ. To do this, we first note that TS/σ^2 is $\chi^2(ab - 1)$. In addition, from Section 11.2, we see that expressions such as RS/σ^2 and CS/σ^2 are chi-square variables, namely, $\chi^2(a - 1)$ and $\chi^2(b - 1)$ by replacing the n_i of Section 11.2 by a and b, respectively. Obviously, $ES \geq 0$, and hence by Theorem 11.1–1 we have that RS/σ^2, CS/σ^2, and ES/σ^2 are independent chi-square variables with $a - 1$, $b - 1$, and $ab - 1 - (a - 1) - (b - 1) = (a - 1)(b - 1)$ degrees of freedom, respectively.

To test the hypothesis $H_R : \alpha_1 = \alpha_2 = \cdots = \alpha_a = 0$, we shall use the row sum of squares RS and the residual sum of squares ES. When H_R is true, RS/σ^2 and ES/σ^2 are independent chi-square variables with $a - 1$ and $(a - 1)(b - 1)$ degrees of freedom, respectively. Thus, $RS/(a - 1)$ and $ES/[(a - 1)(b - 1)]$ are both unbiased estimators of σ^2 when H_R is true. However, $E[RS/(a - 1)] > \sigma^2$ when H_R is not true and hence we would reject H_R when

$$
F_R = \frac{RS/[\sigma^2(a - 1)]}{ES/[\sigma^2(a - 1)(b - 1)]} = \frac{RS/(a - 1)}{ES/[(a - 1)(b - 1)]}
$$

is "too large." Since F_R has an F distribution with $a - 1$ and $(a - 1)(b - 1)$ degrees of freedom when H_R is true, H_R is rejected if the observed value of F_R is greater than c, where c is selected to yield the desired significance level α.

Similarly the test of the hypothesis $H_C : \beta_1 = \beta_2 = \cdots = \beta_b = 0$ against all alternatives can be based on

$$F_C = \frac{CS/[\sigma^2(b - 1)]}{ES/[\sigma^2(a - 1)(b - 1)]} = \frac{CS/(b - 1)}{ES/[(a - 1)(b - 1)]},$$

which has an F distribution with $b - 1$ and $(a - 1)(b - 1)$ degrees of freedom, provided H_C is true.

Example 11.3–1. Each of three cars is driven with each of four different brands of gasoline. The number of miles per gallon driven for each of the $ab = (3)(4) = 12$ different combinations is recorded in Table 11.3–1.

	GASOLINE				
CAR	1	2	3	4	$\bar{X}_{i.}$
1	16	18	21	21	19
2	14	15	18	17	16
3	15	15	18	16	16
$\bar{X}_{.j}$	15	16	19	18	17

TABLE 11.3–1

We would like to test whether we can expect the same mileage for each of these four brands of gasoline. In our notation, we test the hypothesis

$$H_C : \quad \beta_1 = \beta_2 = \beta_3 = \beta_4 = 0$$

against all alternatives. At a 1 % significance level we shall reject H_C if the computed F, namely,

$$\frac{cs/(4 - 1)}{es/[(3 - 1)(4 - 1)]} > 9.78.$$

We have

$$cs = 3[(15 - 17)^2 + (16 - 17)^2 + (19 - 17)^2 + (18 - 17)^2] = 30;$$

$$es = (16 - 19 - 15 + 17)^2 + (14 - 16 - 15 + 17)^2$$

$$+ \cdots + (16 - 16 - 18 + 17)^2 = 4.$$

Hence, the computed F is

$$\frac{30/3}{4/6} = 15 > 9.78,$$

and the hypothesis H_C is rejected. That is, the gasolines seem to give different performances (at least with these three cars).

In a two-way classification problem, particular combinations of the two factors might interact differently from that expected from the additive model. For example, in Example 11.3–1, gasoline 3 seemed to be the best gasoline and car 1 the best car; however, it sometimes happens that the two best do not "mix" well and the joint performance is poor. That is, there might be a strange interaction between this combination of car and gasoline and accordingly the joint performance is not as good as expected. Sometimes it can happen that we get good results from a combination of some of the poorer levels of each factor. This type of thing is called *interaction*, and it frequently occurs in practice (for illustration, in chemistry). In order to test for possible interaction, we shall consider a two-way classification problem in which $c > 1$ independent observations are taken per cell.

Assume that X_{ijk}, $i = 1, 2, \ldots, a$; $j = 1, 2, \ldots, b$; and $k = 1, 2, \ldots, c$, are $n = abc$ random variables that are mutually independent and have normal distributions with a common, but unknown, variance σ^2. The mean of each X_{ijk}, $k = 1, 2, \ldots, c$, is $\mu_{ij} = \mu + \alpha_i + \beta_j + \gamma_{ij}$, where $\sum_1^a \alpha_i = 0$, $\sum_1^b \beta_j = 0$, $\sum_{i=1}^a \gamma_{ij} = 0$, and $\sum_{j=1}^b \gamma_{ij} = 0$. The parameter γ_{ij} is called the *interaction* associated with cell (i, j). That is, the interaction between the ith level of one classification and the jth level of the other classification is γ_{ij}. The reader is asked to determine μ, α_i, β_j, and γ_{ij} for some given μ_{ij} in Exercise 11.3–5.

To test the hypotheses that (a) the row effects are equal to zero, (b) the column effects are equal to zero, and (c) there is no interaction, we shall again decompose the total sum of squares into several component parts. Letting

$$\bar{X}_{ij\cdot} = \frac{1}{c} \sum_{k=1}^c X_{ijk},$$

$$\bar{X}_{i\cdot\cdot} = \frac{1}{bc} \sum_{j=1}^b \sum_{k=1}^c X_{ijk},$$

$$\bar{X}_{\cdot j\cdot} = \frac{1}{ac} \sum_{i=1}^a \sum_{k=1}^c X_{ijk},$$

$$\bar{X}_{\cdots} = \frac{1}{abc} \sum_{i=1}^a \sum_{j=1}^b \sum_{k=1}^c X_{ijk},$$

we have

$$
TS = \sum_{i=1}^{a} \sum_{j=1}^{b} \sum_{k=1}^{c} (X_{ijk} - \overline{X}...)^2
$$

$$
= bc \sum_{i=1}^{a} (\overline{X}_{i..} - \overline{X}...)^2 + ac \sum_{j=1}^{b} (\overline{X}_{.j.} - \overline{X}...)^2
$$

$$
+ c \sum_{i=1}^{a} \sum_{j=1}^{b} (\overline{X}_{ij.} - \overline{X}_{i..} - \overline{X}_{.j.} + \overline{X}...)^2 + \sum_{i=1}^{a} \sum_{j=1}^{b} \sum_{k=1}^{c} (X_{ijk} - \overline{X}_{ij.})^2
$$

$$
= RS + CS + IS + ES,
$$

where RS is the row sum of squares, CS is the column sum of squares, IS is the interaction sum of squares, and ES is the error or residual sum of squares. Again it is easy to show that the cross-product terms sum to zero.

To consider the joint distribution of RS, CS, IS, and ES, let us assume that all the means equal the same value μ. Of course, we know that TS/σ^2 is $\chi^2(abc - 1)$. And by letting the n_i of Section 11.2 equal bc and ac, respectively, we know that RS/σ^2 and CS/σ^2 are $\chi^2(a - 1)$ and $\chi^2(b - 1)$. Moreover,

$$
\frac{\sum_{k=1}^{c} (X_{ijk} - \overline{X}_{ij.})^2}{\sigma^2}
$$

is $\chi^2(c - 1)$; hence ES/σ^2 is the sum of ab independent chi-square variables such as this and thus is $\chi^2[ab(c - 1)]$. Of course $IS \geq 0$; so, according to Theorem 11.1–1, we have that RS/σ^2, CS/σ^2, IS/σ^2, and ES/σ^2 are mutually independent chi-square variables with $a - 1$, $b - 1$, $(a - 1)(b - 1)$, and $ab(c - 1)$ degrees of freedom, respectively.

To test the hypotheses concerning row, column, and interaction effects, we form F statistics in which the numerators are affected by deviations from the respective hypotheses while the denominator is a function of ES, whose distribution depends only on the value of σ^2 and not on the values of the cell means. Hence, ES acts as our measuring stick here.

The statistic for testing the hypothesis

$$
H_I: \quad \gamma_{ij} = 0, \quad i = 1, 2, \ldots, a, \quad j = 1, 2, \ldots, b,
$$

against all alternatives is

$$
F_I = \frac{c \sum_{i=1}^{a} \sum_{j=1}^{b} (\overline{X}_{ij.} - \overline{X}_{i..} - \overline{X}_{.j.} + \overline{X}...)^2 / [\sigma^2(a - 1)(b - 1)]}{\sum_{i=1}^{a} \sum_{j=1}^{b} \sum_{k=1}^{c} (X_{ijk} - \overline{X}_{ij.})^2 / [\sigma^2 ab(c - 1)]}
$$

$$
= \frac{IS/[(a - 1)(b - 1)]}{ES/[ab(c - 1)]}
$$

which has an F distribution with $(a - 1)(b - 1)$ and $ab(c - 1)$ degrees of freedom when H_I is true. Hence, we select a constant d such that $P(F_I > d) = \alpha$, the desired significance level. If the computed $F_I > d$, we reject H_I and say there is a difference among the means since there seems to be interaction. Many statisticians do *not* proceed to test row and column effects if H_I is rejected.

The statistic for testing the hypothesis

$$H_R: \quad \alpha_1 = \alpha_2 = \cdots = \alpha_a = 0$$

against all alternatives is

$$F_R = \frac{bc \sum_{i=1}^{a} (\bar{X}_{i..} - \bar{X}_{...})^2 / [\sigma^2(a-1)]}{\sum_{i=1}^{a} \sum_{j=1}^{b} \sum_{k=1}^{c} (X_{ijk} - \bar{X}_{ij.})^2 / [\sigma^2 ab(c-1)]} = \frac{RS/(a-1)}{ES/[ab(c-1)]}$$

which has an F distribution with $a - 1$ and $ab(c - 1)$ degrees of freedom when H_R is true. The statistic for testing the hypothesis

$$H_C: \quad \beta_1 = \beta_2 = \cdots = \beta_b = 0$$

against all alternatives is

$$F_C = \frac{ac \sum_{j=1}^{b} (\bar{X}_{.j.} - \bar{X}_{...})^2 / [\sigma^2(b-1)]}{\sum_{i=1}^{a} \sum_{j=1}^{b} \sum_{k=1}^{c} (X_{ijk} - \bar{X}_{ij.})^2 / [\sigma^2 ab(c-1)]} = \frac{CS/(b-1)}{ES/[ab(c-1)]}$$

which has an F distribution with $b - 1$ and $ab(c - 1)$ degrees of freedom when H_C is true. Each of these hypotheses is rejected if the observed value of F is greater than a given constant which is selected to yield the desired significance level.

REMARK. It should be noted that when the hypotheses are not true, each of F_I, F_R, and F_C has a noncentral F distribution. Although they are not too common, tables of probabilities for the *non*central F distribution are available in extensive statistical libraries.

===== *Exercises* =====

11.3–1. For the data given in Example 11.3–1, test the hypothesis $H_R: \alpha_1 = \alpha_2 = \alpha_3 = 0$ against all alternatives at the 5% significance level.

11.3–2. With $a = 3$ and $b = 4$, find μ, α_i, and β_j if μ_{ij}, $i = 1, 2, 3$ and $j = 1, 2, 3, 4$, are given by

6	3	7	8
10	7	11	12
8	5	9	10

Please note that in an "additive" model such as this one, one row (column) can be determined by adding a constant value to each of the elements of another row (column).

11.3–3. In order to test whether four brands of gasoline give equal performance in terms of mileage, each of three cars was driven with each of the four brands of gasoline. Then each of the $(3)(4) = 12$ possible combinations was repeated four times. The number of miles per gallon for each of the four repetitions in each cell is recorded in Table 11.3–2. Test each of the hypotheses H_I, H_R, H_C at the 5% significance level.

		BRAND OF GASOLINE		
CAR	**1**	**2**	**3**	**4**
1	21.0 14.9	16.3 20.0	15.8 19.4	17.8 17.3
	16.2 18.8	15.2 21.6	14.5 14.8	18.2 20.4
2	20.6 19.5	15.5 16.8	16.6 13.7	18.1 17.1
	20.8 18.9	17.4 19.4	18.2 16.1	21.5 19.1
3	14.2 13.1	17.4 18.1	15.2 16.7	16.3 16.4
	16.8 17.6	16.4 16.9	17.7 18.1	17.9 18.8

TABLE 11.3–2

11.3–4. Show that the cross-product terms formed from $\bar{X}_{i\cdot} - \bar{X}_{\cdot\cdot}$, $\bar{X}_{\cdot j} - \bar{X}_{\cdot\cdot}$, and $\bar{X}_{ij} - \bar{X}_{i\cdot} - \bar{X}_{\cdot j} + \bar{X}_{\cdot\cdot}$ sum to zero, $i = 1, 2, \ldots, a$ and $j = 1, 2, \ldots, b$. HINT: For example, write

$$\sum_{i=1}^{a} \sum_{j=1}^{b} (\bar{X}_{\cdot j} - \bar{X}_{\cdot\cdot})(X_{ij} - \bar{X}_{i\cdot} - \bar{X}_{\cdot j} + \bar{X}_{\cdot\cdot})$$

$$= \sum_{j=1}^{b} (\bar{X}_{\cdot j} - \bar{X}_{\cdot\cdot}) \sum_{i=1}^{a} [(X_{ij} - \bar{X}_{\cdot j}) - (\bar{X}_{i\cdot} - \bar{X}_{\cdot\cdot})]$$

and sum each term in the inner summation as grouped here to get zero.

11.3–5. With $a = 3$ and $b = 4$, find μ, α_i, β_j, and γ_{ij}, if μ_{ij}, $i = 1, 2, 3$ and $j = 1, 2, 3, 4$, are given by

$$
\begin{array}{cccc}
6 & 7 & 7 & 12 \\
10 & 3 & 11 & 8 \\
8 & 5 & 9 & 10
\end{array}
$$

Note the difference between the layout here and that in Exercise 11.3–2. Does the interaction help explain these differences?

11.3–6. A student in psychology was interested in testing how food consumption by rats would be affected by a particular drug. She used two levels of one attribute, namely, drug and placebo, and four levels of a second attribute, namely, male (M), castrated (C), female (F), and ovarectomized (O). For each cell she observed five rats. The amount of food consumed in grams per 24 hours is listed in Table 11.3–3. Test the hypotheses

(a) H_I: $\gamma_{ij} = 0$, $i = 1, 2$, $j = 1, 2, 3, 4$;
(b) H_R: $\alpha_1 = \alpha_2 = 0$;
(c) H_C: $\beta_1 = \beta_2 = \beta_3 = \beta_4 = 0$.
Let $\alpha = 0.05$.

	M	C	F	O
	22.56	16.54	18.58	18.20
	25.02	24.64	15.44	14.56
DRUG	23.66	24.62	16.12	15.54
	17.22	19.06	16.88	16.82
	22.58	20.12	17.58	14.56
	25.64	22.50	17.82	19.74
	28.84	24.48	15.76	17.48
PLACEBO	26.00	25.52	12.96	16.46
	26.02	24.76	15.00	16.44
	23.24	20.62	19.54	15.70

TABLE 11.3–3

11.3–7. How could you modify the model used in Exercise 11.3–6 so that there are three attributes of classification, each with two levels?

A BRIEF THEORY OF
STATISTICAL INFERENCE

12.1 Transformations of Random Variables

Some additional theory of statistics is presented in this chapter in the hope that the reader will find these facets of the subject interesting and intriguing enough to want to study it further. Obviously, we cannot do justice to these topics in our limited space, and accordingly we use simple, but worthwhile, examples to emphasize the major points.

To go very far in theory or application of statistics, one must know something about transformations of random variables. We saw how important this was in dealing with the normal distribution when we noted that if X is $N(\mu, \sigma^2)$, then $Z = (X - \mu)/\sigma$ is $N(0, 1)$; this simple transformation allows us to use one table for probabilities associated with all normal distributions. In the proof of this, we found that the distribution function of Z was given by the integral

$$G(z) = P\left(\frac{X - \mu}{\sigma} \le z\right) = P(X \le z\sigma + \mu)$$

$$= \int_{-\infty}^{z\sigma + \mu} \frac{1}{\sigma\sqrt{2\pi}} \exp\left[-\frac{(x - \mu)^2}{2\sigma^2}\right] dx.$$

In Section 4.5, we changed variables, $x = w\sigma + \mu$, in the integral to determine the p.d.f. of Z; but let us now simply differentiate the integral with respect to z. If we recall from calculus that the derivative

$$D_z\left[\int_a^{v(z)} f(t)\, dt\right] = f[v(z)]v'(z),$$

then, since certain assumptions are satisfied in our case, we have

$$g(z) = G'(z) = \frac{1}{\sigma\sqrt{2\pi}} \exp\left[-\frac{(z\sigma + \mu - \mu)^2}{2\sigma^2}\right]\frac{d(z\sigma + \mu)}{dz}$$

$$= \frac{1}{\sqrt{2\pi}} \exp\left(-\frac{z^2}{2}\right), \quad -\infty < z < \infty.$$

That is, Z has the p.d.f. of a standard normal distribution.

From the preceding argument we note that, in general, if X is a continuous-type random variable with p.d.f. $f(x)$ on support $a < x < b$ and if $Z = u(X)$ and its inverse $X = v(Z)$ are increasing continuous functions, then the p.d.f. of Z is

$$g(z) = f[v(z)]v'(z)$$

on the support given by $a < v(z) < b$ or, equivalently, $u(a) < z < u(b)$. Moreover, if u and v are decreasing functions, then the p.d.f. is

$$g(z) = f[v(z)][-v'(z)], \quad u(b) < z < u(a).$$

Hence, to cover both cases, we can simply write

$$g(z) = |v'(z)|f[v(z)], \quad c < z < d,$$

where the support $c < z < d$ corresponds to $a < x < b$ through the transformation $x = v(z)$.

Example 12.1–1. Let the positive random variable X have the p.d.f. $f(x) = e^{-x}, 0 < x < \infty$, which is skewed to the right. To find a distribution that is more symmetric than that of X, statisticians frequently use the square root transformation, namely $Z = \sqrt{X}$. Here $z = \sqrt{x}$ corresponds to $x = z^2$, which has derivative $2z$. Thus, the p.d.f. of Z is

$$g(z) = 2z\, e^{-z^2}, \quad 0 < z < \infty,$$

which is of the Weibull type. The graphs of $f(x)$ and $g(z)$ should convince the reader that the latter is more symmetric than the former.

Example 12.1–2. Let X be binomial with parameters n and p. Since X has a discrete distribution, $Z = u(X)$ will also have a discrete distribution with the same probabilities as those in the support of X. For illustration, with $n = 3$, $p = 1/4$, and $Z = X^2$, we have

$$g(z) = \binom{3}{\sqrt{z}}\left(\frac{1}{4}\right)^{\sqrt{z}}\left(\frac{3}{4}\right)^{3-\sqrt{z}}, \quad z = 0, 1, 4, 9.$$

For a more interesting problem with the binomial random variable X, suppose we were to search for a transformation $u(X/n)$ of the relative frequency X/n that would have a variance very little dependent on p itself when n is large. That is, we want the variance of $u(X/n)$ to be essentially a constant. Recall that X/n has variance $p(1 - p)/n$, which created a problem in the search for a confidence interval for p. Of course, the problem was resolved by replacing p in $p(1 - p)/n$ by X/n.

Consider the function $u(X/n)$ and find, using two terms of Taylor's expansion about p, that

$$u\left(\frac{X}{n}\right) \approx u(p) + u'(p)\left(\frac{X}{n} - p\right).$$

Here terms of higher powers can be disregarded if n is large enough so that X/n is close enough to p. Thus,

$$\mathrm{Var}\left[u\left(\frac{X}{n}\right)\right] \approx [u'(p)]^2 \, \mathrm{Var}\left(\frac{X}{n} - p\right) = [u'(p)]^2 \frac{p(1 - p)}{n}.$$

However, if $\mathrm{Var}[u(X/n)]$ is to be constant with respect to p, then

$$[u'(p)]^2 p(1 - p) = k \qquad \text{or} \qquad u'(p) = \frac{c}{\sqrt{p(1 - p)}},$$

where k and c are constants. Clearly

$$u(p) = 2c \arcsin \sqrt{p}$$

is a solution to this differential equation. Thus, with $c = 1/2$, we frequently see, in the literature, use of the arcsine transformation, namely,

$$Z = \arcsin \sqrt{\frac{X}{n}},$$

which, with large n, has an approximate normal distribution with mean

$$\mu = \mathrm{arsin} \sqrt{p}$$

and variance

$$\sigma^2 = [D_p(\arcsin \sqrt{p})]^2 \frac{p(1 - p)}{n} = \frac{1}{4n}.$$

There should be one note of warning here: If the function $Z = u(X)$ does not have a single-valued inverse, the determination of the distribution of Z will not be as simple. We did consider one such example in Section 4.5 by finding the distribution of Z^2, where Z is $N(0, 1)$. In this case, there were "two inverse functions" and special care was exercised. In our examples, we will *not* consider problems with "many inverses"; however, we thought that such a warning should be issued here.

When two or more random variables are involved, many interesting problems can result. In the case of a single-valued inverse, the rule is about the same as that in the one-variable case, with the derivative being replaced by the Jacobian. Namely, if X_1 and X_2 are two continuous-type random variables with joint p.d.f. $f(x_1, x_2)$ and if $Z_1 = u_1(X_1, X_2)$, $Z_2 = u_2(X_1, X_2)$ has the single-valued inverse $X_1 = v_1(Z_1, Z_2)$, $X_2 = v_2(Z_1, Z_2)$, then the joint p.d.f. of Z_1 and Z_2 is

$$g(z_1, z_2) = |J| f[v_1(z_1, z_2), v_2(z_1, z_2)],$$

where the Jacobian J is the determinant

$$J = \begin{vmatrix} \dfrac{\partial x_1}{\partial z_1} & \dfrac{\partial x_1}{\partial z_2} \\[2mm] \dfrac{\partial x_2}{\partial z_1} & \dfrac{\partial x_2}{\partial z_2} \end{vmatrix}.$$

Of course, we find the support of Z_1, Z_2 by considering the mapping of the support of X_1, X_2 under the transformation $z_1 = u_1(x_1, x_2), z_2 = u_2(x_1, x_2)$.

Example 12.1–3. Let X_1 and X_2 have independent gamma distributions with parameters α, θ and β, θ, respectively. That is, the joint p.d.f. of X_1 and X_2 is

$$f(x_1, x_2) = \frac{1}{\Gamma(\alpha)\Gamma(\beta)\theta^{\alpha+\beta}} x_1^{\alpha-1} x_2^{\beta-1} \exp\left(-\frac{x_1 + x_2}{\theta}\right),$$

$$0 < x_1 < \infty, \quad 0 < x_2 < \infty.$$

Consider

$$Z_1 = \frac{X_1}{X_1 + X_2}, \qquad Z_2 = X_1 + X_2$$

or, equivalently,

$$X_1 = Z_1 Z_2, \qquad X_2 = Z_2 - Z_1 Z_2.$$

The Jacobian is

$$J = \begin{vmatrix} z_2 & z_1 \\ -z_2 & 1 - z_1 \end{vmatrix} = z_2(1 - z_1) + z_1 z_2 = z_2.$$

Thus, the joint p.d.f. $g(z_1, z_2)$ of Z_1 and Z_2 is

$$g(z_1, z_2) = |z_2| \frac{1}{\Gamma(\alpha)\Gamma(\beta)\theta^{\alpha+\beta}} (z_1 z_2)^{\alpha-1} (z_2 - z_1 z_2)^{\beta-1} e^{-z_2/\theta},$$

where the support is $0 < z_1 < 1, 0 < z_2 < \infty$. The marginal p.d.f. of Z_1 is

$$g_1(z_1) = \frac{z_1^{\alpha-1}(1 - z_1)^{\beta-1}}{\Gamma(\alpha)\Gamma(\beta)} \int_0^\infty \frac{z_2^{\alpha+\beta-1}}{\theta^{\alpha+\beta}} e^{-z_2/\theta} \, dz_2.$$

But the integral in this expression is that of a gamma p.d.f. with parameters $\alpha + \beta$ and θ, except for $\Gamma(\alpha + \beta)$ in the denominator; hence, the integral equals $\Gamma(\alpha + \beta)$ and

$$g_1(z_1) = \frac{\Gamma(\alpha + \beta)}{\Gamma(\alpha)\Gamma(\beta)}z_1^{\alpha-1}(1 - z_1)^{\beta-1}, \qquad 0 < z_1 < 1.$$

We say that Z_1 has a *beta* p.d.f. with parameters α and β.

Example 12.1–4. (Box–Muller transformation) It is not easy to generate normal random variables using $Y = F(X)$, where Y is uniform $U(0, 1)$ and F is the distribution function of the desired normal distribution, because we cannot express the normal distribution function $F(x)$ in closed form. Consider the following transformation, however, where X_1 and X_2 are items of a random sample from $U(0, 1)$:

$$Z_1 = \sqrt{-2 \ln X_1} \cos(2\pi X_2), \qquad Z_2 = \sqrt{-2 \ln X_1} \sin(2\pi X_2)$$

or, equivalently,

$$X_1 = \exp\left(-\frac{Z_1^2 + Z_2^2}{2}\right) = e^{-q/2}, \qquad X_2 = \frac{1}{2\pi}\arctan\left(\frac{Z_2}{Z_1}\right),$$

which has Jacobian

$$J = \begin{vmatrix} -z_1 e^{-q/2} & -z_2 e^{-q/2} \\[2mm] \dfrac{-z_2}{2\pi(z_1^2 + z_2^2)} & \dfrac{z_1}{2\pi(z_1^2 + z_2^2)} \end{vmatrix} = \frac{-1}{2\pi} e^{-q/2}.$$

Since the joint p.d.f. of X_1 and X_2 is

$$f(x_1, x_2) = 1, \qquad 0 < x_1 < 1, \quad 0 < x_2 < 1,$$

we have that the joint p.d.f. of Z_1 and Z_2 is

$$g(z_1, z_2) = \left| \frac{-1}{2\pi} e^{-q/2} \right| \quad (1)$$

$$= \frac{1}{2\pi} \exp\left(-\frac{z_1^2 + z_2^2}{2}\right), \qquad -\infty < z_1 < \infty, \quad -\infty < z_2 < \infty.$$

(The student should note that there is some difficulty with the definition of the transformation, particularly when $Z_1 = 0$. However, these difficulties occur at events with probability zero and hence cause no problems.) To summarize, from two independent $U(0, 1)$ random variables, we have generated two independent $N(0, 1)$ random variables through this *Box–Muller transformation*.

The techniques described for two random variables can be extended to three or more random variables. We do not give any details here but mention, for illustration, that with three random variables X_1, X_2, X_3 of the continuous type, we need three "new" random variables Z_1, Z_2, Z_3 so that the corresponding Jacobian of the single-valued inverse transformation is the nonzero determinant

$$
J = \begin{vmatrix}
\dfrac{\partial x_1}{\partial z_1} & \dfrac{\partial x_1}{\partial z_2} & \dfrac{\partial x_1}{\partial z_3} \\[2mm]
\dfrac{\partial x_2}{\partial z_1} & \dfrac{\partial x_2}{\partial z_2} & \dfrac{\partial x_2}{\partial z_3} \\[2mm]
\dfrac{\partial x_3}{\partial z_1} & \dfrac{\partial x_3}{\partial z_2} & \dfrac{\partial x_3}{\partial z_3}
\end{vmatrix}.
$$

══ *Exercises* ══

12.1–1. Let the p.d.f. of X be defined by $f(x) = x^3/4, 0 < x < 2$. Find the p.d.f. of $Y = X^2$.

12.1–2. Let the p.d.f. of X be defined $f(x) = 1/\pi,\ -\pi/2 < x < \pi/2$. Find the p.d.f. of $Y = \tan X$. We say that Y has a *Cauchy* distribution.

12.1–3. Let the p.d.f. of X be defined by $f(x) = (3/2)x^2,\ -1 < x < 1$. Find the p.d.f. of $Y = (X^3 + 1)/2$.

12.1–4. If Y has a uniform distribution on the interval $(0, 1)$, find the p.d.f. of

$$X = (2Y - 1)^{1/3}.$$

12.1–5. Let X_1, X_2 denote a random sample from a distribution $\chi^2(2)$. Find the joint p.d.f. of $Y_1 = X_1$ and $Y_2 = X_2 + X_1$. Here note that the support of Y_1, Y_2 is $0 < y_1 < y_2 < \infty$. Also find the marginal p.d.f. of each of Y_1 and Y_2. Are Y_1 and Y_2 independent?

12.1–6. Let X_1, X_2 be a random sample from $N(0, 1)$. Use the transformation defined by polar coordinates: $X_1 = Z_1 \cos Z_2, X_2 = Z_1 \sin Z_2$.
(a) Show that the Jacobian equals z_1. (This explains the factor r of $r\, dr\, d\theta$ in the usual polar coordinate notation.)
(b) Find the joint p.d.f. of Z_1 and Z_2.
(c) Are Z_1 and Z_2 independent?

12.1–7. Let the independent random variables X_1 and X_2 be $N(0, 1)$ and $\chi^2(r)$, respectively. Let $Z_1 = X_1/\sqrt{X_2/r}$ and $Z_2 = X_2$.
(a) Find the joint p.d.f. of Z_1 and Z_2.
(b) Determine the marginal p.d.f. of Z_1, frequently called *Student's t* distribution.

12.1–8. Let X_1 and X_2 be independent chi-square random variables with r_1 and r_2 degrees of freedom, respectively. Let $Z_1 = (X_1/r_1)/(X_2/r_2)$ and $Z_2 = X_2$.
(a) Find the joint p.d.f. of Z_1 and Z_2.
(b) Determine the marginal p.d.f. of Z_1, frequently called the F distribution.

12.1–9. It is true, when sampling from a bivariate normal distribution, that the correlation coefficient R has an approximate $N[\rho, (1 - \rho^2)^2/n]$ if the sample size n is large. Since, for large n, R is close to ρ, use two terms of the Taylor's expansion of $u(R)$ about ρ and determine that function $u(R)$ such that it has a variance (essentially) free of ρ. (The solution of this exercise explains why the transformation $(1/2)\ln[(1 + R)/(1 - R)]$ was suggested in Section 9.7.)

12.1–10. Let X have a beta distribution with parameters α and β. (See Example 12.1–3.) Show that the mean and variance of X are

$$\mu = \frac{\alpha}{\alpha + \beta} \quad \text{and} \quad \sigma^2 = \frac{\alpha\beta}{(\alpha + \beta + 1)(\alpha + \beta)^2}.$$

HINT: Use the fact that

$$\int_0^1 y^{\alpha - 1}(1 - y)^{\beta - 1}\, dy = \frac{\Gamma(\alpha)\Gamma(\beta)}{\Gamma(\alpha + \beta)}$$

to compute $E(X)$ and $E(X^2)$.

12.1–11. Determine the constant c such that $f(x) = cx^3(1 - x)^6, 0 < x < 1$, is a p.d.f.

12.2 Best Critical Regions

In this section, we consider the properties a satisfactory test (or critical region) should possess. To introduce our investigation, we begin with a nonstatistical example.

Example 12.2–1. Say you have α dollars with which to buy books. Suppose further that you are not interested in the books themselves, but only in filling as much of your bookshelves as possible. How do you decide which books to buy? Does the following approach seem reasonable? First of all, take all the available free books. Then start choosing those books for which the cost of filling an inch of bookshelf space is smallest. That is, choose those books for which the ratio c/w is a minimum, where w is the width of the book in inches and c is the cost of the book. Continue choosing books this way until you have spent the α dollars.

To see how Example 12.2–1 provides the background for selecting a good critical region of size α, let us consider a test of the simple hypothesis $H_0: \theta = \theta_0$ against a simple alternative hypothesis $H_1: \theta = \theta_1$. In this discussion we assume that the random variables $X_1, X_2, \ldots, X_n$ under consideration have a joint p.d.f. of the discrete type, which we here denote by $L(\theta; x_1, x_2, \ldots, x_n)$. That is,

$$P(X_1 = x_1, X_2 = x_2, \ldots, X_n = x_n) = L(\theta; x_1, x_2, \ldots, x_n).$$

A critical region C of size α is a set of points $(x_1, x_2, \ldots, x_n)$ with the probability of α when $\theta = \theta_0$. For a good test, this set C of points should have a large probability when $\theta = \theta_1$ because, under $H_1 : \theta = \theta_1$, we wish to reject $H_0 : \theta = \theta_0$. Accordingly, the first point we would place in the critical region C is the one with the smallest ratio

$$\frac{L(\theta_0; x_1, x_2, \ldots, x_n)}{L(\theta_1; x_1, x_2, \ldots, x_n)}.$$

That is, the "cost" in terms of probability under $H_0 : \theta = \theta_0$ is small compared to the probability that we can "buy" if $\theta = \theta_1$. The next point to add to C would be the one with the next smallest ratio. We would continue to add points to C in this manner until the probability of C, under $H_0 : \theta = \theta_0$, equals α. In this way, we have achieved, for the given significance level α, the region C with the largest probability when $H_1 : \theta = \theta_1$. We now formalize this discussion by defining a best critical region and proving the well-known Neyman–Pearson lemma.

DEFINITION 12.2–1.   *Consider the test of the simple null hypothesis $H_0 : \theta = \theta_0$ against the simple alternative hypothesis $H_1 : \theta = \theta_1$. Let C be a critical region of size α; that is, $\alpha = P(C; \theta_0)$. Then C is a best critical region of size α if, for every other critical region D of size $\alpha = P(D; \theta_0)$, we have that*

$$P(C; \theta_1) \geq P(D; \theta_1).$$

That is, when $H_1 : \theta = \theta_1$ is true, the probability of rejecting $H_0 : \theta = \theta_0$ using the critical region C is at least as great as the corresponding probability using any other critical region D of size α.

Thus, a best critical region of size α is the critical region that has the greatest power among all critical regions of size α. The Neyman–Pearson Lemma gives sufficient conditions for a best critical region of size α.

THEOREM 12.2–1.   (Neyman–Pearson Lemma)   *Let $X_1, X_2, \ldots, X_n$ be a random sample of size n from a distribution with p.d.f. $f(x; \theta)$, where θ_0 and θ_1 are two possible values of θ. Denote the joint p.d.f. of $X_1, X_2, \ldots, X_n$ by the likelihood function*

$$L(\theta) = L(\theta; x_1, x_2, \ldots, x_n) = f(x_1; \theta)f(x_2; \theta) \cdots f(x_n; \theta).$$

If there exist a positive constant k and a subset C of the sample space such that

(i) $P\{(X_1, X_2, \ldots, X_n) \in C; \theta_0\} = \alpha$,

(ii) $\dfrac{L(\theta_0)}{L(\theta_1)} \leq k$ *for* $(x_1, x_2, \ldots, x_n) \in C$,

(iii) $\dfrac{L(\theta_0)}{L(\theta_1)} \geq k$ *for* $(x_1, x_2, \ldots, x_n) \in C'$

then C is a best critical region of size α for testing the simple hypothesis $H_0: \theta = \theta_0$ against the alternative simple hypothesis $H_1: \theta = \theta_1$.

Proof: We prove the theorem when the random variables are of the continuous type; for discrete-type random variables, replace the integral signs by summation signs. To simplify the exposition, we shall use the following notation:

$$\int_B L(\theta) = \int \cdots \int_B L(\theta; x_1, x_2, \ldots, x_n) \, dx_1 \, dx_2 \cdots dx_n.$$

Assume that there exists another critical region of size α, say D, such that, in this new notation,

$$\alpha = \int_C L(\theta_0) = \int_D L(\theta_0).$$

So we have

$$0 = \int_C L(\theta_0) - \int_D L(\theta_0)$$

$$= \int_{C \cap D'} L(\theta_0) + \int_{C \cap D} L(\theta_0) - \int_{C \cap D} L(\theta_0) - \int_{C' \cap D} L(\theta_0)$$

and, hence,

$$0 = \int_{C \cap D'} L(\theta_0) - \int_{C' \cap D} L(\theta_0).$$

By hypothesis (ii), $kL(\theta_1) \geq L(\theta_0)$ at each point in C and therefore in $C \cap D'$; thus

$$k \int_{C \cap D'} L(\theta_1) \geq \int_{C \cap D'} L(\theta_0).$$

By hypothesis (iii), $kL(\theta_1) \leq L(\theta_0)$ at each point in C', and therefore in $C' \cap D$; thus, we obtain

$$k \int_{C' \cap D} L(\theta_1) \leq \int_{C' \cap D} L(\theta_0).$$

Therefore,

$$0 = \int_{C \cap D'} L(\theta_0) - \int_{C' \cap D} L(\theta_0) \leq (k) \left\{ \int_{C \cap D'} L(\theta_1) - \int_{C' \cap D} L(\theta_1) \right\}.$$

That is,

$$0 \leq (k) \left\{ \int_{C \cap D'} L(\theta_1) + \int_{C \cap D} L(\theta_1) - \int_{C \cap D} L(\theta_1) - \int_{C' \cap D} L(\theta_1) \right\}$$

or, equivalently,

$$0 \leq (k)\left\{\int_C L(\theta_1) - \int_D L(\theta_1)\right\}.$$

Thus,

$$\int_C L(\theta_1) \geq \int_D L(\theta_1);$$

that is, $P(C; \theta_1) \geq P(D; \theta_1)$. Since that is true for every critical region D of size α, C is a best critical region of size α.

For a realistic application of the Neyman–Pearson Lemma, consider the following in which the test is based on a random sample from a normal distribution.

Example 12.2–2. Let $X_1, X_2, \ldots, X_n$ be a random sample from $N(\mu, 36)$. We shall find the best critical region for testing the simple hypothesis $H_0: \mu = 50$ against the simple alternative hypothesis $H_1: \mu = 55$. Using the ratio of the likelihood functions, namely, $L(50)/L(55)$, we shall find those points in the sample space for which this ratio is less than or equal to some constant k. That is, we shall solve the following inequality:

$$\frac{L(50)}{L(55)} = \frac{(72\pi)^{-n/2} \exp[-(1/72) \sum_1^n (x_i - 50)^2]}{(72\pi)^{-n/2} \exp[-(1/72) \sum_1^n (x_i - 55)^2]}$$

$$= \exp\left[-(1/72)\left(10 \sum_1^n x_i + n50^2 - n55^2\right)\right] \leq k.$$

If we take the natural logarithm of each member of the inequality, we find that

$$-10 \sum_1^n x_i - n50^2 + n55^2 \leq (72) \ln k.$$

Thus,

$$\frac{1}{n} \sum_1^n x_i \geq -\frac{1}{10n} [n50^2 - n55^2 + (72) \ln k]$$

or, equivalently,

$$\bar{x} \geq c,$$

where $c = -(1/10n)[n50^2 - n55^2 + (72) \ln k]$. Thus, $L(50)/L(55) \leq k$ is equivalent to $\bar{x} \geq c$. A best critical region is, according to the Neyman–Pearson Lemma,

$$C = \{(x_1, x_2, \ldots, x_n): \ \bar{x} \geq c\},$$

where c is selected so that the size of the critical region is α. Say $n = 16$ and $c = 53$. Since $\overline{X}$ is $N(50, 36/16)$ under H_0, we have

$$\alpha = P(\overline{X} \geq 53; \mu = 50)$$

$$= P\left(\frac{\overline{X} - 50}{6/4} \geq \frac{3}{6/4}; \mu = 50\right) = 1 - \Phi(2) = 0.0228.$$

This last example illustrates what is often true, namely, the inequality $L(\theta_0)/L(\theta_1) \leq k$ can be expressed in terms of a function $u(x_1, x_2, \ldots, x_n)$, say

$$u(x_1, \ldots, x_n) \leq c_1$$

or

$$u(x_1, \ldots, x_n) \geq c_2$$

where c_1 or c_2 is selected so that the size of the critical region is α. Thus, the test can be based upon the statistic $u(X_1, \ldots, X_n)$. And, for illustration, if we want α to be a given value, say 0.05, we could then choose our c_1 or c_2. In Example 12.2–2, with $\alpha = 0.05$, we want

$$0.05 = P(\overline{X} \geq c; \mu = 50)$$

$$= P\left(\frac{\overline{X} - 50}{6/4} \geq \frac{c - 50}{6/4}; \mu = 50\right) = 1 - \Phi\left(\frac{c - 50}{6/4}\right).$$

Hence, it must be true that $(c - 50)/(3/2) = 1.645$, or, equivalently,

$$c = 50 + \frac{3}{2}(1.645) \approx 52.47.$$

When $H_0: \theta = \theta_0$ and $H_1: \theta = \theta_1$ are both simple hypotheses, a critical region of size α is a best critical region if the probability of rejecting H_0 when H_1 is true is a maximum when compared with all other critical regions of size α. The test using the best critical region is called a *most powerful test* because it has the greatest value of the power function at $\theta = \theta_1$ when compared with that of other tests of significance level α. If H_1 is a composite hypothesis, the power of a test depends on each simple alternative in H_1.

DEFINITION 12.2–2. *A test, defined by a critical region C of size α, is a uniformly most powerful test if it is a most powerful test against each simple alternative in H_1. The critical region C is called a uniformly most powerful critical region of size α.*

Let us reconsider Example 12.2–2 when the alternative hypothesis is composite.

Example 12.2–3. Let $X_1, \ldots, X_n$ be a random sample from $N(\mu, 36)$. We have seen that, when testing $H_0 : \mu = 50$ against $H_1 : \mu = 55$, a best critical region C is defined by $C = \{(x_1, x_2, \ldots, x_n) : \bar{x} \geq c\}$, where c is selected so that the significance level is α. Now consider testing $H_0 : \mu = 50$ against the one-sided composite alternative hypothesis $H_1 : \mu > 50$. For each simple hypothesis in H_1, say $\mu = \mu_1$, the quotient of the likelihood functions is

$$\frac{L(50)}{L(\mu_1)} = \frac{(72\pi)^{-n/2} \exp[-(1/72) \sum_1^n (x_i - 50)^2]}{(72\pi)^{-n/2} \exp[-(1/72) \sum_1^n (x_i - \mu_1)^2]}$$

$$= \exp\left[-\frac{1}{72} \left\{ 2(\mu_1 - 50) \sum_1^n x_i + n(50^2 - u_1^2) \right\} \right].$$

Now $L(50)/L(\mu_1) \leq k$ if and only if

$$\bar{x} \geq \frac{(-72)\ln(k)}{2n(\mu_1 - 50)} + \frac{50 + \mu_1}{2} = c.$$

Thus, the best critical region of size α for testing $H_0 : \mu = 50$ against $H_1 : \mu = \mu_1$, where $\mu_1 > 50$, is given by $C = \{(x_1, x_2, \ldots, x_n); \bar{x} \geq c\}$, where c is selected such that $P(\bar{X} \geq c; H_0 : \mu = 50) = \alpha$. Note that the same value of c can be used for each $\mu_1 > 50$, but (of course) k does not remain the same. Since the critical region C defines a test that is most powerful against each simple alternative $\mu_1 > 50$, this is a uniformly most powerful test and C is a uniformly most powerful critical region of size α. Again if $\alpha = 0.05$, then $c \approx 52.47$.

Exercise 12.2–5 will demonstrate that uniformly most powerful tests do not always exist; in particular, they do not usually exist when the composite alternative hypothesis is two-sided.

═══ *Exercises* ═══

12.2–1. Let $X_1, X_2, \ldots, X_n$ be a random sample from a distribution $N(\mu, 64)$.
(a) Show that $C = \{(x_1, x_2, \ldots, x_n) : \bar{x} < c\}$ is a best critical region for testing $H_0 : \mu = 80$ against $H_1 : \mu = 76$.
(b) Find n and c so that $\alpha = 0.05$ and $\beta = 0.10$, approximately.
HINT: Find equivalent statements for $P(\bar{X} < c; \mu = 80) = 0.05$ and $P(\bar{X} \geq c; \mu = 76) = 0.10$ by using the fact that $(\bar{X} - \mu)/(8/\sqrt{n})$ is $N(0, 1)$.

12.2–2. Let $X_1, X_2, \ldots, X_n$ be a random sample from $N(0, \sigma^2)$.
(a) Show that $C = \{(x_1, x_2, \ldots, x_n) : \sum_1^n x_i^2 \geq c\}$ is a best critical region for testing $H_0 : \sigma^2 = 4$ against $H_1 : \sigma^2 = 9$.
(b) If $n = 15$, find the value of c so that $\alpha = 0.05$. HINT: Recall that $\sum_1^n X_i^2/\sigma^2$ is $\chi^2(n)$.
(c) If $n = 15$ and c is the value found in part (b), find the approximate value of $\beta = P(\sum_1^n X_i^2 \leq c; \sigma^2 = 9)$.

12.2–3. Let X have an exponential distribution with a mean of θ; that is, the p.d.f. of X is $f(x;\theta) = (1/\theta)e^{-x/\theta}, 0 < x < \infty$.

(a) Show that a best critical region for testing $H_0: \theta = 3$ against $H_1: \theta = 5$ can be based on the statistic $\sum_1^n X_i$.

(b) If $n = 12$, use the fact that $(2/\theta)\sum_1^{12} X_i$ is $\chi^2(24)$ to find a best critical region of size $\alpha = 0.10$.

(c) If $n = 12$, find a best critical region of size $\alpha = 0.10$ for testing $H_0: \theta = 3$ against $H_1: \theta = 7$.

(d) If $H_1: \theta > 3$, is the common region found in parts (b) and (c) a uniformly most powerful critical region of size $\alpha = 0.10$?

12.2–4. Let $X_1, X_2, \ldots, X_n$ be a random sample from a distribution $b(1, p)$.

(a) Show that a best critical region for testing $H_0: p = 0.9$ against $H_1: p = 0.8$ can be based on the statistic $Y = \sum_1^n X_i$, which is $b(n, p)$.

(b) If $C = \{(x_1, x_2, \ldots, x_n): \sum_{i=1}^n x_i \le n(0.85)\}$ and $Y = \sum_1^n X_i$, find the value of n such that $\alpha = 0.10 = P[Y \le n(0.85); p = 0.9]$, approximately. HINT: Use the normal approximation for the binomial distribution.

(c) What is the approximate value of $\beta = P[Y > n(0.85); p = 0.8]$ for the test given in part (b)?

12.2–5. Let $X_1, X_2, \ldots, X_n$ be a random sample from $N(\mu, 36)$.

(a) Show that a uniformly most powerful critical region for testing $H_0: \mu = 50$ against $H_1: \mu < 50$ is given by $C_2 = \{\bar{x}: \bar{x} \le c\}$.

(b) With this result and that of Example 12.2–3, argue that a uniformly most powerful test for testing $H_0: \mu = 50$ against $H_1: \mu \ne 50$ does not exist.

12.2–6. Let $X_1, X_2, \ldots, X_n$ be a random sample from $N(\mu, 9)$. To test the hypothesis $H_0: \mu = 80$ against $H_1: \mu \ne 80$, consider the following three critical regions: $C_1 = \{\bar{x}: \bar{x} \ge c_1\}$, $C_2 = \{\bar{x}: \bar{x} \le c_2\}$, and $C_3 = \{\bar{x}: |\bar{x} - 80| \ge c_3\}$.

(a) If $n = 16$, find the values of c_1, c_2, c_3 such that the size of each critical region is 0.05. That is, find c_1, c_2, c_3 such that

$$0.05 = P(\bar{X} \in C_1; \mu = 80) = P(\bar{X} \in C_2; \mu = 80) = P(\bar{X} \in C_3; \mu = 80).$$

(b) Sketch, on the same graph paper, the power functions for these three critical regions.

12.2–7. Let $X_1, X_2, \ldots, X_{10}$ be a random sample of size 10 from a Poisson distribution with mean μ.

(a) Show that a uniformly most powerful critical region for testing $H_0: \mu = 0.5$ against $H_1: \mu > 0.5$ can be defined using the statistic $\sum_1^{10} X_i$.

(b) What is a uniformly most powerful critical region of size $\alpha = 0.068$? Recall that $\sum_1^{10} X_i$ has a Poisson distribution with mean 10μ.

(c) Sketch the power function of this test.

12.3 Likelihood Ratio Tests

In this section, we consider a general test-construction method that is applicable when both the null and alternative hypotheses, say H_0 and H_1, are composite.

We continue to assume that the functional form of the p.d.f. is known but that it depends on an unknown parameter or unknown parameters. That is, we assume that the p.d.f. of X is $f(x; \theta)$, where θ represents one or more unknown parameters. We let Ω denote the total parameter space, that is, the set of all possible values of the parameter θ given by either H_0 or H_1. These hypotheses will be stated as follows:

$$H_0: \quad \theta \in \omega, \qquad H_1: \quad \theta \in \omega',$$

where ω is a subset of Ω and ω' is the complement of ω with respect to Ω. The test will be constructed by using a ratio of likelihood functions that have been maximized in ω and Ω, respectively. In a sense, this is a natural generalization of the ratio appearing in the Neyman–Pearson Lemma when the two hypotheses were simple.

DEFINITION 12.3–1.   *The* likelihood ratio *is the quotient*

$$\lambda = \frac{L(\hat{\omega})}{L(\hat{\Omega})},$$

where $L(\hat{\omega})$ is the maximum of the likelihood function with respect to θ when $\theta \in \omega$ and $L(\hat{\Omega})$ is the maximum of the likelihood function with respect to θ when $\theta \in \Omega$.

Because λ is the quotient of nonnegative functions, $\lambda \geq 0$. In addition, since $\omega \subset \Omega$, we have that $L(\hat{\omega}) \leq L(\hat{\Omega})$ and hence $\lambda \leq 1$. Thus, $0 \leq \lambda \leq 1$. If the maximum of L in ω is much smaller than that in Ω, it would seem that the data $x_1, x_2, \ldots, x_n$ do not support the hypothesis $H_0: \theta \in \omega$. That is, a small value of the ratio $\lambda = L(\hat{\omega})/L(\hat{\Omega})$ would lead to the rejection of H_0. On the other hand, a value of the ratio λ that is close to one would support the null hypothesis H_0. This leads us to the following definition.

DEFINITION 12.3–2.   *To test $H_0 : \theta \in \omega$ against $H_1 : \theta \in \omega'$, the* critical region *for the* likelihood ratio test *is the set of points in the sample space for which*

$$\lambda = \frac{L(\hat{\omega})}{L(\hat{\Omega})} \leq k,$$

where $0 < k < 1$ and k is selected so that the test has a desired significance level α.

The following example will illustrate these definitions.

Example 12.3–1.   Assume that the weight X in ounces of a "10-pound bag" of sugar is $N(\mu, 5)$. We shall test the hypothesis $H_0: \mu = 160$ against the alternative hypothesis $H_1: \mu \neq 160$. Thus, $\Omega = \{\mu: \ -\infty < \mu < \infty\}$ and $\omega = \{160\}$. To find the likelihood ratio, we need $L(\hat{\omega})$ and $L(\hat{\Omega})$. When H_0

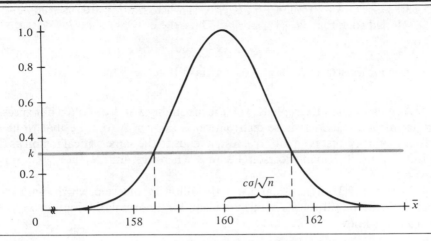

FIGURE 12.3–1

is true, μ can take on only one value, namely, $\mu = 160$. Thus, $L(\hat{\omega}) = L(160)$. To find $L(\hat{\Omega})$, we must find the value of μ that maximizes $L(\mu)$. We recall that $\hat{\mu} = \bar{x}$ is the maximum likelihood estimate of μ. Thus, $L(\hat{\Omega}) = L(\bar{x})$ and the likelihood ratio $\lambda = L(\hat{\omega})/L(\hat{\Omega})$ is given by

$$\lambda = \frac{(10\pi)^{-n/2} \exp[-(1/10) \sum_1^n (x_i - 160)^2]}{(10\pi)^{-n/2} \exp[-(1/10) \sum_1^n (x_i - \bar{x})^2]}$$

$$= \frac{\exp[-(1/10) \sum_1^n (x_i - \bar{x})^2 - (n/10)(\bar{x} - 160)^2]}{\exp[-(1/10) \sum_1^n (x_i - \bar{x})^2]}$$

$$= \exp\left[-\frac{n}{10}(\bar{x} - 160)^2\right].$$

A value of $\bar{x}$ close to 160 would tend to support H_0 and in that case λ is close to 1. On the other hand, an $\bar{x}$ that differs from 160 by too much would tend to support H_1. See Figure 12.3–1 for the graph of this likelihood ratio when $n = 5$.

A likelihood ratio critical region is given by $\lambda \leq k$, where k is selected so that the significance level of the test is α. Using this criterion and simplifying the inequality as we do using the Neyman–Pearson Lemma, we have that $\lambda \leq k$ is equivalent to each of the following inequalities:

$$-(n/10)(\bar{x} - 160)^2 \leq \ln k,$$

$$(\bar{x} - 160)^2 \geq -(10/n) \ln k,$$

$$\frac{|\bar{x} - 160|}{\sigma/\sqrt{n}} \geq \frac{\sqrt{-(10/n) \ln k}}{\sigma/\sqrt{n}} = c.$$

Since $Z = (\bar{X} - 160)/(\sigma/\sqrt{n})$ is $N(0, 1)$ when $H_0 : \mu = 160$ is true, let c be selected such that $P(|Z| \geq c) = \alpha$. Thus, the critical region is

$$C = \{\bar{x} : \ |\bar{x} - 160| \geq c\sigma/\sqrt{n}\}$$

and, for illustration, if $\alpha = 0.05$, we have that $c = 1.96$.

As illustrated in Example 12.3–1, the inequality $\lambda \leq k$ can often be expressed in terms of a statistic whose distribution is known. Also, note that while the likelihood ratio test is an intuitive test, it leads to the same critical region as that given by the Neyman–Pearson Lemma when H_0 and H_1 are both simple hypotheses.

Example 12.3–1 is a special case of the following theorem, which summarizes some likelihood ratio tests about the mean for a normal distribution when the variance is known.

THEOREM 12.3–1. *Let $X_1, X_2, \ldots, X_n$ be a random sample of size n from $N(\mu, \sigma^2)$. Assume that the variance σ^2 is known. Table 12.3–1 lists the critical region C as specified by the likelihood ratio test criterion for testing H_0 against H_1. In each case, the constant z_0 is selected to yield the desired significance level α.*

H_0	H_1	C
$\mu = \mu_0$	$\mu > \mu_0$	$\bar{x} \geq \mu_0 + t_0(s/\sqrt{n-1})$
$\mu = \mu_0$	$\mu < \mu_0$	$\bar{x} \leq \mu_0 - t_0(s/\sqrt{n-1})$
$\mu = \mu_0$	$\mu \neq \mu_0$	$|\bar{x} - \mu_0| \geq t_0(s/\sqrt{n-1})$

TABLE 12.3–2

The proof of this theorem is left as an exercise. Note that each of the first two tests is uniformly most powerful, whereas the third is not.

It is often true that when the hypotheses concern the unknown mean of a normal distribution, the variance is also unknown. When this is the case, the variance is estimated and a ratio, with a t distribution, is involved in the test. The next theorem summarizes some tests about the mean of a normal distribution when the variance is unknown.

H_0	H_1	C
$\mu = \mu_0$	$\mu > \mu_0$	$\bar{x} \geq \mu_0 + t_0(s/\sqrt{n-1})$
$\mu = \mu_0$	$\mu < \mu_0$	$\bar{x} \leq \mu_0 - t_0(s/\sqrt{n-1})$
$\mu = \mu_0$	$\mu \neq \mu_0$	$|\bar{x} - \mu_0| \geq t_0(s/\sqrt{n-1})$

TABLE 12.3–2

THEOREM 12.3–2. *Let $X_1, X_2, \ldots, X_n$ be a random sample of size n from $N(\mu, \sigma^2)$, where μ and σ^2 are both unknown. Table 12.3–2 lists the critical region C as specified by the likelihood ratio test criterion for testing H_0 against H_1. In each case, the constant t_0 is selected to yield the desired significance level α.*

Proof: We shall give the proof for the last test only, that is, when H_1 is $\mu \neq \mu_0$. For this test

$$\omega = \{(\mu, \sigma^2): \quad \mu = \mu_0, 0 < \sigma^2 < \infty\}$$

and

$$\Omega = \{(\mu, \sigma^2): \quad -\infty < \mu < \infty, 0 < \sigma^2 < \infty\}.$$

If $(\mu, \sigma^2) \in \Omega$, the observed maximum likelihood estimates are $\hat{\mu} = \bar{x}$ and $\widehat{\sigma^2} = (1/n) \sum_1^n (x_i - \bar{x})^2$. Thus,

$$L(\hat{\Omega}) = \left[\frac{1}{2\pi(1/n)\sum_1^n (x_i - \bar{x})^2} \right]^{n/2} \exp\left[-\frac{\sum_1^n (x_i - \bar{x})^2}{(2/n)\sum_1^n (x_i - \bar{x})^2} \right]$$

$$= \left[\frac{ne^{-1}}{2\pi \sum_1^n (x_i - \bar{x})^2} \right]^{n/2}.$$

Similarly, if $(\mu, \sigma^2) \in \omega$, the observed maximum likelihood estimates are $\hat{\mu} = \mu_0$ and $\widehat{\sigma^2} = (1/n) \sum_1^n (x_i - \mu_0)^2$. Thus,

$$L(\hat{\omega}) = \left[\frac{1}{2\pi(1/n)\sum_1^n (x_i - \mu_0)^2} \right]^{n/2} \exp\left[-\frac{\sum_1^n (x_i - \mu_0)^2}{(2/n)\sum_1^n (x_i - \mu_0)^2} \right]$$

$$= \left[\frac{ne^{-1}}{2\pi \sum_1^n (x_i - \mu_0)^2} \right]^{n/2}.$$

The likelihood ratio $\lambda = L(\hat{\omega})/L(\hat{\Omega})$ for this test is

$$\lambda = \frac{\{ne^{-1}/[2\pi \sum_1^n (x_i - \mu_0)^2]\}^{n/2}}{\{ne^{-1}/[2\pi \sum_1^n (x_i - \bar{x})^2]\}^{n/2}} = \{[\sum_1^n (x_i - \bar{x})^2]/[\sum_1^n (x_i - \mu_0)^2]\}^{n/2}.$$

However, note that

$$\sum_1^n (x_i - \mu_0)^2 = \sum_1^n (x_i - \bar{x} + \bar{x} - \mu_0)^2 = \sum_1^n (x_i - \bar{x})^2 + n(\bar{x} - \mu_0)^2.$$

If this substitution is made in the denominator of λ, we have

$$\lambda = \left[\frac{\sum_1^n (x_i - \bar{x})^2}{\sum_1^n (x_i - \bar{x})^2 + n(\bar{x} - \mu_0)^2} \right]^{n/2}$$

$$= \left\{ \frac{1}{1 + [n(\bar{x} - \mu_0)^2]/[\sum_1^n (x_i - \bar{x})^2]} \right\}^{n/2}.$$

Note that λ is close to one when $\bar{x}$ is close to μ_0 and its is small when $\bar{x}$ and μ_0 differ by a great deal. The likelihood ratio test, given by the inequality $\lambda \leq k$, is the same as

$$\frac{1}{1 + n(\bar{x} - \mu_0)^2/\sum_1^n (x_i - \bar{x})^2} \leq k^{2/n}$$

or, equivalently,

$$\frac{n(\bar{x} - \mu_0)^2}{\sum_1^n (x_i - \bar{x})^2/(n - 1)} \geq (n - 1)(k^{-2/n} - 1).$$

When H_0 is true, $\sqrt{n}(\bar{X} - \mu_0)/\sigma$ is $N(0, 1)$ and $\sum_1^n (X_i - \bar{X})^2/\sigma^2$ has an independent distribution $\chi^2(n - 1)$. Hence, under H_0,

$$T = \frac{\sqrt{n}(\bar{X} - \mu_0)/\sigma}{\sqrt{\sum_1^n (X_i - \bar{X})^2/[\sigma^2(n - 1)]}} = \frac{\sqrt{n}(\bar{X} - \mu_0)}{\sqrt{\sum_1^n (X_i - \bar{X})^2/(n - 1)}} = \frac{\bar{X} - \mu_0}{S/\sqrt{n - 1}}$$

has a t distribution with $r = n - 1$ degrees of freedom. In accordance with the likelihood ratio test criterion, H_0 is rejected if the observed

$$T^2 \geq (n - 1)(k^{-2/n} - 1).$$

That is, we reject $H_0: \mu = \mu_0$ and accept $H_1: \mu \neq \mu_0$ if the observed $|T| \geq t_0$, where t_0 is selected from the t tables so that $P(|T| \geq t_0) = \alpha$.

Example 12.3–2. Assume that the length X of a bluegill caught by a particular fisherman is $N(\mu, \sigma^2)$. The mean length of fish caught in the past has been $\mu = 6.30$ inches. It is claimed that a new and more expensive bait will enable the fisherman to catch longer fish. If $n = 10$ fish caught with this new bait have an average length of $\bar{x} = 6.81$ inches and a standard deviation of $s = 1.05$ inches, is the claim support by these data? Say we use $\alpha = 0.10$. Note that we can formalize this by testing

$$H_0: \quad \mu = 6.30 \quad \text{against} \quad H_1: \quad \mu > 6.30.$$

The critical region for $\alpha = 0.10$ is given by

$$C = \left\{ \bar{x}: \quad \bar{x} \geq 6.30 + 1.383\left(\frac{1.05}{\sqrt{9}}\right) = 6.784 \right\}.$$

Because $6.81 > 6.784$, H_0 is rejected at an $\alpha = 0.10$ significance level in favor of the alternative hypothesis. Thus, we accept the claim.

Rather than give many additional examples, we simply note that all of the following, which involve underlying normal assumptions, are actually likelihood ratio tests: tests about differences of means, ratio of variances, and the correlation coefficient, in addition to those in the analysis of variance chapter. Thus, we can see that the likelihood ratio criterion does produce some tests that are standard and extremely good.

Exercises

12.3–1. In Example 12.3–1, if $n = 20$ and $\bar{x} = 159.1$, is H_0 accepted at a significance level of size α **(a)** if $\alpha = 0.10$? **(b)** if $\alpha = 0.05$?

12.3–2. Assume that the weight X in ounces of a "10-ounce box" of corn flakes is $N(\mu, 0.25)$.
(a) To test the hypothesis $H_0 : \mu \geq 10$ against the alternative hypothesis $H_1 : \mu < 10$, what is the critical region of size $\alpha = 0.05$ specified by the likelihood ratio test criterion? HINT: If $\mu \geq 10$ and $\bar{x} < 10$, note that $\hat{\mu} = 10$.
(b) If a random sample of $n = 50$ boxes yielded a sample mean of $\bar{x} = 9.87$, is H_0 rejected? HINT: Find the critical value z_0 when H_0 is true by taking $\mu = 10$, which is the extreme value in $\mu \geq 10$.

12.3–3. Let X be $N(\mu, 100)$.
(a) To test $H_0 : \mu = 230$ against $H_1 : \mu > 230$, what is the critical region specified by the likelihood ratio test criterion?
(b) Is this test uniformly most powerful?
(c) If a random sample of $n = 16$ yielded $\bar{x} = 232.6$, is H_0 accepted at a significance level of $\alpha = 0.10$?

12.3–4. Let X be $N(\mu, 225)$.
(a) To test $H_0 : \mu = 59$ against $H_1 : \mu \neq 59$, what is the critical region of size $\alpha = 0.05$ specified by the likelihood ratio test criterion?
(b) If a sample of size $n = 100$ yielded $\bar{x} = 56.13$, is H_0 accepted?

12.3–5. In Example 12.3–2, is H_0 rejected if $\alpha = 0.05$?

12.3–6. It is desired to test the hypothesis $H_0 : \mu = 30$ against the alternative hypothesis $H_1 : \mu \neq 30$, where μ is the mean of a normal distribution and σ^2 is unknown. If a random sample of size $n = 10$ has a mean of $\bar{x} = 32.8$ and a standard deviation of $s = 4$, is H_0 accepted at an $\alpha = 0.05$ significance level?

12.3–7. To test $H_0 : \mu = 335$ against $H_1 : \mu < 335$, under normal assumptions, a random sample of size 17 yielded $\bar{x} = 324.8$ and $s = 40$. Is H_0 accepted at an $\alpha = 0.10$ significance level?

12.3–8. Let X have a normal distribution, in which μ and σ^2 are both unknown. It is desired to test $H_0 : \mu = 1.80$ against $H_1 : \mu > 1.80$ at an $\alpha = 0.10$ significance level. If a random sample of size $n = 121$ yielded $\bar{x} = 1.89$ and $s = 0.20$, is H_0 accepted or rejected? HINT: Here n is so large that Theorem 12.3–1 can be used, replacing σ by s.

12.4 Sufficient Statistics

We first define a sufficient statistic $Y = u(X_1, X_2, \ldots, X_n)$ for a parameter θ using a statement that, in most books, is given as a necessary and sufficient condition for sufficiency, namely, the well-known Fisher–Neyman Factorization

Theorem. Using this as a definition, we will note, by examples, the implications of this definition, one of which is sometimes used as the definition.

DEFINITION 12.4–1. (Factorization Theorem) *Let $X_1, X_2, \ldots, X_n$ denote random variables with joint p.d.f. $f(x_1, \ldots, x_n; \theta)$, which depends upon the parameter θ. The statistic $Y = u(X_1, X_2, \ldots, X_n)$ is* sufficient *for θ if and only if*

$$f(x_1, x_2, \ldots, x_n; \theta) = \phi[u(x_1, \ldots, x_n); \theta]h(x_1, \ldots, x_n),$$

where ϕ depends upon $x_1, \ldots, x_n$ only through $u(x_1, \ldots, x_n)$ and $h(x_1, \ldots, x_n)$ does not depend upon θ.

Let us consider several important examples and consequences of this definition. We first note, however, that in all instances in this book the random variables $X_1, X_2, \ldots, X_n$ will be items of a random sample and hence their joint p.d.f. will be of the form

$$f(x_1; \theta)f(x_2; \theta) \cdots f(x_n; \theta).$$

Example 12.4–1. Let $X_1, X_2, \ldots, X_n$ denote a random sample from a Poisson distribution with parameter $\lambda > 0$. Then

$$f(x_1; \lambda)f(x_2; \lambda) \cdots f(x_n; \lambda) = \frac{\lambda^{\Sigma x_i}e^{-n\lambda}}{x_1!x_2!\cdots x_n!} = (\lambda^{n\bar{x}}e^{-n\lambda})\left(\frac{1}{x_1!x_2!\cdots x_n!}\right),$$

where $\bar{x} = (1/n)\sum x_i$. Thus, from the Factorization Theorem (definition), it is clear that the sample mean $\bar{X}$ is a sufficient statistic for λ. It can easily be shown that the maximum likelihood estimator for λ is also $\bar{X}$ (see Exercise 12.4–1).

In Example 12.4–1, if we replace $n\bar{x}$ by $\sum x_i$, it is quite obvious that the sum $\sum X_i$ is also a sufficient statistic for λ. This certainly agrees with our intuition because if we know one of the statistics, $\bar{X}$ or $\sum X_i$, we can easily find the other. If we generalize this, we see that if Y is sufficient for a parameter θ, then every single-valued function of Y, not involving θ but with a single-valued inverse, is also a sufficient statistic for θ. Again the reason is that knowing either Y or that function of Y we know the other. More formally, if $Z = v(Y) = v[u(X_1, \ldots, X_n)]$ is that function and $Y = v^{-1}(Z)$ is the single-valued inverse, then the display of the Factorization Theorem can be written as

$$f(x_1, \ldots, x_n; \theta) = \phi[v^{-1}\{v[u(x_1, \ldots, x_n)]\}; \theta]h(x_1, \ldots, x_n).$$

The first factor of the right-hand member of this equation depends upon $x_1, x_2, \ldots, x_n$ through $v[u(x_1, \ldots, x_n)]$, so $Z = v[u(X_1, \ldots, X_n)]$ is a sufficient statistic for θ. We illustrate this fact and the Factorization Theorem with an underlying distribution of the continuous type.

Example 12.4–2. Let $X_1, X_2, \ldots, X_n$ be a random sample from $N(\mu, 1)$, $-\infty < \mu < \infty$. The joint p.d.f. of these random variables is

$$(2\pi)^{-n/2} \exp\left[-\frac{\sum_{i=1}^{n} (x_i - \mu)^2}{2} \right]$$

$$= (2\pi)^{-n/2} \exp\left\{ -\frac{\sum_{i=1}^{n} [(x_i - \bar{x}) + (\bar{x} - \mu)]^2}{2} \right\}$$

$$= \left\{ \exp\left[-\frac{n(\bar{x} - \mu)^2}{2} \right] \right\} \left\{ (2\pi)^{-n/2} \exp\left[-\frac{\sum (x_i - \bar{x})^2}{2} \right] \right\}.$$

From the Factorization Theorem, we see that $\bar{X}$ is sufficient for μ. Now $\bar{X}^3$ is also sufficient for θ because knowing $\bar{X}^3$ is equivalent to having knowledge of the value of $\bar{X}$. However, $\bar{X}^2$ does not have this property and it is not sufficient for θ.

One consequence of the sufficiency of a statistic Y is that the conditional probability of any given event A in the support of $X_1, X_2, \ldots, X_n$, given $Y = y$, does not depend upon θ. This is sometimes used as the definition of sufficiency and is illustrated by the following example.

Example 12.4–3. Let $X_1, X_2, \ldots, X_n$ be a random sample from a distribution with p.d.f.

$$f(x; p) = p^x (1 - p)^{1-x}, \qquad x = 0, 1,$$

where the parameter p is between zero and one. We know that

$$Y = X_1 + X_2 + \cdots + X_n$$

is $b(n, p)$ and Y is sufficient for p because the joint p.d.f. of $X_1, X_2, \ldots, X_n$ is

$$p^{x_1}(1 - p)^{1-x_1} \cdots p^{x_n}(1 - p)^{1-x_n} = [p^{\sum x_i}(1 - p)^{n - \sum x_i}](1),$$

where $\phi(u; p) = p^u (1 - p)^{n-u}$ and $h(x_1, x_2, \ldots, x_n) = 1$. What then is the conditional probability $P(X_1 = x_1, \ldots, X_n = x_n | Y = y)$, where $y = 0$, $1, \ldots, n - 1$, or n? Unless the sum of the nonnegative integers $x_1, x_2, \ldots, x_n$ equals y, this conditional probability is obviously equal to zero, which does not depend upon p. Hence, it is only interesting to consider the solution when $y = x_1 + \cdots + x_n$. From the definition of conditional probability, we have

$$P(X_1 = x_1, \ldots, X_n = x_n | Y = y) = \frac{P(X_1 = x_1, \ldots, X_n = x_n)}{P(Y = y)},$$

$$= \frac{p^{x_1}(1 - p)^{1-x_1} \cdots p^{x_n}(1 - p)^{1-x_n}}{\binom{n}{y} p^y (1 - p)^{n-y}}$$

$$= \frac{1}{\binom{n}{y}},$$

where $y = x_1 + \cdots + x_n$. Since y equals the number of ones in the collection $x_1, x_2, \ldots, x_n$, this answer is only the probability of selecting a particular arrangement, namely, $x_1, x_2, \ldots, x_n$ of y ones and $n - y$ zeros, and does not depend upon the parameter p. That is, given that the sufficient statistic $Y = y$, the conditional probability of $X_1 = x_1, X_2 = x_2, \ldots, X_n = x_n$ does not depend upon the parameter p.

It is interesting to observe that the underlying p.d.f. in Examples 12.4–1, 12.4–2, and 12.4–3 can be written in the exponential form

$$f(x; \theta) = \exp[K(x)p(\theta) + S(x) + q(\theta)],$$

where the support is free of θ. That is, we have, respectively,

$$\frac{e^{-\lambda}\lambda^x}{x!} = \exp\{x \ln \lambda - \ln x! - \lambda\}, \qquad x = 0, 1, 2, \ldots,$$

$$\frac{1}{\sqrt{2\pi}} e^{-(x-\mu)^2/2} = \exp\left\{x\mu - \frac{x^2}{2} - \frac{\mu^2}{2} - \frac{1}{2}\ln(2\pi)\right\}, \qquad -\infty < x < \infty,$$

and

$$p^x(1 - p)^{1-x} = \exp\left\{x \ln\left(\frac{p}{1-p}\right) + \ln(1 - p)\right\}, \qquad x = 0, 1.$$

In each of these examples, the sum $\sum X_i$ of the items of the random sample was the sufficient statistic for the parameter. This is generalized by Theorem 12.4–1.

THEOREM 12.4–1. *Let $X_1, X_2, \ldots, X_n$ be a random sample from a distribution with a p.d.f. of the exponential form*

$$f(x; \theta) = \exp[K(x)p(\theta) + S(x) + q(\theta)]$$

on a support free of θ. The statistic $\sum_{i=1}^{n} K(X_i)$ is sufficient for θ.

Proof: The joint p.d.f. of $X_1, X_2, \ldots, X_n$ is

$$\exp\left[p(\theta) \sum_1^n K(x_i) + \sum_1^n S(x_1) + nq(\theta) \right]$$

$$= \left\{\exp\left[p(\theta) \sum_1^n K(x_i) + nq(\theta)\right]\right\}\left\{\exp\left[\sum_1^n S(x_i)\right]\right\}.$$

In accordance with the Factorization Theorem, the $\sum K(X_i)$ is sufficient for θ.

In many cases Theorem 12.4–1 permits the student to find the sufficient statistic for the parameter with very little effort, as shown by the following example.

Example 12.4–4. Let $X_1, X_2, \ldots, X_n$ be a random sample from an exponential distribution with p.d.f.

$$f(x; \theta) = \frac{1}{\theta} e^{-x/\theta} = \exp\left[x\left(-\frac{1}{\theta}\right) - \ln \theta \right], \qquad 0 < x < \infty,$$

provided $0 < \theta < \infty$. Here $K(x) = x$. Thus, $\sum_1^n X_i$ is sufficient for θ; of course, $\bar{X} = \sum_1^n X_i/n$ is also sufficient.

Another consequence of the sufficiency of the statistic Y for a parameter θ is that the conditional distribution of any other statistic, say Z, given $Y = y$, must be free of θ. This fact is illustrated by the next example.

Example 12.4–5. If the sample arises from $N(\mu, 1)$, $-\infty < \mu < \infty$, we have seen that $Y = \sum X_i$ is a sufficient statistic for μ. Let $n = 2$. We find the conditional p.d.f. of $Z = X_1 + 2X_2$, given $Y = y$, by first finding the joint p.d.f. of

$$Y = X_1 + X_2, \qquad Z = X_1 + 2X_2.$$

Equivalently, we have that

$$X_1 = 2Y - Z, \qquad X_2 = Z - Y,$$

which has Jacobian

$$J = \begin{vmatrix} 2 & -1 \\ -1 & 1 \end{vmatrix} = 1.$$

Thus, the joint p.d.f. of Y and Z is

$$\frac{1}{2\pi} \exp\left[-\frac{(2y - z - \mu)^2}{2} - \frac{(z - y - \mu)^2}{2} \right],$$

where $-\infty < y < \infty$, $-\infty < z < \infty$. Now we know that the marginal p.d.f. of $Y = X_1 + X_2$ is $N(2\mu, 2)$. Thus, the conditional p.d.f. Z, given $Y = y$, is

$$\frac{(1/2\pi)\exp[-(2y - z - \mu)^2/2 - (z - y - \mu)^2/2]}{(1/\sqrt{2\pi}\sqrt{2})\exp[-(y - 2\mu)^2/2(2)]}$$

$$= \frac{\sqrt{2}}{\sqrt{2\pi}} \exp\left[-\frac{(z - 3y/2)^2}{2(1/2)} \right].$$

That is, the conditional distribution of Z, given $Y = y$, is normal with mean $3y/2$ and variance $1/2$, and it does *not* depend upon μ.

We close this section by noting the importance of sufficient statistics in estimation and tests of hypotheses. The Rao–Blackwell Theorem can be used to emphasize that in point estimation.

THEOREM 12.4–2. (Rao–Blackwell Theorem)   *Let V and Y be two random variables such that V has mean $E(V) = \theta$ and positive finite variance. Let $E(V | Y = y) = w(y)$. Then the random variable $W = w(Y)$ is such that $E(W) = \theta$ and $\mathrm{Var}(W) \leq \mathrm{Var}(V)$.*

Proof: We present the proof for random variables of the continuous type; the discrete case can be handled by replacing integrals by summations. The joint, marginal, and conditional probability density functions are given by $g(y, v)$, $g_1(y)$, $g_2(v)$, and $h(v | y)$, respectively. The mean of W is given by

$$E(W) = E[w(Y)] = \int_{-\infty}^{\infty} E(V | y) g_1(y)\, dy$$

$$= \int_{-\infty}^{\infty} \left\{ \int_{-\infty}^{\infty} v\, \frac{g(y, v)}{g_1(y)}\, dv \right\} g_1(y)\, dy$$

$$= \int_{-\infty}^{\infty} \int_{-\infty}^{\infty} v\, \frac{g(y, v)}{g_1(y)}\, g_1(y)\, dy\, dv$$

$$= \int_{-\infty}^{\infty} \int_{-\infty}^{\infty} v g(y, v)\, dy\, dv = \int_{-\infty}^{\infty} v g_2(v)\, dv = \theta.$$

The last equality follows because $E(V) = \theta$. Hence we have proved that $E(W) = \theta$.

The variance of V is given by

$$\mathrm{Var}(V) = E[(V - \theta)^2] = E[(V - W + W - \theta)^2]$$
$$= E[(V - W)^2] + 2E[(V - W)(W - \theta)] + \mathrm{Var}(W),$$

since $E[(W - \theta)^2] = \mathrm{Var}(W)$. If we can show that $E[(V - W)(W - \theta)] = 0$, then we will have established that $\mathrm{Var}(W) \leq \mathrm{Var}(V)$ because $E[(V - W)^2] \geq 0$. We have that

$$E[(V - W)(W - \theta)] = \int_{-\infty}^{\infty} \int_{-\infty}^{\infty} [v - w(y)][w(y) - \theta] g(y, v)\, dy\, dv$$

$$= \int_{-\infty}^{\infty} \int_{-\infty}^{\infty} [v - w(y)][w(y) - \theta] h(v | y) g_1(y)\, dv\, dy$$

$$= \int_{-\infty}^{\infty} [w(y) - \theta] g_1(y) \left\{ \int_{-\infty}^{\infty} [v - w(y)] h(v | y)\, dv \right\} dy.$$

However,

$$\int_{-\infty}^{\infty} [v - w(y)]h(v|y)\, dv = 0$$

because $E(V|y) = w(y)$ by definition. This completes the proof.

Theorem 12.4–2 means that if a sufficient statistic for θ exists, say Y, we may limit our search for a minimum variance unbiased estimator to functions of Y. Why? Because, given another unbiased estimator, say V, we can always find a better estimator, based on Y, that is unbiased and has a smaller variance. This can be accomplished by computing $E(V|y) = w(y)$, which does not depend upon θ, and by using $w(Y)$ as an unbiased estimator whose variance is at least as small as that of V.

After studying the Rao–Blackwell Theorem and the preceding discussion, many students feel that they must first find another unbiased estimator V of θ in order to find a function of the sufficient statistic Y that is an unbiased estimator of θ. This is not the case at all! The theorem and the preceding remarks mean that we can go directly to the consideration of functions of Y when looking for the best unbiased estimator. Thus, the correct procedure is first to find the sufficient statistic Y and then consider expectations of Y and functions of Y in the search for the best unbiased estimator. In many important cases, only one function of Y is an unbiased estimator of θ.

Example 12.4–6.   From Example 12.4–4 we know that the sum $\sum_1^n X_i$ of the items of a random sample is a sufficient statistic for the parameter θ of the exponential p.d.f. $f(x;\theta) = (1/\theta)e^{-x/\theta}, 0 < x < \infty, 0 < \theta < \infty$.

$$E\left(\sum_1^n X_i\right) = \sum_1^n E(X_i) = n\theta.$$

Thus,

$$E\left(\frac{1}{n}\sum_1^n X_i\right) = \theta.$$

That is, $\overline{X}$ is that function of the sufficient statistic for θ which is unbiased. Of course, $\overline{X}$ is also sufficient for θ. Here we accept the fact that $\overline{X}$ itself is the only function of $\overline{X}$ that is an unbiased estimator of θ.

We should also note that if there is a sufficient statistic for the parameter under consideration and if the maximum likelihood estimator of this parameter is unique, then the maximum likelihood estimator is a function of the sufficient statistic. To see this heuristically, consider the following. If a sufficient statistic exists, then the likelihood function is

$$L(\theta) = f(x_1, x_2, \ldots, x_n; \theta) = \phi[u(x_1, \ldots, x_n); \theta]h(x_1, \ldots, x_n).$$

Since $h(x_1, \ldots, x_n)$ does not depend upon θ, we maximize $L(\theta)$ by maximizing $\phi[u(x_1, \ldots, x_n); \theta]$. But ϕ is a function of $x_1, x_2, \ldots, x_n$ only through $u(x_1, \ldots, x_n)$. Thus, if there is a unique value of θ that maximizes ϕ, then it must be a function of $u(x_1, \ldots, x_n)$. That is, $\hat{\theta}$ is a function of the sufficient statistic $u(X_1, X_2, \ldots, X_n)$. This fact was alluded to in Example 12.4–1, but it could be checked using other examples and exercises.

From the factorization of $L(\theta)$ when a sufficient statistic $u(X_1, X_2, \ldots, X_n)$ for θ exists, we see that the Neyman–Pearson Lemma would require a best critical region to be based upon that sufficient statistic. The reason for this is that

$$\frac{L(\theta_0)}{L(\theta_1)} = \frac{\phi[u(x_1, x_2, \ldots, x_n); \theta_0]}{\phi[u(x_1, x_2, \ldots, x_n); \theta_1]} \leq k$$

results in an inequality involving $x_1, x_2, \ldots, x_n$ only through the function $u(x_1, x_2, \ldots, x_n)$.

Finally, it should be mentioned that the concept of a single sufficient statistic for one parameter can be extended to joint sufficient statistics for several parameters. There is a Factorization Theorem in these cases too, and it is still true that unique maximum likelihood estimators must be functions of the joint sufficient statistics, provided they exist. Because of these extensions, we immediately note that the likelihood ratio test

$$\lambda = \frac{L(\hat{\omega})}{L(\hat{\Omega})} \leq k$$

must be based upon these sufficient statistics. Thus, in estimating and testing, we see that the good procedures are based on sufficient statistics, provided they exist.

═══ *Exercises* ═══

12.4–1. Show that the sample mean $\bar{X}$ is the maximum likelihood estimator of λ, the parameter of the Poisson distribution. That is, here, the sufficient statistic for λ is the maximum likelihood estimator.

12.4–2. Let $X_1, X_2, \ldots, X_n$ be a random sample from the distribution with p.d.f. $f(x; p) = p(1 - p)^x$, $x = 0, 1, 2, \ldots$, where $0 < p < 1$.
   (a)   Show that $Y = \sum_{i=1}^{n} X_i$ is a sufficient statistic for p.
   (b)   Find a function of $Y = \sum_{i=1}^{n} X_i$ that is an unbiased estimator for p.

12.4–3. Let $X_1, X_2, \ldots, X_n$ be a random sample from a Poisson distribution with mean $\lambda > 0$. Find the conditional probability $P(X_1 = x_1, \ldots, X_n = x_n | Y = y)$, where $Y = X_1 + \cdots + X_n$ and the nonnegative integers $x_1, x_2, \ldots, x_n$ sum to y.

12.4–4. Let X_1, X_2 be a random sample of size 2 from a distribution with p.d.f. $f(x; \theta) = (1/\theta)e^{-x/\theta}, 0 < x < \infty, 0 < \theta < \infty$.

   (a) Find the joint p.d.f. of the sufficient statistic $Y = X_1 + X_2$ and $Y_2 = X_2$.

   (b) Show that Y_2 is an unbiased estimator for θ with variance θ^2.

   (c) Show that the conditional p.d.f. of Y_2, given $Y_1 = y_1$, does not depend upon θ.

   (d) Find $E(Y_2 | y_1) = w(y_1)$ and the variance of $w(Y_1)$.

12.4–5. Let $X_1, X_2, \ldots, X_n$ be a random sample from a distribution with p.d.f. $f(x; \theta) = \theta x^{\theta - 1}, 0 < x < 1$, where $0 < \theta$.

   (a) Find the sufficient statistic Y for θ.

   (b) Show that the maximum likelihood estimator $\hat{\theta}$ is a function of Y.

   (c) Argue that $\hat{\theta}$ is also sufficient for θ.

12.4–6. Let $X_1, X_2, \ldots, X_n$ be a random sample from $N(0, \sigma^2)$.

   (a) Find the sufficient statistic Y for σ^2.

   (b) Show that the maximum likelihood estimator for σ^2 is a function of Y.

   (c) Is the maximum likelihood estimator for σ^2 unbiased?

12.5 Decision Theory and Bayesian Methods

Certainly it would not be appropriate to discuss interesting topics in statistics without at least mentioning decision theory and Bayesian methods, a very attractive way of solving some of the related problems. Statisticians who support these methods are often called Bayesians, and many of them believe that Bayesian procedures comprise the most important area in statistical inference.

Since it is easier to discuss decision theory and Bayesian methods in the estimation framework, we consider only the problem of estimating one parameter, say θ. Moreover, in each of our examples, a sufficient statistic Y for θ does exist, so we will start directly with this sufficient statistic in each case. However, we must emphasize that we could have started more naturally with the sample items $X_1, X_2, \ldots, X_n$ and obtained exactly the same results. That is, in Bayesian procedures, sufficient statistics, if they exist, arise naturally in the analysis. We only begin with a sufficient statistic Y to simplify the exposition.

Let us use $w(Y)$ as that function of the sufficient statistic Y which is to serve as a point estimator of θ. Sometimes w is called the *decision function* because it decides, from the data through Y, the value of our estimate of θ. Of course, if the observed $w(y)$ is close to the unknown θ, then $w(y)$ is a good estimate. On the other hand, if $w(y)$ differs much from θ, we would find it a less attractive estimate. To measure the seriousness of the difference between $w(y)$ and θ, we introduce a nonnegative *loss function* $L[w(y), \theta]$. Frequently, the loss function $L[w(y), \theta] = [w(y) - \theta]^2$ is used; that is, the loss is equal to the square of the error or, equivalently, the square of the difference between the estimate $w(y)$ and θ.

In some experiments using the decision function $w(Y)$, the actual losses can often be small but at other times large. Accordingly, we find it desirable to compute the average loss (or mean loss) by determining the expectation

$$R(w, \theta) = E\{L[w(Y), \theta]\} = \int_{-\infty}^{\infty} L[w(y), \theta]g(y; \theta)\, dy,$$

where $g(y; \theta)$ is the p.d.f. of Y (which here we have assumed to be of the continuous type by using an integral). This expectation $R(w, \theta)$ is called the *risk function*. For example, if the square error loss $[w(y) - \theta]^2$ is used, then $R(w, \theta)$ is referred to as the *mean square error* of $w(Y)$. Ideally we would like to select that decision function w which minimizes $R(w, \theta)$ for all possible values of θ. This, however, is impossible unless w is restricted to some class of decision functions, like the class of all unbiased estimators of θ; this fact is now illustrated by the following example.

Example 12.5–1. Let us consider n Bernoulli trials, each of which has probability θ of success. Let Y be the number of successes in the n trials; of course, Y has a binomial distribution and is a sufficient statistic for θ. Let us take $w_1(Y) = Y/n$ and take L equal to the square error loss function. Thus, the risk function is

$$R(w_1, \theta) = E\left[\left(\frac{Y}{n} - \theta\right)^2\right] = \frac{\theta(1 - \theta)}{n},$$

which is the variance of Y/n and depends upon θ. Suppose, however, someone proposed the use of the decision function $w_2(Y)$, which is the average of the relative frequency Y/n and the fraction $1/2$; that is,

$$w_2(Y) = \frac{1}{2}\left(\frac{Y}{n} + \frac{1}{2}\right) = \frac{2Y + n}{4n}.$$

To see whether w_1 or w_2 is better, let us compute the risk function associated with w_2. Since $E(Z^2) = \mu_Z^2 + \sigma_Z^2$ for all random variables Z with finite variance, we have

$$R(w_2, \theta) = E\left[\left(\frac{2Y + n}{4n} - \theta\right)^2\right]$$

$$= \left[E\left(\frac{2Y + n}{4n} - \theta\right)\right]^2 + \text{Var}\left(\frac{Y}{2n}\right)$$

$$= \left(\frac{2n\theta + n}{4n} - \theta\right)^2 + \frac{1}{4}\frac{\theta(1 - \theta)}{n}$$

$$= \frac{n(1 - 2\theta)^2 + 4\theta(1 - \theta)}{16n}.$$

For some values of θ, $R(w_2, \theta)$ is actually less than $R(w_1, \theta)$, but for the other values of θ the inequality is reversed. To be explicit, suppose that the sample size $n = 4$, then

$$\frac{4(1 - 2\theta)^2 + 4\theta(1 - \theta)}{(16)(4)} < \frac{\theta(1 - \theta)}{4}$$

is equivalent to

$$(1 - 2\theta)^2 < 3\theta(1 - \theta),$$

which means that

$$\frac{1}{2} - \frac{\sqrt{21}}{14} < \theta < \frac{1}{2} + \frac{\sqrt{21}}{14}.$$

For θ values in this interval, w_2 has a smaller risk than the usual maximum likelihood estimator $w_1(Y) = Y/n$ when $n = 4$. The latter, however, has a smaller risk when $|\theta - 1/2| > \sqrt{21}/14$. Thus, this example clearly illustrates that we cannot find one risk function that is smaller than another for all possible values of θ. That is, in this example, each of w_1 and w_2 is better than the other for certain values of θ. Please note that if we had restricted the decision functions to unbiased ones, $w_2(Y)$ could not have been used and $w_1(Y)$ is actually the one with smallest risk in that restricted class (this can be proved in a more advanced course). However, in general, it would be foolish to restrict our search in such a way, and thus we must have some way of ordering the worths of w_1 and w_2.

If we have several competing decision functions, say $w_1, w_2, \ldots, w_k$, how do we order their respective risk functions to select the best one among these k decision functions? Some statisticians suggest finding the maximum of each of the risk functions, and then selecting the decision function with the smallest maximum. This *minimax criterion* is usually very conservative, and a more appealing scheme is that which compares the "average" values of the risk functions instead.

To obtain the average of $R(w, \theta)$, a function of θ, we need to weight it properly for various values of θ. Suppose, as in Example 12.5–1, θ is restricted to the interval $0 < \theta < 1$. Certainly

$$\int_0^1 R(w, \theta) \, d\theta$$

would be an average of $R(w, \theta)$, but so would

$$\int_0^1 R(w, \theta) h(\theta) \, d\theta,$$

where $h(\theta)$ has the properties of a p.d.f. on the support $0 < \theta < 1$. Note the first integral of these last two is just a special case of the second when $h(\theta) = 1$, $0 < \theta < 1$.

To determine the correct average to use, each investigator must somehow assign these prior weights $h(\theta)$. Since $h(\theta)$ has the properties of a p.d.f., we can think of this process as assigning prior probabilities to the parameter. For illustration, suppose that we selected $h(\theta)$ such that

$$\int_0^{1/4} h(\theta)\, d\theta = \frac{1}{10}.$$

This would mean that our prior probability, which is subjective but based (we hope) upon past experience, of $0 < \theta < 1/4$ is $1/10$. That is, if we are not opposed (for some moral reason) to gambling on the position of θ, we would give "nine to one odds" that θ is outside the interval $0 < \theta < 1/4$. Moreover, if we really believed that $1/10$ is the correct prior probability, then we would be willing to take either side of this "nine to one" bet: Bet on $0 < \theta < 1/4$ and win nine units if it happens (and lose one if it does not), or bet on $1/4 < \theta < 1$ and win one unit if it occurs (and lose nine if it does not).

The Bayesians have developed several schemes for assigning prior probability; we certainly cannot do justice to these procedures in this short section and have only alluded to them in the previous paragraph. Thus, we accept the fact here that somehow we can obtain the *prior p.d.f.* $h(\theta)$, which is usually very subjective. The average risk is

$$\int_{-\infty}^{\infty} R(w, \theta) h(\theta)\, d\theta = \int_{-\infty}^{\infty} \left[\int_{-\infty}^{\infty} L[w(y), \theta] g(y; \theta)\, dy \right] h(\theta)\, d\theta,$$

where here we have used $-\infty < \theta < \infty$ as the support of $h(\theta)$. At this point, $g(y; \theta)$ can be thought of as the conditional p.d.f. of Y, given θ; and henceforth we write $g(y; \theta) = g(y|\theta)$. If we let

$$g(y|\theta) h(\theta) = k(y, \theta),$$

and treat $k(y, \theta)$ as a joint p.d.f. of the sufficient statistic and the parameter, then

$$\frac{k(y, \theta)}{k_1(y)} = \frac{g(y|\theta) h(\theta)}{k_1(y)} = k(\theta|y)$$

would serve as the conditional p.d.f. of the parameter, given $Y = y$. This is essentially Bayes' Theorem and $k(\theta|y)$ is called the *posterior p.d.f.* of θ, given $Y = y$. Therefore, replacing $g(y|\theta) h(\theta)$ by $k(\theta|y) k_1(y)$ in the expression for the average risk and changing the order of integration, we have that

$$\int_{-\infty}^{\infty} \left[\int_{-\infty}^{\infty} L[w(y), \theta] k(\theta|y)\, d\theta \right] k_1(y)\, dy$$

is our expression for the average risk. The decision function $w(y)$ that minimizes the interior integral

$$\int_{-\infty}^{\infty} L[w(y), \theta] k(\theta|y)\, d\theta = E_\theta[L|y],$$

which can be thought of as the conditional expectation with respect to θ of the loss $L[w(y), \theta]$, given $Y = y$, is called the *Bayes' solution*. Please note that if $E_\theta[L \,|\, y]$ is minimized by $w(y)$, it is obvious that $w(y)$ must also minimize the average risk. Hence, the Bayes' solution presents the best decision function $w(y)$ as judged by the average values of the risk functions.

Example 12.5–2. In Example 12.5–1, let us take the prior p.d.f. of the parameter to be the beta p.d.f.

$$h(\theta) = \frac{\Gamma(\alpha + \beta)}{\Gamma(\alpha)\Gamma(\beta)} \, \theta^{\alpha - 1}(1 - \theta)^{\beta - 1}, \qquad 0 < \theta < 1.$$

Thus, the joint p.d.f. is a product of a binomial p.d.f. with parameters n and θ and this beta p.d.f., namely

$$k(y, \theta) = \binom{n}{y} \frac{\Gamma(\alpha + \beta)}{\Gamma(\alpha)\Gamma(\beta)} \, \theta^{y + \alpha - 1}(1 - \theta)^{n - y + \beta - 1},$$

on the support given by $y = 0, 1, 2, \ldots, n$ and $0 < \theta < 1$. Suppose we find

$$k_1(y) = \int_0^1 k(y, \theta) \, d\theta$$

$$= \binom{n}{y} \frac{\Gamma(\alpha + \beta)}{\Gamma(\alpha)\Gamma(\beta)} \frac{\Gamma(\alpha + y)\Gamma(n + \beta - y)}{\Gamma(n + \alpha + \beta)}$$

on the support $y = 0, 1, 2, \ldots, n$. Therefore,

$$k(\theta \,|\, y) = \frac{k(y, \theta)}{k_1(y)}$$

$$= \frac{\Gamma(n + \alpha + \beta)}{\Gamma(\alpha + y)\Gamma(n + \beta - y)} \, \theta^{y + \alpha - 1}(1 - \theta)^{n - y + \beta - 1}, \qquad 0 < \theta < 1,$$

which is a beta p.d.f. with parameters $y + \alpha$ and $n - y + \beta$. With the square error loss function we must minimize with respect to $w(y)$, the integral

$$\int_0^1 [\theta - w(y)]^2 k(\theta \,|\, y) \, d\theta$$

to obtain the Bayes' solution. But, in general, if Z is a random variable with a second moment, $E[(Z - b)^2]$ is minimized by $b = E(Z)$. In the preceding display, θ is like the Z with p.d.f. $k(\theta \,|\, y)$ and $w(y)$ is like the b, so the minimization is accomplished by taking

$$w(y) = E(\theta \,|\, y) = \frac{\alpha + y}{\alpha + \beta + n},$$

which is the mean of the beta distribution with parameters $y + \alpha$ and $n - y + \beta$. It is very instructive to note that this Bayes' solution can be written as

$$w(y) = \left(\frac{n}{\alpha + \beta + n}\right)\left(\frac{y}{n}\right) + \left(\frac{\alpha + \beta}{\alpha + \beta + n}\right)\left(\frac{\alpha}{\alpha + \beta}\right),$$

which is a weighted average of the maximum likelihood estimate y/n of θ and the mean $\alpha/(\alpha + \beta)$ of the prior p.d.f. of the parameter. Moreover the respective weights are $n/(\alpha + \beta + n)$ and $(\alpha + \beta)/(\alpha + \beta + n)$. Thus, we see that α and β should be selected so that not only is $\alpha/(\alpha + \beta)$ the desired prior mean, but the sum $\alpha + \beta$ plays a role corresponding to a sample size. That is, if we want our prior opinion to have as much weight as a sample size of 20, we would take $\alpha + \beta = 20$. So if our prior mean is 3/4; we have that α and β are selected such that $\alpha = 15$ and $\beta = 5$.

In Example 12.5–2, it is extremely convenient to note that it is not really necessary to determine $k_1(y)$ to find $k(\theta|y)$. If we divide $k(y, \theta)$ by $k_1(y)$ we get the product of a factor, which depends upon y but does *not* depend upon θ, say $c(y)$, and

$$\theta^{y+\alpha-1}(1 - \theta)^{n-y+\beta-1}.$$

That is,

$$k(\theta|y) = c(y)\theta^{y+\alpha-1}(1 - \theta)^{n-y+\beta-1}, \qquad 0 < \theta < 1.$$

However, $c(y)$ must be that "constant" needed to make $k(\theta|y)$ a p.d.f.,namely,

$$c(y) = \frac{\Gamma(n + \alpha + \beta)}{\Gamma(y + \alpha)\Gamma(n - y + \beta)}.$$

Accordingly, Bayesians frequently write that $k(\theta|y)$ is proportional to $k(y, \theta) = g(y|\theta)h(\theta)$; that is,

$$k(\theta|y) \propto g(y|\theta)h(\theta).$$

Then to actually form the p.d.f. $k(\theta|y)$, they simply find a "constant," which is some function of y, so that the expression integrates to one.

Example 12.5–3.   Suppose that $Y = \bar{X}$ is the mean of a random sample of size n that arises from the normal distribution $N(\theta, \sigma^2)$, where σ^2 is known. Then $g(y|\theta)$ is $N(\theta, \sigma^2/n)$. Further suppose that we are able to assign prior weights to θ through a prior p.d.f. $h(\theta)$, which is $N(\theta_0, \sigma_0^2)$. Then we have that

$$k(\theta|y) \propto \frac{1}{\sqrt{2\pi}(\sigma/\sqrt{n})}\frac{1}{\sqrt{2\pi}\sigma_0}\exp\left[-\frac{(y - \theta)^2}{2(\sigma^2/n)} - \frac{(\theta - \theta_0)^2}{2\sigma_0^2}\right].$$

If we eliminate all constant factors (including factors involving y only), then we have

$$k(\theta|y) \propto \exp\left[-\frac{(\sigma_0^2 + \sigma^2/n)\theta^2 - 2(y\sigma_0^2 + \theta_0\sigma^2/n)\theta}{2(\sigma^2/n)\sigma_0^2}\right].$$

This can be simplified, by completing the square, to read (after eliminating factors not involving θ)

$$k(\theta|y) \propto \exp\left\{-\frac{[\theta - (y\sigma_0^2 + \theta_0\sigma^2/n)/(\sigma_0^2 + \sigma^2/n)]^2}{[2(\sigma^2/n)\sigma_0^2]/[\sigma_0^2 + (\sigma^2/n)]}\right\}.$$

That is, the posterior p.d.f. of the parameter is obviously normal with mean

$$\frac{y\sigma_0^2 + \theta_0\sigma^2/n}{\sigma_0^2 + \sigma^2/n} = \left(\frac{\sigma_0^2}{\sigma_0^2 + \sigma^2/n}\right)y + \left(\frac{\sigma^2/n}{\sigma_0^2 + \sigma^2/n}\right)\theta_0$$

and variance $(\sigma^2/n)\sigma_0^2/(\sigma_0^2 + \sigma^2/n)$. If the square error loss function is used, then this posterior mean is the Bayes' solution. Again note that it is a weighted average of the maximum likelihood estimate $y = \bar{x}$ and the prior mean θ_0. The Bayes' solution $w(y)$ will always be a value between the prior judgment and the usual estimate. Also, please note here and in Example 12.5–2 that the Bayes' solution gets closer to the maximum likelihood estimate as n increases. Thus, the Bayesian procedures permit the decision maker to enter his or her prior opinions into the solution in a very formal way so that the influence of those prior notions will be less and less as n increases.

In Bayesian statistics, all the information is contained in the posterior p.d.f. $k(\theta|y)$. In Examples 12.5–2 and 12.5–3, we found Bayesian point estimates using the square error loss function. It should be noted that if $L[w(y), \theta] = |w(y) - \theta|$, the absolute value of the error, then the Bayes' solution would be the median of the posterior distribution of the parameter, which is given by $k(\theta|y)$. Hence, the Bayes' solution changes, *as it should*, with different loss functions.

Finally, if an interval estimate of θ is desired, we would find two functions of y, say $u(y)$ and $v(y)$, such that

$$\int_{u(y)}^{v(y)} k(\theta|y)\,d\theta = 1 - \alpha,$$

where α is small, say $\alpha = 0.05$. Then the observed interval $u(y)$ to $v(y)$ would serve as an interval estimate for the parameter in the sense that the posterior probability of the parameter being in that interval is $1 - \alpha$. In Example 12.5–3 where the posterior p.d.f. of the parameter was normal, the interval

$$\frac{y\sigma_0^2 + \theta_0\sigma^2/n}{\sigma_0^2 + \sigma^2/n} \pm 1.96\sqrt{\frac{(\sigma^2/n)\sigma_0^2}{\sigma_0^2 + \sigma^2/n}}$$

serves as an interval estimate for θ with posterior probability of 0.95.

═══ *Exercises* ═══

12.5–1. Let $X_1, X_2, \ldots, X_n$ be a random sample from a distribution $b(1, \theta)$. Let the prior p.d.f. of θ be a beta one with parameters α and β. Show that the posterior p.d.f.

$$k(\theta | x_1, x_2, \ldots, x_n)$$

is exactly the same as $k(\theta | y)$ given in Example 12.5–2. This demonstrates that we get exactly the same result whether we begin with the sufficient statistic or with the sample items. HINT: Note that $k(\theta | x_1, x_2, \ldots, x_n)$ is proportional to the product of the joint p.d.f. of $X_1, X_2, \ldots, X_n$ and the prior p.d.f. of θ.

12.5–2. Let Y be the sum of the items of a random sample from a Poisson distribution with mean θ. Let the prior p.d.f. of θ be a gamma one with parameters α and β.
(a) Find the posterior p.d.f. of θ, given $Y = y$.
(b) If the loss function is $[w(y) - \theta]^2$, find the Bayesian point estimate $w(y)$.
(c) Show that this $w(y)$ is a weighted average of the maximum likelihood estimate y/n and the prior mean $\alpha\beta$, with respective weights of $n/(n + 1/\beta)$ and $(1/\beta)/(n + 1/\beta)$.

12.5–3. In Example 12.5–2, take $n = 30$, $\alpha = 15$, and $\beta = 5$.
(a) Using the square error loss, compute the risk function associated with the Bayes' solution $w(Y)$.
(b) The risk function associated with the usual estimator Y/n is of course $\theta(1 - \theta)/30$. Find those values of θ for which the risk function in part (a) is less than $\theta(1 - \theta)/30$. In particular, note that if the prior mean $\alpha/(\alpha + \beta) = 3/4$ is a reasonable guess, then the risk function in part (a) is the better of the two (that is, it is smaller in a neighborhood of $\theta = 3/4$).

12.5–4. Let $f(x_1, x_2, \ldots, x_n | \theta)$ represent the conditional p.d.f. of $X_1, X_2, \ldots, X_n$, given θ, in a situation in which the simple hypothesis $H_0 : \theta = \theta_0$ is tested against the simple alternative $H_1 : \theta = \theta_1$. The two prior probabilities $h(\theta_0)$ and $h(\theta_1)$ must sum to one; that is, $h(\theta_0) + h(\theta_1) = 1$. The nonnegative loss function is such that

$$L(w = \theta_0, \theta_0) = L(w = \theta_1, \theta_1) = 0,$$

but $L(w = \theta_1, \theta_0) > 0$ and $L(w = \theta_0, \theta_1) > 0$.
(a) If $w = \theta_0$, show that the conditional expected value of L, given $X_1 = x_1, \ldots, X_n = x_n$, equals

$$\frac{L(\theta_0, \theta_1) f(x_1, \ldots, x_n | \theta_1) h(\theta_1)}{f(x_1, \ldots, x_n | \theta_0) h(\theta_0) + f(x_1, \ldots, x_n | \theta_1) h(\theta_1)}.$$

(b) If $w = \theta_1$, find the conditional expected value of L, given $X_1 = x_1, \ldots, X_n = x_n$.
(c) If the expected loss in part (a) is less than that of part (b), show that the Bayesian decision would be $w = \theta_0$ (that is, H_0) if

$$\frac{f(x_1, \ldots, x_n | \theta_1)}{f(x_1, \ldots, x_n | \theta_0)} < \frac{L(\theta_1, \theta_0) h(\theta_0)}{L(\theta_0, \theta_1) h(\theta_1)}.$$

Thus, we see by the Neyman–Pearson Lemma that the Bayes' solution leads to a best test.

References

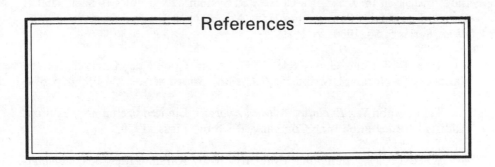

Box, G. E. P., and M. E. Muller, "A Note on the Generation of Random Normal Deviates," *Ann. Math. Statist.*, **29** (1958), p. 610.

Cramér, H., *Mathematical Methods of Statistics*, Princeton University Press, Princeton, N.J., 1946.

Crisman, Roger, "Shortest Confidence Interval for the Standard Deviation of a Normal Distribution," *J. Undergrad. Math.*, **7**, 2 (1975), p. 57.

Feller, W., *An Introduction to Probability Theory and Its Applications*, Vol. I, 3rd ed., John Wiley & Sons, Inc., New York, 1968.

Ferguson, T. S., *Mathematical Statistics*, Academic Press, Inc., New York, 1967.

Guenther, William C., "Shortest Confidence Intervals," *Amer. Statist.*, **23**, 1 (1969), p. 22.

Hogg, R. V., and A. T. Craig, *Introduction to Mathematical Statistics*, 3rd ed., Macmillan Publishing Co., Inc., New York, 1970.

Hogg, R. V., and A. T. Craig, "On the Decomposition of Certain Chi-Square Variables," *Ann. Math. Statist.*, **29** (1958), p. 608.

Lindgren, B. W., *Statistical Theory*, 3rd ed., Macmillan Publishing Co., Inc., New York, 1976.

Mood, A. M., F. A. Graybill, and D. C. Boes, *Introduction to the Theory of Statistics*, 3rd ed., McGraw-Hill Book Company, Inc., New York, 1974.

Pearson, K., "On the Criterion That a Given System of Deviations from the Probable in the Case of a Correlated System of Variables Is Such That It Can Be Reasonably Supposed to Have Arisen from Random Sampling," *Phil. Mag.*, Series 5, **50** (1900), p. 157.

Tate, R. F., and G. W. Klett, "Optimum Confidence Intervals for the Variance of a Normal Distribution," *J. Amer. Statist. Assoc.*, **54** (1959), p. 674.

Tukey, John W., *Exploratory Data Analysis*, Limited Preliminary Edition, Addison-Wesley Publishing Company, Reading, Mass., 1970.

Wilcoxon, F., "Individual Comparisons by Ranking Methods," *Biometrics Bull.*, **1** (1945), p. 80.

Wilks, S. S., *Mathematical Statistics*, John Wiley & Sons, Inc., New York, 1962.

Appendix

TABLE I

THE BINOMIAL DISTRIBUTION

$$P(X \leq x) = \sum_{k=0}^{x} \frac{n!}{k!(n-k)!} p^k (1-p)^{n-k}$$

n	x	\(p\) 0.05	0.10	0.15	0.20	0.25	0.30	0.35	0.40	0.45	0.50
2	0	0.9025	0.8100	0.7225	0.6400	0.5625	0.4900	0.4225	0.3600	0.3025	0.2500
	1	0.9975	0.9900	0.9775	0.9600	0.9375	0.9100	0.8775	0.8400	0.7975	0.7500
	2	1.0000	1.0000	1.0000	1.0000	1.0000	1.0000	1.0000	1.0000	1.0000	1.0000
3	0	0.8574	0.7290	0.6141	0.5120	0.4219	0.3430	0.2746	0.2160	0.1664	0.1250
	1	0.9928	0.9720	0.9392	0.8960	0.8438	0.7840	0.7182	0.6480	0.5748	0.5000
	2	0.9999	0.9990	0.9966	0.9920	0.9844	0.9730	0.9571	0.9360	0.9089	0.8750
	3	1.0000	1.0000	1.0000	1.0000	1.0000	1.0000	1.0000	1.0000	1.0000	1.0000
4	0	0.8145	0.6561	0.5220	0.4096	0.3164	0.2401	0.1785	0.1296	0.0915	0.0625
	1	0.9860	0.9477	0.8905	0.8192	0.7383	0.6517	0.5630	0.4752	0.3910	0.3125
	2	0.9995	0.9963	0.9880	0.9728	0.9492	0.9163	0.8735	0.8208	0.7585	0.6875
	3	1.0000	0.9999	0.9995	0.9984	0.9961	0.9919	0.9850	0.9744	0.9590	0.9375
	4	1.0000	1.0000	1.0000	1.0000	1.0000	1.0000	1.0000	1.0000	1.0000	1.0000
5	0	0.7738	0.5905	0.4437	0.3277	0.2373	0.1681	0.1160	0.0778	0.0503	0.0312
	1	0.9774	0.9185	0.8352	0.7373	0.6328	0.5282	0.4284	0.3370	0.2562	0.1875
	2	0.9988	0.9914	0.9734	0.9421	0.8965	0.8369	0.7648	0.6826	0.5931	0.5000
	3	1.0000	0.9995	0.9978	0.9933	0.9844	0.9692	0.9460	0.9130	0.8688	0.8125
	4	1.0000	1.0000	0.9999	0.9997	0.9990	0.9976	0.9947	0.9898	0.9815	0.9688
	5	1.0000	1.0000	1.0000	1.0000	1.0000	1.0000	1.0000	1.0000	1.0000	1.0000

n	r										
6	0	0.0156	0.0277	0.0467	0.0754	0.1176	0.1780	0.2621	0.3771	0.5314	0.7351
	1	0.1094	0.1636	0.2333	0.3191	0.4202	0.5339	0.6553	0.7765	0.8857	0.9672
	2	0.3438	0.4415	0.5443	0.6471	0.7443	0.8306	0.9011	0.9527	0.9842	0.9978
	3	0.6562	0.7447	0.8208	0.8826	0.9295	0.9624	0.9830	0.9941	0.9987	0.9999
	4	0.8906	0.9308	0.9590	0.9777	0.9891	0.9954	0.9984	0.9996	0.9999	1.0000
	5	0.9844	0.9917	0.9959	0.9982	0.9993	0.9998	0.9999	1.0000	1.0000	1.0000
	6	1.0000	1.0000	1.0000	1.0000	1.0000	1.0000	1.0000	1.0000	1.0000	1.0000
7	0	0.0078	0.0152	0.0280	0.0490	0.0824	0.1335	0.2097	0.3206	0.4783	0.6983
	1	0.0625	0.1024	0.1586	0.2338	0.3294	0.4449	0.5767	0.7166	0.8503	0.9556
	2	0.2266	0.3164	0.4199	0.5323	0.6471	0.7564	0.8520	0.9262	0.9743	0.9962
	3	0.5000	0.6083	0.7102	0.8002	0.8740	0.9294	0.9667	0.9879	0.9973	0.9998
	4	0.7734	0.8471	0.9037	0.9444	0.9712	0.9871	0.9953	0.9988	0.9998	1.0000
	5	0.9375	0.9643	0.9812	0.9910	0.9962	0.9987	0.9996	0.9999	1.0000	1.0000
	6	0.9922	0.9963	0.9984	0.9994	0.9998	0.9999	1.0000	1.0000	1.0000	1.0000
	7	1.0000	1.0000	1.0000	1.0000	1.0000	1.0000	1.0000	1.0000	1.0000	1.0000
8	0	0.0039	0.0084	0.0168	0.0319	0.0576	0.1001	0.1678	0.2725	0.4305	0.6634
	1	0.0352	0.0632	0.1064	0.1691	0.2553	0.3671	0.5033	0.6572	0.8131	0.9428
	2	0.1445	0.2201	0.3154	0.4278	0.5518	0.6785	0.7969	0.8948	0.9619	0.9942
	3	0.3633	0.4770	0.5941	0.7064	0.8059	0.8862	0.9437	0.9786	0.9950	0.9996
	4	0.6367	0.7396	0.8263	0.8939	0.9420	0.9727	0.9896	0.9971	0.9996	1.0000
	5	0.8555	0.9115	0.9502	0.9747	0.9887	0.9958	0.9988	0.9998	1.0000	1.0000
	6	0.9648	0.9819	0.9915	0.9964	0.9987	0.9996	0.9999	1.0000	1.0000	1.0000
	7	0.9961	0.9983	0.9993	0.9998	0.9999	1.0000	1.0000	1.0000	1.0000	1.0000
	8	1.0000	1.0000	1.0000	1.0000	1.0000	1.0000	1.0000	1.0000	1.0000	1.0000
9	0	0.0020	0.0046	0.0101	0.0207	0.0404	0.0751	0.1342	0.2316	0.3874	0.6302
	1	0.0195	0.0385	0.0705	0.1211	0.1960	0.3003	0.4362	0.5995	0.7748	0.9288
	2	0.0898	0.1495	0.2318	0.3373	0.4628	0.6007	0.7382	0.8591	0.9470	0.9916
	3	0.2539	0.3614	0.4826	0.6089	0.7297	0.8343	0.9144	0.9661	0.9917	0.9994
	4	0.5000	0.6214	0.7334	0.8283	0.9012	0.9511	0.9804	0.9944	0.9991	1.0000
	5	0.7461	0.8342	0.9006	0.9464	0.9747	0.9900	0.9969	0.9994	0.9999	1.0000
	6	0.9102	0.9502	0.9750	0.9888	0.9957	0.9987	0.9997	1.0000	1.0000	1.0000
	7	0.9805	0.9909	0.9962	0.9986	0.9996	0.9999	1.0000	1.0000	1.0000	1.0000
	8	0.9980	0.9992	0.9997	0.9999	1.0000	1.0000	1.0000	1.0000	1.0000	1.0000
	9	1.0000	1.0000	1.0000	1.0000	1.0000	1.0000	1.0000	1.0000	1.0000	1.0000

TABLE I (continued)

n	x		0.05	0.10	0.15	0.20	0.25	0.30	0.35	0.40	0.45	0.50
		p										
10	0		0.5987	0.3487	0.1969	0.1074	0.0563	0.0282	0.0135	0.0060	0.0025	0.0010
	1		0.9139	0.7361	0.5443	0.3758	0.2440	0.1493	0.0860	0.0464	0.0233	0.0107
	2		0.9885	0.9298	0.8202	0.6778	0.5256	0.3828	0.2616	0.1673	0.0996	0.0547
	3		0.9990	0.9872	0.9500	0.8791	0.7759	0.6496	0.5138	0.3823	0.2660	0.1719
	4		0.9999	0.9984	0.9901	0.9672	0.9219	0.8497	0.7515	0.6331	0.5044	0.3770
	5		1.0000	0.9999	0.9986	0.9936	0.9803	0.9527	0.9051	0.8338	0.7384	0.6230
	6		1.0000	1.0000	0.9999	0.9991	0.9965	0.9894	0.9740	0.9452	0.8980	0.8281
	7		1.0000	1.0000	1.0000	0.9999	0.9996	0.9984	0.9952	0.9877	0.9726	0.9453
	8		1.0000	1.0000	1.0000	1.0000	1.0000	0.9999	0.9995	0.9983	0.9955	0.9893
	9		1.0000	1.0000	1.0000	1.0000	1.0000	1.0000	1.0000	0.9999	0.9997	0.9990
	10		1.0000	1.0000	1.0000	1.0000	1.0000	1.0000	1.0000	1.0000	1.0000	1.0000
11	0		0.5688	0.3138	0.1673	0.0859	0.0422	0.0198	0.0088	0.0036	0.0014	0.0005
	1		0.8981	0.6974	0.4922	0.3221	0.1971	0.1130	0.0606	0.0302	0.0139	0.0059
	2		0.9848	0.9104	0.7788	0.6174	0.4552	0.3127	0.2001	0.1189	0.0652	0.0327
	3		0.9984	0.9815	0.9306	0.8389	0.7133	0.5696	0.4256	0.2963	0.1911	0.1133
	4		0.9999	0.9972	0.9841	0.9496	0.8854	0.7897	0.6683	0.5328	0.3971	0.2744
	5		1.0000	0.9997	0.9973	0.9883	0.9657	0.9218	0.8513	0.7535	0.6331	0.5000
	6		1.0000	1.0000	0.9997	0.9980	0.9924	0.9784	0.9499	0.9006	0.8262	0.7256
	7		1.0000	1.0000	1.0000	0.9998	0.9988	0.9957	0.9878	0.9707	0.9390	0.8867
	8		1.0000	1.0000	1.0000	1.0000	0.9999	0.9994	0.9980	0.9941	0.9852	0.9673
	9		1.0000	1.0000	1.0000	1.0000	1.0000	1.0000	0.9998	0.9993	0.9978	0.9941
	10		1.0000	1.0000	1.0000	1.0000	1.0000	1.0000	1.0000	1.0000	0.9998	0.9995
	11		1.0000	1.0000	1.0000	1.0000	1.0000	1.0000	1.0000	1.0000	1.0000	1.0000

n	x										
12	0	0.0002	0.0008	0.0022	0.0057	0.0138	0.0317	0.0687	0.1422	0.2824	0.5404
	1	0.0032	0.0083	0.0196	0.0424	0.0850	0.1584	0.2749	0.4435	0.6590	0.8816
	2	0.0193	0.0421	0.0834	0.1513	0.2528	0.3907	0.5583	0.7358	0.8891	0.9804
	3	0.0730	0.1345	0.2253	0.3467	0.4925	0.6488	0.7946	0.9078	0.9744	0.9978
	4	0.1938	0.3044	0.4382	0.5833	0.7237	0.8424	0.9274	0.9761	0.9957	0.9998
	5	0.3872	0.5269	0.6652	0.7873	0.8822	0.9456	0.9806	0.9954	0.9995	1.0000
	6	0.6128	0.7393	0.8418	0.9154	0.9614	0.9857	0.9961	0.9993	0.9999	1.0000
	7	0.8062	0.8883	0.9427	0.9745	0.9905	0.9972	0.9994	0.9999	1.0000	1.0000
	8	0.9270	0.9644	0.9847	0.9944	0.9983	0.9996	0.9999	1.0000	1.0000	1.0000
	9	0.9807	0.9921	0.9972	0.9992	0.9998	1.0000	1.0000	1.0000	1.0000	1.0000
	10	0.9968	0.9989	0.9997	0.9999	1.0000	1.0000	1.0000	1.0000	1.0000	1.0000
	11	0.9998	0.9999	1.0000	1.0000	1.0000	1.0000	1.0000	1.0000	1.0000	1.0000
	12	1.0000	1.0000	1.0000	1.0000	1.0000	1.0000	1.0000	1.0000	1.0000	1.0000
13	0	0.0001	0.0004	0.0013	0.0037	0.0097	0.0238	0.0550	0.1209	0.2542	0.5133
	1	0.0017	0.0049	0.0126	0.0296	0.0637	0.1267	0.2336	0.3983	0.6213	0.8646
	2	0.0112	0.0269	0.0579	0.1132	0.2025	0.3326	0.5017	0.6920	0.8661	0.9755
	3	0.0461	0.0929	0.1686	0.2783	0.4206	0.5843	0.7473	0.8820	0.9658	0.9969
	4	0.1334	0.2279	0.3530	0.5005	0.6543	0.7940	0.9009	0.9658	0.9935	0.9997
	5	0.2905	0.4268	0.5744	0.7159	0.8346	0.9198	0.9700	0.9924	0.9991	1.0000
	6	0.5000	0.6437	0.7712	0.8705	0.9376	0.9757	0.9930	0.9987	0.9999	1.0000
	7	0.7095	0.8212	0.9023	0.9538	0.9818	0.9944	0.9988	0.9998	1.0000	1.0000
	8	0.8666	0.9302	0.9679	0.9874	0.9960	0.9990	0.9998	1.0000	1.0000	1.0000
	9	0.9539	0.9797	0.9922	0.9975	0.9993	0.9999	1.0000	1.0000	1.0000	1.0000
	10	0.9888	0.9959	0.9987	0.9997	0.9999	1.0000	1.0000	1.0000	1.0000	1.0000
	11	0.9983	0.9995	0.9999	1.0000	1.0000	1.0000	1.0000	1.0000	1.0000	1.0000
	12	0.9999	1.0000	1.0000	1.0000	1.0000	1.0000	1.0000	1.0000	1.0000	1.0000
	13	1.0000	1.0000	1.0000	1.0000	1.0000	1.0000	1.0000	1.0000	1.0000	1.0000

TABLE I (continued)

n	x	0.05	0.10	0.15	0.20	0.25	0.30	0.35	0.40	0.45	0.50
14	0	0.4877	0.2288	0.1028	0.0440	0.0178	0.0068	0.0024	0.0008	0.0002	0.0001
	1	0.8470	0.5846	0.3567	0.1979	0.1010	0.0475	0.0205	0.0081	0.0029	0.0009
	2	0.9699	0.8416	0.6479	0.4481	0.2811	0.1608	0.0839	0.0398	0.0170	0.0065
	3	0.9958	0.9559	0.8535	0.6982	0.5213	0.3552	0.2205	0.1243	0.0632	0.0287
	4	0.9996	0.9908	0.9533	0.8702	0.7415	0.5842	0.4227	0.2793	0.1672	0.0898
	5	1.0000	0.9985	0.9885	0.9561	0.8883	0.7805	0.6405	0.4859	0.3373	0.2120
	6	1.0000	0.9998	0.9978	0.9884	0.9617	0.9067	0.8164	0.6925	0.5461	0.3953
	7	1.0000	1.0000	0.9997	0.9976	0.9897	0.9685	0.9247	0.8499	0.7414	0.6047
	8	1.0000	1.0000	1.0000	0.9996	0.9978	0.9917	0.9757	0.9417	0.8811	0.7880
	9	1.0000	1.0000	1.0000	1.0000	0.9997	0.9983	0.9940	0.9825	0.9574	0.9102
	10	1.0000	1.0000	1.0000	1.0000	1.0000	0.9998	0.9989	0.9961	0.9886	0.9713
	11	1.0000	1.0000	1.0000	1.0000	1.0000	1.0000	0.9999	0.9994	0.9978	0.9935
	12	1.0000	1.0000	1.0000	1.0000	1.0000	1.0000	1.0000	0.9999	0.9997	0.9991
	13	1.0000	1.0000	1.0000	1.0000	1.0000	1.0000	1.0000	1.0000	1.0000	0.9999
	14	1.0000	1.0000	1.0000	1.0000	1.0000	1.0000	1.0000	1.0000	1.0000	1.0000
15	0	0.4633	0.2059	0.0874	0.0352	0.0134	0.0047	0.0016	0.0005	0.0001	0.0000
	1	0.8290	0.5490	0.3186	0.1671	0.0802	0.0353	0.0142	0.0052	0.0017	0.0005
	2	0.9638	0.8159	0.6042	0.3980	0.2361	0.1268	0.0617	0.0271	0.0107	0.0037
	3	0.9945	0.9444	0.8227	0.6482	0.4613	0.2969	0.1727	0.0905	0.0424	0.0176
	4	0.9994	0.9873	0.9383	0.8358	0.6865	0.5155	0.3519	0.2173	0.1204	0.0592
	5	0.9999	0.9978	0.9832	0.9389	0.8516	0.7216	0.5643	0.4032	0.2608	0.1509
	6	1.0000	0.9997	0.9964	0.9819	0.9434	0.8689	0.7548	0.6098	0.4522	0.3036
	7	1.0000	1.0000	0.9994	0.9958	0.9827	0.9500	0.8868	0.7869	0.6535	0.5000
	8	1.0000	1.0000	0.9999	0.9992	0.9958	0.9848	0.9578	0.9050	0.8182	0.6964
	9	1.0000	1.0000	1.0000	0.9999	0.9992	0.9963	0.9876	0.9662	0.9231	0.8491
	10	1.0000	1.0000	1.0000	1.0000	0.9999	0.9993	0.9972	0.9907	0.9745	0.9408
	11	1.0000	1.0000	1.0000	1.0000	1.0000	0.9999	0.9995	0.9981	0.9937	0.9824
	12	1.0000	1.0000	1.0000	1.0000	1.0000	1.0000	0.9999	0.9997	0.9989	0.9963
	13	1.0000	1.0000	1.0000	1.0000	1.0000	1.0000	1.0000	1.0000	0.9999	0.9995
	14	1.0000	1.0000	1.0000	1.0000	1.0000	1.0000	1.0000	1.0000	1.0000	1.0000
	15	1.0000	1.0000	1.0000	1.0000	1.0000	1.0000	1.0000	1.0000	1.0000	1.0000

TABLE II

THE POISSON DISTRIBUTION

$$P(X \leq x) = \sum_{k=0}^{x} \frac{\lambda^k e^{-\lambda}}{k!}$$

				$\lambda = E(X)$						
x	0.1	0.2	0.3	0.4	0.5	0.6	0.7	0.8	0.9	1.0
0	0.905	0.819	0.741	0.670	0.607	0.549	0.497	0.449	0.407	0.368
1	0.995	0.982	0.963	0.938	0.910	0.878	0.844	0.809	0.772	0.736
2	1.000	0.999	0.996	0.992	0.986	0.977	0.966	0.953	0.937	0.920
3	1.000	1.000	1.000	0.999	0.998	0.997	0.994	0.991	0.987	0.981
4	1.000	1.000	1.000	1.000	1.000	1.000	0.999	0.999	0.998	0.996
5	1.000	1.000	1.000	1.000	1.000	1.000	1.000	1.000	1.000	0.999
6	1.000	1.000	1.000	1.000	1.000	1.000	1.000	1.000	1.000	1.000

x	1.1	1.2	1.3	1.4	1.5	1.6	1.7	1.8	1.9	2.0
0	0.333	0.301	0.273	0.247	0.223	0.202	0.183	0.165	0.150	0.135
1	0.699	0.663	0.627	0.592	0.558	0.525	0.493	0.463	0.434	0.406
2	0.900	0.879	0.857	0.833	0.809	0.783	0.757	0.731	0.704	0.677
3	0.974	0.966	0.957	0.946	0.934	0.921	0.907	0.891	0.875	0.857
4	0.995	0.992	0.989	0.986	0.981	0.976	0.970	0.964	0.956	0.947
5	0.999	0.998	0.998	0.997	0.996	0.994	0.992	0.990	0.987	0.983
6	1.000	1.000	1.000	0.999	0.999	0.999	0.998	0.997	0.997	0.995
7	1.000	1.000	1.000	1.000	1.000	1.000	1.000	0.999	0.999	0.999
8	1.000	1.000	1.000	1.000	1.000	1.000	1.000	1.000	1.000	1.000

x	2.2	2.4	2.6	2.8	3.0	3.2	3.4	3.6	3.8	4.0
0	0.111	0.091	0.074	0.061	0.050	0.041	0.033	0.027	0.022	0.018
1	0.355	0.308	0.267	0.231	0.199	0.171	0.147	0.126	0.107	0.092
2	0.623	0.570	0.518	0.469	0.423	0.380	0.340	0.303	0.269	0.238
3	0.819	0.779	0.736	0.692	0.647	0.603	0.558	0.515	0.473	0.433
4	0.928	0.904	0.877	0.848	0.815	0.781	0.744	0.706	0.668	0.629
5	0.975	0.964	0.951	0.935	0.916	0.895	0.871	0.844	0.816	0.785
6	0.993	0.988	0.983	0.976	0.966	0.955	0.942	0.927	0.909	0.889
7	0.998	0.997	0.995	0.992	0.988	0.983	0.977	0.969	0.960	0.949
8	1.000	0.999	0.999	0.998	0.996	0.994	0.992	0.988	0.984	0.979
9	1.000	1.000	1.000	0.999	0.999	0.998	0.997	0.996	0.994	0.992
10	1.000	1.000	1.000	1.000	1.000	1.000	0.999	0.999	0.998	0.997
11	1.000	1.000	1.000	1.000	1.000	1.000	1.000	1.000	0.999	0.999
12	1.000	1.000	1.000	1.000	1.000	1.000	1.000	1.000	1.000	1.000

TABLE II (continued)

x	\u03bb = E(X)									
	4.2	4.4	4.6	4.8	5.0	5.2	5.4	5.6	5.8	6.0
0	0.015	0.012	0.010	0.008	0.007	0.006	0.005	0.004	0.003	0.002
1	0.078	0.066	0.056	0.048	0.040	0.034	0.029	0.024	0.021	0.017
2	0.210	0.185	0.163	0.143	0.125	0.109	0.095	0.082	0.072	0.062
3	0.395	0.359	0.326	0.294	0.265	0.238	0.213	0.191	0.170	0.151
4	0.590	0.551	0.513	0.476	0.440	0.406	0.373	0.342	0.313	0.285
5	0.753	0.720	0.686	0.651	0.616	0.581	0.546	0.512	0.478	0.446
6	0.867	0.844	0.818	0.791	0.762	0.732	0.702	0.670	0.638	0.606
7	0.936	0.921	0.905	0.887	0.867	0.845	0.822	0.797	0.771	0.744
8	0.972	0.964	0.955	0.944	0.932	0.918	0.903	0.886	0.867	0.847
9	0.989	0.985	0.980	0.975	0.968	0.960	0.951	0.941	0.929	0.916
10	0.996	0.994	0.992	0.990	0.986	0.982	0.977	0.972	0.965	0.957
11	0.999	0.998	0.997	0.996	0.995	0.993	0.990	0.988	0.984	0.980
12	1.000	0.999	0.999	0.999	0.998	0.997	0.996	0.995	0.993	0.991
13	1.000	1.000	1.000	1.000	0.999	0.999	0.999	0.998	0.997	0.996
14	1.000	1.000	1.000	1.000	1.000	1.000	0.999	0.999	0.999	0.999
15	1.000	1.000	1.000	1.000	1.000	1.000	1.000	1.000	1.000	0.999
16	1.000	1.000	1.000	1.000	1.000	1.000	1.000	1.000	1.000	1.000

x	6.5	7.0	7.5	8.0	8.5	9.0	9.5	10.0	10.5	11.0
0	0.002	0.001	0.001	0.000	0.000	0.000	0.000	0.000	0.000	0.000
1	0.011	0.007	0.005	0.003	0.002	0.001	0.001	0.000	0.000	0.000
2	0.043	0.030	0.020	0.014	0.009	0.006	0.004	0.003	0.002	0.001
3	0.112	0.082	0.059	0.042	0.030	0.021	0.015	0.010	0.007	0.005
4	0.224	0.173	0.132	0.100	0.074	0.055	0.040	0.029	0.021	0.015
5	0.369	0.301	0.241	0.191	0.150	0.116	0.089	0.067	0.050	0.038
6	0.527	0.450	0.378	0.313	0.256	0.207	0.165	0.130	0.102	0.079
7	0.673	0.599	0.525	0.453	0.386	0.324	0.269	0.220	0.179	0.143
8	0.792	0.729	0.662	0.593	0.523	0.456	0.392	0.333	0.279	0.232
9	0.877	0.830	0.776	0.717	0.653	0.587	0.522	0.458	0.397	0.341
10	0.933	0.901	0.862	0.816	0.763	0.706	0.645	0.583	0.521	0.460
11	0.966	0.947	0.921	0.888	0.849	0.803	0.752	0.697	0.639	0.579
12	0.984	0.973	0.957	0.936	0.909	0.876	0.836	0.792	0.742	0.689
13	0.993	0.987	0.978	0.966	0.949	0.926	0.898	0.864	0.825	0.781
14	0.997	0.994	0.990	0.983	0.973	0.959	0.940	0.917	0.888	0.854
15	0.999	0.998	0.995	0.992	0.986	0.978	0.967	0.951	0.932	0.907
16	1.000	0.999	0.998	0.996	0.993	0.989	0.982	0.973	0.960	0.944
17	1.000	1.000	0.999	0.998	0.997	0.995	0.991	0.986	0.978	0.968
18	1.000	1.000	1.000	0.999	0.999	0.998	0.996	0.993	0.988	0.982
19	1.000	1.000	1.000	1.000	0.999	0.999	0.998	0.997	0.994	0.991
20	1.000	1.000	1.000	1.000	1.000	1.000	0.999	0.998	0.997	0.995
21	1.000	1.000	1.000	1.000	1.000	1.000	1.000	0.999	0.999	0.998
22	1.000	1.000	1.000	1.000	1.000	1.000	1.000	1.000	0.999	0.999
23	1.000	1.000	1.000	1.000	1.000	1.000	1.000	1.000	1.000	1.000

TABLE III

THE CHI–SQUARE DISTRIBUTION

$$P(X \leq x) = \int_0^x \frac{1}{\Gamma(r/2)2^{r/2}} w^{r/2-1} e^{-w/2} \, dw$$

r	$P(X \leq x)$							
	0.010	0.025	0.050	0.100	0.900	0.950	0.975	0.990
1	0.000	0.001	0.004	0.016	2.706	3.841	5.024	6.635
2	0.020	0.051	0.103	0.211	4.605	5.991	7.378	9.210
3	0.115	0.216	0.352	0.584	6.251	7.815	9.348	11.34
4	0.297	0.484	0.711	1.064	7.779	9.488	11.14	13.28
5	0.554	0.831	1.145	1.610	9.236	11.07	12.83	15.09
6	0.872	1.237	1.635	2.204	10.64	12.59	14.45	16.81
7	1.239	1.690	2.167	2.833	12.02	14.07	16.01	18.48
8	1.646	2.180	2.733	3.490	13.36	15.51	17.54	20.09
9	2.088	2.700	3.325	4.168	14.68	16.92	19.02	21.67
10	2.558	3.247	3.940	4.865	15.99	18.31	20.48	23.21
11	3.053	3.816	4.575	5.578	17.28	19.68	21.92	24.72
12	3.571	4.404	5.226	6.304	18.55	21.03	23.34	26.22
13	4.107	5.009	5.892	7.042	19.81	22.36	24.74	27.69
14	4.660	5.629	6.571	7.790	21.06	23.68	26.12	29.14
15	5.229	6.262	7.261	8.547	22.31	25.00	27.49	30.58
16	5.812	6.908	7.962	9.312	23.54	26.30	28.84	32.00
17	6.408	7.564	8.672	10.08	24.77	27.59	30.19	33.41
18	7.015	8.231	9.390	10.86	25.99	28.87	31.53	34.80
19	7.633	8.907	10.12	11.65	27.20	30.14	32.85	36.19
20	8.260	9.591	10.85	12.44	28.41	31.41	34.17	37.57
21	8.897	10.28	11.59	13.24	29.62	32.67	35.48	38.93
22	9.542	10.98	12.34	14.04	30.81	33.92	36.78	40.29
23	10.20	11.69	13.09	14.85	32.01	35.17	38.08	41.64
24	10.86	12.40	13.85	15.66	33.20	36.42	39.36	42.98
25	11.52	13.12	14.61	16.47	34.38	37.65	40.65	44.31
26	12.20	13.84	15.38	17.29	35.56	38.88	41.92	45.64
27	12.88	14.57	16.15	18.11	36.74	40.11	43.19	46.96
28	13.56	15.31	16.93	18.94	37.92	41.34	44.46	48.28
29	14.26	16.05	17.71	19.77	39.09	42.56	45.72	49.59
30	14.95	16.79	18.49	20.60	40.26	43.77	46.98	50.89
40	22.16	24.43	26.51	29.05	51.80	55.76	59.34	63.69
50	29.71	32.36	34.76	37.69	63.17	67.50	71.42	76.15
60	37.48	40.48	43.19	46.46	74.40	79.08	83.30	88.38
70	45.44	48.76	51.74	55.33	85.53	90.53	95.02	100.4
80	53.34	57.15	60.39	64.28	96.58	101.9	106.6	112.3

This table is abridged and adapted from Table III in *Biometrika Tables for Statisticians*, edited by E. S. Pearson and H. O. Hartley. It is published here with the kind permission of the *Biometrika* Trustees.

TABLE IV

THE NORMAL DISTRIBUTION

$$P(Z \leq z) = \Phi(z) = \int_{-\infty}^{z} \frac{1}{\sqrt{2\pi}} e^{-w^2/2} \, dw$$

$$[\Phi(-z) = 1 - \Phi(z)]$$

z	0.00	0.01	0.02	0.03	0.04	0.05	0.06	0.07	0.08	0.09
0.0	0.5000	0.5040	0.5080	0.5120	0.5160	0.5199	0.5239	0.5279	0.5319	0.5359
0.1	0.5398	0.5438	0.5478	0.5517	0.5557	0.5596	0.5636	0.5675	0.5714	0.5753
0.2	0.5793	0.5832	0.5871	0.5910	0.5948	0.5987	0.6026	0.6064	0.6103	0.6141
0.3	0.6179	0.6217	0.6255	0.6293	0.6331	0.6368	0.6406	0.6443	0.6480	0.6517
0.4	0.6554	0.6591	0.6628	0.6664	0.6700	0.6736	0.6772	0.6808	0.6844	0.6879
0.5	0.6915	0.6950	0.6985	0.7019	0.7054	0.7088	0.7123	0.7157	0.7190	0.7224
0.6	0.7257	0.7291	0.7324	0.7357	0.7389	0.7422	0.7454	0.7486	0.7517	0.7549
0.7	0.7580	0.7611	0.7642	0.7673	0.7703	0.7734	0.7764	0.7794	0.7823	0.7852
0.8	0.7881	0.7910	0.7939	0.7967	0.7995	0.8023	0.8051	0.8078	0.8106	0.8133
0.9	0.8159	0.8186	0.8212	0.8238	0.8264	0.8289	0.8315	0.8340	0.8365	0.8389
1.0	0.8413	0.8438	0.8461	0.8485	0.8508	0.8531	0.8554	0.8577	0.8599	0.8621
1.1	0.8643	0.8665	0.8686	0.8708	0.8729	0.8749	0.8770	0.8790	0.8810	0.8830
1.2	0.8849	0.8869	0.8888	0.8907	0.8925	0.8944	0.8962	0.8980	0.8997	0.9015
1.3	0.9032	0.9049	0.9066	0.9082	0.9099	0.9115	0.9131	0.9147	0.9162	0.9177
1.4	0.9192	0.9207	0.9222	0.9236	0.9251	0.9265	0.9279	0.9292	0.9306	0.9319
1.5	0.9332	0.9345	0.9357	0.9370	0.9382	0.9394	0.9406	0.9418	0.9429	0.9441
1.6	0.9452	0.9463	0.9474	0.9484	0.9495	0.9505	0.9515	0.9525	0.9535	0.9545
1.7	0.9554	0.9564	0.9573	0.9582	0.9591	0.9599	0.9608	0.9616	0.9625	0.9633
1.8	0.9641	0.9649	0.9656	0.9664	0.9671	0.9678	0.9686	0.9693	0.9699	0.9706
1.9	0.9713	0.9719	0.9726	0.9732	0.9738	0.9744	0.9750	0.9756	0.9761	0.9767

z										
2.0	0.9772	0.9778	0.9783	0.9788	0.9793	0.9798	0.9803	0.9808	0.9812	0.9817
2.1	0.9821	0.9826	0.9830	0.9834	0.9838	0.9842	0.9846	0.9850	0.9854	0.9857
2.2	0.9861	0.9864	0.9868	0.9871	0.9875	0.9878	0.9881	0.9884	0.9887	0.9890
2.3	0.9893	0.9896	0.9898	0.9901	0.9904	0.9906	0.9909	0.9911	0.9913	0.9916
2.4	0.9918	0.9920	0.9922	0.9925	0.9927	0.9929	0.9931	0.9932	0.9934	0.9936
2.5	0.9938	0.9940	0.9941	0.9943	0.9945	0.9946	0.9948	0.9949	0.9951	0.9952
2.6	0.9953	0.9955	0.9956	0.9957	0.9959	0.9960	0.9961	0.9962	0.9963	0.9964
2.7	0.9965	0.9966	0.9967	0.9968	0.9969	0.9970	0.9971	0.9972	0.9973	0.9974
2.8	0.9974	0.9975	0.9976	0.9977	0.9977	0.9978	0.9979	0.9979	0.9980	0.9981
2.9	0.9981	0.9982	0.9982	0.9983	0.9984	0.9984	0.9985	0.9985	0.9986	0.9986
3.0	0.9987	0.9987	0.9987	0.9988	0.9988	0.9989	0.9989	0.9989	0.9990	0.9990

z	1.282	1.645	1.960	2.326	2.576
$1 - \Phi(z)$	0.100	0.050	0.025	0.010	0.005

TABLE V

THE F DISTRIBUTION

$$P(F \leq f) = \int_0^f \frac{\Gamma[(r_1 + r_2)/2](r_1/r_2)^{r_1/2} w^{r_1/2 - 1}}{\Gamma(r_1/2)\Gamma(r_2/2)(1 + r_1 w/r_2)^{(r_1 + r_2)/2}} \, dw$$

$P(F \leq f)$	r_2	r_1										
		1	2	3	4	5	6	7	8	9	10	12	15
0.95	1	161	200	216	225	230	234	237	239	241	242	244	246
0.975		648	800	864	900	922	937	948	957	963	969	977	985
0.99		4052	4999	5403	5625	5764	5859	5928	5982	6023	6056	6106	6157
0.95	2	18.51	19.00	19.16	19.25	19.30	19.33	19.35	19.37	19.38	19.40	19.41	19.43
0.975		38.51	39.00	39.17	39.25	39.30	39.33	39.36	39.37	39.39	39.40	39.41	39.43
0.99		98.50	99.00	99.17	99.25	99.30	99.33	99.36	99.37	99.39	99.40	99.42	99.43
0.95	3	10.13	9.55	9.28	9.12	9.01	8.94	8.89	8.85	8.81	8.79	8.74	8.70
0.975		17.44	16.04	15.44	15.10	14.88	14.73	14.62	14.54	14.47	14.42	14.34	14.25
0.99		34.12	30.82	29.46	28.71	28.24	27.91	27.67	27.49	27.35	27.23	27.05	26.87
0.95	4	7.71	6.94	6.59	6.39	6.26	6.16	6.09	6.04	6.00	5.96	5.91	5.86
0.975		12.22	10.65	9.98	9.60	9.36	9.20	9.07	8.98	8.90	8.84	8.75	8.66
0.99		21.20	18.00	16.69	15.98	15.52	15.21	14.98	14.80	14.66	14.55	14.37	14.20
0.95	5	6.61	5.79	5.41	5.19	5.05	4.95	4.88	4.82	4.77	4.74	4.68	4.62
0.975		10.01	8.43	7.76	7.39	7.15	6.98	6.85	6.76	6.68	6.62	6.52	6.43
0.99		16.26	13.27	12.06	11.39	10.97	10.67	10.46	10.29	10.16	10.05	9.89	9.72
0.95	6	5.99	5.14	4.76	4.53	4.39	4.28	4.21	4.15	4.10	4.06	4.00	3.94
0.975		8.81	7.26	6.60	6.23	5.99	5.82	5.70	5.60	5.52	5.46	5.37	5.27
0.99		13.75	10.92	9.78	9.15	8.75	8.47	8.26	8.10	7.98	7.87	7.72	7.56

7	0.95	5.59	4.74	4.35	4.12	3.97	3.87	3.79	3.73	3.68	3.64	3.57	3.51
	0.975	8.07	6.54	5.89	5.52	5.29	5.12	4.99	4.90	4.82	4.76	4.67	4.57
	0.99	12.25	9.55	8.45	7.85	7.46	7.19	6.99	6.84	6.72	6.62	6.47	6.31
8	0.95	5.32	4.46	4.07	3.84	3.69	3.58	3.50	3.44	3.39	3.35	3.28	3.22
	0.975	7.57	6.06	5.42	5.05	4.82	4.65	4.53	4.43	4.36	4.30	4.20	4.10
	0.99	11.26	8.65	7.59	7.01	6.63	6.37	6.18	6.03	5.91	5.81	5.67	5.52
9	0.95	5.12	4.26	3.86	3.63	3.48	3.37	3.29	3.23	3.18	3.14	3.07	3.01
	0.975	7.21	5.71	5.08	4.72	4.48	4.32	4.20	4.10	4.03	3.96	3.87	3.77
	0.99	10.56	8.02	6.99	6.42	6.06	5.80	5.61	5.47	5.35	5.26	5.11	4.96
10	0.95	4.96	4.10	3.71	3.48	3.33	3.22	3.14	3.07	3.02	2.98	2.91	2.85
	0.975	6.94	5.46	4.83	4.47	4.24	4.07	3.95	3.85	3.78	3.72	3.62	3.52
	0.99	10.04	7.56	6.55	5.99	5.64	5.39	5.20	5.06	4.94	4.85	4.71	4.56
12	0.95	4.75	3.89	3.49	3.26	3.11	3.00	2.91	2.85	2.80	2.75	2.69	2.62
	0.975	6.55	5.10	4.47	4.12	3.89	3.73	3.61	3.51	3.44	3.37	3.28	3.18
	0.99	9.33	6.93	5.95	5.41	5.06	4.82	4.64	4.50	4.39	4.30	4.16	4.01
15	0.95	4.54	3.68	3.29	3.06	2.90	2.79	2.71	2.64	2.59	2.54	2.48	2.40
	0.975	6.20	4.77	4.15	3.80	3.58	3.41	3.29	3.20	3.12	3.06	2.96	2.86
	0.99	8.68	6.36	5.42	4.89	4.56	4.32	4.14	4.00	3.89	3.80	3.67	3.52

This table is abridged and adapted from Table IV in *Biometrika Tables for Statisticians*, edited by E. S. Pearson and H. O. Hartley. It is published here with the kind permission of the *Biometrika* Trustees.

TABLE VI

THE t DISTRIBUTION

$$P(T \leq t) = \int_{-\infty}^{t} \frac{\Gamma[(r + 1)/2]}{\sqrt{\pi r}\,\Gamma(r/2)(1 + w^2/r)^{(r+1)/2}}\, dw$$

$$[P(T \leq -t) = 1 - P(T \leq t)]$$

	$P(T \leq t)$				
r	0.90	0.95	0.975	0.99	0.995
1	3.078	6.314	12.706	31.821	63.657
2	1.886	2.920	4.303	6.965	9.925
3	1.638	2.353	3.182	4.541	5.841
4	1.533	2.132	2.776	3.747	4.604
5	1.476	2.015	2.571	3.365	4.032
6	1.440	1.943	2.447	3.143	3.707
7	1.415	1.895	2.365	2.998	3.499
8	1.397	1.860	2.306	2.896	3.355
9	1.383	1.833	2.262	2.821	3.250
10	1.372	1.812	2.228	2.764	3.169
11	1.363	1.796	2.201	2.718	3.106
12	1.356	1.782	2.179	2.681	3.055
13	1.350	1.771	2.160	2.650	3.012
14	1.345	1.761	2.145	2.624	2.977
15	1.341	1.753	2.131	2.602	2.947
16	1.337	1.746	2.120	2.583	2.921
17	1.333	1.740	2.110	2.567	2.898
18	1.330	1.734	2.101	2.552	2.878
19	1.328	1.729	2.093	2.539	2.861
20	1.325	1.725	2.086	2.528	2.845
21	1.323	1.721	2.080	2.518	2.831
22	1.321	1.717	2.074	2.508	2.819
23	1.319	1.714	2.069	2.500	2.807
24	1.318	1.711	2.064	2.492	2.797
25	1.316	1.708	2.060	2.485	2.787
26	1.315	1.706	2.056	2.479	2.779
27	1.314	1.703	2.052	2.473	2.771
28	1.313	1.701	2.048	2.467	2.763
29	1.311	1.699	2.045	2.462	2.756
30	1.310	1.697	2.042	2.457	2.750

This table is taken from Table III of Fisher and Yates: *Statistical Tables for Biological, Agricultural, and Medical Research*, published by Longman Group Ltd., London (previously published by Oliver and Boyd, Edinburgh), by permission of the authors and publishers.

TABLE VII

KOLMOGOROV–SMIRNOV ACCEPTANCE LIMITS

$$D_n = \sup_x [\,|F_n(x) - F_0(x)|\,]$$

$$\alpha = 1 - P(D_n \leq d)$$

	α			
n	0.20	0.10	0.05	0.01
1	0.90	0.95	0.98	0.99
2	0.68	0.78	0.84	0.93
3	0.56	0.64	0.71	0.83
4	0.49	0.56	0.62	0.73
5	0.45	0.51	0.56	0.67
6	0.41	0.47	0.52	0.62
7	0.38	0.44	0.49	0.58
8	0.36	0.41	0.46	0.54
9	0.34	0.39	0.43	0.51
10	0.32	0.37	0.41	0.49
11	0.31	0.35	0.39	0.47
12	0.30	0.34	0.38	0.45
13	0.28	0.32	0.36	0.43
14	0.27	0.31	0.35	0.42
15	0.27	0.30	0.34	0.40
16	0.26	0.30	0.33	0.39
17	0.25	0.29	0.32	0.38
18	0.24	0.28	0.31	0.37
19	0.24	0.27	0.30	0.36
20	0.23	0.26	0.29	0.35
25	0.21	0.24	0.26	0.32
30	0.19	0.22	0.24	0.29
35	0.18	0.21	0.23	0.27
40	0.17	0.19	0.21	0.25
45	0.16	0.18	0.20	0.24
Large n	$\dfrac{1.07}{\sqrt{n}}$	$\dfrac{1.22}{\sqrt{n}}$	$\dfrac{1.36}{\sqrt{n}}$	$\dfrac{1.63}{\sqrt{n}}$

TABLE VIII

RANDOM NUMBERS ON THE INTERVAL (0, 1)

3407	1440	6960	8675	5649	5793	1514
5044	9859	4658	7779	7986	0520	6697
0045	4999	4930	7408	7551	3124	0527
7536	1448	7843	4801	3147	3071	4749
7653	4231	1233	4409	0609	6448	2900
6157	1144	4779	0951	3757	9562	2354
6593	8668	4871	0946	3155	3941	9662
3187	7434	0315	4418	1569	1101	0043
4780	1071	6814	2733	7968	8541	1003
9414	6170	2581	1398	2429	4763	9192
1948	2360	7244	9682	5418	0596	4971
1843	0914	9705	7861	6861	7865	7293
4944	8903	0460	0188	0530	7790	9118
3882	3195	8287	3298	9532	9066	8225
6596	9009	2055	4081	4842	7852	5915
4793	2503	2906	6807	2028	1075	7175
2112	0232	5334	1443	7306	6418	9639
0743	1083	8071	9779	5973	1141	4393
8856	5352	3384	8891	9189	1680	3192
8027	4975	2346	5786	0693	5615	2047
3134	1688	4071	3766	0570	2142	3492
0633	9002	1305	2256	5956	9256	8979
8771	6069	1598	4275	6017	5946	8189
2672	1304	2186	8279	2430	4896	3698
3136	1916	8886	8617	9312	5070	2720
6490	7491	6562	5355	3794	3555	7510
8628	0501	4618	3364	6709	1289	0543
9270	0504	5018	7013	4423	2147	4089
5723	3807	4997	4699	2231	3193	8130
6228	8874	7271	2621	5746	6333	0345
7645	3379	8376	3030	0351	8290	3640
6842	5836	6203	6171	2698	4086	5469
6126	7792	9337	7773	7286	4236	1788
4956	0215	3468	8038	6144	9753	3131
1327	4736	6229	8965	7215	6458	3937
9188	1516	5279	5433	2254	5768	8718
0271	9627	9442	9217	4656	7603	8826
2127	1847	1331	5122	8332	8195	3322
2102	9201	2911	7318	7670	6079	2676
1706	6011	5280	5552	5180	4630	4747
7501	7635	2301	0889	6955	8113	4364
5705	1900	7144	8707	9065	8163	9846
3234	2599	3295	9160	8441	0085	9317
5641	4935	7971	8917	1978	5649	5799
2127	1868	3664	9376	1984	6315	8396

TABLE IX

DIVISORS FOR THE CONFIDENCE INTERVAL FOR
σ OF MINIMUM LENGTH

Let $X_1, X_2, \ldots, X_n$ be a random sample of size n from $N(\mu, \sigma^2)$; let $\bar{x} = (1/n) \sum_{i=1}^{n} x_i$ and $s^2 = (1/n) \sum_{i=1}^{n} (x_i - \bar{x})^2$. A $(100\gamma)\%$ confidence interval for σ which has minimum length is given by $[\sqrt{ns^2/b}, \sqrt{ns^2/a}]$. In the table, a is the upper number, b the lower number, $r = n - 1$ the number of degrees of freedom, and $\gamma = 1 - \alpha$ the confidence level.

r	γ 0.90	0.95	0.99	r	γ 0.90	0.95	0.99
2	0.206	0.101	0.020	16	8.774	7.604	5.649
	12.521	15.111	20.865		29.233	32.072	38.097
3	0.565	0.345	0.114	17	9.505	8.282	6.226
	13.153	15.589	20.973		30.480	33.362	39.469
4	1.020	0.692	0.294	18	10.242	8.969	6.814
	14.180	16.573	21.838		31.721	34.647	40.835
5	1.535	1.109	0.546	19	10.986	9.663	7.413
	15.350	17.743	22.985		32.959	35.927	42.195
6	2.093	1.578	0.857	20	11.736	10.365	8.021
	16.581	18.996	24.262		34.192	37.202	43.550
7	2.683	2.085	1.214	21	12.492	11.073	8.638
	17.839	20.286	25.602		35.420	38.472	44.899
8	3.298	2.623	1.611	22	13.253	11.788	9.264
	19.110	21.595	26.975		36.646	39.738	46.243
9	3.934	3.187	2.039	23	14.019	12.509	9.898
	20.385	22.912	28.364		37.867	41.000	47.586
10	4.588	3.773	2.496	24	14.790	13.236	10.539
	21.660	24.230	29.760		39.084	42.257	48.914
11	5.257	4.377	2.976	25	15.565	13.968	11.186
	22.933	25.548	31.158		40.299	43.510	50.243
12	5.940	4.997	3.477	26	16.344	14.704	11.841
	24.202	26.862	32.554		41.509	44.760	51.566
13	6.634	5.631	3.997	27	17.127	15.446	12.501
	25.467	28.172	33.947		42.717	46.006	52.886
14	7.338	6.278	4.533	28	17.914	16.192	13.168
	26.727	29.477	35.336		43.922	47.248	54.200
15	8.052	6.936	5.084	29	18.705	16.942	13.840
	27.982	30.777	36.719		45.123	48.487	55.511

Answers to Selected Exercises

CHAPTER 1

1.1–1 (a) $S = \{H, T\}$, 1/2;

(b) $S = \{1, 2, 3, 4, 5, 6\}$, 2/3;

(c) S is the set of 52 cards, 1/4;

(d) $S = \{(x, y): \ 0 \le x \le 1, 0 \le y \le 1\}$, 9/32.

1.1–3 1/4.

1.1–5 $1 - \sqrt{3}/2$.

1.1–7 (a) 1/4; (b) 1/4; (c) 3/4.

1.1–9 (a) $S = \{1, 2; 1, 3; 1, 4; 1, 5; 2, 3; 2, 4; 2, 5; 3, 4; 3, 5; 4, 5\}$;

(b) 1/10; 2/10; 5/10.

1.2–1 (a) $\{1, 2, 4, 6\}$; (b) $\{4\}$; (c) $\{1, 3, 5\}$; (d) $\{3, 5\}$;

(e) $\{3, 4, 5\}$; (f) S; (g) $\{1\}$; (h) $\{1\}$; (i) C;

(j) $\{4\}$.

1.2–3 (a) $S = \{DD, GDD, DGD, GGDD, GDGD, DGGD, GGGDD,$
$GGDGD, GDGGD, DGGGD\}$, $\{2, 3, 4, 5\}$;

(b) $\{GGDD, GDGD, DGGD\}$;

(c) $\{GDD, DGD\}$;

(d) $\{GGDD, GDGD, DGGD, GDD, DGD\}$, ϕ, $\{DD, GDD,$
$DGD, GGGDD, GGDGD, GDGGD, DGGGD\}$.

1.2–5 (a) $A \cup B \cup C$; (b) $A \cap B \cap C$;

(c) $(A \cap B' \cap C') \cup (A' \cap B \cap C') \cup (A' \cap B' \cap C)$;

(d) $(A \cup B \cup C)'$; (e) $A \cap B \cap C'$.

1.2–6   (a) $(0, 1)$;   (b) ϕ;   (c) $(0, 1]$;   (d) $[0, 1]$.

1.2–9 $A \cap B$.

1.3–1   (a) $12/52$;   (b) $2/52$;   (c) $16/52$;   (d) 1;   (e) 0.

1.3–3   (a) 0.6;   (b) 0.1;   (c) 0.7.

1.3–5   (a) 0.9;   (b) 0.8.

1.3–7   0.6.

1.3–8   44.

1.3–10   (a) $1 - (1/2)^{10}$;   (b) $1 - (1/2)^{20}$;   (c) $1 - (1/2)^{20}$;   (d) $1 - (1/2)^{10}$;   (e) $(1/2)^{10} - (1/2)^{20}$;   (f) $(1/2)^{20}$.

1.4–1 $6^3 = 216$.

1.4–3   (a) 6,760,000;   (b) 17,576,000.

1.4–5   151,200.

1.4–7   (a) 24;   (b) 256.

1.4–9   34,650.

1.4–11   4,096.

1.5–1   (a) 1,000;   (b) 720.

1.5–2   50,400.

1.5–3   (a) $\dbinom{3}{1}\dbinom{47}{9}\bigg/\dbinom{50}{10}$;   (b) $\left[\dbinom{3}{0}\dbinom{47}{10} + \dbinom{3}{1}\dbinom{47}{9}\right]\bigg/\dbinom{50}{10}$.

1.5–4 $1 - \dbinom{4}{0}\dbinom{96}{10}\bigg/\dbinom{100}{10}$.

1.5–7   (a) 365^r;   (b) $P(365, r)$;   (c) $P(365, r)/365^r$;   (d) 23.

1.5–8   635,013,559,600.

1.5–9 $\dbinom{39}{13}\bigg/\dbinom{52}{13}$.

1.5–11 $\dbinom{13}{3}\dbinom{13}{2}\dbinom{13}{4}\dbinom{13}{4}\bigg/\dbinom{52}{13}$.

1.5–14 $52!/[13!13!13!13!]$.

1.5–17 $\dbinom{4}{2}\dbinom{48}{3}\bigg/\dbinom{52}{5}$.

1.5–18   (a) 0.00024;   (b) 0.00144;   (c) 0.02113;   (d) 0.04754;   (e) 0.42257.

1.5–20 $\binom{5}{3}\binom{10}{2}\binom{10}{2}\Big/\binom{25}{7}$.

1.5–22 $\dfrac{12!}{4!3!2!3!}\,a^4b^3c^2d^3$.

1.6–1 (a) $1/17$; (b) $13/204$; (c) $1/52$.

1.6–3 0.1665.

1.6–5 0.776.

1.6–7 $7/55$.

1.6–9 (a) $1/56$; (b) $3/28$; (c) $3/8$.

1.6–11 $1/5$

1.6–13 (a) $6/22$; (b) $5/13$; (c) $3/7$.

1.7–1 (a) $8/21$; (b) $1/2$.

1.7–3 (a) $107/156$; (b) $8/107$.

1.7–5 $5/16$.

1.8–1 Yes.

1.8–4 (a) $1/6$; (b) $1/12$; (c) $1/4$; (d) $1/4$; (e) $1/2$.

1.8–7 0.398.

1.8–11 (a) $8/243$; (b) $8/243$; (c) $8/243$; (d) $10(8/243)$.

1.8–12 $1 - (0.05)^3 = 0.999875$.

1.8–13 (a) $\left(\dfrac{7}{15}\right)^3\left(\dfrac{3}{15}\right)\left(\dfrac{5}{15}\right)^2$; (b) $\left(\dfrac{7}{15}\right)^3\left(\dfrac{3}{15}\right)\left(\dfrac{5}{15}\right)^2$;

(c) $\dfrac{6!}{3!1!2!}\left(\dfrac{7}{15}\right)^3\left(\dfrac{3}{15}\right)\left(\dfrac{5}{15}\right)^2$.

CHAPTER 2

2.1–1 (a) $f(0) = 10/28, f(1) = 15/28, f(2) = 3/28$.

2.1–3 (a) 10; (b) $1/55$; (c) 2; (d) $1/30$.

2.1–5 (a) $g(y) = \dfrac{6 - |7 - y|}{36}$, $y = 2, 3, \ldots, 12$; (b) $0, 3/36, 3/36$.

2.2–1 $\$5$.

2.2–3 $\$11.20$.

2.2–5 7.

2.2–7 $-30.33¢.$

2.2–9 $\sum_{k=1}^{n} \frac{1}{k}\left(\frac{1}{n}\right); 0.4567; \frac{1}{100}\left(\frac{\ln 101 + 1 + \ln 100}{2}\right).$

2.2–11 22.

2.3–1 1.4; 1.44; 1.2.

2.3–3 3/2.

2.3–5 15; 12.5.

2.3–7 0; 1.

2.3–9 40/49.

2.4–1 (a) $\frac{1}{3}e^t + \frac{1}{3}e^{2t} + \frac{1}{3}e^{3t};$ (b) $e^{5t};$ (c) $\left(\frac{2}{3} + \frac{1}{3}e^t\right)^5.$

2.4–3 $2; 0.8; f(1) = 2/5, f(2) = 1/5, f(3) = 2/5.$

2.4–4 $e^{5t}; f(5) = 1.$

2.5–2 $\left(\frac{7}{18}\right)^x\left(\frac{11}{18}\right)^{1-x}, x = 0, 1; \frac{11}{18} + \frac{7}{18}e^t; \frac{7}{18}; \frac{77}{324}.$

2.5–3 $f(1) = \frac{7}{18}, f(-1) = \frac{11}{18}; \frac{11}{18}e^{-t} + \frac{7}{18}e^t; -\frac{4}{18}; \frac{308}{324}.$

2.5–5 (a) 0.3087; (b) 0.8319; (c) 1.5, 1.05, $(0.7 + 0.3e^t)^5.$

2.5–7 (a) $b(2{,}000, \pi/4);$ (b) 1,570.80, 337.10, 18.36.

2.5–9 (a) 9; (b) 2.25; (c) 0.3907.

2.5–10 (a) 0.2090; (b) 0.2082.

2.5–12 (a) 0.6513; (b) 0.7941.

2.5–13 7; 29.

2.5–15 0.75.

2.5–17 0.1268.

2.6–1 (a) 1/8; (b) 1/4; (c) 7/8.

2.6–2 (a) 27/256; (b) 175/256.

2.6–5 1/4.

2.6–7 $(0.99)^{99}.$

2.6–9 (a) $1/p;$ (b) 6.

2.6–10　　**(b)**　$5/(6 - i)$;　　**(c)**　11.417.

2.6–12　　**(a)**　27/256;　　**(b)**　67/256;　　**(c)**　27/32.

2.7–1　　**(a)**　0.693;　　**(b)**　0.762;　　**(c)**　0.433.

2.7–3　　0.315.

2.7–5　　0.558.

2.7–7　　**(a)**　0.082;　　**(b)**　0.488.

2.7–9　　**(a)**　0.217;　　**(b)**　0.971.

CHAPTER 3

3.1–1　　**(a)**　$2e^{-7} = 0.0018$;　　**(b)**　$7e^{-7} = 0.0064$.

3.1–2　　$(80/81)^4$.

3.1–5　　**(b)**　0.4, 0.5, 0.1.

3.1–9　　**(b)**　2/4.

3.1–10　　**(b)**　0.5;　　**(c)**　0.512.

3.2–1　　Class frequencies: 1, 2, 2, 7, 14, 5, 6, 4, 4, 5.

3.2–3　　**(a)**　Class frequencies: 2, 4, 9, 5, 14, 7, 3, 0, 1, 3;
　　　　(b)　Class frequencies: 3, 12, 11, 15, 3, 1, 3.

3.2–5　　**(a)**　Class frequencies: 1, 3, 3, 8, 10, 12, 12, 15, 21, 15;
　　　　(c)　$f(x) = x/2, 0 < x < 2$;
　　　　(d)　$F(x) = x^2/4, 0 < x < 2$.

3.3–1　　86; 45.6; 6.75; 17; 83; 86.5.

3.3–4　　**(a)**　4.0, 5.3, 6.0, 7.2, 4.5;　　**(b)**　$-0.4, 2.2, 3.6, 6.0, 0.6$;
　　　　(c)　2.6, 5.2, 6.6, 9.0, 3.6.

3.3–5　　2; 0.833.

3.3–7　　0.49; 0.078.

3.3–9　　5.59; 4.94.

CHAPTER 4

4.1–1　　**(a)**　3/2;　　**(b)**　$(y^3 + 1)/2, -1 \le y < 1$;　　**(c)**　1/2, 1/2, 0, 9/16.

4.1–3　　**(a)**　2;　　**(b)**　2.

4.1–6　　**(a)**　0, 3/5;　　**(b)**　0, 1/3;　　**(c)**　0, 1/6.

4.1–7 $(1 - 10t)^{-1}, t < \dfrac{1}{10}; 10; 100.$

4.1–9   (a) 0;   (b) -0.5;   (c) 0.80.

4.2–2   0, 1/3.

4.2–4   (a) $b(1, 1/4)$;   (c) $b(4, 1/4)$.

4.2–9   (a) $x^2, 0 \le x < 1.$

4.2–10   (a) $x/(1 - x), 0 \le x < 1/2.$

4.2–12   (a) $(w - a)/(b - a), a \le w < b$;   (b) $U(a, b).$

4.2–13   (a) 4.5;   (b) 1/12;   (c) 0.5.

4.3–1   (a) $e^{-1/2} - e^{-3/2}$;   (b) $e^{-3/2}$;   (c) $e^{-3/2}.$

4.3–3   (a) $(1/3)e^{-x/3}, 0 \le x < \infty$;   (b) $3e^{-3x}, 0 \le x < \infty.$

4.3–4   (a) $(2/3)e^{-2x/3}, 0 \le x < \infty$;   (b) $e^{-4/3}.$

4.3–8   (a) $e^{-1.2}.$

4.3–9 $\theta \ln 2; 0.8125.$

4.4–2 $1 - (9/4)e^{-5/4}.$

4.4–3   0.05.

4.4–5   (a) $\dfrac{14.7^{100}}{\Gamma(100)} w^{99} e^{-14.7w}, 0 \le w < \infty; 6.80; 0.463;$

      (b) 6.74, 0.443;
      (c) 0.36.

4.4–7   (a) 0.025;   (b) 0.05;   (c) 0.94.

4.4–9   (a) 0.94;   (b) 11.69, 38.08;   (c) 23, 46.

4.4–11   0.0746.

4.5–1   (a) 0.2784;   (b) 0.7209;   (c) 0.3007;   (d) 0.9616;
      (e) 0.0019;   (f) 0.95;   (g) 0.6826;   (h) 0.9544;
      (i) 0.9974.

4.5–3   (a) 1.96;   (b) 1.645;   (c) 1.645.

4.5–5   (a) 0.6326;   (b) 50.

4.5–7   0.025.

4.5–9   (c) 0.3, 0.3143.

4.6–1   (a) 0.5;   (b) 0;   (c) 0.25;   (d) 0.75;   (e) 0.625.

4.6–3   **(b)**   31/24, 167/576;

   (c)   (i) 15/64;   (ii) 1/4;   (iii) 0;   (iv) 11/16.

4.6–5   **(a)** $1.25.

CHAPTER 5

5.1–1   1/36, 1/9, 5/18, 1/3, 1/4; 14/3; 10/9.

5.1–3 $\exp(-2t + 25t^2/2)$; $N(-2, 25)$.

5.1–5   0.6554.

5.2–1   1; 61.

5.2–2   0.4772; 0.8560.

5.2–4 $\exp[7(e^t - 1)]$; 0.800.

5.2–6   0.7257.

5.2–7   0.8849.

5.2–11   **(a)**   1/36, 4/36, 10/36, 12/36, 9/36;
   (c)   1/1296, 1/162, 1/36, 13/162, 107/648, 13/54, 1/4, 1/6, 1/16.

5.3–1   0.4772.

5.3–3   0.8185.

5.3–5   **(a)** $\chi^2(18)$;   **(b)**   0.0756, 0.9974.

5.4–1   **(a)**   0.3721;   **(b)**   0.5468;   **(c)**   0.1747.

5.4–3   **(a)**   0.5548;   **(b)**   0.3823;   **(c)**   0.5802.

5.4–5   **(a)**   0.3802;   **(b)**   0.7066.

5.4–7   **(a)**   0.9984;   **(b)**   0.998.

5.5–1   **(b)**   18.3, 26.1;   **(c)**   14.15, 23.9.

5.5–3   **(a)**   0.2553;   **(b)**   0.7483.

5.5–5   **(a)**   0.8697;   **(b)**   0.8634.

5.5–7   **(b)** $r(n - r + 1)/[(n + 1)^2(n + 2)]$.

CHAPTER 6

6.1–1   **(a)**   0.7812;   **(b)**   0.7844;   **(c)**   0.4528.

6.1–3   0.143; 0.316; 0.670; ($y_{41} = 0.394$, $y_{60} = 0.564$).

6.1–5   (15.40, 17.05)

6.2–1   [2.03, 2.15].

6.2–3   [48.47, 72.27].

6.2–5   [29.30, 118.56]; $(y_{12} = 11.57, y_{24} = 65.68)$.

6.3–2   [0.007, 0.093].

6.3–3   (a) [0.789, 0.891].

6.3–5   [0.487, 0.691].

6.3–7   [−0.059, 0.099].

6.4–1   208.

6.4–3   384.

6.4–5   (a) 38;   (b) [0.621, 0.845].

CHAPTER 7

7.1–4   (a) $-n/\ln(X_1 X_2 \cdots X_n)$;   (b) $\bar{X}/2$;   (c) median.

7.1–5   33.43; 5.10.

7.2–1   [71.35, 76.25].

7.2–3   [−59.7, −43.3].

7.2–5   [−22.96, 40.30].

7.3–1   (a) [7.06, 27.05];   (b) [2.66, 5.20] or [2.54, 4.98].

7.3–4   (a) 0.05;   (b) 0.99;   (c) 0.01;   (d) 0.94.

7.3–5   (a) 0.279, 4.15;   (b) 0.157, 8.10.

7.3–8   [0.091, 12.6].

7.3–10   [0.047, 2.189].

7.4–1   (a) 0.025;   (b) 0.975;   (c) 0.05;   (d) 0.94;   (e) 0.09.

7.4–4   2.120.

7.4–5   [106.55, 118.89].

7.4–7   (a) 1.75σ;   (b) 2.33σ.

7.4–9   [−74.54, 63.70].

7.4–11   [−4.16, −0.84].

7.5–1   (d) 3.48.

7.5–5 $\sqrt{n/2} \; \Gamma[(n-1)/2]/\Gamma(n/2)$.

7.6–3 $\bar{X}, \bar{Y}, (nS_X^2 + mS_Y^2)/(n+m)$.

7.7–1   (a) $86.8 + 10.2(x - 74.5)$.

7.7–3   (a) $125.73 + 4.49(x - 26.6)$.

7.7–5 $[53.16, 80.84]; [-0.72, 4.92]; [164.57, 1321.09]$.

CHAPTER 8

8.1–1 $0.7 < 1.645$, accept; yes, $[-0.378, 0.938]$.

8.1–3 $1.19 < 2.12$, accept.

8.1–5 $1.43 < 1.96$, accept; yes, $[-0.24, 0.04]$.

8.2–1   (a) $1.4 < 1.645$, accept;
             (b) $1.4 > 1.282$, reject.

8.2–3 $0.59 < 1.645$, accept.

8.2–5 $-6.06 < -1.734$, reject.

8.3–1 $4.35 > 3.941$, accept.

8.3–3 $1.17 < 2.64$, accept.

8.3–5 $0.843 < 1.717$, accept.

8.3–7 $-2.221 < -1.734$, reject.

8.4–1   (a) $C = \{2, 3\}$;   (b) $\alpha = \beta = 7/27$.

8.4–4   (a) 0.0122;   (b) $1 - \Phi[(11.5 - \mu)/(2/3)]$;
             (c) $0.2266, 0.5, 0.7734$.

8.4–5 $24, 1.6$.

8.4–7 $66, 4.55$.

8.5–1 $C = \{y: \; y \le 4 \text{ or } y \ge 11\}, \alpha = 0.1184$, accept.

8.5–3 $C = \{w: \; w \le 7 \text{ or } w \ge 17\}, \alpha \approx 0.0662$, accept.

8.5–5 $C = \{y: \; 86 < y\}, \alpha \approx 0.0521$, reject.

8.6–1 $z = -21, |0.543| < 1.645$, accept.

8.6–3   Reject; accept; accept.

8.7–1   Reject; $-3.02 < -1.645$, reject; accept.

8.7–3 $2.136 > 1.645$, reject.

8.8–1   Reject.

8.8–3 $-2.08 < -1.96$; reject at $\alpha = 0.025$.

8.8–5   Reject at $\alpha = 0.025$.

8.9–1   Accept.

8.9–3   **(a)** $d_{25} \geq 0.24$;   **(c)** accept.

8.9–5   Accept.

CHAPTER 9

9.1–1   **(a)** $(2x + 5)/16$, $x = 1, 2$;   **(b)** $(2y + 3)/32$, $y = 1, 2, 3, 4$;   **(c)** $3/32$;   **(d)** $9/32$;   **(e)** $3/16$;   **(f)** $1/4$;   **(g)** dependent.

9.1–3   **(b)** 0.0034;   **(c)** 0.296;   **(d)** no.

9.1–5   **(d)** 0.043.

9.2–1 $25/16, 45/16, 63/256, 295/256$; -0.037.

9.2–2 $1/\sqrt{3}$; $y = x/2$.

9.2–5   **(a)** $1/2$;   **(b)** $3/2$;   **(c)** $5/12$;   **(d)** $3/4$;   **(e)** $-1/4$;   **(f)** $-1/\sqrt{5}$.

9.3–1   **(c)** $20/7$;   **(d)** $25/9$;   **(e)** $9/14$;   **(f)** $55/49$.

9.3–3   **(b)** $b(30 - y, 1/5)$;   **(c)** 20.

9.3–5   **(a)** $1/[10(10 - x)]$, $x = 0, 1, \ldots, 9$, $y = x, x + 1, \ldots, 9$;   **(c)** $(x + 9)/2$.

9.4–1 $2e^{-2x}, 0 < x < \infty$; $2e^{-y}(1 - e^{-y}), 0 < y < \infty$; no.

9.4–3 $1/2, 0 \leq x \leq 2$; $1/2, 0 \leq y \leq 2$; yes.

9.4–5   **(a)** $1/2$;   **(b)** e^{-2};   **(c)** 0;   **(d)** $1 - e^{-2}$.

9.4–7   **(a)** $1/(2x^2), 0 < x < 2, 0 < y < x^2$;

         (b) $(2 - \sqrt{y})/(4\sqrt{y}), 0 < y < 4$;

         (c) $\ln(2/\sqrt{y}) \div (2 - \sqrt{y})/(2\sqrt{y})$.

9.4–9 $1 - e^{-1}$.

9.5–2   **(a)** 0.6006;   **(b)** 0.7888;   **(c)** 0.8185;   **(d)** 0.9371.

9.5–4   **(a)** 0.5746;   **(b)** 0.7357.

9.5–5   **(a)** 0.5998;   **(b)** 0.8957.

9.6–1 (a) $4, 5, 8/3, 8/3, 1/2, x/2 + 3$.

9.6–3 (a) $2.667, 4.50, 1.556, 1.25, 0.598, 0.536x + 3.07$;
        (c) $2.53, 4.47, 1.97, 1.97, 0.48, 0.48x + 3.25$.

9.6–5 (b) $0.89x + 0.02$.

9.6–7 (b) $0.48x + 0.94$.

9.7–1 $-2.52 < -2.06$, reject.

9.7–3 $[0.419, 0.802]$.

CHAPTER 10

10.1–1 $10.2 < 11.07$, no.

10.1–3 $3.65 < 7.815$, accept.

10.1–5 $14.6 > 14.07$, reject.

10.2–1 $A_1 = \{0, 1\}, A_8 = \{8, 9, \ldots\}, 9.43 < 18.48$, accept.

10.2–3 $2.365 < 16.92$, accept.

10.2–5 Accept.

10.3–2 $0.000067; 0.054$.

10.3–4 $1 - \exp[-(a/b)e^{bw} - cw + (a/b)]$.

10.4–1 $3.23 < 11.07$, accept.

10.4–3 $2.4 < 5.991$, accept.

10.4–5 $6.04 < 9.488$, accept.

10.5–1 $9.18 > 5.991$, reject.

10.5–3 $23.78 > 21.03$, reject.

CHAPTER 11

11.1–3 $1/4, 1/4, 1/4$.

11.2–1 $7.88 > 4.26$, reject.

11.2–3 $13.77 > 4.07$, reject.

11.2–5 $5.17 > 3.24$, reject.

11.3–1 $18 > 5.14$, reject.

11.3–2 $8; -2, 2, 0; 0, -3, 1, 2.$

11.3–5 $8; 0, 0, 0; 0, -3, 1, 2; -2, 2, -2, 2, 2, -2, 2, -2, 0, 0, 0, 0.$

CHAPTER 12

12.1–1 $y/8, 0 < y < 4.$

12.1–2 $1/[\pi(1 + y^2)], -\infty < y < \infty.$

12.1–3 $1, 0 < y < 1.$

12.1–5 $(1/4) \exp(-y_2/2), 0 < y_1 < y_2 < \infty; (1/2) \exp(-y_1/2), 0 < y_1 < \infty;$
$(y_2/4) \exp(-y_2/2), 0 < y_2 < \infty;$ no.

12.1–9 $c \ln[(1 + R)/(1 - R)].$

12.1–11 $1/840.$

12.2–1   **(b)** $34; 77.8.$

12.2–3   **(b)** $\bar{x} > 4.15;$ **(c)** $\bar{x} > 4.15;$ **(d)**   yes.

12.2–6   **(a)** $81.23, 78.77, 1.47.$

12.2–7   **(b)** $y > 8.$

12.3–1   **(a)** $0.90 > 0.822,$ reject;
(b) $0.90 < 0.98,$ accept.

12.3–3   **(a)** $\bar{x} \geq 230 + 10c/\sqrt{n};$
(b)   yes;
(c) $232.6 < 233.2,$ accept.

12.3–4   **(b)** $2.87 < 2.94,$ accept.

12.3–7 $324.8 > 321.63,$ accept.

12.4–2   **(b)** $\bar{X} + 1.$

12.4–3 $[y!/(x_1!x_2! \cdots x_n!)][1/n]^y.$

12.4–5   **(a)** $\ln(X_1 X_2 \cdots X_n);$ **(b)** $-n/\ln(X_1 X_2 \cdots X_n).$

12.5–2   **(a)** $\text{gamma}[y + \alpha, \beta/(n\beta + 1)];$ **(b)** $(y + \alpha)\beta/(n\beta + 1).$

12.5–3   **(a)** $(74\theta^2 - 114\theta + 45)/500;$ **(b)** 0.57 to $0.87.$

INDEX